P9-DVH-868

To

ERIKA

R

Walter Gautschi

Numerical Analysis
An Introduction

1997

Birkhäuser
Boston • Basel • Berlin

West Lafayette, IN 47907-1398
USA

Library of Congress Cataloging-in-Publication Data

Gautschi, Walter.
Numerical analysis : an introduction / Walter Gautschi.
p. cm.
Includes bibliographical references (p. 451- 481).
ISBN 0-8176-3895-4 (alk. paper). -- ISBN 3-7643-3895-4 (Basel : alk. paper)
1. Numerical analysis. I. Title.
QA297.G35 1997
519.4--dc21 97-186
CIP

Printed on acid-free paper

ISBN 0-8176-3895-4
ISBN 3-7643-3895-4
Typeset by the Author in LATEX.
Cover design by Dutton & Sherman Design, New Haven, CT.
Printed and bound by Quinn-Woodbine, Woodbine, NJ.
Printed in the U.S.A.

9 8 7 6 5 4 3 2 1

CONTENTS

PREFACE

The book is designed for use in a graduate program in Numerical Analysis that is structured so as to include a basic introductory course and subsequent more specialized courses. The latter are envisaged to cover such topics as numerical linear algebra, the numerical solution of ordinary and partial differential equations, and perhaps additional topics related to complex analysis, to multidimensional analysis, in particular optimization, and to functional analysis and related functional equations. Viewed in this context, the first four chapters of our book could serve as a text for the basic introductory course, and the remaining three chapters (which indeed are at a distinctly higher level) could provide a text for an advanced course on the numerical solution of ordinary differential equations. In a sense, therefore, the book breaks with tradition in that it does no longer attempt to deal with all major topics of numerical mathematics. It is felt by the author that some of the current subdisciplines, particularly those dealing with linear algebra and partial differential equations, have developed into major fields of study that have attained a degree of autonomy and identity that justifies their treatment in separate books and separate courses on the graduate level. The term "Numerical Analysis" as used in this book, therefore, is to be taken in the narrow sense of the numerical analogue of Mathematical Analysis, comprising such topics as machine arithmetic, the approximation of functions, approximate differentiation and integration, and the approximate solution of nonlinear equations and of ordinary differential equations.

What is being covered, on the other hand, is done so with a view toward stressing basic principles and maintaining simplicity and student-friendliness as far as possible. In this sense, the book is "An Introduction". Topics that, even though important and of current interest, require a level of technicality that transcends the bounds of simplicity striven for, are referenced in detailed bibliographic notes at the end of each chapter. It is hoped, in this way, to place the material treated in proper context and to help, indeed encourage, the reader to pursue advanced modern topics in more depth.

A significant feature of the book is the large collection of exercises that are designed to help the student develop problem-solving skills, and to provide interesting extensions of topics treated in the text. Particular attention is given to machine assignments, where the student is encouraged to implement numerical techniques on the computer and to make use of modern software packages.

The author has taught the basic introductory course, and the advanced course on ordinary differential equations regularly at Purdue University for the last 30 years or so. The former, typically, was offered both in the fall and spring semesters, to a mixed audience consisting of graduate (and some good undergraduate) students in mathematics, computer science, and engineering, while the latter was taught only in the fall, to a smaller but also mixed audience. Written notes began to materialize in the 1970s, when the author taught the basic course repeatedly in summer courses on Mathematics held in Perugia, Italy. Indeed, for some time, these notes existed only in the Italian language. Over the years, they were progressively expanded, updated, and transposed into English, and along with that, notes for the advanced course were developed. This, briefly, is how the present book evolved.

A long gestation period such as this, of course, is not without dangers, the most notable one being a tendency for the material to become dated. The author tried to counteract this by constantly updating and revising the notes, adding newer developments when deemed appropriate. There are, however, benefits as well: over time, one develops a sense for what is likely to stand the test of time and what may only be of temporary interest, and one selects and deletes accordingly. Another benefit is the steady accumulation of exercises and the opportunity to have them tested on a large and diverse student population.

The purpose of academic teaching, in the author's view, is twofold: to transmit knowledge, and, perhaps more important, to kindle interest and even enthusiasm in the student. Accordingly, the author did not strive for comprehensiveness — even within the boundaries delineated — but rather tried to concentrate on what is essential, interesting and intellectually pleasing, and teachable. In line with this, an attempt has been made to keep the text uncluttered with numerical examples and other illustrative material. Being well aware, however, that mastery of a subject does not come from studying alone, but from active participation, the author provided many exercises, including machine projects. Attributions of results to specific authors and citations to the literature have been deliberately omitted from the body of the text. Each chapter, as already mentioned, has a set of appended notes that help the reader to pursue related topics in more depth and to consult the specialized literature. It is here where attributions and historical remarks are made, and where citations to the literature — both textbook and research — appear.

The main text is preceded by a Prologue (Chapter 0), which is intended to place the book in proper perspective. In addition to other textbooks on the subject, and information on software, it gives a detailed list of topics not treated in this book, but definitely belonging to the vast area of computational mathematics, and it provides ample references to relevant texts. A list of numerical analysis journals is also included.

The reader is expected to have a good background in calculus and advanced calculus. Some passages of the text require a modest degree of acquaintance with linear algebra, complex analysis, or differential equations. These passages, however, can easily be skipped, without loss of continuity, by a student who is not familiar with these subjects.

It is a pleasure to thank the publisher for his interest in this book and his cooperation in producing it. The author is also grateful to Soren Jensen and Manil Suri, who taught from this text, and to an anonymous reader; they all made many helpful suggestions on improving the presentation. He is particularly indebted to Professor Jensen for substantially helping in preparing the exercises to Chapter 7. The author further acknowledges assistance from Carl de Boor in preparing the notes to Chapter 2, and from Werner C. Rheinboldt for helping with the notes to Chapter 4. Last but not least, he owes a measure of gratitude to Connie Wilson for typing a preliminary version of the text, and to Adam Hammer for assisting the author with the more intricate aspects of LaTeX.

January, 1997 *Walter Gautschi*

CHAPTER 0

PROLOGUE

§0.1. **Overview.** *Numerical Analysis* is the branch of mathematics that provides tools and methods for solving mathematical problems in numerical form. The objective is to develop detailed computational procedures, capable of being implemented on electronic computers, and to study their performance characteristics. Related fields are *Scientific Computation*, which explores the application of numerical techniques and computer architectures to concrete problems arising in the sciences and engineering; *Complexity Theory*, which analyzes the number of "operations" and the amount of computer memory required to solve a problem; and *Parallel Computation*, which is concerned with organizing computational procedures in a manner that allows running various parts of the procedures simultaneously on different processors.

The problems dealt with in computational mathematics come from virtually all branches of pure and applied mathematics. There are computational aspects in number theory, combinatorics, abstract algebra, linear algebra, approximation theory, geometry, statistics, optimization, complex analysis, nonlinear equations, differential and other functional equations, and so on. It is clearly impossible to deal with all these topics in a single text of reasonable size. Indeed, the tendency today is to develop specialized texts dealing with one or the other of these topics. In the present text we concentrate on subject matters that are basic to problems in approximation theory, nonlinear equations, and differential equations. Accordingly, we have chapters on machine arithmetic, approximation and interpolation, numerical differentiation and integration, nonlinear equations, one-step and multistep methods for ordinary differential equations, and boundary value problems in ordinary differential equations. Important topics not covered in this text are computational number theory, algebra, and geometry; constructive methods in optimization and complex analysis; numerical linear algebra; and the numerical solution of problems involving partial differential equations and integral equations. Selected texts for these areas are listed in §0.3.

We now describe briefly the topics treated in this text. Chapter 1 deals with the basic facts of life regarding machine computation. It recognizes that, although present-day computers are extremely powerful in terms of computational speed, reliability, and amount of memory available, they are less than ideal — unless supplemented by appropriate software — when it comes to the precision available, and accuracy attainable, in the execution of elementary arithmetic operations. This raises serious questions as to how arithmetic errors, either present in the input data of a problem or committed during the execution of a solution algorithm, affect the accuracy of the desired results. Concepts and tools required to answer such questions are the topic of this introductory chapter. In Chapter 2, the central theme is the approximation of functions by simpler functions, typically, polynomials and piecewise polynomial functions. Approximation in the sense of least squares provides an opportunity to introduce orthogonal polynomials, which are relevant also in connection with problems of numerical integration treated in Chapter 3. A large part of the chapter, however, deals with polynomial interpolation and associated error estimates, which are basic to many numerical procedures for integrating functions and differential equations. Also discussed briefly is inverse interpolation, an idea useful in solving equations.

First applications of interpolation theory are given in Chapter 3, where the tasks presented are the computation of derivatives and definite integrals. Although the formulae developed for derivatives are subject to the detrimental effects of machine arithmetic, they are useful, nevertheless, for purposes of discretizing differential operators. The treatment of numerical integration includes routine procedures, such as the trapezoidal and Simpson's rules, appropriate for well-behaved integrands, as well as the more sophisticated procedures based on Gaussian quadrature to deal with singularities. It is here where orthogonal polynomials reappear. The method of undetermined coefficients is another technique for developing integration formulae. It is applied in order to approximate general linear functionals, the Peano representation of linear functionals providing an important tool for estimating the error. The chapter ends with a discussion of extrapolation techniques; although applicable to more general problems, they are inserted here since the composite trapezoidal rule together with the Euler-Maclaurin formula provides the best-known application — Romberg integration.

Chapter 4 deals with iterative methods for solving nonlinear equations and systems thereof, the pièce de résistance being Newton's method. The emphasis here lies in the study of, and the tools necessary to analyze, convergence. The special case of algebraic equations is also briefly given attention.

Chapter 5 is the first of three chapters devoted to the numerical solution of ordinary differential equations. It concerns itself with one-step methods for solving initial value problems, such as the Runge-Kutta method, and gives a detailed analysis of local and global errors. Also included is a brief introduction to stiff equations and special methods to deal with them. Multistep methods and, in particular, Dahlquist's theory of stability and its applications, is the subject of Chapter 6. The final Chapter 7 is devoted to boundary value problems and their solution by shooting methods, finite difference techniques, and variational methods.

§0.2. **Numerical analysis software**. There are many software packages available, both in the public domain and distributed commercially, that deal with numerical analysis algorithms. A widely used source of numerical software is Netlib, which can be accessed either on the World Wide Web at the URL `http://www.netlib.org`, or by e-mail at the address (in the US) `netlib@netlib.org`.

If you wish to see an index of all packages contained in the Netlib repository, you may, on the Web, open up the highlighted entry `Browse the Netlib repository`. This will display on the screen a list of all software packages, which may then be examined more closely by clicking on `A more descriptive version of the list`. Instructions as to how files are retrieved can be found under `Frequently Asked Questions about Netlib (FAQ)` on the home page of Netlib. Via e-mail, you can get an index, as well as instructions regarding retrieval of an individual package or file(s) from a package, by sending the message `send index` to the e-mail address given here.

Large collections of general-purpose numerical algorithms are contained in sources such as Slatec and TOMS (ACM Transactions on Mathematical Software). Specialized packages relevant to the topics in the chapters ahead are identified in the "Notes" to each chapter. Likewise, specific files needed to do some of the machine assignments in the Exercises are identified as part of the exercise.

Among the commercial software packages we mention the visual numerics (formerly IMSL) package, the NAG library, and MLAB (Modeling LABoratory). Interactive systems include HiQ, MACSYMA, MAPLE, Mathcad, Mathematica, and MATLAB. Many of these packages, in addition to numerical computation, have symbolic computation and graphics capabilities. Further information is available in the Netlib file `commercial`. For more libraries, and for the interactive systems, also see Lozier and Olver [1994, §3].

§0.3. **Textbooks and monographs**. We provide here an annotated list (ordered alphabetically with respect to authors) of other textbooks on numerical analysis, written at about the same, or higher, level as the present one. Following this we also mention books and monographs dealing with topics in computational mathematics not covered in our (and many other) books on numerical analysis. Additional books dealing with specialized subject areas, as well as other literature, are referenced in the "Notes" to the individual chapters. We generally restrict ourselves to books written in English and, with a few exceptions, published within the last 15 years or so. Even so, we have had to be selective. (No value judgment is to be implied by our selections or omissions.) A reader with access to the AMS (American Mathematical Society) MathSci CD-ROM will have no difficulty in retrieving a more complete list of relevant items, including older texts.

Selected Textbooks on Numerical Analysis

Atkinson [1989]

A comprehensive in-depth treatment of standard topics short of partial differential equations; includes an appendix describing some of the better-known software packages.

Bruce, Giblin, and Rippon [1990]

A collection of interesting mathematical problems, ranging from number theory and computer-aided design to differential equations, that require the use of computers for their solution.

Cheney and Kincaid [1994]

Although an undergraduate text, it covers a broad area, has many examples from science and engineering as well as computer programs; there are many exercises, including machine assignments.

Conte and de Boor [1980]

A widely used text for upper-division undergraduate students; written for a broad audience, with algorithmic concerns in the foreground; has FORTRAN subroutines for many algorithms discussed in the text.

Deuflhard and Hohmann [1995]

An introductory text with emphasis on machine computation and algorithms; includes a discussion of three-term recurrence relations (not usually found in textbooks), but no differential equations.

Fröberg [1985]

A thorough and exceptionally lucid exposition of all major topics of numerical analysis exclusive of algorithms and computer programs.

Hämmerlin and Hoffmann [1991]

Similar to Stoer and Bulirsch [1993] in its emphasis on mathematical theory; has more on approximation theory and multivariate interpolation and integration, but nothing on differential equations.

Isaacson and Keller [1994]

One of the older but still eminently readable texts, stressing the mathematical analysis of numerical methods.

Kincaid and Cheney [1996]

Related to Cheney and Kincaid [1985] but more mathematically oriented and unusually rich in exercises and bibliographic items.

Rutishauser [1990]

An annotated translation from the German of an older text based on posthumous notes by one of the pioneers of numerical analysis; although the subject matter reflects the state of the art in the early 1970s, the treatment is highly original and is supplemented by translator's notes to each chapter pointing to more recent developments.

Schwarz [1989]

A mathematically oriented treatment of all major areas of numerical analysis, including ordinary and partial differential equations.

Stoer and Bulirsch [1993]

Fairly comprehensive in coverage; written in a style appealing more to mathematicians than engineers and computer scientists; has many exercises and bibliographic references; serves not only as a textbook, but also as a reference work.

Todd [1980, 1977]

Rather unique books, emphasizing problem solving in areas often not covered in other books on numerical analysis.

A collection of outstanding survey papers on specialized topics in numerical analysis is being assembled by Ciarlet and Lions [1990,1991,1994] in handbooks of numerical analysis; three volumes have appeared so far. Another source of surveys on a variety of topics is *Acta numerica*, an annual series of books edited by Iserles [1992–1996], of which five volumes have so far been published. For an authoritative account of the history of numerical analysis, the reader is referred to the book by Goldstine [1977].

The related areas of *Scientific Computing* and *Parallel Computing* are rather more recent fields of study, and currently are most actively pursued in proceedings of conferences and workshops. Nevertheless, a few textbooks have also appeared, notably in the linear algebra context and in connection with ordinary and partial differential equations, for example Schendel [1984], Ortega and Voigt [1985], Ortega [1989], Golub and Ortega [1992], [1993], Van de Velde [1994], and Heath [1997], but also in optimization, Pardalos, Phillips, and Rosen [1992], computational geometry, Akl and Lyons [1993], and other miscellaneous areas, Crandall [1994], Köckler [1994], and Bellomo and Preziosi [1995]. Interesting historical essays are contained in Nash [1990]. Matters regarding the *Complexity* of numerical algorithms are discussed in an abstract framework in books by Traub and Woźniakowski [1980] and Traub, Wasilkowski, and Woźniakowski [1983], [1988], with applications to the numerical integration of functions and nonlinear equations, and similarly, applied to elliptic partial differential equations and integral equations, in the book by Werschulz [1991]. Other treatises are those by Kronsjö [1987], Ko [1991], Bini and Pan [1994], and Wang, Xu, and Gao [1994]. For an in-depth complexity analysis of Newton's method, the reader is encouraged to study Smale's [1987] lecture.

Material on *Computational Number Theory* can be found, at the undergraduate level, in the book by Rosen [1993], which also contains applications to cryptography and computer science, and in Allenby and Redfern [1989], and at a more advanced level in the books by Niven, Zuckerman, and Montgomery [1991], Cohen [1993], and Bach and Shallit [1996], the first volume of a projected two-volume set. Computational methods of factorization are dealt with in the book by Riesel [1994]. Other useful sources are the set of lecture notes by Pohst [1993] on algebraic number theory algorithms, and the proceedings volumes edited by Pomerance [1990] and Gautschi [1994a, Part II]. For algorithms in *Combinatorics*, see the books by Nijenhuis and Wilf [1978], Hu [1982], and Cormen, Leiserson, and Rivest [1990]. Various aspects of *Computer Algebra* are treated in the books by Cox, Little, and O'Shea [1992], Geddes, Czapor, and Labahn [1992], Mignotte [1992], Davenport, Siret, and Tournier [1993], Heck [1996], and Mishra [1993].

Other relatively new disciplines are *Computational Geometry* and *Computer-Aided Design*, for which relevant texts are Preparata and Shamos [1985], Edelsbrunner [1987], Mäntylä [1988], and Taylor [1992]; and Farin [1995], [1997] and Hoschek and Lasser [1993], respectively. *Statistical Computing* is covered in general textbooks such as Kennedy and Gentle [1980], Anscombe [1981], Maindonald [1984], and Thisted [1988]. More specialized texts are Devroye [1986] on the generation of nonuniform random variables, Späth [1992] on regression analysis, Heiberger [1989] on the design of experiments, Stewart [1994] on Markov chains, and Fang and Wang [1994] on the application of number-theoretic methods. Numerical techniques in *Optimization* (including optimal control problems) are discussed in Evtushenko [1985]. An introductory book on unconstrained optimization is Wolfe [1978]; among the more advanced and broader texts on optimization techniques we mention Gill, Murray, and Wright [1981], Fletcher [1987], and Ciarlet [1989]. Linear programming is treated in Nazareth [1987] and Panik [1996], linear and quadratic problems in Sima [1996], and the application of conjugate direction methods to problems in optimization in Hestenes [1980]. The most comprehensive text on (numerical and applied) *Complex Analysis* is the three-volume treatise by Henrici [1988, 1991, 1986]. Numerical methods for conformal mapping are also treated in Schinzinger and Laura [1991]. For approximation in the complex domain, the standard text is Gaier [1987]; Stenger [1993] deals with approximation by sinc functions. The book by Iserles and Nørsett [1991] contains interesting discussions on the interface between complex rational approximation and the stability theory of discretized differential equations. The impact of high-precision computation on problems and conjectures involving complex approximation is beautifully illustrated in the set of lectures by Varga [1990].

For an in-depth treatment of many of the preceding topics, also see the three-volume work of Knuth [1975, 1981, 1973].

Perhaps the most significant topic omitted in our book is numerical linear algebra and its application to solving partial differential equations by finite difference or finite element methods. Fortunately, there are good treatises available that address these areas. For *Numerical Linear Algebra*, we refer to the book by Golub and Van Loan [1996] and the classic work of Wilkinson [1988]. Other general texts are Watkins [1991], Jennings and McKeown [1992], and Datta [1995]; Higham [1996] has a comprehensive treatment of error and stability analyses. The solution of sparse linear systems, and the special data structures and pivoting strategies required in direct methods, are treated in Østerby and Zlatev [1983], Duff, Erisman, and Reid [1989], and Zlatev [1991], whereas iterative techniques are discussed in the classic texts by Varga [1962] and Young [1971], and more recently in Hageman and Young [1981], Il'in [1992], Hackbusch [1994], Saad [1996], and Weiss [1996]. The books by Branham [1990] and Björck [1996] are devoted especially to least squares problems. For eigenvalues, see Chatelin [1983], [1993], and for a good introduction to the numerical analysis of symmetric eigenvalue problems, see Parlett [1980]. The currently very active investigation of large sparse symmetric and nonsymmetric eigenvalue problems and their solution by Lanczos-type methods has given rise to many books, for example, Cullum and Willoughby [1985], Meyer [1987], Sehmi [1989], and Saad [1992]. For readers wishing to test their algorithms on specific matrices, the collection of test matrices in Gregory and Karney [1978], and the recently established "matrix market" on the Web (`http://math.nist.gov./MatrixMarket`), are useful sources.

Even more extensive is the textbook literature on the numerical solution of *Partial Differential Equations*. The field has grown so much that there are currently only a few books that attempt to cover the subject as a whole. Among these are Birkhoff and Lynch [1984] (for elliptic problems), Sewell [1988], Hall and Porsching [1990], Ames [1992], Celia and Gray [1992], Morton and Mayers [1994], and Quarteroni and Valli [1994]. Variational and finite element methods seem to have attracted the most attention. An early and still frequently cited reference is the book by Ciarlet [1978]; among the more recent texts we mention the Texas Finite Element Series (Becker, Carey, and Oden [1981], Carey and Oden [1983], [1984], [1986], Oden [1983], Oden and Carey [1984]), Axelsson and Barker [1984], Wait and Mitchell [1985], Sewell [1985] (with a slant toward software), White [1985], Girault and Raviart [1986] (focusing on Navier-Stokes equations), Burnett [1987], Hughes [1987], Johnson [1987], Schwarz [1988], Beltzer [1990] (using symbolic computation), Křížek and Neittaanmäki [1990], Brezzi and Fortin [1991], and Brenner and Scott [1994]. Finite difference methods are treated in Godunov and Ryaben'kiĭ [1987], Strikwerda [1989], Ashyralyev and Sobolevskiĭ [1994], Gustafsson, Kreiss, and Oliger [1995], and Thomas [1995], the method of lines in Schiesser [1991], and the more refined techniques of multigrids and domain decomposition in Hackbusch [1985], Briggs [1987], McCormick [1989], [1992], Bramble [1993], Shaĭdurov [1995], and Smith, Bjørstad, and Gropp [1996]. Problems in potential theory and elasticity are often approached via boundary element methods, for which representative texts are Banerjee and Butterfield [1981], Brebbia [1984], Hartmann [1989], Chen and Zhou [1992], and Hall [1994]. A discussion of conservation laws is given in the classic monograph by Lax [1973] and more recently in LeVeque [1992] and Godlewski and Raviart [1996]. Spectral methods (i.e., expansions in (typically) orthogonal polynomials), applied to a variety of problems, were pioneered in the monograph by Gottlieb and Orszag [1977] and have received extensive treatments in more recent texts by Canuto, Hussaini, Quarteroni, and Zang [1988], Mercier [1989], and Fornberg [1996]. The numerical solution of elliptic boundary value problems nowadays is greatly facilitated thanks to the software package ELLPACK, which is described in Rice and Boisvert [1985].

Early, but still relevant, texts on the numerical solution of *Integral Equations* are Atkinson [1976] and Baker [1977]. A more recent introduction to the subject is Delves and Mohamed [1988]. Volterra integral equations are discussed in Linz [1985] and, more extensively, in Brunner and van der Houwen [1986], whereas singular integral equations are the subject of Prössdorf and Silbermann [1991].

§0.4. **Journals**. Here we list the major journals (in alphabetical order) covering the areas of numerical analysis and mathematical software.

ACM Transactions on Mathematical Software
Applied Numerical Mathematics
BIT Numerical Mathematics
Calcolo
Chinese Journal of Numerical Mathematics and Applications
Computational Mathematics and Mathematical Physics
Computing
IMA Journal on Numerical Analysis
Journal of Computational and Applied Mathematics
Mathematical Modelling and Numerical Analysis
Mathematics of Computation
Numerische Mathematik
SIAM Journal on Numerical Analysis

CHAPTER 1

MACHINE ARITHMETIC AND RELATED MATTERS

The questions addressed in this introductory chapter are fundamental in the sense that they are relevant in any situation that involves numerical machine computation, regardless of the kind of problem that gave rise to these computations. In the first place, one has to be aware of the rather primitive type of number system available on computers. It is basically a finite system of numbers of finite length, thus a far cry from the idealistic number system familiar to us from mathematical analysis. The passage from a real number to a machine number entails *rounding*, and thus small errors, called *roundoff errors*. Additional errors are introduced when the individual arithmetic operations are carried out on the computer. In themselves, these errors are harmless, but acting in concert and propagating through a lengthy computation, they can have significant — even disastrous — effects.

Most problems involve input data not representable exactly on the computer. Therefore, even before the solution process starts, simply by storing the input in computer memory, the problem is already slightly perturbed, owing to the necessity of rounding the input. It is important, then, to estimate how such small perturbations in the input affect the output, the solution of the problem. This is the question of the (numerical) *condition of a problem*: the problem is called well-conditioned if the changes in the solution of the problem are of the same order of magnitude as the perturbations in the input that caused those changes. If, on the other hand, they are much larger, the problem is called ill-conditioned. It is desirable to measure by a single number — the *condition number* of the problem — the extent to which the solution is sensitive to perturbations in the input. The larger this number, the more ill-conditioned the problem.

Once the solution process starts, additional rounding errors will be committed, which also contaminate the solution. The resulting errors, in contrast to those caused by input errors, depend on the particular solution algorithm. It makes sense, therefore, to also talk about the *condition of an algorithm*, although its analysis is usually quite a bit harder. The quality of the computed solution is then determined by both (essentially the product of) the condition of the problem and the condition of the algorithm.

§1. Real Numbers, Machine Numbers, and Rounding

We begin with the number system commonly used in mathematical analysis and confront it with the more primitive number system available to us on any particular computer. We identify the basic constant (the machine precision) that determines the level of precision attainable on such a computer.

§1.1. **Real numbers**. One can introduce real numbers in many different ways. Mathematicians favor the axiomatic approach, which leads them to define the set of real numbers as a "complete Archimedean ordered field." Here we adopt a more pedestrian attitude and consider the set of real numbers $\mathbb{R}$ to consist of positive and negative numbers represented in some appropriate number system and manipulated in the usual manner known from elementary arithmetic. We adopt here the *binary number system*, since it is the one most commonly used on computers. Thus,

$$x \in \mathbb{R} \text{ iff } x = \pm\,(b_n 2^n + b_{n-1} 2^{n-1} + \cdots + b_0 + b_{-1} 2^{-1} + b_{-2} 2^{-2} + \cdots). \quad (1.1)$$

Here $n \geq 0$ is some integer, and the "binary digits" b_i are either 0 or 1,

$$b_i = 0 \quad \text{or} \quad b_i = 1 \quad \text{for all} \quad i. \quad (1.2)$$

It is important to note that in general we need infinitely many binary digits to represent a real number. We conveniently write such a number in the abbreviated form (familiar from the decimal number system)

$$x = \pm\,(b_n b_{n-1} \cdots b_0 \,.\, b_{-1} b_{-2} b_{-3} \cdots)_2, \quad (1.3)$$

where the subscript 2 at the end is to remind us that we are dealing with a binary number. (Without this subscript, the number could also be read as a decimal number, which would be a source of ambiguity.) The dot in (1.3) — appropriately called the binary point — separates the integer part on the left from the fractional part on the right. Note that the representation (1.3) is not unique; for example, $(.011\bar{1}\ldots)_2 = (.1)_2$. We regain uniqueness if we always insist on a finite representation, if one exists.

Examples.

(1) $(10011.01)_2 = 2^4 + 2^1 + 2^0 + 2^{-2} = 16 + 2 + 1 + \frac{1}{4} = (19.25)_{10}$

(2) $(.010\overline{1}\overline{01}\ldots)_2 = \sum_{\substack{k=2 \\ (k \text{ even})}}^{\infty} 2^{-k} = \sum_{m=1}^{\infty} 2^{-2m} = \frac{1}{4}\sum_{m=0}^{\infty}\left(\frac{1}{4}\right)^m$

$= \frac{1}{4}\,\frac{1}{1-\frac{1}{4}} = \frac{1}{3} = (.33\overline{3}\ldots)_{10}$

(3) $\frac{1}{5} = (.2)_{10} = (.0011\overline{0011}\ldots)_2$

To determine the binary digits on the right, one keeps multiplying by 2 and observing the integer part in the result; if it is zero, the binary digit in question is 0, otherwise 1. In the latter case, the integral part is removed and the process repeated.

The last example is of interest insofar as it shows that to a finite decimal number there may correspond a (nontrivial) infinite binary representation. One cannot assume, therefore, that a finite decimal number is exactly representable on a binary computer. Conversely, however, to a finite binary number there always corresponds a finite decimal representation. (Why?)

§1.2. **Machine numbers**. There are two kinds of machine numbers: floating-point and fixed-point. The first corresponds to the "scientific notation" in the decimal system, whereby a number is written as a decimal fraction times an integral power of 10. The second allows only for fractions. On a binary computer, one consistently uses powers of 2 instead of 10. More important, the number of binary digits, both in the fraction and in the exponent of 2 (if any), is finite and cannot exceed certain limits that are characteristics of the particular computer at hand.

(a) *Floating-point numbers.* We denote by t the number of binary digits allowed by the computer in the fractional part, and by s the number of binary digits in the exponent. Then the set of (real) floating-point numbers on that computer will be denoted by $\mathbb{R}(t,s)$. Thus,

$$x \in \mathbb{R}(t,s) \quad \text{iff} \quad x = f \cdot 2^e, \tag{1.4}$$

where, in the notation of (1.3),

$$f = \pm\,(.\,b_{-1}b_{-2}\cdots b_{-t})_2, \quad e = \pm\,(c_{s-1}c_{s-2}\cdots c_0.)_2. \tag{1.5}$$

Here all b_i and c_j are binary digits, that is, either zero or one. The binary fraction f is usually referred to as the *mantissa* of x, and the integer e as the *exponent* of x. The number x in (1.4) is said to be *normalized* if in its fraction f we have $b_{-1} = 1$. We assume that all numbers in $\mathbb{R}(t, s)$ are normalized (with the exception of $x = 0$, which is treated as a special number). If $x \neq 0$ were not normalized, we could multiply f by an appropriate power of 2, to normalize it, and adjust the exponent accordingly. This is always possible as long as the adjusted exponent is still in the admissible range.

We can think of a floating-point number (1.4) as being accommodated in a machine register as shown in Figure 1.1.1. The figure does not quite correspond to reality, but is close enough to it for our purposes.

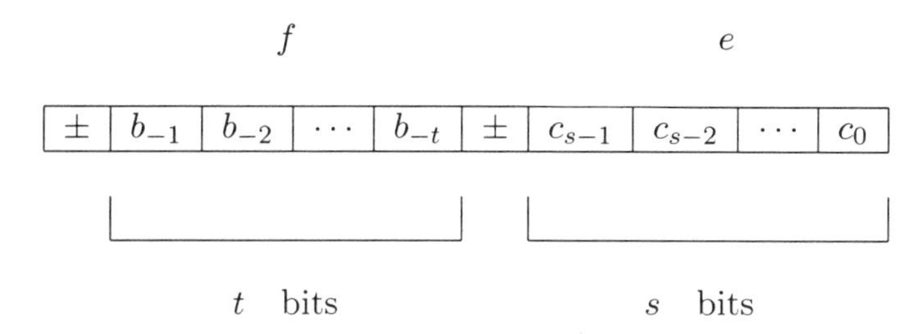

FIGURE 1.1.1. Packing of a floating-point number in a machine register

Note that the set (1.4) of normalized floating-point numbers is finite, and is thus represented by a finite set of points on the real line. What is worse, these points are not uniformly distributed (cf. Ex. 1). This, then, is all we have to work with!

It is immediately clear from (1.4) and (1.5) that the largest and smallest magnitude of a (normalized) floating-point number is given, respectively, by

$$\max_{x \in \mathbb{R}(t,s)} |x| = (1 - 2^{-t})\, 2^{2^s - 1}, \quad \min_{x \in \mathbb{R}(t,s)} |x| = 2^{-2^s}. \tag{1.6}$$

On a SUN SPARC workstation, for example, one has $t = 23$, $s = 7$, so that the maximum and minimum in (1.6) are 1.70×10^{38} and 2.94×10^{-39}, respectively. (Because of an asymmetric internal hardware representation of the exponent on these computers, the true range of floating-point numbers is slightly shifted, more like from 1.18×10^{-38} to 3.40×10^{38}.)

A real nonzero number whose modulus is not in the range determined by (1.6) cannot be represented on this particular computer. If such a number is produced during the course of a computation, one says that *overflow* has occurred if its modulus is larger than the maximum in (1.6), and *underflow* if it is smaller than the minimum in (1.6). The occurrence of overflow is fatal, and the machine (or its operating system) usually prompts the computation to be interrupted. Underflow is less serious, and one may get away with replacing the delinquent number by zero. However, this is not foolproof. Imagine that at the next step the number that underflowed is to be multiplied by a huge number. If the replacement by zero has been made, the result will always be zero.

In order to increase the precision, one can use two machine registers to represent a machine number. In effect, one then embeds $\mathbb{R}(t,s) \subset \mathbb{R}(2t,s)$, and calls $x \in \mathbb{R}(2t,s)$ a *double-precision* number.

(b) *Fixed-point numbers.* This is the case (1.4) where $e = 0$. That is, fixed-point numbers are binary fractions, $x = f$, hence $|f| < 1$. We can therefore only deal with numbers that are in the interval (–1,1). This, in particular, requires extensive scaling and rescaling to make sure that all initial data, as well as all intermediate and final results, lie in that interval. Such a complication can only be justified in special circumstances where machine time and/or precision are at a premium. Note that on the same computer as considered before, we do not need to allocate space for the exponent in the machine register, and thus have in effect $s + t$ binary digits available for the fraction f, hence more precision; cf. Figure 1.1.2.

$\pm$	b_{-1}	b_{-2}	$\cdots$	b_{-t}	$b_{-(t+1)}$	$\cdots$	$b_{-(t+s)}$

FIGURE 1.1.2. Packing of a fixed-point number in a machine register

(c) *Other data structures for numbers.* Complex floating-point numbers consist of pairs of real floating-point numbers, the first of the pair representing the real part and the second the imaginary part. To avoid rounding errors in arithmetic operations altogether, one can employ rational arithmetic, in which each (rational) number is represented by a pair of extended-precision integers — the numerator and denominator of the rational number. The Euclidean algorithm is used to remove common factors. A device that allows keeping track of error propagation and the influence of data errors is interval arithmetic involving intervals guaranteed to contain the desired numbers. In complex arithmetic one employs rectangular or circular domains.

§1.3. **Rounding.** A machine register acts much like the infamous Procrustes bed in Greek mythology. Procrustes was the innkeeper whose inn had only beds of one size. If a fellow came along who was too tall to fit into his beds, he cut off his feet. If the fellow was too short, he stretched him. In the same way, if a real number comes along that is too long, its tail end (not the head!) is cut off; if it is too short, it is padded by zeros at the end.

More specifically, let

$$x \in \mathbb{R}, \quad x = \pm \left(\sum_{k=1}^{\infty} b_{-k} 2^{-k} \right) 2^{e} \tag{1.7}$$

be the "exact" real number (in normalized floating-point form), and

$$x^* \in \mathbb{R}(t,s), \quad x^* = \pm \left(\sum_{k=1}^{t} b^*_{-k} 2^{-k} \right) 2^{e^*} \tag{1.8}$$

the rounded number. One then distinguishes between two methods of rounding, the first being Procrustes' method.

(a) *Chopping.* One takes

$$x^* = \operatorname{chop}(x), \quad e^* = e, \quad b^*_{-k} = b_{-k} \quad \text{for} \quad k = 1, 2, \ldots, t. \tag{1.9}$$

(b) *Symmetric rounding.* This corresponds to the familiar rounding up or rounding down in decimal arithmetic, based on the first discarded decimal digit: if it is larger than or equal to 5, one rounds up; if it is less than 5, one rounds down. In binary arithmetic, the procedure is somewhat simpler, since there are only two possibilities: either the first discarded binary digit is 1, in which case one rounds up, or it is 0, in which case one rounds down. We can write the procedure very simply in terms of the chop operation in (1.9):

$$x^* = \operatorname{rd}(x), \quad \operatorname{rd}(x) := \operatorname{chop}\left(x + \frac{1}{2} \cdot 2^{-t} \cdot 2^{e} \right). \tag{1.10}$$

There is a small error incurred in rounding, which is most easily estimated in the case of chopping. Here the *absolute error* $|x - x^*|$ is

$$\begin{aligned} |x - \operatorname{chop}(x)| &= \left| \pm \sum_{k=t+1}^{\infty} b_{-k} 2^{-k} \right| 2^{e} \\ &\leq \sum_{k=t+1}^{\infty} 2^{-k} \cdot 2^{e} = 2^{-t} \cdot 2^{e}. \end{aligned}$$

It depends on e (i.e., the magnitude of x), which is the reason why one prefers the *relative error* $|(x - x^*)/x|$ (if $x \neq 0$), which, for normalized x, can be estimated as

$$\left|\frac{x - \text{chop}(x)}{x}\right| \leq \frac{2^{-t} \cdot 2^e}{\left|\pm\sum_{k=1}^{\infty} b_{-k} 2^{-k}\right| 2^e} \leq \frac{2^{-t} \cdot 2^e}{\frac{1}{2} \cdot 2^e} = 2 \cdot 2^{-t}. \tag{1.11}$$

Similarly, in the case of symmetric rounding, one finds (cf. Ex. 6)

$$\left|\frac{x - \text{rd}(x)}{x}\right| \leq 2^{-t}. \tag{1.12}$$

The number on the right is an important, machine-dependent quantity, called the *machine precision*,

$$\text{eps} = 2^{-t}; \tag{1.13}$$

it determines the level of precision of any large-scale floating-point computation. On the SUN SPARC workstation, where $t = 23$, we have eps $\approx 1.19 \times 10^{-7}$, corresponding to a precision of 6 to 7 significant decimal digits.

Since it is awkward to work with inequalities, one prefers writing (1.12) equivalently as an equality,

$$\text{rd}(x) = x(1 + \varepsilon), \quad |\varepsilon| \leq \text{eps}, \tag{1.14}$$

and defers dealing with the inequality (for ε) to the very end.

§2. Machine Arithmetic

The arithmetic used on computers unfortunately does not respect the laws of ordinary arithmetic. Each elementary floating-point operation, in general, generates a small error that may then propagate through subsequent machine operations. As a rule, this error propagation is harmless, except in the case of subtraction, where cancellation effects may seriously compromise the accuracy of the results.

§2.1. **A model of machine arithmetic**. Any of the four basic arithmetic operations, when applied to two machine numbers, may produce a result no longer representable on the computer. We have therefore errors also associated with arithmetic operations. Barring the occurrence of overflow or underflow, we may assume as a *model of machine arithmetic* that each arithmetic operation $\circ$ $(= +, -, \times, /)$ produces a correctly rounded result. Thus, if x, $y \in \mathbb{R}(t,s)$ are floating-point machine numbers, and $\mathrm{fl}(x \circ y)$ denotes the machine-produced result of the arithmetic operation $x \circ y$, then

$$\mathrm{fl}(x \circ y) = x \circ y\,(1+\varepsilon), \quad |\varepsilon| \leq \mathrm{eps}. \tag{2.1}$$

This can be interpreted in a number of ways; for example, in the case of multiplication,

$$\mathrm{fl}(x \times y) = [x(1+\varepsilon)] \times y = x \times [y(1+\varepsilon)] = (x\sqrt{1+\varepsilon}) \times (y\sqrt{1+\varepsilon}) = \cdots .$$

In each equation we identify the computed result as the exact result on data that are slightly perturbed, whereby the respective relative perturbations can be estimated, for example, by $|\varepsilon| \leq \mathrm{eps}$ in the first two equations, and $\sqrt{1+\epsilon} \approx 1 + \frac{1}{2}\varepsilon$, $\left|\frac{1}{2}\varepsilon\right| \leq \frac{1}{2}\mathrm{eps}$ in the third. These are elementary examples of *backward error analysis*, a powerful tool for estimating errors in machine computation.

Even though a single arithmetic operation causes a small error that can be neglected, a succession of arithmetic operations can well result in a significant error, owing to *error propagation*. It is like the small microorganisms that we all carry in our bodies: if our defense mechanism is in good order, the microorganisms cause no harm, in spite of their large presence. If for some reason our defenses are weakened, then all of a sudden they can play havoc with our health. The same is true in machine computation: the rounding errors, although widespread, will cause little harm unless our computations contain some weak spots that allow rounding errors to take over to the point of completely invalidating the results. We learn about one such weak spot (indeed the only one) in the next subsection.[1]

[1]Rounding errors can also have significant implications in real life. One example, taken from politics, concerns the problem of apportionment: how should the representatives in an assembly, such as the US House of Representatives or the Electoral College, be constituted to fairly reflect the size of population in the various states? If the total number of representatives in the assembly is given, say, A, the total population of the US is P, and the population of State i is p_i, then State i should be allocated

$$r_i = \frac{p_i}{P} A$$

representatives. The problem is that r_i is not an integer, in general. How then should r_i

§2.2. **Error propagation in arithmetic operations; cancellation error.** We now study the extent to which the basic arithmetic operations propagate errors already present in their operands. Previously, in §2.1, we assumed the operands to be exact machine-representable numbers and discussed the errors due to imperfect execution of the arithmetic operations by the computer. We now change our viewpoint and assume that the operands themselves are contaminated by errors, but the arithmetic operations are carried out exactly. (We already know what to do, cf. (2.1), when we are dealing with machine operations.) Our interest is in the errors in the results caused by errors in the data.

(a) *Multiplication.* We consider values $x(1+\varepsilon_x)$ and $y(1+\varepsilon_y)$ of x and y contaminated by relative errors ε_x and ε_y, respectively. What is the relative error in the product? We assume ε_x, ε_y sufficiently small so that quantities of second order, ε_x^2, $\varepsilon_x\varepsilon_y$, ε_y^2 — and even more so, quantities of still higher order — can be neglected against the epsilons themselves. Then

$$x(1+\varepsilon_x)\cdot y(1+\varepsilon_y) = x\cdot y\,(1+\varepsilon_x+\varepsilon_y+\varepsilon_x\varepsilon_y) \approx x\cdot y\,(1+\varepsilon_x+\varepsilon_y).$$

Thus, the relative error $\varepsilon_{x\cdot y}$ in the product is given (at least approximately) by

$$\varepsilon_{x\cdot y} = \varepsilon_x + \varepsilon_y; \tag{2.2}$$

that is, the (relative) errors in the data are being added to produce the (relative) error in the result. We consider this to be acceptable error propagation, and in this sense, multiplication is a *benign* operation.

(b) *Division.* Here we have similarly (if $y \neq 0$)

$$\begin{aligned}\frac{x(1+\varepsilon_x)}{y(1+\varepsilon_y)} &= \frac{x}{y}(1+\varepsilon_x)(1-\varepsilon_y+\varepsilon_y^2-+\cdots)\\ &\approx \frac{x}{y}(1+\varepsilon_x-\varepsilon_y);\end{aligned}$$

---

be rounded to an integer r_i^*? One can think of three natural criteria to be imposed: (i) r_i^* should be one of the two integers closest to r_i ("quota condition"). (ii) If A is increased, all other things being the same, then r_i^* should not decrease ("house monotonicity"). (iii) If p_i is increased, the other p_j remaining constant, then r_i^* should not decrease ("population monotonicity"). Unfortunately, there is no apportionment method that satisfies all three criteria. There is indeed a case in US history when Samuel J. Tilden lost his bid for the presidency in 1876 in favor of Rutherford B. Hayes, purely on the basis of the apportionment method adopted on that occasion (which, incidentally, was not the one prescribed by law at the time).

that is,

$$\varepsilon_{x/y} = \varepsilon_x - \varepsilon_y. \tag{2.3}$$

Also division is a benign operation.

(c) *Addition and subtraction.* Since x and y can be numbers of arbitrary signs, it suffices to look at addition. We have

$$\begin{aligned} x(1+\varepsilon_x) + y(1+\varepsilon_y) &= x + y + x\varepsilon_x + y\varepsilon_y \\ &= (x+y)\left(1 + \frac{x\varepsilon_x + y\varepsilon_y}{x+y}\right), \end{aligned}$$

assuming $x + y \neq 0$. Therefore,

$$\varepsilon_{x+y} = \frac{x}{x+y}\varepsilon_x + \frac{y}{x+y}\varepsilon_y. \tag{2.4}$$

As before, the error in the result is a linear combination of the errors in the data, but now the coefficients are no longer ± 1 but can assume values that are arbitrarily large. Note first, however, that when x and y have the same sign, then both coefficients are positive and bounded by 1, so that

$$|\varepsilon_{x+y}| \leq |\varepsilon_x| + |\varepsilon_y| \quad (x \cdot y > 0); \tag{2.5}$$

addition, in this case, is again a benign operation. It is only when x and y have opposite signs that the coefficients in (2.4) can be arbitrarily large, namely, when $|x + y|$ is arbitrarily small compared to $|x|$ and $|y|$. This happens when x and y are almost equal in absolute value, but opposite in sign. The large magnification of error then occurring in (2.4) is referred to as *cancellation error*. It is the only serious weakness — the Achilles heel, as it were — of numerical computation, and it should be avoided whenever possible. In particular, one should be prepared to encounter cancellation effects not only in single devastating amounts, but also repeatedly over a long period of time involving "small doses" of cancellation. Either way, the end result can be disastrous.

We illustrate the cancellation phenomenon schematically in Figure 1.2.1, where b, b', b'' stand for binary digits that are reliable, and the gs represent binary digits contaminated by error; these are often called "garbage" digits. Note in Figure 1.2.1 that "garbage – garbage = garbage," but, more important, that the final normalization of the result moves the first garbage digit from the 12th position to the 3rd.

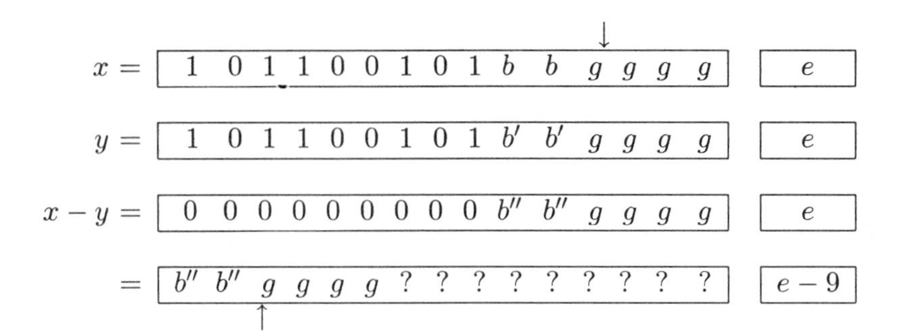

FIGURE 1.2.1. The cancellation phenomenon

Cancellation is such a serious matter that we wish to give a number of elementary examples, not only of its occurrence, but also of how it might be avoided.

Examples.

(1) An algebraic identity: $(a-b)^2 = a^2-2ab+b^2$. Although this is a valid identity in algebra, it is no longer valid in machine arithmetic. Thus, on a 2-decimal-digit computer, with $a = 1.8$, $b = 1.7$, we get, using symmetric rounding,

$$\mathrm{fl}(a^2 - 2ab + b^2) = 3.2 - 6.2 + 2.9 = -.10$$

instead of the true result .010, which we obtain also on our 2-digit computer if we use the left-hand side of the identity. The expanded form of the square thus produces a result which is off by one order of magnitude and on top has the wrong sign!

(2) Quadratic equation: $x^2 - 56x + 1 = 0$. The usual formula for a quadratic gives, in 5-decimal arithmetic,

$$x_1 = 28 - \sqrt{783} = 28 - 27.982 = .018000,$$

$$x_2 = 28 + \sqrt{783} = 28 + 27.982 = 55.982.$$

This should be contrasted with the exact roots .0178628... and 55.982137... . As can be seen, the smaller of the two is obtained to only two correct decimal digits, owing to cancellation. An easy way out, of course, is to compute x_2 first, which involves a benign addition, and then to compute $x_1 = 1/x_2$ by Vieta's formula, which again involves a benign operation — division. In this way we obtain both roots to full machine accuracy.

(3) Compute $y = \sqrt{x+\delta} - \sqrt{x}$, where $x > 0$ and $|\delta|$ is very small. Clearly, the formula as written causes severe cancellation errors, since each square root has to be rounded. Writing instead

$$y = \frac{\delta}{\sqrt{x+\delta} + \sqrt{x}}$$

completely removes the problem.

(4) Compute $y = \cos(x + \delta) - \cos x$, where $|\delta|$ is very small. Here cancellation can be avoided by writing y in the equivalent form

$$y = -2 \sin \frac{\delta}{2} \sin \left(x + \frac{\delta}{2} \right).$$

(5) Compute $y = f(x + \delta) - f(x)$, where $|\delta|$ is very small and f a given function. Special tricks, such as those used in the two preceding examples, can no longer be played, but if f is sufficiently smooth in the neighborhood of x, we can use Taylor expansion:

$$y = f'(x)\delta + \tfrac{1}{2} f''(x)\delta^2 + \cdots .$$

The terms in this series decrease rapidly when $|\delta|$ is small, so that cancellation is no longer a problem.

Addition is an example of a potentially ill-conditioned function (of two variables). It naturally leads us to study the condition of more general functions.

§3. The Condition of a Problem

A problem typically has an input and an output. The input consists of a set of data, say, the coefficients of some equation, and the output of another set of numbers uniquely determined by the input, say, all the roots of the equation in some prescribed order. If we collect the input in a vector $x \in \mathbb{R}^m$ (assuming the data consist of real numbers), and the output in the vector $y \in \mathbb{R}^n$ (also assumed real), we have the black box situation shown in Figure 1.3.1, where the box P accepts some input x and then solves the problem for this input to produce the output y.

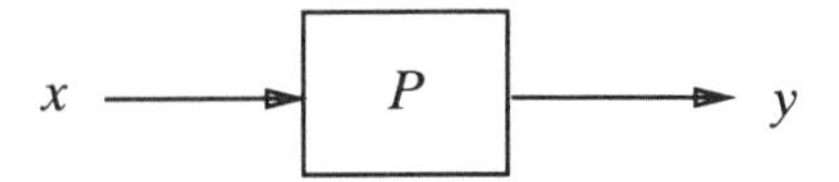

FIGURE 1.3.1. Black box representation of a problem

We may thus think of a problem as a map f, given by

$$f: \quad \mathbb{R}^m \to \mathbb{R}^n, \quad y = f(x). \tag{3.1}$$

(One or both of the spaces $\mathbb{R}^m$, $\mathbb{R}^n$ could be complex spaces without changing in any essential way the discussion that follows.) What we are interested in is the sensitivity of the map f at some given point x to a small perturbation of x, that is, how much bigger (or smaller) the perturbation in y is compared to the perturbation in x. In particular, we wish to measure the degree of sensitivity by a single number — the *condition number* of the map f at the point x. We emphasize that, as we perturb x, the function f is always assumed to be evaluated exactly, with infinite precision. The condition of f, therefore, is an inherent property of the map f and does not depend on any algorithmic considerations concerning its implementation.

This is not to say that knowledge of the condition of a problem is irrelevant to any algorithmic solution of the problem. On the contrary! The reason is that quite often the *computed* solution y^* of (3.1) (computed in floating-point machine arithmetic, using a specific algorithm) can be demonstrated to be the *exact* solution to a "nearby" problem; that is,

$$y^* = f(x^*), \tag{3.2}$$

where x^* is a vector close to the given data x,

$$x^* = x + \delta, \tag{3.3}$$

and moreover, the distance $\|\delta\|$ of x^* to x can be estimated in terms of the machine precision. Therefore, if we know how strongly (or weakly) the map f reacts to a small perturbation, such as δ in (3.3), we can say something about the error $y^* - y$ in the solution caused by this perturbation. This, indeed, is an important technique of error analysis — known as *backward error analysis* — which was pioneered in the 1950s by J. W. Givens, C. Lanczos, and, above all, J. H. Wilkinson.

Maps f between more general spaces (in particular, function spaces) have also been considered from the point of view of conditioning, but eventually, these spaces have to be reduced to finite-dimensional spaces for practical implementation.

§3.1. **Condition numbers**. We start with the simplest case of a single function of one variable.

The case $m = n = 1$: $y = f(x)$. Assuming first $x \neq 0$, $y \neq 0$, and denoting by Δx a small perturbation of x, we have for the corresponding perturbation Δy by Taylor's formula

$$\Delta y = f(x + \Delta x) - f(x) \approx f'(x)\Delta x, \tag{3.4}$$

assuming that f is differentiable at x. Since our interest is in *relative* errors, we write this in the form

$$\frac{\Delta y}{y} \approx \frac{xf'(x)}{f(x)} \cdot \frac{\Delta x}{x}. \tag{3.5}$$

The approximate equality becomes a true equality in the limit as $\Delta x \to 0$. This suggests that the condition of f at x be defined by the quantity

$$(\text{cond}\, f)(x) := \left| \frac{xf'(x)}{f(x)} \right|. \tag{3.6}$$

This number tells us how much larger the relative perturbation in y is compared to the relative perturbation in x.

If $x = 0$ and $y \neq 0$, it is more meaningful to consider the absolute error measure for x and for y still the relative error. This leads to the condition number $|f'(x)/f(x)|$. Similarly for $y = 0$, $x \neq 0$. If $x = y = 0$, the condition number by (3.4) would then simply be $|f'(x)|$.

The case of arbitrary m, n: Here we write

$$x = [x_1, x_2, \ldots, x_m]^T \in \mathbb{R}^m, \quad y = [y_1, y_2, \ldots, y_n]^T \in \mathbb{R}^n$$

and exhibit the map f in component form

$$y_\nu = f_\nu(x_1, x_2, \ldots, x_m), \quad \nu = 1, 2, \ldots, n. \tag{3.7}$$

We assume again that each function f_ν has partial derivatives with respect to all m variables at the point x. Then the most detailed analysis departs from considering each component y_ν as a function of one single variable, x_μ. In other words, we subject only one variable, x_μ, to a small change and observe the resulting change in just one component, y_ν. Then we can apply (3.6) and obtain

$$\gamma_{\nu\mu}(x) := (\mathrm{cond}_{\nu\mu} f)(x) := \left| \frac{x_\mu \frac{\partial f_\nu}{\partial x_\mu}}{f_\nu(x)} \right| . \tag{3.8}$$

This gives us a whole matrix $\Gamma(x) = [\gamma_{\nu\mu}(x)] \in \mathbb{R}_+^{n\times m}$ of condition numbers. To obtain a single condition number, we can take any convenient measure of the "magnitude" of the matrix $\Gamma(x)$ such as one of the matrix norms defined in (3.11),

$$(\mathrm{cond}\, f)(x) = \|\Gamma(x)\|, \quad \Gamma(x) = [\gamma_{\nu\mu}(x)]. \tag{3.9}$$

The condition so defined, of course, depends on the choice of norm, but the order of magnitude (and that is all that counts) should be more or less the same for any reasonable norm.

If a component of x, or of y, vanishes, one modifies (3.8) as discussed earlier.

A less refined analysis can be modeled after the one-dimensional case by defining the relative perturbation of $x \in \mathbb{R}^m$ to mean

$$\frac{\|\Delta x\|_{\mathbb{R}^m}}{\|x\|_{\mathbb{R}^m}}, \quad \Delta x = [\Delta x_1, \Delta x_2, \ldots, \Delta x_m]^T, \tag{3.10}$$

where Δx is a perturbation vector whose components Δx_μ are small compared to x_μ, and where $\|\cdot\|_{\mathbb{R}^m}$ is some vector norm in $\mathbb{R}^m$. For the perturbation Δy caused by Δx, one defines similarly the relative perturbation $\|\Delta y\|_{\mathbb{R}^n}/\|y\|_{\mathbb{R}^n}$, with a suitable vector norm $\|\cdot\|_{\mathbb{R}^n}$ in $\mathbb{R}^n$. One then tries to relate the relative perturbation in y to the one in x.

To carry this out, one needs to define a matrix norm for matrices $A \in \mathbb{R}^{n\times m}$. We choose the so-called "operator norm,"

$$\|A\|_{\mathbb{R}^{n\times m}} := \max_{\substack{x\in\mathbb{R}^m \\ x\neq 0}} \frac{\|Ax\|_{\mathbb{R}^n}}{\|x\|_{\mathbb{R}^m}} . \tag{3.11}$$

In the following we take for the vector norms the "uniform" (or infinity) norm,

$$\|x\|_{\mathbb{R}^m} = \max_{1\le\mu\le m} |x_\mu| =: \|x\|_\infty, \quad \|y\|_{\mathbb{R}^n} = \max_{1\le\nu\le n} |y_\nu| =: \|y\|_\infty . \tag{3.12}$$

It is then easy to show that (cf. Ex. 30)

$$\|A\|_{\mathbb{R}^{n\times m}} =: \|A\|_\infty = \max_{1\le\nu\le n} \sum_{\mu=1}^{m} |a_{\nu\mu}|, \quad A = [a_{\nu\mu}] \in \mathbb{R}^{n\times m}. \tag{3.13}$$

Now in analogy to (3.4), we have

$$\Delta y_\nu = f_\nu(x+\Delta x) - f_\nu(x) \approx \sum_{\mu=1}^{m} \frac{\partial f_\nu}{\partial x_\mu} \Delta x_\mu.$$

Therefore, at least approximately,

$$\begin{aligned} |\Delta y_\nu| &\le \sum_{\mu=1}^{m} \left|\frac{\partial f_\nu}{\partial x_\mu}\right| |\Delta x_\mu| \le \max_\mu |\Delta x_\mu| \cdot \sum_{\mu=1}^{m} \left|\frac{\partial f_\nu}{\partial x_\mu}\right| \\ &\le \max_\mu |\Delta x_\mu| \cdot \max_\nu \sum_{\mu=1}^{m} \left|\frac{\partial f_\nu}{\partial x_\mu}\right|. \end{aligned}$$

Since this holds for each $\nu = 1, 2, \ldots, n$, it also holds for $\max_\nu |\Delta y_\nu|$, giving, in view of (3.12) and (3.13),

$$\|\Delta y\|_\infty \le \|\Delta x\|_\infty \left\|\frac{\partial f}{\partial x}\right\|_\infty. \tag{3.14}$$

Here

$$\frac{\partial f}{\partial x} = \begin{bmatrix} \frac{\partial f_1}{\partial x_1} & \frac{\partial f_1}{\partial x_2} & \cdots & \frac{\partial f_1}{\partial x_m} \\ \frac{\partial f_2}{\partial x_1} & \frac{\partial f_2}{\partial x_2} & \cdots & \frac{\partial f_2}{\partial x_m} \\ \cdot & \cdot & \cdots & \cdot \\ \frac{\partial f_n}{\partial x_1} & \frac{\partial f_n}{\partial x_2} & \cdots & \frac{\partial f_n}{\partial x_m} \end{bmatrix} \in \mathbb{R}^{n\times m} \tag{3.15}$$

is the *Jacobian matrix* of f. (This is the analogue of the first derivative for *systems* of functions of *several* variables.) From (3.14) one now immediately obtains for the relative perturbations

$$\frac{\|\Delta y\|_\infty}{\|y\|_\infty} \le \frac{\|x\|_\infty \, \|\partial f/\partial x\|_\infty}{\|f(x)\|_\infty} \cdot \frac{\|\Delta x\|_\infty}{\|x\|_\infty}.$$

Although this is an inequality, it is sharp in the sense that equality can be achieved for a suitable perturbation Δx. We are justified, therefore, in defining a global condition number by

$$(\operatorname{cond} f)(x) := \frac{\|x\|_\infty \|\partial f/\partial x\|_\infty}{\|f(x)\|_\infty} . \tag{3.16}$$

Clearly, in the case $m = n = 1$, the definition (3.16) reduces precisely to the definition (3.6) (as well as (3.9)) given earlier. In higher dimensions (m and/or n larger than 1), however, the condition number in (3.16) is much cruder than the one in (3.9). This is because norms tend to destroy detail: if x, for example, has components of vastly different magnitudes, then $\|x\|_\infty$ is simply equal to the largest of these components, and all the others are ignored. For this reason, some caution is required when using (3.16).

To give an example, consider

$$f(x) = \begin{bmatrix} \dfrac{1}{x_1} + \dfrac{1}{x_2} \\ \dfrac{1}{x_1} - \dfrac{1}{x_2} \end{bmatrix}, \quad x = \begin{bmatrix} x_1 \\ x_2 \end{bmatrix}.$$

The components of the condition matrix $\Gamma(x)$ in (3.8) are then

$$\gamma_{11} = \left| \frac{x_2}{x_1 + x_2} \right|, \quad \gamma_{12} = \left| \frac{x_1}{x_1 + x_2} \right|, \quad \gamma_{21} = \left| \frac{x_2}{x_2 - x_1} \right|, \quad \gamma_{22} = \left| \frac{x_1}{x_2 - x_1} \right|,$$

indicating ill-conditioning if either $x_1 \approx x_2$ or $x_1 \approx -x_2$ and $|x_1|$ (hence also $|x_2|$) is not small. The global condition number (3.16), on the other hand, since

$$\frac{\partial f}{\partial x}(x) = -\frac{1}{x_1^2 x_2^2} \begin{bmatrix} x_2^2 & x_1^2 \\ x_2^2 & -x_1^2 \end{bmatrix},$$

becomes, when L_1 vector and matrix norms are used (cf. Ex. 31),

$$(\operatorname{cond} f)(x) = \frac{\|x\|_1 \cdot \dfrac{2}{x_1^2 x_2^2} \max(x_1^2,\ x_2^2)}{\dfrac{1}{|x_1 x_2|}(|x_1 + x_2| + |x_1 - x_2|)} = 2 \frac{|x_1| + |x_2|}{|x_1 x_2|} \frac{\max(x_1^2,\ x_2^2)}{|x_1 + x_2| + |x_1 - x_2|}.$$

Here $x_1 \approx x_2$ or $x_1 \approx -x_2$ yield $(\operatorname{cond} f)(x) \approx 2$, which is obviously misleading.

§3.2. **Examples**. We illustrate the idea of numerical condition in a number of examples, some of which are of considerable interest in applications.

(1) Compute $I_n = \int_0^1 \frac{t^n}{t+5} dt$ for some fixed integer $n \geq 1$. As it stands, the example here deals with a map from the integers to reals, and therefore does not fit our concept of "problem" in (3.1). However, we propose to compute I_n recursively by relating I_k to I_{k-1} and noting that

$$I_0 = \int_0^1 \frac{dt}{t+5} = \ln(t+5)\Big|_0^1 = \ln\frac{6}{5}\,. \tag{3.17}$$

To find the recursion, observe that

$$\frac{t}{t+5} = 1 - \frac{5}{t+5}\,.$$

Thus, multiplying both sides by t^{k-1} and integrating from 0 to 1 yields

$$I_k = -5I_{k-1} + \frac{1}{k}, \quad k = 1, 2, \ldots, n. \tag{3.18}$$

We see that I_k is a solution of the (linear, inhomogeneous, first-order) difference equation

$$y_k = -5y_{k-1} + \frac{1}{k}, \quad k = 1, 2, 3, \ldots\,. \tag{3.19}$$

We now have what appears to be a practical scheme to compute I_n: start with $y_0 = I_0$ given by (3.17), and then apply in succession (3.19) for $k = 1, 2, \ldots, n$; then $y_n = I_n$. The recursion (3.19), for any starting value y_0, defines a function,

$$y_n = f_n(y_0). \tag{3.20}$$

We have the black box in Figure 1.3.2 and thus a problem $f_n: \mathbb{R} \to \mathbb{R}$.

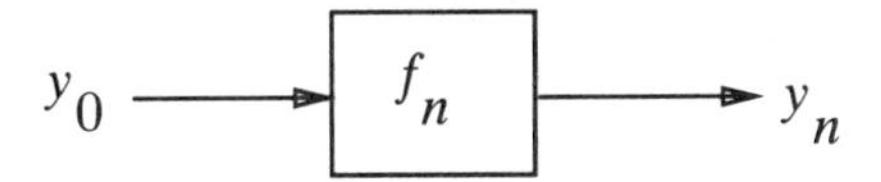

FIGURE 1.3.2. Black box for the recursion (3.19)

(Here n is a parameter.) We are interested in the condition of f_n at the point $y_0 = I_0$ given by (3.17). Indeed, I_0 in (3.17) is not machine-representable, and must be rounded to I_0^* before the recursion (3.19) can be employed. Even if no further errors are introduced during the recursion, the final result will not be exactly I_n, but some approximation $I_n^* = f_n(I_0^*)$, and we have, at least approximately (actually exactly; see the remark after (3.27)),

$$\left|\frac{I_n^* - I_n}{I_n}\right| = (\text{cond}\, f_n)(I_0)\left|\frac{I_0^* - I_0}{I_0}\right| . \tag{3.21}$$

To compute the condition number, note that f_n is a linear function of y_0. Indeed, if $n = 1$, then

$$y_1 = f_1(y_0) = -5y_0 + 1.$$

If $n = 2$, then

$$y_2 = f_2(y_0) = -5y_1 + \frac{1}{2} = (-5)^2 y_0 - 5 + \frac{1}{2} ,$$

and so on. In general,

$$y_n = f_n(y_0) = (-5)^n y_0 + p_n,$$

where p_n is some number (independent of y_0). There follows

$$(\text{cond}\, f_n)(y_0) = \left|\frac{y_0 f_n'(y_0)}{y_n}\right| = \left|\frac{y_0(-5)^n}{y_n}\right| . \tag{3.22}$$

Now, if $y_0 = I_0$, then $y_n = I_n$, and from the definition of I_n as an integral it is clear that I_n decreases monotonically in n (and indeed converges monotonically to zero as $n \to \infty$). Therefore,

$$(\text{cond}\, f_n)(I_0) = \frac{I_0 \cdot 5^n}{I_n} > \frac{I_0 \cdot 5^n}{I_0} = 5^n. \tag{3.23}$$

We see that $f_n(y_0)$ is severely ill-conditioned at $y_0 = I_0$, the more so the larger n.

We could have anticipated this result by just looking at the recursion (3.19): we keep multiplying by (–5), which tends to make things bigger, whereas they should get smaller! Thus, there will be continuous cancellation occurring throughout the recursion.

How can we avoid this ill-conditioning? The clue comes from the remark just made: instead of multiplying by a large number, we would prefer dividing by a large number, especially if the results get bigger at the same time. This is accomplished by reversing the recurrence (3.19), that is, by choosing an $\nu > n$ and computing

$$y_{k-1} = \frac{1}{5}\left(\frac{1}{k} - y_k\right), \quad k = \nu, \nu - 1, \ldots, n+1. \tag{3.24}$$

The problem then, of course, is how to compute the starting value y_ν. Before we deal with this, let us observe that we now have a new black box, as shown in Figure 1.3.3.

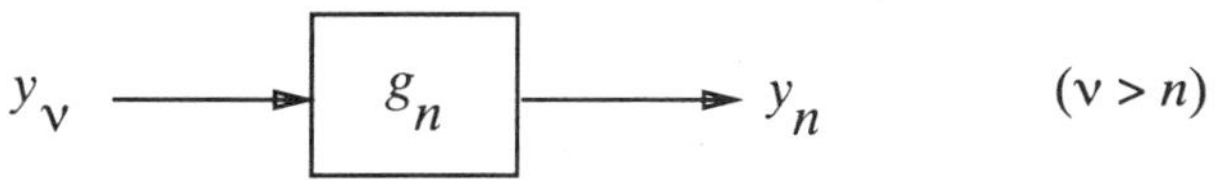

$y_\nu \longrightarrow \boxed{g_n} \longrightarrow y_n \qquad (\nu > n)$

FIGURE 1.3.3. Black box for the recursion (3.24)

As before, the function involved, g_n, is a linear function of y_ν, and an argument similar to the one leading to (3.22) then gives

$$(\operatorname{cond} g_n)(y_\nu) = \left|\frac{y_\nu\left(-\frac{1}{5}\right)^{\nu-n}}{y_n}\right|, \quad \nu > n. \tag{3.25}$$

For $y_\nu = I_\nu$, we get, again by the monotonicity of I_n,

$$(\operatorname{cond} g_n)(I_\nu) < \left(\frac{1}{5}\right)^{\nu-n}, \quad \nu > n. \tag{3.26}$$

In analogy to (3.21), we now have

$$\left|\frac{I_n^* - I_n}{I_n}\right| = (\operatorname{cond} g_n)(I_\nu)\left|\frac{I_\nu^* - I_\nu}{I_\nu}\right| < \left(\frac{1}{5}\right)^{\nu-n}\left|\frac{I_\nu^* - I_\nu}{I_\nu}\right|, \tag{3.27}$$

where I_ν^* is some approximation of I_ν. Actually, I_ν^* does not even have to be close to I_ν for (3.27) to hold, since the function g_n is linear. Thus, we may take $I_\nu^* = 0$, committing a 100% error in the starting value, yet obtaining I_n^* with a relative error

$$\left|\frac{I_n^* - I_n}{I_n}\right| < \left(\frac{1}{5}\right)^{\nu-n}, \quad \nu > n. \tag{3.28}$$

The bound on the right can be made arbitrarily small, say, $\leq \varepsilon$, if we choose ν large enough; for example,

$$\nu \geq n + \frac{\ln \frac{1}{\varepsilon}}{\ln 5} \,. \tag{3.29}$$

The final procedure, therefore, is: given the desired relative accuracy ε, choose ν to be the smallest integer satisfying (3.29), and then compute

$$\begin{aligned} I_\nu^* &= 0, \\ I_{k-1}^* &= \frac{1}{5}\left(\frac{1}{k} - I_k^*\right), \quad k = \nu, \nu - 1, \ldots, n+1. \end{aligned} \tag{3.30}$$

This will produce a sufficiently accurate $I_n^* \approx I_n$, even in the presence of rounding errors committed in (3.30): they, too, will be consistently attenuated.

Similar ideas can be applied to the more important problem of computing solutions to second-order linear recurrence relations such as those satisfied by Bessel functions and many other special functions of mathematical physics. The procedure of backward recurrence is then closely tied up with the theory of continued fractions.

(2) *Algebraic equations*: these are equations involving a polynomial of given degree n,

$$p(x) = 0, \;\; p(x) = x^n + a_{n-1}x^{n-1} + \cdots + a_1 x + a_0, \;\; a_0 \neq 0. \tag{3.31}$$

Let ξ be some fixed root of the equation, which we assume to be simple,

$$p(\xi) = 0, \quad p'(\xi) \neq 0. \tag{3.32}$$

The problem then is to find ξ, given p. The data vector $a = [a_0, a_1, \ldots, a_{n-1}]^T \in \mathbb{R}^n$ consists of the coefficients of the polynomial p, and the result is ξ, a real or complex number. Thus, we have

$$\xi : \;\; \mathbb{R}^n \to \mathbb{C}, \quad \xi = \xi(a_0, a_1, \ldots, a_{n-1}). \tag{3.33}$$

What is the condition of ξ? We adopt the detailed approach of (3.8) and first define

$$\gamma_\nu = (\text{cond}_\nu\, \xi)(a) = \left| \frac{a_\nu \frac{\partial \xi}{\partial a_\nu}}{\xi} \right|, \quad \nu = 0, 1, \ldots, n-1. \tag{3.34}$$

Then we take a convenient norm, say, the L_1 norm $\|\gamma\|_1 := \sum_{\nu=0}^{n-1} |\gamma_\nu|$ of the vector $\gamma = [\gamma_0, \ldots, \gamma_{n-1}]^T$, to define

$$(\operatorname{cond} \xi)(a) = \sum_{\nu=0}^{n-1} (\operatorname{cond}_\nu \xi)(a). \tag{3.35}$$

To determine the partial derivative of ξ with respect to a_ν, observe that we have the identity

$$[\xi(a_0, a_1, \ldots, a_n)]^n + a_{n-1}[\xi(\cdots)]^{n-1} + \cdots + a_\nu[\xi(\cdots)]^\nu + \cdots + a_0 \equiv 0.$$

Differentiating this with respect to a_ν, we get

$$n[\xi(a_0, a_1, \ldots, a_n)]^{n-1} \frac{\partial \xi}{\partial a_\nu} + a_{n-1}(n-1)[\xi(\cdots)]^{n-2} \frac{\partial \xi}{\partial a_\nu} + \cdots$$

$$+ a_\nu \nu [\xi(\cdots)]^{\nu-1} \frac{\partial \xi}{\partial a_\nu} + \cdots + a_1 \frac{\partial \xi}{\partial a_\nu} + [\xi(\cdots)]^\nu \equiv 0,$$

where the last term comes from differentiating the first factor in the product $a_\nu \xi^\nu$. The last identity can be written as

$$p'(\xi) \frac{\partial \xi}{\partial a_\nu} + \xi^\nu = 0.$$

Since $p'(\xi) \neq 0$, we can solve for $\partial \xi / \partial a_\nu$ and insert the result in (3.34) and (3.35) to obtain

$$(\operatorname{cond} \xi)(a) = \frac{1}{|\xi p'(\xi)|} \sum_{\nu=0}^{n-1} |a_\nu| \, |\xi|^\nu. \tag{3.36}$$

We illustrate (3.36) by considering the polynomial p of degree n that has the zeros $1, 2, \ldots, n$,

$$p(x) = \prod_{\nu=1}^{n} (x - \nu) = x^n + a_{n-1} x^{n-1} + \cdots + a_0. \tag{3.37}$$

This is a famous example due to J. H. Wilkinson, who discovered the ill-conditioning of some of the zeros almost by accident. If we let $\xi_\mu = \mu$, $\mu = 1, 2, \ldots, n$, it can be shown that

$$\min_\mu \operatorname{cond} \xi_\mu = \operatorname{cond} \xi_1 \sim n^2 \quad \text{as} \quad n \to \infty,$$

$$\max_{\mu} \operatorname{cond} \xi_\mu \sim \frac{1}{\left(2-\sqrt{2}\right)\pi n} \left(\frac{\sqrt{2}+1}{\sqrt{2}-1}\right)^n \quad \text{as} \quad n \to \infty.$$

The worst-conditioned root is ξ_{μ_0} with μ_0 the integer closest to $n/\sqrt{2}$, when n is large. Its condition number grows like $(5.828\ldots)^n$, thus exponentially fast in n. For example, when $n = 20$, then $\operatorname{cond} \xi_{\mu_0} = .540 \times 10^{14}$.

The example teaches us that the roots of an algebraic equation written in the form (3.31) can be extremely sensitive to small changes in the coefficients a_ν. It would, therefore, be ill-advised to express every polynomial in terms of powers, as in (3.37) and (3.31). This is particularly true for characteristic polynomials of matrices. It is much better here to work with the matrices themselves and try to reduce them (by similarity transformations) to a form that allows the eigenvalues — the roots of the characteristic equation — to be read off relatively easily.

(3) *Systems of linear algebraic equations*: given a nonsingular square matrix $A \in \mathbb{R}^{n\times n}$, and a vector $b \in \mathbb{R}^n$, the problem now discussed is solving the system

$$Ax = b. \tag{3.38}$$

Here the data are the elements of A and b, and the result the vector x. The map in question is thus $\mathbb{R}^{n^2+n} \to \mathbb{R}^n$. To simplify matters, let us assume that A is a fixed matrix not subject to change, and only the vector b is undergoing perturbations. We then have a map $f: \mathbb{R}^n \to \mathbb{R}^n$ given by

$$x = f(b) := A^{-1}b.$$

It is in fact a linear map. Therefore, $\partial f/\partial b = A^{-1}$, and we get, using (3.16),

$$(\operatorname{cond} f)(b) = \frac{\|b\| \, \|A^{-1}\|}{\|A^{-1}b\|}, \tag{3.39}$$

where we may take any vector norm in $\mathbb{R}^n$ and associated matrix norm (cf. (3.11)). We can write (3.39) alternatively in the form

$$(\operatorname{cond} f)(b) = \frac{\|Ax\| \, \|A^{-1}\|}{\|x\|} \quad (\text{where } Ax = b),$$

and since there is a one-to-one correspondence between x and b, we find for the worst condition number

$$\max_{\substack{b\in\mathbb{R}^n \\ b\neq 0}} (\operatorname{cond} f)(b) = \max_{\substack{x\in\mathbb{R}^n \\ x\neq 0}} \frac{\|Ax\|}{\|x\|} \cdot \|A^{-1}\| = \|A\| \cdot \|A^{-1}\|,$$

by definition of the norm of A. The number on the far right no longer depends on the particular system (i.e., on b) and is called the *condition number* of the matrix A. We denote it by

$$\text{cond } A := \|A\| \cdot \|A^{-1}\| \ . \tag{3.40}$$

It should be clearly understood, though, that it measures the condition of a linear system with coefficient matrix A, and not the condition of other quantities that may depend on A, such as eigenvalues.

Although we have considered only perturbations in the right-hand vector b, it turns out that the condition number in (3.40) is also relevant when perturbations in the matrix A are allowed, provided they are sufficiently small (so small, for example, that $\|\Delta A\| \cdot \|A^{-1}\| < 1$).

We illustrate (3.40) by several examples.

(i) Hilbert[2] matrix:

$$H_n = \begin{bmatrix} 1 & \frac{1}{2} & \cdots & \frac{1}{n} \\ \frac{1}{2} & \frac{1}{3} & \cdots & \frac{1}{n+1} \\ \cdots & \cdots & \cdots & \cdots \\ \frac{1}{n} & \frac{1}{n+1} & \cdots & \frac{1}{2n-1} \end{bmatrix} \in \ \mathbb{R}^{n \times n}. \tag{3.41}$$

This is clearly a symmetric matrix, and it is also positive definite. Some numerical values for the condition number of H_n, computed with the Euclidean norm,[3] are shown in Table 1.3.1. Their rapid growth is devastating.

---

[2]David Hilbert (1862–1943) was the most prominent member of the Göttingen school of mathematics. Hilbert's fundamental contributions to almost all parts of mathematics — algebra, number theory, geometry, integral equations, calculus of variations, and foundations — and in particular the 23 now famous problems he proposed in 1900 at the International Congress of Mathematicians in Paris, gave a new impetus, and new directions, to 20th-century mathematics. Hilbert is also known for his work in mathematical physics, where among other things he formulated a variational principle for Einstein's equations in the theory of relativity.

[3]We have $\text{cond}_2 H_n = \lambda_{\max}(H_n) \cdot \lambda_{\max}(H_n^{-1})$, where $\lambda_{\max}(A)$ denotes the largest eigenvalue of the (symmetric, positive definite) matrix A. We computed all eigenvalues of H_n and H_n^{-1}, using the appropriate Eispack routine. The inverse of H_n was computed from its well-known explicit form (not by inversion!). The total computing time (on a CDC 6500 computer in the 1980s) was 45 sec, at a cost of \$1.04

n	$\text{cond}_2 H_n$
10	1.60×10^{13}
20	2.45×10^{28}
40	7.65×10^{58}

TABLE 1.3.1. The condition of Hilbert matrices

A system of order $n = 10$, for example, cannot be solved with any reliability in single precision on a 14-decimal computer. Double precision will be "exhausted" by the time we reach $n = 20$. The Hilbert matrix thus is a prototype of an ill-conditioned matrix. From a result of G. Szegő it can be seen that

$$\text{cond}_2 H_n \sim \frac{\left(\sqrt{2}+1\right)^{4n+4}}{2^{15/4}\sqrt{\pi n}} \quad \text{as} \quad n \to \infty.$$

(ii) Vandermonde[4] matrices: these are matrices of the form

$$V_n = \begin{bmatrix} 1 & 1 & \cdots & 1 \\ t_1 & t_2 & \cdots & t_n \\ \cdot & \cdot & & \cdot \\ \cdot & \cdot & & \cdot \\ \cdot & \cdot & & \cdot \\ t_1^{n-1} & t_2^{n-1} & \cdots & t_n^{n-1} \end{bmatrix} \in \mathbb{R}^{n\times n}, \tag{3.42}$$

where $t_1, t_2, \ldots, t_n$ are parameters, here assumed real. The condition number of these matrices, in the ∞-norm, has been studied at length. Here are some sample results: if the parameters are equally spaced in [–1,1], that is,

$$t_\nu = 1 - \frac{2(\nu-1)}{n-1}, \quad \nu = 1, 2, \ldots, n,$$

then

$$\text{cond}_\infty V_n \sim \frac{1}{\pi} e^{-\pi/4} e^{n\left(\frac{\pi}{4}+\frac{1}{2}\ln 2\right)}, \quad n \to \infty.$$

Numerical values are shown in Table 1.3.2.

[4] Alexandre Théophile Vandermonde (1735–1796), the author of only four mathematical papers, was elected to the French Academy of Sciences before he even wrote his first paper, apparently as a result of influential acquaintances. Nevertheless, his papers, especially the first, made important contributions to the then emerging theory of equations. By virtue of his fourth paper, he is regarded as the founder of the theory of determinants. What today is referred to as the "Vandermonde determinant," however, does not seem to appear anywhere in his writings. As a member of the Academy, he was appointed to the committee that in 1799 was to define the unit of length — the meter.

n	$\text{cond}_\infty V_n$
10	1.36×10^4
20	1.05×10^9
40	6.93×10^{18}
80	3.15×10^{38}

TABLE 1.3.2. The condition of Vandermonde matrices

They are not growing quite as fast as those for the Hilbert matrix, but still exponentially fast. Worse than exponential growth is observed if one takes harmonic numbers as parameters,

$$t_\nu = \frac{1}{\nu}, \quad \nu = 1, 2, \ldots, n.$$

Then indeed

$$\text{cond}_\infty V_n > n^{n+1}.$$

Fortunately, there are not many matrices occurring naturally in applications that are *that* ill-conditioned, but moderately to severely ill-conditioned matrices are no rarity in real-life applications.

§4. The Condition of an Algorithm

We again assume that we are dealing with a problem f given by

$$f: \quad \mathbb{R}^m \to \mathbb{R}^n, \quad y = f(x). \tag{4.1}$$

Along with the problem f, we are also given an algorithm A that "solves" the problem. That is, given a machine vector $x \in \mathbb{R}^m(t,s)$, the algorithm A produces a vector y_A (in machine arithmetic) that is supposed to approximate $y = f(x)$. Thus, we have another map f_A describing how the problem f is solved by the algorithm A,

$$f_A: \quad \mathbb{R}^m(t,s) \to \mathbb{R}^n(t,s), \quad y_A = f_A(x). \tag{4.2}$$

In order to be able to analyze f_A in these general terms, we must make a basic assumption, namely, that

$$\begin{aligned} &\text{for every} \quad x \in \mathbb{R}^m(t,s), \quad \text{there holds} \\ &f_A(x) = f(x_A) \quad \text{for some} \quad x_A \in \mathbb{R}^m. \end{aligned} \tag{4.3}$$

That is, the computed solution corresponding to some input x is the exact solution for some different input x_A (not necessarily a machine vector and not necessarily uniquely determined) that we hope is close to x. The closer we can find an x_A to x, the more confidence we should place in the algorithm A. We therefore define the condition of A in terms of the x_A closest to x (if there is more than one), by comparing its relative error with the machine precision eps:

$$(\text{cond } A)(x) = \inf_{x_A} \frac{\|x_A - x\|}{\|x\|} / \text{eps}. \tag{4.4}$$

Here the infimum is over all x_A satisfying $y_A = f(x_A)$. In practice one can take any such x_A and then obtain an upper bound for the condition number:

$$(\text{cond } A)(x) \leq \frac{\|x_A - x\|}{\|x\|} / \text{eps}. \tag{4.5}$$

The vector norm in (4.4), respectively, (4.5), can be chosen as seems convenient.

Here are some very elementary examples.

(1) Suppose a library routine for the logarithm function furnishes $y = \ln x$, for any positive machine number x, by producing a y_A satisfying $y_A = [\ln x](1 + \varepsilon)$, $|\varepsilon| \leq 5\,\text{eps}$. What can we say about the condition of the underlying algorithm A? We clearly have

$$y_A = \ln x_A \quad \text{where,} \quad x_A = x^{1+\varepsilon} \quad \text{(uniquely)}.$$

Consequently,

$$\left| \frac{x_A - x}{x} \right| = \left| \frac{x^{1+\varepsilon} - x}{x} \right| = |x^\varepsilon - 1| \approx |\varepsilon \ln x| \leq 5\,|\ln x| \cdot \text{eps},$$

and, therefore, $(\text{cond } A)(x) \leq 5\,|\ln x|$. The algorithm A is well-conditioned, except in the immediate right-hand vicinity of $x = 0$ and for x very large. (In the latter case, however, x is likely to overflow before A becomes seriously ill-conditioned.)

(2) Consider the problem

$$f: \quad \mathbb{R}^n \to \mathbb{R}, \quad y = x_1 x_2 \cdots x_n.$$

We solve the problem by the obvious algorithm

$$A: \quad \begin{aligned} &p_1 = x_1, \\ &p_k = \mathrm{fl}(x_k p_{k-1}), \quad k = 2, 3, \ldots, n, \\ &y_A = p_n. \end{aligned}$$

Note that x_1 is machine-representable, since for the algorithm A we assume $x \in \mathbb{R}^n(t, s)$.

Now using the basic law of machine arithmetic (cf. (2.1)), we get

$$\begin{aligned} &p_1 = x_1, \\ &p_k = x_k p_{k-1}(1 + \varepsilon_k), \quad k = 2, 3, \ldots, n, \quad |\varepsilon_k| \leq \text{eps}, \end{aligned}$$

from which

$$p_n = x_1 x_2 \cdots x_n (1 + \varepsilon_2)(1 + \varepsilon_3) \cdots (1 + \varepsilon_n).$$

Therefore, we can take, for example (there is no uniqueness),

$$x_A = [x_1, x_2(1 + \varepsilon_2), \ldots, x_n(1 + \varepsilon_n)]^T.$$

This gives, using the ∞-norm,

$$\frac{\|x_A - x\|_\infty}{\|x\|_\infty \text{eps}} = \frac{\|[0, x_2\varepsilon_2, \ldots, x_n\varepsilon_n]^T\|_\infty}{\|x\|_\infty \text{eps}} \leq \frac{\|x\|_\infty \text{eps}}{\|x\|_\infty \text{eps}} = 1,$$

and so, by (4.5), $(\text{cond}\, A)(x) \leq 1$ for any $x \in \mathbb{R}^n(t, s)$. Our algorithm, to nobody's surprise, is perfectly well-conditioned.

§5. Computer Solution of a Problem; Overall Error

The problem to be solved is again

$$f: \quad \mathbb{R}^m \to \mathbb{R}^n, \quad y = f(x). \tag{5.1}$$

This is the mathematical (idealized) problem, where the data are exact real numbers, and the solution is the mathematically exact solution.

When solving such a problem on a computer, in floating-point arithmetic with precision eps, and using some algorithm A, one first of all rounds the data, and then applies to these rounded data not f, but f_A:

$$\begin{aligned} &x^* = \text{rounded data}, \quad \frac{\|x^* - x\|}{\|x\|} = \varepsilon, \\ &y_A^* = f_A(x^*). \end{aligned} \tag{5.2}$$

Here ε represents the rounding error in the data. (The error ε could also be due to sources other than rounding, e.g., measurement.) The total error that we wish to estimate is then

$$\frac{\|y_A^* - y\|}{\|y\|} . \tag{5.3}$$

By the basic assumption (4.3) made on the algorithm A, and choosing x_A^* optimally, we have

$$f_A(x^*) = f(x_A^*), \quad \frac{\|x_A^* - x^*\|}{\|x^*\|} = (\text{cond}\, A)(x^*) \cdot \text{eps}. \tag{5.4}$$

Let $y^* = f(x^*)$. Then, using the triangle inequality, we have

$$\frac{\|y_A^* - y\|}{\|y\|} \le \frac{\|y_A^* - y^*\|}{\|y\|} + \frac{\|y^* - y\|}{\|y\|} \approx \frac{\|y_A^* - y^*\|}{\|y^*\|} + \frac{\|y^* - y\|}{\|y\|} ,$$

where we have used the (harmless) approximation $\|y\| \approx \|y^*\|$. By virtue of (5.4), we now have for the first term on the right,

$$\begin{aligned} \frac{\|y_A^* - y^*\|}{\|y^*\|} = \frac{\|f_A(x^*) - f(x^*)\|}{\|f(x^*)\|} &= \frac{\|f(x_A^*) - f(x^*)\|}{\|f(x^*)\|} \\ &\le (\text{cond}\, f)(x^*) \cdot \frac{\|x_A^* - x^*\|}{\|x^*\|} \\ &= (\text{cond}\, f)(x^*) \cdot (\text{cond}\, A)(x^*) \cdot \text{eps}. \end{aligned}$$

For the second term we have

$$\frac{\|y^* - y\|}{\|y\|} = \frac{\|f(x^*) - f(x)\|}{\|f(x)\|} \le (\text{cond}\, f)(x) \cdot \frac{\|x^* - x\|}{\|x\|} = (\text{cond}\, f)(x) \cdot \varepsilon.$$

Assuming finally that $(\operatorname{cond} f)(x^*) \approx (\operatorname{cond} f)(x)$, we get

$$\frac{\|y_A^* - y\|}{\|y\|} \leq (\operatorname{cond} f)(x)\{\varepsilon + (\operatorname{cond} A)(x^*) \cdot \mathrm{eps}\}. \tag{5.5}$$

This shows how the data error and machine precision contribute toward the total error: both are amplified by the condition of the problem, but the latter is further amplified by the condition of the algorithm.

NOTES TO CHAPTER 1

In addition to rounding errors in the data and those committed during the execution of arithmetic operations, there may be other sources of errors not considered in this introductory chapter. One such source of error, which is not entirely dismissible, is a faulty design of the computer chip that executes arithmetic operations. This was brought home in a recent incident when it was discovered in 1994 (by Thomas Nicely in the course of number-theoretic computations involving reciprocals of twin primes) that the Pentium floating-point divide chip manufactured by Intel can produce erroneous results for certain (extremely rare) bit patterns in the divisor. The incident — rightly so — has stirred up considerable concern, and prompted not only remedial actions, but also careful analysis of the phenomenon; some relevant articles are those by Coe, Mathisen, Moler, and Pratt [1995] and Edelman [preprint].

Neither should the occurrence of overflow and proper handling thereof be taken lightly, especially not in real-time applications. Again, a case in point is the failure of the French rocket Ariane 5, which on June 4, 1996, less than a minute into its flight, self-destructed. The failure was eventually traced to an overflow in a floating-point to integer conversion and lack of protection against this occurrence in the rocket's on-board software (cf. Anonymous [1996]).

§1.1. The abstract notion of the real number system is discussed in most texts on real analysis, for example, Hewitt and Stromberg [1975, Ch.1, §5] or Rudin [1976, Ch.1]. The development of the concept of real (and complex) numbers has had a long and lively history, extending from pre-Hellenic times to the recent past. Many of the leading thinkers over time contributed to this development. A reader interested in a detailed historical account (and who knows German) is referred to the monograph by Gericke [1970].

§1.2 (a) The notion of the floating-point number system and associated arithmetic, including interval arithmetic, can also be phrased in abstract algebraic terms; for this, see, for example, Kulisch and Miranker [1981]. A more elementary, but detailed, discussion of floating-point numbers and arithmetic is given in Sterbenz [1974]. There the reader will learn, for example, that computing the average of two floating-point numbers, or solving a quadratic equation, can be fairly intricate tasks if they are to be made foolproof. The quadratic equation problem is also considered at some length in Young and Gregory [1988, §3.4], where further references are given to earlier work of W. Kahan and G. E. Forsythe.

The basic standard for binary floating-point arithmetic, used on all contemporary computers, is the ANSI/IEEE Standard 754 established in IEEE [1985]. It provides for $t = 23$ bits in the mantissa and $s = 7$ bits in the exponent, in single-precision arithmetic, and has $t = 52$, $s = 11$ in double precision. There is also an "extended precision" for which $t = 63$, $s = 14$, allowing for a number range of approx. 10^{-4964} to 10^{+4964}.

(c) Rational arithmetic is available in all major symbolic computation packages such as Mathematica and MACSYMA.

Interval arithmetic has evolved to become an important tool in computations that strive at obtaining guaranteed and sharp inclusion regions for the results of mathematical problems. The basic texts on interval analysis are Moore [1966], [1979] and Alefeld and Herzberger [1983]. Specific applications such as computing inclusions of the range of functions, of global extrema of functions of one and several variables, and of solutions to systems of linear and nonlinear equations are studied, respectively, in Ratschek and Rokne [1984], [1988], Hansen [1992], and Neumaier [1990]. Concrete algorithms and codes (in Pascal and C^{++}) for "verified computing" are contained in Hammer, Hocks, Kulisch, and Ratz [1993], [1995]. Interval arithmetic has been most widely used in processes involving finite-dimensional spaces; for applications to infinite-dimensional problems, notably differential equations, see, however, Eijgenraam [1981] and Kaucher and Miranker [1984].

§2. The fact that thoughtless use of mathematical formulae and numerical methods, or inherent sensitivities in a problem, can lead to disastrous results, has been known since the early days of computers; see, for example, the old but still relevant papers by Stegun and Abramowitz [1956] and Forsythe [1970]. Nearby singularities can also cause the accuracy to deteriorate unless corrective measures are taken; Forsythe [1958] has an interesting discussion of this.

§2.1. For the implications of rounding in the problem of apportionment, mentioned in Footnote 1, a good reference is Garfunkel and Steen [1988, Ch.12, pp.230–249].

§3.1. An early but basic reference for ideas of conditioning and error analysis in algebraic processes is Wilkinson [1963]. An impressive continuation of this work, containing copious references to the literature, is Higham [1996]. It analyzes the behavior in floating-point arithmetic of virtually all the algebraic processes in current use. Problems of conditioning specifically involving polynomials are discussed in Gautschi [1984]. The condition of general (differentiable) maps has been studied as early as 1966 in Rice [1966].

§3.2. (1) For a treatment of stability aspects of more general difference equations, and systems thereof, including nonlinear ones, the reader is referred to the monograph by Wimp [1984]. This also contains many applications to special functions. Another relevant text is Lakshmikantham and Trigiante [1988].

(2) The condition of algebraic equations, although considered already in Wilkinson's book [1963], has been further analyzed by Gautschi [1973]. The circumstances that led to Wilkinson's example (3.37), which he himself describes as "the most traumatic experience in [his] career as a numerical analyst," are related in the essay by Wilkinson [1984, §2]. This reference also deals with errors committed in the evaluation and deflation of polynomials. For the latter, also see Cohen [1994]. The asymptotic estimates for the best- and worst-conditioned roots in Wilkinson's example are from Gautschi [1973]. For the computation of eigenvalues of matrices, the classic treatment is Wilkinson [1988]; more recent accounts are Parlett [1980] for symmetric matrices, and Golub and Van Loan [1996, Ch. 7–9] for general matrices.

(3) A more complete analysis of the condition of linear systems, that also allows for perturbations of the matrix, can be found, for example, in the very readable books by Forsythe and Moler [1967, Ch.8] and Stewart [1973, Ch.4, §3]. The asymptotic result of Szegő cited in connection with the Euclidean condition number of the Hilbert matrix is taken from Szegő [1936]. For the explicit inverse of the Hilbert matrix, referred to in Footnote 3, see Todd [1954]. The condition of Vandermonde and Vandermonde-like matrices has been studied in a series of papers by the author; for a summary, see Gautschi [1990].

§§4 and 5. The treatment of the condition of algorithms and of the overall error in computer solutions of problems, as given in these sections, seems to be more or less original. Similar ideas, however, can be found in the book by Dahlquist and Björck [1974, Ch.2, §4].

EXERCISES AND MACHINE ASSIGNMENTS TO CHAPTER 1

EXERCISES

1. Represent all elements of $\mathbb{R}_+(3,2) = \{x \in \mathbb{R}(3,2) : \ x > 0, \ x \text{ normalized}\}$ as dots on the real axis. For clarity, draw two axes, one from 0 to 8, the other from 0 to $\frac{1}{2}$.

2. (a) What is the distance $d(x)$ of a positive normalized floating-point number $x \in \mathbb{R}(t,s)$ to its next larger floating-point number:

$$d(x) = \min_{\substack{y \in \mathbb{R}(t,s) \\ y > x}} (y - x)\,?$$

   (b) Determine the relative distance $r(x) = d(x)/x$, with x as in (a), and give upper and lower bounds for it.

3. The identity $\text{fl}(1+x) = 1$, $x \geq 0$, is true for $x = 0$ and for x sufficiently small. What is the largest machine number x for which the identity still holds?

4. Consider a miniature binary computer whose floating-point words consist of 4 binary digits for the mantissa and 3 binary digits for the exponent (plus sign bits). Let

$$x = (.1011)_2 \times 2^0, \qquad y = (.1100)_2 \times 2^0.$$

   Mark in the following table whether the machine operation indicated (with the result z assumed normalized) is exact, rounded (i.e., subject to a nonzero rounding error), overflows, or underflows.

operation	exact	rounded	overflow	underflow
$z = \text{fl}(x - y)$				
$z = \text{fl}((y - x)^{10})$				
$z = \text{fl}(x + y)$				
$z = \text{fl}(y + (x/4))$				
$z = \text{fl}(x + (y/4))$				

5. The following algorithm (attributed to Cleve Moler) estimates eps:

```
a=4./3.
b=a-1.
c=b+b+b
eps=abs(c-1.).
```

Run the program with the corresponding double-precision statements appended to it and print the single- and double-precision eps.

6. Prove (1.12).

7. A set S of elements, or pairs of elements, is said to possess a metric if there is defined a distance function $d(x, y)$ for any two elements $x, y \in S$ that has the following properties:

   (i) $d(x, y) \geq 0$ and $d(x, y) = 0$ if and only if $x = y$ (positive definiteness);

   (ii) $d(x, y) = d(y, x)$ (symmetry);

   (iii) $d(x, y) \leq d(x, z) + d(z, y)$ (triangle inequality).

   Discuss which of the following error measures is a distance function on what set S (of real numbers, or pairs of real numbers):

   (a) absolute error: $\mathrm{ae}(x, y) = |x - y|$;

   (b) relative error: $\mathrm{re}(x, y) = \left|\frac{x-y}{x}\right|$;

   (c) relative precision (F.W.J. OLVER, 1978): $\mathrm{rp}(x, y) = |\ln|x| - \ln|y||$;

   If $y = x(1 + \varepsilon)$, show that $\mathrm{rp}(x, y) = O(\varepsilon)$ as $\varepsilon \to 0$.

8. Assume that x_1^*, x_2^* are approximations to x_1, x_2 with relative errors E_1 and E_2, respectively, and that $|E_i| \leq E$, $i = 1, 2$. Assume further that $x_1 \neq x_2$.

   (a) How small must E be in order to ensure that $x_1^* \neq x_2^*$?

   (b) Taking $\dfrac{1}{x_1^* - x_2^*}$ to approximate $\dfrac{1}{x_1 - x_2}$, obtain a bound on the relative error committed, assuming (i) exact arithmetic; (ii) machine arithmetic with machine precision eps. (Neglect higher-order terms in E_1, E_2, eps.)

9. Consider the quadratic equation $x^2 + px + q = 0$ with roots x_1, x_2. As seen in Example (2) of §2.2, the absolutely larger root must be computed first, whereupon the other can be accurately obtained from $x_1 x_2 = q$. Suppose one incorporates this idea in a program such as

```
x1=abs(p/2)+sqrt(p*p/4-q)
if(p.gt.0.) x1=-x1
x2=q/x1.
```

Find three serious faults with this program as a "general-purpose quadratic equation solver." Take into consideration that the program will be executed in floating-point machine arithmetic. Be specific and support your arguments by examples, if necessary.

10. Let $f(x) = \sqrt{1+x^2} - 1$.

    (a) Explain the difficulty of computing $f(x)$ for a small value of $|x|$ and show how it can be circumvented.

    (b) Compute $(\operatorname{cond} f)(x)$ and discuss the conditioning of $f(x)$ for small $|x|$.

    (c) How can the answers to (a) and (b) be reconciled?

11. The nth power of some positive (machine) number x can be computed

    (i) either by repeated multiplication by x, or

    (ii) as $x^n = e^{n \ln x}$.

    In each case, derive bounds for the relative error due to machine arithmetic, neglecting higher powers of the machine precision against the first power. Based on these bounds, state a criterion (involving x and n) for (i) to be better than (ii).

12. Let $f(x) = (1 - \cos x)/x$, $x \neq 0$.

    (a) Show that direct evaluation of f is inaccurate if $|x|$ is small; assume $\mathrm{fl}(f(x)) = \mathrm{fl}((1 - \mathrm{fl}(\cos x))/x)$, where $\mathrm{fl}(\cos x) = (1 + \varepsilon_c)\cos x$, and estimate the relative error of $\mathrm{fl}(f(x))$ as $x \to 0$.

    (b) A mathematically equivalent form of f is $f(x) = \sin^2 x/(x(1 + \cos x))$. Carry out a similar analysis as in (a), based on $\mathrm{fl}(f(x)) = \mathrm{fl}([\mathrm{fl}(\sin x)]^2/(x(1 + \mathrm{fl}(\cos x))))$, assuming $\mathrm{fl}(\cos x) = (1 + \varepsilon_c)\cos x$, $\mathrm{fl}(\sin x) = (1 + \varepsilon_s)\sin x$ and retaining only first-order terms in ε_s and ε_c. Discuss the result.

    (c) Determine the condition of $f(x)$. Indicate for what values of x (if any) $f(x)$ is ill-conditioned. ($|x|$ is no longer small, necessarily.)

13. If $z = x + iy$, then $\sqrt{z} = \left(\frac{r+x}{2}\right)^{1/2} + i\left(\frac{r-x}{2}\right)^{1/2}$, where $r = (x^2 + y^2)^{1/2}$. Alternatively, $\sqrt{z} = u + iv$, $u = \left(\frac{r+x}{2}\right)^{1/2}$, $v = y/2u$. Discuss the computational merits of these two (mathematically equivalent) expressions when $x > 0$. Illustrate with $z = 4.5 + .025i$, using 8 significant decimal places. How would you deal with $x < 0$?

14. Consider the numerical evaluation of

$$f(t) = \sum_{n=0}^{\infty} \frac{1}{1 + n^4(t-n)^2(t-n-1)^2},$$

say, for $t = 20$, and 7-digit accuracy. Discuss the danger involved.

15. Let X_+ be the largest positive machine-representable number, and X_- the absolute value of the smallest negative one (so that $-X_- \le x \le X_+$ for any machine number x). Determine, approximately, all intervals on $\mathbb{R}$ on which the tangent function overflows.

16. Consider a decimal computer with 3 (decimal) digits in the floating-point mantissa.

    (a) Estimate the relative error committed in symmetric rounding.

    (b) Let $x_1 = .982$, $x_2 = .984$ be two machine numbers. Calculate in machine arithmetic the mean $m = \frac{1}{2}(x_1 + x_2)$. Is the computed number between x_1 and x_2?

    (c) Derive sufficient conditions for $x_1 < \mathrm{fl}(m) < x_2$ to hold, where x_1, x_2 are two machine numbers with $0 < x_1 < x_2$.

17. For this problem, assume a binary computer with 12 bits in the floating-point mantissa.

    (a) What is the machine precision eps?

    (b) Let $x = 6/7$ and x^* be the correctly rounded machine approximation to x (symmetric rounding). Exhibit x and x^* as binary numbers.

    (c) Determine (exactly!) the relative error ε of x^* as an approximation to x, and calculate the ratio $|\varepsilon|/\text{eps}$.

18. The associative law of algebra states that

$$(a + b)c = ac + bc.$$

Discuss to what extent this is violated in machine arithmetic. Assume a computer with machine precision eps and assume that a, b, c are machine-representable numbers.

    (a) Let y_1 be the floating-point number obtained by evaluating $(a + b)c$ (as written) in floating-point arithmetic, and let $y_1 = (a + b)c(1 + e_1)$. Estimate $|e_1|$ in terms of eps (neglecting second-order terms in eps).

    (b) Let y_2 be the floating-point number obtained by evaluating $ac + bc$ (as written) in floating-point arithmetic, and let $y_2 = (a + b)c(1 + e_2)$. Estimate $|e_2|$ (neglecting second-order terms in eps) in terms of eps (and a, b, and c).

(c) Identify conditions (if any) under which one of the two ys is significantly less accurate than the other.

19. Let $x_1, x_2, \ldots, x_n$ be machine numbers. Their product can be computed by the algorithm

$$p_1 = x_1,$$
$$p_k = \mathrm{fl}(x_k p_{k-1}), \quad k = 2, 3, \ldots, n.$$

(a) Find an upper bound for the relative error $(p_n - x_1 x_2 \cdots x_n)/(x_1 x_2 \cdots x_n)$ in terms of the machine precision eps and n.

(b) For any integer $r \geq 1$ small enough to satisfy $r \cdot \mathrm{eps} < \frac{1}{10}$, show that

$$(1 + \mathrm{eps})^r - 1 < (1.06) r \cdot \mathrm{eps}.$$

Hence simplify the answer given in (a). {*Hint*: Use the binomial theorem.}

20. Analyze the error propagation in exponentiation, x^α $(x > 0)$:

(a) assuming x exact and α subject to a small relative error ε_α;

(b) assuming α exact and x subject to a small relative error ε_x.

Discuss the possibility of any serious loss of accuracy.

21. Indicate how you would accurately compute

$$(x + y)^{1/4} - y^{1/4}, \quad x > 0, \quad y > 0.$$

22. (a) Let $a = .23371258 \times 10^{-4}$, $b = .33678429 \times 10^2$, $c = -.33677811 \times 10^2$. Assuming an 8-decimal-digit computer, determine the sum $s = a + b + c$ either as (i) $\mathrm{fl}(s) = \mathrm{fl}(\mathrm{fl}(a + b) + c)$ or as (ii) $\mathrm{fl}(s) = \mathrm{fl}(a + \mathrm{fl}(b + c))$. Explain the discrepancy between the two answers.

(b) For arbitrary machine numbers a, b, c, on a computer with machine precision eps, find a criterion on a, b, c for the result of (ii) in (a) to be more accurate than the result of (i). {*Hint*: Compare bounds on the relative errors, neglecting higher-order terms in eps and assuming $a + b + c \neq 0$.}

23. Write the expression $a^2 - 2ab\cos\gamma + b^2$ $(a > 0, b > 0)$ as the sum of two positive terms in order to avoid cancellation errors. Illustrate the advantage gained in the case $a = 16.5$, $b = 15.7$, $\gamma = 5°$, using 3-decimal-digit arithmetic. Is the method foolproof?

24. Determine the condition number for the following functions.

(a) $f(x) = \ln x, \quad x > 0;$ (b) $f(x) = \cos x, \quad |x| < \frac{1}{2}\pi;$

(c) $f(x) = \sin^{-1} x, \quad |x| < 1;$ (d) $f(x) = \sin^{-1} \dfrac{x}{\sqrt{1+x^2}}\,.$

Indicate the possibility of ill-conditioning.

25. Compute the condition number of the following functions, and discuss any possible ill-conditioning.

(a) $f(x) = x^{1/n} \quad (x > 0,\ n > 0$ an integer)
(b) $f(x) = x - \sqrt{x^2 - 1} \quad (x > 1)$
(c) $f(x_1, x_2) = \sqrt{x_1^2 + x_2^2}$
(d) $f(x_1, x_2) = x_1 + x_2$

26. (a) Consider the composite function $h(t) = g(f(t))$. Express the condition of h in terms of the condition of g and f. Be careful to state at which points the various condition numbers are to be evaluated.
(b) Illustrate (a) with $h(t) = \frac{1+\sin t}{1-\sin t}$, $t = \frac{1}{4}\pi$.

27. Show that $(\text{cond}\, f \cdot g)(x) \le (\text{cond}\, f)(x) + (\text{cond}\, g)(x)$.

28. Let $f : \mathbb{R}^2 \to \mathbb{R}$ be given by $y = x_1 + x_2$. Define $(\text{cond}\, f)(x) = (\text{cond}_1 f)(x) + (\text{cond}_2 f)(x)$, where $\text{cond}_i f$ is the condition number of f considered a function of x_i only $(i = 1, 2)$.

(a) Derive a formula for $\kappa(x_1, x_2) = (\text{cond}\, f)(x)$.
(b) Show that $\kappa(x_1, x_2)$ as a function of x_1, x_2 is symmetric with respect to both bisectors b_1 and b_2 (see figure).

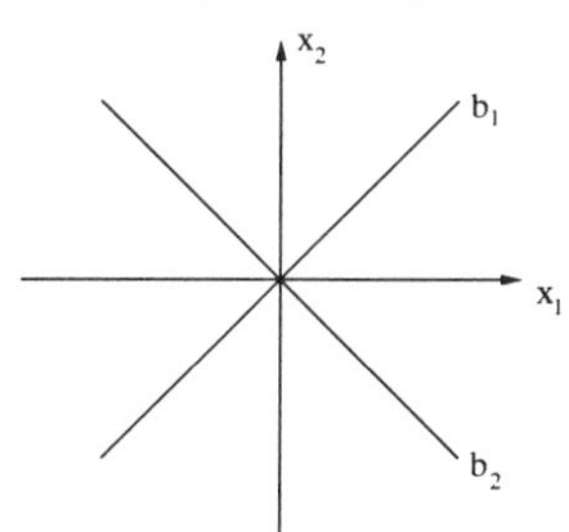

(c) Determine the lines (or domains) in $\mathbb{R}^2$ on which $\kappa(x_1, x_2) = c$, $c \ge 1$ a constant. (Simplify the analysis by using symmetry; cf. part (b).)

29. Let $\|\cdot\|$ be a vector norm in $\mathbb{R}^n$ and denote by the same symbol the associated matrix norm. Show for arbitrary matrices A, $B \in \mathbb{R}^{n\times n}$ that

    (a) $\|AB\| \le \|A\|\,\|B\|$;
    (b) $\mathrm{cond}(AB) \le \mathrm{cond}\,A \cdot \mathrm{cond}\,B$.

30. Prove (3.13). {*Hint*: Let $m_\infty = \max_\nu \sum_\mu |a_{\nu\mu}|$. Show that $\|A\|_\infty \le m_\infty$ as well as $\|A\|_\infty \ge m_\infty$, the latter by taking a special vector x in (3.11).}

31. Let the L_1 norm of a vector $y = [y_\lambda]$ be defined by $\|y\|_1 = \sum_\lambda |y_\lambda|$. For a matrix $A \in \mathbb{R}^{n\times m}$, show that

$$\|A\|_1 := \max_{\substack{x\in\mathbb{R}^m \\ x\neq 0}} \frac{\|Ax\|_1}{\|x\|_1} = \max_\mu \sum_\nu |a_{\nu\mu}|;$$

    that is, $\|A\|_1$ is the "maximum column sum." {*Hint*: Let $m_1 = \max_\mu \sum_\nu |a_{\nu\mu}|$. Show that $\|A\|_1 \le m_1$ as well as $\|A\|_1 \ge m_1$, the latter by taking for x in (3.11) an appropriate coordinate vector.}

32. Let a, q be linearly independent vectors in $\mathbb{R}^n$ of (Euclidean) length 1. Define $b(\rho) \in \mathbb{R}^n$ as follows:

$$b(\rho) = a - \rho q, \quad \rho \in \mathbb{R}.$$

    Compute the condition of the angle $\alpha(\rho)$ between $b(\rho)$ and q at the value $\rho = \rho_0 = q^T a$. (Then $b(\rho_0) \perp q$; see figure.) Discuss the answer.

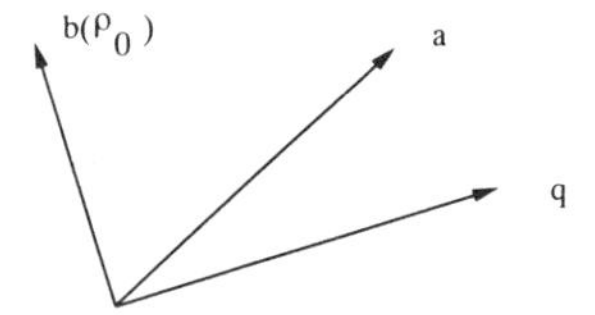

33. The area Δ of a triangle ABC is given by $\Delta = \frac{1}{2}ab\sin\gamma$ (see figure). Discuss the condition of Δ.

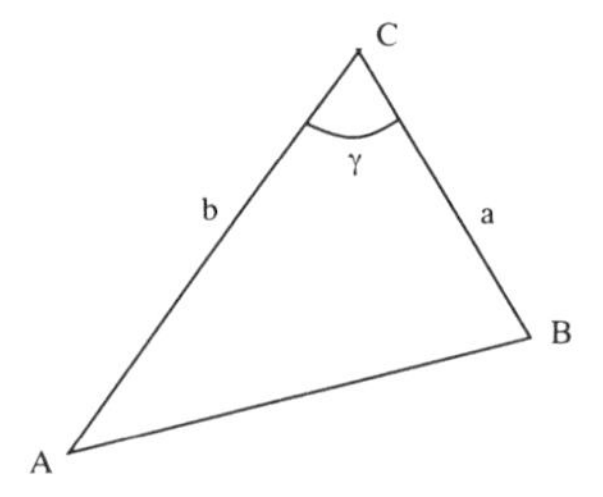

34. Define, for $x \neq 0$,

$$f_n = f_n(x) = (-1)^n \frac{d^n}{dx^n}\left(\frac{e^{-x}}{x}\right), \quad n = 0, 1, 2, \ldots .$$

(a) Show that $\{f_n\}$ satisfies the recursion

$$y_k = \frac{k}{x}\, y_{k-1} + \frac{e^{-x}}{x}, \quad k = 1, 2, 3, \ldots\,; \quad y_0 = \frac{e^{-x}}{x}.$$

{*Hint*: Differentiate k times the identity $e^{-x} = x \cdot (e^{-x}/x)$.}

(b) Why do you expect the recursion in (a), without doing any analysis, to be numerically stable if $x > 0$? How about $x < 0$?

(c) Support and discuss your answer to (b) by showing

$$(\mathrm{cond}\, y_n)(f_0) = \frac{1}{|e_n(x)|},$$

where $e_n(x) = 1 + x + x^2/2! + \cdots + x^n/n!$ is the nth partial sum of the exponential series. {*Hint*: Use Leibniz's formula to evaluate f_n.}

35. Consider the algebraic equation

$$x^n + ax - 1 = 0, \quad a > 0, \quad n \geq 2.$$

(a) Show that the equation has exactly one positive root $\xi(a)$.

(b) Obtain a formula for $(\mathrm{cond}\ \xi)(a)$.

(c) Obtain (good) upper and lower bounds for $(\mathrm{cond}\ \xi)(a)$.

36. Consider the algebraic equation

$$x^n + x^{n-1} - a = 0, \quad a > 0, \quad n \geq 2.$$

(a) Show that there is exactly one positive root $\xi(a)$.

(b) Show that $\xi(a)$ is well-conditioned as a function of a. Indeed, prove

$$(\mathrm{cond}\, \xi)(a) < \frac{1}{n-1}.$$

37. Consider the equation

$$xe^x = a$$

for real values of x and a.

(a) Show graphically that the equation has exactly one root $\xi(a) \geq 0$ if $a \geq 0$, exactly two roots $\xi_2(a) < \xi_1(a) < 0$ if $-1/e < a < 0$, and none if $a < -1/e$.

(b) Discuss the condition of $\xi(a)$, $\xi_1(a)$, $\xi_2(a)$ as a varies in the respective intervals.

38. Given the natural number n, let $\xi = \xi(a)$ be the unique positive root of the equation $x^n = ae^{-x}$ $(a > 0)$. Determine the condition of ξ as a function of the parameter a; simplify the answer as much as possible. In particular, show that $(\operatorname{cond} \xi)(a) < 1/n$.

39. Let $f(x_1, x_2) = x_1 + x_2$ and consider the algorithm A given as follows,

$$f_A : \mathbb{R}^2(t,s) \to \mathbb{R}(t,s) \qquad y_A = \mathrm{fl}(x_1 + x_2).$$

Estimate $\gamma(x_1, x_2) = (\operatorname{cond} A)(x)$, using any of the norms

$$\|x\|_1 = |x_1| + |x_2|, \quad \|x\|_2 = \sqrt{x_1^2 + x_2^2}, \quad \|x\|_\infty = \max\,(|x_1|, |x_2|).$$

Discuss the answer in the light of the conditioning of f.

40. This problem deals with the function $f(x) = \sqrt{1-x} - 1$, $-\infty < x < 1$.

(a) Compute the condition number $(\operatorname{cond} f)(x)$.

(b) Let A be the algorithm that evaluates $f(x)$ in floating-point arithmetic on a computer with machine precision eps, given an (error-free) floating-point number x. Let ε_1, ε_2, ε_3 be the relative errors due, respectively, to the subtraction in $1 - x$, to taking the square root, and to the final subtraction of 1. Assume $|\varepsilon_i| \le \mathrm{eps}$ $(i = 1, 2, 3)$. Letting $f_A(x)$ be the value of $f(x)$ so computed, write $f_A(x) = f(x_A)$ and $x_A = x(1 + \varepsilon_A)$. Express ε_A in terms of x, ε_1, ε_2, ε_3 (neglecting terms of higher order in the ε_i). Then determine an upper bound for $|\varepsilon_A|$ in terms of x and eps, and finally an estimate of $(\operatorname{cond} A)(x)$.

(c) Sketch a graph of $(\operatorname{cond} f)(x)$ (found in (a)) and a graph of the estimate of $(\operatorname{cond} A)(x)$ (found in (b)) as functions of x on $(-\infty, 1)$. Discuss your results.

41. Consider the function $f(x) = 1 - e^{-x}$ on the interval $0 \le x \le 1$.

(a) Show that $(\operatorname{cond} f)(x) \le 1$ on [0,1].

(b) Let A be the algorithm that evaluates $f(x)$ for the machine number x in floating-point arithmetic (with machine precision eps). Assume that the exponential routine returns a correctly rounded answer. Estimate $(\operatorname{cond} A)(x)$ for $0 \le x \le 1$, neglecting terms of $O(\mathrm{eps}^2)$. {*Point of information*: $\ln(1 + \varepsilon) = \varepsilon + O(\varepsilon^2)$, $\varepsilon \to 0$.}

(c) Plot $(\operatorname{cond} f)(x)$ and your estimate of $(\operatorname{cond} A)(x)$ as functions of x on [0,1]. Comment on the results.

42. (a) Suppose A is an algorithm that computes the (smooth) function $f(x)$ for a given machine number x, producing $f_A(x) = f(x)(1 + \varepsilon_f)$, where $|\varepsilon_f| \le \varphi(x)\text{eps}$ (eps = machine precision). Show that

$$(\text{cond}\, A)(x) \le \frac{\varphi(x)}{(\text{cond}\, f)(x)}$$

if second-order terms in eps are neglected. {*Hint*: Set $f_A(x) = f(x_A)$, $x_A = x(1 + \varepsilon_A)$, and expand in powers of ε_A, keeping only the first.}

(b) Apply the result of (a) to $f(x) = \frac{1-\cos x}{\sin x}$, $0 < x < \frac{1}{2}\pi$, when evaluated as shown. (You may assume that $\cos x$ and $\sin x$ are computed within a relative error of eps.) Discuss the answer.

(c) Do the same as (b), but for the (mathematically equivalent) function $f(x) = \frac{\sin x}{1+\cos x}$, $0 < x < \frac{1}{2}\pi$.

MACHINE ASSIGNMENTS

1. Let $x = 1 + \pi/10^6$ and compute x^n for $n = 1, 2, \ldots, 10^6$, both in single precision and double precision. Use the double-precision results to observe the rounding errors ρ_n in the single-precision results (relative errors). Print (in e-format with 5 decimal digits after the decimal point, except for the integer n): n, x^n, ρ_n, $\rho_n/(n \times \text{eps})$ for $n = k \times 10^5$, $k = 1, 2, \ldots, 10$, where eps is the machine precision. What should x^n be, approximately, when $n = 10^6$? Comment on the results. (Compute the number π in a machine-independent manner, using an elementary function routine.)

2. Compute the derivative dy/dx of the exponential function $y = e^x$ at $x = 0$ from difference quotients $(e^{x+h} - e^x)/h$ with decreasing h. Use

   (a) $h = 2^{-i}$, $i = 1, 2, \ldots, 20$;

   (b) $h = (2.2)^{-i}$, $i = 1, 2, \ldots, 20$,

   and print 8 decimal digits after the decimal point. Explain what you observe.

3. Consider the following procedure for determining the limit $\lim_{h \to 0} (e^h - 1)/h$ on a computer. Let

$$d_n = \text{fl}\left(\frac{e^{2^{-n}} - 1}{2^{-n}}\right) \quad \text{for } n = 0, 1, 2, \ldots$$

and accept as the machine limit the first value satisfying $d_n = d_{n-1}$ $(n \ge 1)$.

   (a) Run the procedure on a computer.

(b) In floating-point arithmetic for $\mathbb{R}(t, s)$, with rounding by chopping, for what value of n will the correct limit be reached, assuming no underflow (of 2^{-n}) occurs? {*Hint*: Use $e^h = 1 + h + \frac{1}{2}h^2 + \cdots$.} Compare with the experiment made in (a).

(c) On what kind of computer (i.e., under what conditions on s and t) will underflow occur before the limit is reached?

4. Euler's constant $\gamma = .57721566490153286\ldots$ is defined as the limit

$$\gamma = \lim_{n \to \infty} \gamma_n, \quad \text{where} \quad \gamma_n = 1 + \frac{1}{2} + \frac{1}{3} + \cdots + \frac{1}{n} - \ln n.$$

Assuming that $\gamma - \gamma_n \sim cn^{-d}$, $n \to \infty$, for some constants c and $d > 0$, try to determine c and d experimentally on the computer.

5. Letting $\Delta u_n = u_{n+1} - u_n$, one has the easy formula

$$\sum_{n=1}^{N} \Delta u_n = u_{N+1} - u_1.$$

With $u_n = \ln(1 + n)$, compute each side (as it stands) for $N = 1000(1000)$ 10,000, the right-hand side in double precision. Print the relative discrepancy of the two sides. Repeat with $\sum_{n=1}^{N} u_n$, computed in single and double precision. Explain the results.

6. (a) Write a program that computes

$$S_N = \sum_{n=1}^{N} \left[\frac{1}{n} - \frac{1}{n+1}\right],$$

where the summation is carried out as written.

(a1) What is the exact sum S_N?

(a2) Run your program for $N = 10, 100, 1000, \ldots, 1,000,000$ and print out the error $|\mathrm{fl}(S_N) - S_N|$. Comment on the answers obtained.

(b) Do the same as in part (a), but for the product

$$p_N = \prod_{n=1}^{N} \frac{n}{n+1}.$$

Comment on the results.

7. (a) Suppose x, y, z are floating-point numbers with $x > y > z > 0$. Estimate the relative errors in

$$\mathrm{fl}(\mathrm{fl}(x + y) + z) \quad \text{and} \quad \mathrm{fl}(\mathrm{fl}(z + y) + x).$$

Which result is more accurate? Explain. Adapt your discussion so it applies to the sum s_n in Part (b).

(b) It is known that

$$\pi = 4 \lim_{n\to\infty} s_n, \qquad s_n = \sum_{k=0}^{n} \frac{(-1)^k}{2k+1}.$$

Write a program computing s_n for $n = 10^7$, using forward as well as backward summation. Compute the respective errors (in double precision as `err=abs(sngl(4.d0*dble(sn)-dpi))`, where `dpi` is the double-precision value of π and `sn` the single-precision value of s_n). Interpret the results.

8. In the theory of Fourier series the numbers

$$\lambda_n = \frac{1}{2n+1} + \frac{2}{\pi} \sum_{k=1}^{n} \frac{1}{k} \tan \frac{k\pi}{2n+1}, \quad n = 1, 2, 3, \ldots ,$$

known as *Lebesgue constants*, are of some importance.

(a) Show that the terms in the sum increase monotonically with k. How do the terms for k near n behave when n is large?

(b) Compute λ_n for $n = 1, 10, 10^2, \ldots, 10^5$ in single and double precision and compare the results. Explain what you observe.

9. Sum the series

$$\text{(a)} \quad \sum_{n=0}^{\infty} (-1)^n/n!^2 \qquad \text{(b)} \quad \sum_{n=0}^{\infty} 1/n!^2$$

until there is no more change in the partial sums to within the machine precision. Generate the terms recursively. Print the number of terms required and the value of the sum. (Answers in terms of Bessel functions: (a) $J_0(2)$; (b) $I_0(2)$.)

10. (P.J. Davis, 1993) Consider the series $\sum_{k=1}^{\infty} \frac{1}{k^{3/2} + k^{1/2}}$. Try to compute the sum to three correct decimal digits.

11. We know from calculus that

$$\lim_{n\to\infty} \left(1 + \frac{1}{n}\right)^n = e.$$

What is the "machine limit"? Explain.

12. Let $f(x) = (n+1)x - 1$. The iteration

$$x_k = f(x_{k-1}), \quad k = 1, 2, \ldots, K; \quad x_0 = \frac{1}{n}$$

in exact arithmetic converges to the fixed point $1/n$ in one step (why?). What happens in machine arithmetic? Run a program with $n = 1(1)5$ and $K = 10(10)50$ and explain quantitatively what you observe.

13. Compute the integral $\int_0^1 e^x dx$ from Riemann sums with n equal subintervals, evaluating the integrand at the midpoint of each. Print the Riemann sums for $n = 10, 20, 30, \ldots, 200$ (showing 6 decimal digits after the decimal point), together with the exact answers. Comment on the results.

14. Let $y_n = \int_0^1 t^n e^{-t} dt$, $n = 0, 1, 2, \ldots$.

    (a) Use integration by parts to obtain a recurrence formula relating y_k to y_{k-1} for $k = 1, 2, 3, \ldots$, and determine the starting value y_0.

    (b) Write and run in single precision a program that generates $y_0, y_1, \ldots, y_{12}$, using the recurrence of (a), and prints the results in `e15.7` format. Explain in detail (quantitatively, using mathematical analysis) what is happening.

    (c) Use the recursion in (a) in reverse order, starting (arbitrarily) with $y_N = 0$. Print in four consecutive columns (in `e15.7` format) the values $y_0^{(N)}, y_1^{(N)}, \ldots, y_{12}^{(N)}$ thus obtained for $N = 14, 16, 18, 20$. Explain in detail (quantitatively) what you observe.

CHAPTER 2

APPROXIMATION AND INTERPOLATION

The present chapter is basically concerned with the approximation of functions. The functions in question may be functions defined on a continuum — typically a finite interval — or functions defined only on a finite set of points. The first instance arises, for example, in the context of special functions (elementary or transcendental) that one wishes to evaluate as part of a subroutine. Since any such evaluation must be reduced to a finite number of arithmetic operations, we must ultimately approximate the function by means of a polynomial or a rational function. The second instance is frequently encountered in the physical sciences when measurements are taken of a certain physical quantity as a function of some other physical quantity (such as time). In either case one wants to approximate the given function "as well as possible" in terms of other simpler functions.

The general scheme of approximation can be described as follows. We are given the function f to be approximated, along with a class Φ of "approximating functions" φ and a "norm" $\|\cdot\|$ measuring the overall magnitude of functions. We are looking for an approximation $\hat{\varphi} \in \Phi$ of f such that

$$\|f - \hat{\varphi}\| \leq \|f - \varphi\| \quad \text{for all} \quad \varphi \in \Phi. \tag{0.1}$$

The function $\hat{\varphi}$ is called the *best approximation* to f from the class Φ, relative to the norm $\|\cdot\|$.

The class Φ is called a (real) *linear space* if with any two functions φ_1, $\varphi_2 \in \Phi$ it also contains $\varphi_1 + \varphi_2$ and $c\varphi_1$ for any $c \in \mathbb{R}$, hence also any (finite) linear combination of functions $\varphi_i \in \Phi$. Given n "basis functions" $\pi_j \in \Phi$, $j = 1, 2, \ldots, n$, we can define a linear space of finite dimension n by

$$\Phi = \Phi_n = \{\varphi : \quad \varphi(t) = \sum_{j=1}^{n} c_j \pi_j(t), \; c_j \in \mathbb{R}\}. \tag{0.2}$$

Examples of linear spaces Φ

(1) $\Phi = \mathbb{P}_m$: polynomials of degree $\leq m$. A basis for $\mathbb{P}_m$ is, for example, $\pi_j(t) = t^{j-1}$, $j = 1, 2, \ldots, m+1$, so that $n = m+1$. Polynomials are the most frequently used "general-purpose" approximants for dealing with functions on bounded domains (finite intervals or finite sets of points). One reason

is *Weierstrass's theorem*, which states that any continuous function can be approximated on a finite interval as closely as one wishes by a polynomial of sufficiently high degree.

(2) $\Phi = \mathbf{S}_m^k(\Delta)$: (polynomial) spline functions of degree m and smoothness class k on the subdivision

$$\Delta: \quad a = t_1 < t_2 < t_3 < \cdots < t_{N-1} < t_N = b$$

of the interval $[a, b]$. These are piecewise polynomials of degree $\leq m$ pieced together at the "joints" $t_2, \ldots, t_{N-1}$ in such a way that all derivatives up to and including the kth are continuous on the whole interval $[a, b]$, including the joints:

$$\mathbf{S}_m^k(\Delta) = \{s \in C^k[a,b]: \ s\Big|_{[t_i,t_{i+1}]} \in \mathbb{P}_m, \quad i = 1, 2, \ldots, N-1\}.$$

We assume here $0 \leq k < m$; otherwise, we are back to polynomials $\mathbb{P}_m$ (see Ex. 62). We set $k = -1$ if we allow discontinuities at the joints. The dimension of $\mathbf{S}_m^k(\Delta)$ is $n = (m - k) \cdot (N - 2) + m + 1$ (see Ex. 65), but to find a basis is a nontrivial task; for $m = 1$, see §3.2.

(3) $\Phi = \mathbf{T}_m\ [0, 2\pi]$: trigonometric polynomials of degree $\leq m$ on $[0,\ 2\pi]$. These are linear combinations of the basic harmonics up to and including the mth one, that is,

$$\pi_k(t) = \cos\ (k-1)t, \quad k = 1, 2, \ldots, m+1;$$
$$\pi_{m+1+k}(t) = \sin kt, \quad k = 1, 2, \ldots, m,$$

where now $n = 2m + 1$. Such approximants are a natural choice when the function f to be approximated is periodic with period 2π. (If f has period p, one makes a preliminary change of variables $t \mapsto t \cdot p/2\pi$.)

(4) $\Phi = \mathbf{E}_n$: exponential sums. For given distinct $\alpha_j > 0$, one takes $\pi_j(t) = e^{-\alpha_j t}$, $j = 1, 2, \ldots, n$. Exponential sums are often employed on the half-infinite interval $\mathbb{R}_+$: $0 \leq t < \infty$, especially if one knows that f decays exponentially as $t \to \infty$.

Note that the important class of rational functions,

$$\Phi = \mathbb{R}_{r,s} = \{\varphi: \quad \varphi = p/q,\ p \in \mathbb{P}_r,\ q \in \mathbb{P}_s\},$$

is *not* a linear space. (Why not?)

Possible choices of norm — both for continuous and discrete functions — and the type of approximation they generate are summarized in Table 2.0.1. The continuous case involves an interval $[a, b]$ and a "weight function"

continuous norm	approximation	discrete norm
$\|u\|_\infty = \max_{a \le t \le b} \|u(t)\|$	L_∞ uniform Chebyshev	$\|u\|_\infty = \max_{1 \le i \le N} \|u(t_i)\|$
$\|u\|_1 = \int_a^b \|u(t)\| dt$	L_1	$\|u\|_1 = \sum_{i=1}^N \|u(t_i)\|$
$\|u\|_{1,w} = \int_a^b \|u(t)\| w(t) dt$	weighted L_1	$\|u\|_{1,w} = \sum_{i=1}^N w_i \|u(t_i)\|$
$\|u\|_{2,w} = \left(\int_a^b \|u(t)\|^2 w(t) dt \right)^{\frac{1}{2}}$	weighted L_2 least squares	$\|u\|_{2,w} = \left(\sum_{i=1}^N w_i \|u(t_i)\|^2 \right)^{\frac{1}{2}}$

TABLE 2.0.1. Types of approximation and associated norms

$w(t)$ (possibly $w(t) \equiv 1$) defined on $[a, b]$ and positive except for isolated zeros. The discrete case involves a set of N distinct points $t_1, t_2, \ldots, t_N$ along with positive weight factors $w_1, w_2, \ldots, w_N$ (possibly all equal to 1). The interval $[a, b]$ may be unbounded if the weight function w is such that the integral extended over $[a, b]$, which defines the norm, makes sense.

Hence, we may take any one of the norms in Table 2.0.1 and combine it with any of the preceding linear spaces Φ to arrive at a meaningful best approximation problem (0.1). In the continuous case, the given function f, and the functions φ of the class Φ, of course, must be defined on $[a, b]$ and such that the norm $\|f - \varphi\|$ makes sense. Likewise, f and φ must be defined at the points t_i in the discrete case.

Note that if the best approximant $\hat{\varphi}$ in the discrete case is such that $\|f - \hat{\varphi}\| = 0$, then $\hat{\varphi}(t_i) = f(t_i)$ for $i = 1, 2, \ldots, N$. We then say that $\hat{\varphi}$ *interpolates* f at the points t_i and we refer to this kind of approximation problem as an *interpolation problem*.

The simplest approximation problems are the least squares problem and the interpolation problem, and the easiest space Φ to work with the space of polynomials of given degree. These are indeed the problems we concentrate on in this chapter. In the case of the least squares problem, however, we admit general linear spaces Φ of approximants, and also in the case of the interpolation problem, we include polynomial splines in addition to straight polynomials.

Before we start with the least squares problem, we introduce a notational device that allows us to treat the continuous and the discrete case simultaneously. We define, in the continuous case,

$$\lambda(t) = \begin{cases} 0 \text{ if } t < a \text{ (whenever } -\infty < a), \\ \int_a^t w(\tau)d\tau \text{ if } a \le t \le b, \\ \int_a^b w(\tau)d\tau \text{ if } t > b \text{ (whenever } b < \infty). \end{cases} \tag{0.3}$$

Then we can write, for any (say, continuous) function u,

$$\int_{\mathbb{R}} u(t)d\lambda(t) = \int_a^b u(t)w(t)dt, \tag{0.4}$$

since $d\lambda(t) \equiv 0$ "outside" $[a, b]$, and $d\lambda(t) = w(t)dt$ inside. We call $d\lambda$ a *continuous* (positive) *measure*. The *discrete measure* (also called "Dirac measure") associated with the point set $\{t_1, t_2, \ldots, t_N\}$ is a measure $d\lambda$ that is nonzero only at the points t_i and has the value w_i there. Thus, in this case,

$$\int_{\mathbb{R}} u(t)d\lambda(t) = \sum_{i=1}^{N} w_i u(t_i). \tag{0.4$'$}$$

(A more precise definition can be given in terms of Stieltjes integrals, if we define $\lambda(t)$ to be a *step function* having jump w_i at t_i.) In particular, we can define the L_2 norm as

$$\|u\|_{2,d\lambda} = \left(\int_{\mathbb{R}} |u(t)|^2 d\lambda(t) \right)^{\frac{1}{2}}, \tag{0.5}$$

and obtain the continuous or the discrete norm depending on whether λ is taken to be as in (0.3), or a step function, as in (0.4$'$).

We call the *support* of $d\lambda$ — and denote it by $\operatorname{supp} d\lambda$ — the interval $[a, b]$ in the continuous case (assuming w positive on $[a, b]$ except for isolated zeros), and the set $\{t_1, t_2, \ldots, t_N\}$ in the discrete case. We say that the set of functions $\pi_j(t)$ in (0.2) is *linearly independent* on the support of $d\lambda$ if

$$\sum_{j=1}^{n} c_j \pi_j(t) \equiv 0 \text{ for all } t \in \operatorname{supp} d\lambda \text{ implies } c_1 = c_2 = \cdots = c_n = 0. \quad (0.6)$$

Example: The powers $\pi_j(t) = t^{j-1}$, $j = 1, 2, \ldots, n$.

Here $\sum_{j=1}^{n} c_j \pi_j(t) = p_{n-1}(t)$ is a polynomial of degree $\leq n-1$. Suppose, first, that $\operatorname{supp} d\lambda = [a, b]$. Then the identity in (0.6) says that $p_{n-1}(t) \equiv 0$ on $[a, b]$. Clearly, this implies $c_1 = c_2 = \cdots = c_n = 0$, so that the powers are linearly independent on $\operatorname{supp} d\lambda = [a, b]$. If, on the other hand, $\operatorname{supp} d\lambda = \{t_1, t_2, \ldots, t_N\}$, then the premise in (0.6) says that $p_{n-1}(t_i) = 0$, $i = 1, 2, \ldots, N$; that is, p_{n-1} has N distinct zeros t_i. This implies $p_{n-1} \equiv 0$ only if $N \geq n$. Otherwise, $p_{n-1}(t) = \prod_{i=1}^{N}(t - t_i) \in \mathbb{P}_{n-1}$ would satisfy $p_{n-1}(t_i) = 0$, $i = 1, 2, \ldots, N$, without being identically zero. Thus, we have linear independence on $\operatorname{supp} d\lambda = \{t_1, t_2, \ldots, t_N\}$ if and only if $N \geq n$.

§1. Least Squares Approximation

We specialize the best approximation problem (0.1) by taking as norm the L_2 norm

$$\|u\|_{2,d\lambda} = \left(\int_{\mathbb{R}} |u(t)|^2 d\lambda(t) \right)^{\frac{1}{2}}, \quad (1.1)$$

where $d\lambda$ is either a continuous measure (cf. (0.3)) or a discrete measure (cf. (0.4′)), and by using approximants φ from an n-dimensional linear space

$$\Phi = \Phi_n = \left\{ \varphi : \ \varphi(t) = \sum_{j=1}^{n} c_j \pi_j(t), \ c_j \in \mathbb{R} \right\}. \quad (1.2)$$

Here the basis functions π_j are assumed linearly independent on $\operatorname{supp} d\lambda$ (cf. (0.6)). We furthermore assume, of course, that the integral in (1.1) is meaningful whenever $u = \pi_j$ or $u = f$, the given function to be approximated.

The solution of the least squares problem is most easily expressed in terms of orthogonal systems π_j relative to an appropriate inner product. We therefore begin with a discussion of inner products.

§1.1. **Inner products**. Given a continuous or discrete measure $d\lambda$, as introduced earlier, and given any two functions u, v having a finite norm (1.1), we can define the *inner product*

$$(u, v) = \int_{\mathbb{R}} u(t)v(t)d\lambda(t). \tag{1.3}$$

(Schwarz's inequality $|(u,v)| \leq \|u\|_{2,d\lambda} \cdot \|v\|_{2,d\lambda}$ tells us that the integral in (1.3) is well defined.) The inner product (1.3) has the following obvious (but useful) properties.

(i) symmetry: $(u, v) = (v, u)$,

(ii) homogeneity: $(\alpha u, v) = \alpha(u, v)$, $\alpha \in \mathbb{R}$,

(iii) additivity: $(u + v, w) = (u, w) + (v, w)$, and

(iv) positive definiteness: $(u, u) \geq 0$, with equality holding if and only if $u \equiv 0$ on $\operatorname{supp} d\lambda$.

Homogeneity and additivity together give *linearity*,

$$(\alpha_1 u_1 + \alpha_2 u_2, v) = \alpha_1(u_1, v) + \alpha_2(u_2, v) \tag{1.4}$$

in the first variable and, by symmetry, also in the second. Moreover, (1.4) easily extends to linear combinations of arbitrary finite length. Note also that

$$\|u\|^2_{2,d\lambda} = (u, u). \tag{1.5}$$

We say that u and v are *orthogonal* if

$$(u, v) = 0. \tag{1.6}$$

This is always trivially true if either u or v vanishes identically on $\operatorname{supp} d\lambda$.

It is now a simple exercise, for example, to prove the *Theorem of Pythagoras*:

$$\text{if} \quad (u, v) = 0, \quad \text{then} \quad \|u + v\|^2 = \|u\|^2 + \|v\|^2, \tag{1.7}$$

where $\| \cdot \| = \| \cdot \|_{2,d\lambda}$. (From now on we use this abbreviated notation for the norm.) Indeed,

$$\begin{aligned}\|u+v\|^2 &= (u+v, u+v) = (u,u) + (u,v) + (v,u) + (v,v) \\ &= \|u\|^2 + 2(u,v) + \|v\|^2 = \|u\|^2 + \|v\|^2,\end{aligned}$$

where the first equality is a definition, the second follows from additivity, the third from symmetry, and the last from orthogonality. Interpreting functions u, v as "vectors," we can picture the configuration of u, v (orthogonal) and $u + v$ as in Figure 2.1.1.

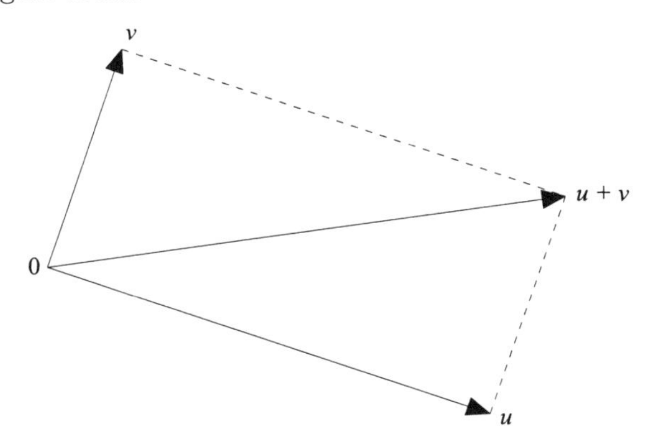

FIGURE 2.1.1. Orthogonal vectors and their sum

More generally, we may consider an *orthogonal system* $\{u_k\}_{k=1}^n$:

$$\begin{aligned}(u_i, u_j) \;=\; 0 \text{ if } i \neq j, \quad u_k \not\equiv 0 \text{ on } \operatorname{supp} d\lambda; \\ i, j = 1, 2, \ldots, n; \quad k = 1, 2, \ldots, n.\end{aligned} \tag{1.8}$$

For such a system we have the *Generalized Theorem of Pythagoras*,

$$\left\| \sum_{k=1}^{n} \alpha_k u_k \right\|^2 = \sum_{k=1}^{n} |\alpha_k|^2 \|u_k\|^2. \tag{1.9}$$

The proof is essentially the same as before. An important consequence of (1.9) is that *every orthogonal system is linearly independent* on the support of $d\lambda$. Indeed, if the left-hand side of (1.9) vanishes, then so does the right-hand side, and this, since $\|u_k\|^2 > 0$ by assumption, implies $\alpha_1 = \alpha_2 = \cdots = \alpha_n = 0$.

§1.2. **The normal equations.** We are now in a position to solve the least squares approximation problem. By (1.5), we can write the L_2 error, or rather its square, in the form

$$E^2[\varphi] := \|\varphi - f\|^2 = (\varphi - f, \varphi - f) = (\varphi, \varphi) - 2(\varphi, f) + (f, f).$$

Inserting φ here from (1.2) gives

$$E^2[\varphi] = \int_{\mathbb{R}} \left(\sum_{j=1}^{n} c_j \pi_j(t) \right)^2 d\lambda(t) - 2 \int_{\mathbb{R}} \left(\sum_{j=1}^{n} c_j \pi_j(t) \right) f(t) d\lambda(t) + \int_{\mathbb{R}} f^2(t) d\lambda(t). \tag{1.10}$$

The squared L_2 error, therefore, is a quadratic function of the coefficients $c_1, c_2, \ldots, c_n$ of φ. The problem of best L_2 approximation thus amounts to minimizing a quadratic function of n variables. This is a standard problem of calculus and is solved by setting all partial derivatives equal to zero. This yields a system of *linear* algebraic equations. Indeed, differentiating partially with respect to c_i under the integral sign in (1.10) gives

$$\frac{\partial}{\partial c_i} E^2[\varphi] = 2 \int_{\mathbb{R}} \left(\sum_{j=1}^{n} c_j \pi_j(t) \right) \pi_i(t) d\lambda(t) - 2 \int_{\mathbb{R}} \pi_i(t) f(t) d\lambda(t),$$

and setting this equal to zero, interchanging integration and summation in the process, we get

$$\sum_{j=1}^{n} (\pi_i, \pi_j) c_j = (\pi_i, f), \quad i = 1, 2, \ldots, n. \tag{1.11}$$

These are called the *normal equations* for the least squares problem. They form a linear system of the form

$$Ac = b, \tag{1.11'}$$

where the matrix A and the vector b have elements

$$A = [a_{ij}], \; a_{ij} = (\pi_i, \pi_j); \quad b = [b_i], \; b_i = (\pi_i, f). \tag{1.11''}$$

By symmetry of the inner product, A is a symmetric matrix. Moreover, A is positive definite; that is,

$$x^T A x = \sum_{i=1}^{n} \sum_{j=1}^{n} a_{ij} x_i x_j > 0 \;\text{ if }\; x \neq [0, 0, \ldots, 0]^T. \tag{1.12}$$

The quadratic function in (1.12) is called a *quadratic form* (since it is homogeneous of degree 2). Positive definiteness of A thus says that the quadratic form whose coefficients are the elements of A is always nonnegative, and zero only if all variables x_i vanish.

To prove (1.12), all we have to do is insert the definition of the a_{ij} and use the elementary properties (i) through (iv) of the inner product:

$$x^T Ax = \sum_{i=1}^{n}\sum_{j=1}^{n} x_i x_j (\pi_i, \pi_j) = \sum_{i=1}^{n}\sum_{j=1}^{n} (x_i\pi_i, x_j\pi_j) = \left\| \sum_{i=1}^{n} x_i \pi_i \right\|^2 .$$

This clearly is nonnegative. It is zero only if $\sum_{i=1}^{n} x_i\pi_i \equiv 0$ on $\operatorname{supp} d\lambda$, which, by the assumption of linear independence of the π_i, implies $x_1 = x_2 = \cdots = x_n = 0$.

Now it is a well-known fact of linear algebra that a symmetric positive definite matrix A is nonsingular. Indeed, its determinant, as well as all its leading principal minor determinants, are strictly positive. It follows that the system (1.11) of normal equations has a unique solution. Does this solution correspond to a minimum of $E[\varphi]$ in (1.10)? Calculus tells us that for this to be the case, the Hessian matrix $H = [\partial^2 E^2/\partial c_i \partial c_j]$ has to be positive definite. But $H = 2A$, since E^2 is a quadratic function. Therefore, H, with A, is indeed positive definite, and the solution of the normal equations gives us the desired minimum. The least squares approximation problem thus has a unique solution, given by

$$\hat{\varphi}(t) = \sum_{j=1}^{n} \hat{c}_j \pi_j(t), \tag{1.13}$$

where $\hat{c} = [\hat{c}_1, \hat{c}_2, \ldots, \hat{c}_n]^T$ is the solution vector of the normal equation (1.11).

This completely settles the least squares approximation problem in theory. How about in practice?

Assuming a general set of (linearly independent) basis functions, we can see the following possible difficulties.

(1) The system (1.11) may be *ill-conditioned.* A simple example is provided by $\operatorname{supp} d\lambda = [0,1]$, $d\lambda(t) = dt$ on [0,1], and $\pi_j(t) = t^{j-1}$, $j = 1, 2, \ldots, n$. Then

$$(\pi_i, \pi_j) = \int_0^1 t^{i+j-2}\, dt = \frac{1}{i+j-1}\,, \quad i, j = 1, 2, \ldots, n;$$

that is, the matrix A in (1.11) is precisely the Hilbert matrix. (Cf. Ch. 1, (3.41).) The resulting severe ill-conditioning of the normal equations in this example is entirely due to an unfortunate choice of basis functions — the powers. These become almost linearly dependent, more so the larger the exponent (cf. Ex. 35). Another source of degradation lies in the element $b_j = \int_0^1 \pi_j(t)f(t)dt$ of the right-hand vector b in (1.11). When j is large, the power $\pi_j = t^{j-1}$ behaves very much like a discontinuous function on [0,1]: it is practically zero for much of the interval until it shoots up to the value 1 at the right endpoint. This has the unfortunate consequence that a good deal of information about f is lost when one forms the integral defining b_j. A polynomial π_j that oscillates rapidly on [0,1] would seem to be preferable from this point of view, since it would "engage" the function f more vigorously over all of the interval [0,1].

(2) The second disadvantage is the fact that all coefficients $\hat{c}_j$ in (1.13) depend on n; that is, $\hat{c}_j = \hat{c}_j^{(n)}$, $j = 1, 2, \ldots, n$. Increasing n, for example, will give an enlarged system of normal equations with a completely new solution vector. We refer to this as the *nonpermanence* of the coefficients $\hat{c}_j$.

Both defects (1) and (2) can be eliminated (or at least attenuated in the case of (1)) in one stroke: select for the basis functions π_j an orthogonal system,

$$(\pi_i, \pi_j) = 0 \ \text{ if } \ i \neq j; \quad (\pi_j, \pi_j) = \|\pi_j\|^2 > 0. \tag{1.14}$$

Then the system of normal equations becomes diagonal and is solved immediately by

$$\hat{c}_j = \frac{(\pi_j, f)}{(\pi_j, \pi_j)}, \quad j = 1, 2, \ldots, n. \tag{1.15}$$

Clearly, each of these coefficients $\hat{c}_j$ is independent of n, and once computed, remains the same for any larger n. We now have *permanence* of the coefficients. Also, we do not have to go through the trouble of solving a linear system of equations, but instead can use the formula (1.15) directly. This does not mean that there are no numerical problems associated with (1.15). Indeed, it is typical that the denominators $\|\pi_j\|^2$ in (1.15) decrease rapidly with increasing j, whereas the integrand in the numerator (or the individual terms in the case of a discrete inner product) are of the same magnitude as f. Yet the coefficients $\hat{c}_j$ also are expected to decrease rapidly. Therefore, cancellation errors must occur when one computes the inner product in the numerator. The cancellation problem can be alleviated somewhat by computing $\hat{c}_j$ in the alternative form

$$\hat{c}_j = \frac{1}{(\pi_j, \pi_j)} \left(f - \sum_{k=1}^{j-1} \hat{c}_k \pi_k, \pi_j \right), \quad j = 1, 2, \ldots, n, \tag{1.15'}$$

where the empty sum (when $j = 1$) is taken to be zero, as usual. Clearly, by orthogonality of the π_j, (1.15′) is equivalent to (1.15) mathematically, but not necessarily numerically.

An algorithm for computing $\hat{c}_j$ from (1.15′), and at the same time $\hat{\varphi}(t)$, is as follows:

$$\begin{aligned} & s_0 = 0, \\ & \text{for } \ j = 1, 2, \ldots, n \ \text{ do} \\ & \left[\begin{aligned} \hat{c}_j &= \frac{1}{\|\pi_j\|^2} (f - s_{j-1}, \pi_j) \\ s_j &= s_{j-1} + \hat{c}_j \pi_j(t). \end{aligned} \right. \end{aligned}$$

This produces the coefficients $\hat{c}_1, \hat{c}_2, \ldots, \hat{c}_n$ as well as $\hat{\varphi}(t) = s_n$.

Any system $\{\hat{\pi}_j\}$ that is linearly independent on the support of $d\lambda$ can be orthogonalized (with respect to the measure $d\lambda$) by a device known as the *Gram-Schmidt procedure*. One takes

$$\pi_1 = \hat{\pi}_1$$

and, for $j = 2, 3, \ldots$, recursively forms

$$\pi_j = \hat{\pi}_j - \sum_{k=1}^{j-1} c_k \pi_k, \quad c_k = \frac{(\hat{\pi}_j, \pi_k)}{(\pi_k, \pi_k)} .$$

Then each π_j so determined is orthogonal to all preceding ones.

§1.3. **Least squares error; convergence**. We have seen in §1.2 that if the class $\Phi = \Phi_n$ consists of n functions π_j, $j = 1, 2, \ldots, n$, that are linearly independent on the support of some measure $d\lambda$, then the least squares problem for this measure,

$$\min_{\varphi \in \Phi_n} \|f - \varphi\|_{2,d\lambda} = \|f - \hat{\varphi}\|_{2,d\lambda}, \tag{1.16}$$

has a unique solution $\hat{\varphi} = \hat{\varphi}_n$ given by (1.13). There are many ways we can select a basis π_j in Φ_n and, therefore, many ways the solution $\hat{\varphi}_n$ can be represented. Nevertheless, it is always one and the same function. The least squares error — the quantity on the right of (1.16) — therefore is independent of the choice of basis functions (although the calculation of the least squares solution, as mentioned previously, is not). In studying this error, we may thus assume, without restricting generality, that the basis π_j is an orthogonal system. (Every linearly independent system can be orthogonalized by the Gram-Schmidt orthogonalization procedure; cf. §1.2.) We then have (cf. (1.15))

$$\hat{\varphi}_n(t) = \sum_{j=1}^{n} \hat{c}_j \pi_j(t), \quad \hat{c}_j = \frac{(\pi_j, f)}{(\pi_j, \pi_j)} \,. \tag{1.17}$$

We first note that the error $f - \hat{\varphi}_n$ is orthogonal to the space Φ_n; that is,

$$(f - \hat{\varphi}_n, \varphi) = 0 \quad \text{for all} \quad \varphi \in \Phi_n, \tag{1.18}$$

where the inner product is the one in (1.3). Since φ is a linear combination of the π_k, it suffices to show (1.18) for each $\varphi = \pi_k$, $k = 1, 2, \ldots, n$. Inserting $\hat{\varphi}_n$ from (1.17) in the left of (1.18), and using orthogonality, we find indeed

$$(f - \hat{\varphi}_n, \pi_k) = \left(f - \sum_{j=1}^{n} \hat{c}_j \pi_j, \pi_k \right) = (f, \pi_k) - \hat{c}_k(\pi_k, \pi_k) = 0,$$

the last equation following from the formula for $\hat{c}_k$ in (1.17). The result (1.18) has a simple geometric interpretation. If we picture functions as vectors, and the space Φ_n as a plane, then for any f that "sticks out" of the plane Φ_n, the least squares approximant $\hat{\varphi}_n$ is the *orthogonal projection* of f onto Φ_n; see Figure 2.1.2.

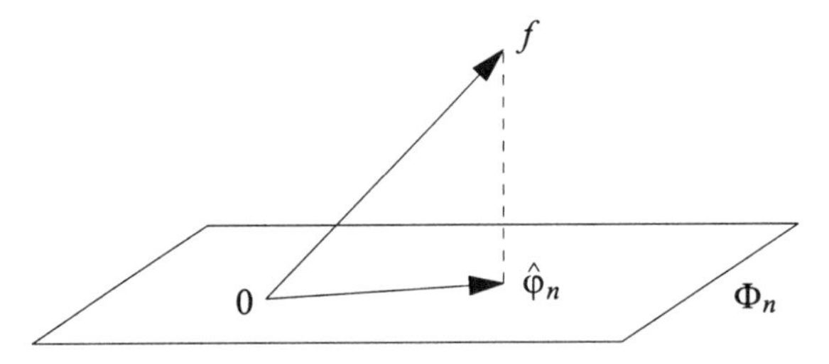

FIGURE 2.1.2. Least squares approximation as orthogonal projection

In particular, choosing $\varphi = \hat{\varphi}_n$ in (1.18), we get

$$(f - \hat{\varphi}_n, \hat{\varphi}_n) = 0$$

and, therefore, since $f = (f - \hat{\varphi}_n) + \hat{\varphi}_n$, by the Theorem of Pythagoras (cf. (1.7)) and its generalization (cf. (1.9)),

$$\begin{aligned} \|f\|^2 &= \|f - \hat{\varphi}_n\|^2 + \|\hat{\varphi}_n\|^2 \\ &= \|f - \hat{\varphi}_n\|^2 + \left\| \sum_{j=1}^n \hat{c}_j \pi_j \right\|^2 \\ &= \|f - \hat{\varphi}_n\|^2 + \sum_{j=1}^n |\hat{c}_j|^2 \|\pi_j\|^2. \end{aligned}$$

Solving for the first term on the right, we get

$$\|f - \hat{\varphi}_n\| = \left\{ \|f\|^2 - \sum_{j=1}^n |\hat{c}_j|^2 \|\pi_j\|^2 \right\}^{\frac{1}{2}}, \quad \hat{c}_j = \frac{(\pi_j, f)}{(\pi_j, \pi_j)}. \tag{1.19}$$

Note that the expression in braces must necessarily be nonnegative.

The formula (1.19) for the error is interesting theoretically, but of limited practical use. Note, indeed, that as the error approaches the level of the machine precision eps, computing the error from the right-hand side of (1.19) cannot produce anything smaller than $\sqrt{\text{eps}}$ because of inevitable rounding errors committed during the subtraction in the radicand. (They may even produce a negative result for the radicand.) Using instead the definition,

$$\|f - \hat{\varphi}_n\| = \left\{ \int_{\mathbb{R}} [f(t) - \hat{\varphi}_n(t)]^2 d\lambda(t) \right\}^{\frac{1}{2}},$$

along, perhaps, with a suitable (positive) quadrature rule (cf. Ch. 3, §2), is guaranteed to produce a nonnegative result that may potentially be as small as $O(\text{eps})$.

If we are now given a sequence of linear spaces Φ_n, $n = 1, 2, 3, \ldots$, as defined in (0.2), then clearly

$$\|f - \hat{\varphi}_1\| \geq \|f - \hat{\varphi}_2\| \geq \|f - \hat{\varphi}_3\| \geq \cdots,$$

which follows not only from (1.19), but more directly from the fact that $\Phi_1 \subset \Phi_2 \subset \Phi_3 \subset \cdots$. If there are infinitely many such spaces, then the sequence of L_2 errors, being monotonically decreasing, must converge to a limit. Is this limit zero? If so, we say that the least squares approximation process *converges* (in the mean) as $n \to \infty$. It is obvious from (1.19) that a necessary and sufficient condition for this is

$$\sum_{j=1}^{\infty} |\hat{c}_j|^2 \|\pi_j\|^2 = \|f\|^2. \tag{1.20}$$

An equivalent way of stating convergence is as follows: given any f with $\|f\| < \infty$, that is, any f in the $L_{2,d\lambda}$ space, and given any $\varepsilon > 0$, no matter how small, there exists an integer $n = n_\varepsilon$ and a function $\varphi^* \in \Phi_n$ such that $\|f - \varphi^*\| \leq \varepsilon$. A class of spaces Φ_n having this property is said to be *complete* with respect to the norm $\|\cdot\| = \|\cdot\|_{2,d\lambda}$. One therefore calls (1.20) also the *completeness relation.*

For a finite interval $[a, b]$, one can define completeness of $\{\Phi_n\}$ also for the uniform norm $\|\cdot\| = \|\cdot\|_\infty$ on $[a, b]$. One then assumes $f \in C[a, b]$ and also $\pi_j \in C[a, b]$ for all basis functions in all classes Φ_n, and one calls $\{\Phi_n\}$ complete in the norm $\|\cdot\|_\infty$ if for any $f \in C[a, b]$ and any $\varepsilon > 0$ there is an $n = n_\varepsilon$ and a $\varphi^* \in \Phi_n$ such that $\|f - \varphi^*\|_\infty \leq \varepsilon$. It is easy to see that completeness of $\{\Phi_n\}$ in the norm $\|\cdot\|_\infty$ (on $[a, b]$) implies completeness of $\{\Phi_n\}$ in the L_2 norm $\|\cdot\|_{2,d\lambda}$, where $\operatorname{supp} d\lambda = [a, b]$, and hence convergence of the least squares approximation process. Indeed, let $\varepsilon > 0$ be arbitrary and let n and $\varphi^* \in \Phi_n$ be such that

$$\|f - \varphi^*\|_\infty \leq \frac{\varepsilon}{\left(\int_{\mathbb{R}} d\lambda(t)\right)^{\frac{1}{2}}}.$$

This is possible by assumption. Then

$$\|f - \varphi^*\|_{2,d\lambda} = \left(\int_{\mathbb{R}} [f(t) - \varphi^*(t)]^2 d\lambda(t)\right)^{\frac{1}{2}}$$

$$\leq \|f - \varphi^*\|_\infty \left(\int_{\mathbb{R}} d\lambda(t)\right)^{\frac{1}{2}} \leq \frac{\varepsilon}{\left(\int_{\mathbb{R}} d\lambda(t)\right)^{\frac{1}{2}}} \left(\int_{\mathbb{R}} d\lambda(t)\right)^{\frac{1}{2}} = \varepsilon,$$

as claimed.

Example: $\Phi_n = \mathbb{P}_{n-1}$. Here completeness of $\{\Phi_n\}$ in the norm $\|\cdot\|_\infty$ (on a finite interval $[a, b]$) is a consequence of Weierstrass's Approximation

Theorem. Thus, polynomial least squares approximation on a finite interval always converges (in the mean).

§1.4. **Examples of orthogonal systems**. There are many orthogonal systems in use. The prototype of them all is the system of trigonometric functions known from Fourier analysis. Other widely used systems involve algebraic polynomials. We restrict ourselves here to these two particular examples of orthogonal systems.

(1) *Trigonometric functions*: $1,\ \cos t,\ \cos 2t,\ \cos 3t\ ,\ldots,\ \sin t,\ \sin 2t,$ $\sin 3t, \ldots$. These are the basic harmonics; they are mutually orthogonal on the interval $[0, 2\pi]$ with respect to the equally weighted measure on $[0,2\pi]$,

$$d\lambda(t) = \begin{cases} dt & \text{on } [0, 2\pi], \\ 0 & \text{otherwise}. \end{cases} \tag{1.21}$$

We verify this for the sine functions: for $k,\ \ell = 1, 2, 3, \ldots$ we have

$$\int_0^{2\pi} \sin kt \cdot \sin \ell t\ dt = -\frac{1}{2} \int_0^{2\pi} [\cos(k+\ell)t - \cos(k-\ell)t]\ dt.$$

The right-hand side is equal to

$$-\frac{1}{2}\left[\frac{\sin(k+\ell)t}{k+\ell} - \frac{\sin(k-\ell)t}{k-\ell}\right]_0^{2\pi} = 0,$$

when $k \neq \ell$, and equal to π otherwise. Thus,

$$\int_0^{2\pi} \sin kt \cdot \sin \ell t\ dt = \begin{cases} 0 & \text{if } k \neq \ell, \\ \pi & \text{if } k = \ell, \end{cases} \qquad k, \ell = 1, 2, 3, \ldots . \tag{1.22ss}$$

Similarly, one shows that

$$\int_0^{2\pi} \cos kt \cdot \cos \ell t\ dt = \begin{cases} 0 & \text{if } k \neq \ell, \\ 2\pi & \text{if } k = \ell = 0, \\ \pi & \text{if } k = \ell > 0, \end{cases} \qquad k, \ell = 0, 1, 2, \ldots \tag{1.22cc}$$

and

$$\int_0^{2\pi} \sin kt \cdot \cos \ell t \, dt = 0, \quad k = 1, 2, 3, \ldots, \quad \ell = 0, 1, 2, \ldots \,. \tag{1.22sc}$$

The theory of *Fourier series* is concerned with the expansion of a given 2π-periodic function in terms of these trigonometric functions,

$$f(t) = \sum_{k=0}^{\infty} a_k \cos kt + \sum_{k=1}^{\infty} b_k \sin kt. \tag{1.23}$$

Using (1.22), one formally obtains

$$\begin{aligned} a_0 &= \frac{1}{2\pi}\int_0^{2\pi} f(t)\, dt, \quad a_k = \frac{1}{\pi}\int_0^{2\pi} f(t)\cos kt \, dt, \quad k = 1, 2, \ldots\,, \\ b_k &= \frac{1}{\pi}\int_0^{2\pi} f(t)\sin kt \, dt, \quad k = 1, 2, \ldots\,, \end{aligned} \tag{1.24}$$

which are known as *Fourier coefficients* of f. They are precisely the coefficients (1.15) for the system π_j consisting of our trigonometric functions. By extension, one therefore calls the coefficients $\hat{c}_j$ in (1.15), for *any* orthogonal system π_j, the Fourier coefficients of f relative to this system. In particular, we now recognize the truncated Fourier series (the series on the right of (1.23) truncated at $k = m$, with a_k, b_k given by (1.24)) as the best L_2 approximation to f from the class of trigonometric polynomials of degree $\leq m$ relative to the norm (cf. (1.21))

$$\|u\|_2 = \left(\int_0^{2\pi} |u(t)|^2 dt\right)^{\frac{1}{2}}.$$

(2) *Orthogonal polynomials.* Given a measure $d\lambda$ as introduced in (0.3) to (0.4′), we know from the example immediately following (0.6) that any finite number of consecutive powers 1, t, $t^2, \ldots$ are linearly independent on $[a, b]$, if $\operatorname{supp} d\lambda = [a, b]$, whereas the finite set $1, t, \ldots, t^{N-1}$ is linearly independent on $\operatorname{supp} d\lambda = \{t_1, t_2, \ldots, t_N\}$. Since a linearly independent set can be orthogonalized by Gram-Schmidt (cf. §1.2), any measure $d\lambda$ of the type considered generates a unique set of (monic) polynomials $\pi_j(t) =$

$\pi_j(t; d\lambda)$, $j = 0, 1, 2, \dots$, satisfying

$$\deg \pi_j = j, \quad j = 0, 1, 2, \dots ,$$
$$\int_{\mathbb{R}} \pi_k(t) \pi_\ell(t) d\lambda(t) = 0 \quad \text{if} \quad k \neq \ell. \tag{1.25}$$

These are called *orthogonal polynomials* relative to the measure $d\lambda$. (We slightly deviate from the notation in §§1.2 and 1.3 by letting the index j start from zero.) The set π_j is infinite if $\operatorname{supp} d\lambda = [a, b]$, and consists of exactly N polynomials $\pi_0, \pi_1, \dots, \pi_{N-1}$ if $\operatorname{supp} d\lambda = \{t_1, t_2, \dots, t_N\}$. The latter are referred to as *discrete orthogonal polynomials.*

It is an important fact that three consecutive orthogonal polynomials are linearly related. Specifically, there are real constants $\alpha_k = \alpha_k(d\lambda)$ and positive constants $\beta_k = \beta_k(d\lambda)$ (depending on the measure $d\lambda$) such that

$$\pi_{k+1}(t) = (t - \alpha_k)\pi_k(t) - \beta_k \pi_{k-1}(t), \quad k = 0, 1, 2, \dots ,$$
$$\pi_{-1}(t) = 0, \quad \pi_0(t) = 1. \tag{1.26}$$

(It is understood that (1.26) holds for all integers $k \geq 0$ if $\operatorname{supp} d\lambda = [a, b]$, and only for $0 \leq k < N - 1$ if $\operatorname{supp} d\lambda = \{t_1, t_2, \dots, t_N\}$.)

To prove (1.26) and, at the same time identify the coefficients α_k, β_k, we note that

$$\pi_{k+1}(t) - t\pi_k(t)$$

is a polynomial of degree $\leq k$, since the leading terms cancel (the polynomials π_j are assumed monic). Since an orthogonal system is linearly independent (cf. the remark after (1.9)), we can express this polynomial as a linear combination of $\pi_0, \pi_1, \dots, \pi_k$. We choose to write this linear combination in the form

$$\pi_{k+1}(t) - t\pi_k(t) = -\alpha_k \pi_k(t) - \beta_k \pi_{k-1}(t) + \sum_{j=0}^{k-2} \gamma_{k,j} \pi_j(t) \tag{1.27}$$

(with the understanding that empty sums are zero). Now multiply both sides of (1.27) by π_k in the sense of the inner product $(\cdot\, , \cdot)$ defined in (1.3). By orthogonality, this gives $(-t\pi_k, \pi_k) = -\alpha_k(\pi_k, \pi_k)$; that is,

$$\alpha_k = \frac{(t\pi_k, \pi_k)}{(\pi_k, \pi_k)}, \quad k = 0, 1, 2, \dots . \tag{1.28}$$

Similarly, forming the inner product of (1.27) with π_{k-1} gives $(-t\pi_k, \pi_{k-1}) = -\beta_k(\pi_{k-1}, \pi_{k-1})$. Since $(t\pi_k, \pi_{k-1}) = (\pi_k, t\pi_{k-1})$ and $t\pi_{k-1}$ differs from π_k by a polynomial of degree $< k$, we obtain by orthogonality $(t\pi_k, \pi_{k-1}) = (\pi_k, \pi_k)$; hence

$$\beta_k = \frac{(\pi_k, \pi_k)}{(\pi_{k-1}, \pi_{k-1})}, \quad k = 1, 2, \dots . \tag{1.29}$$

Finally, multiplication of (1.27) by π_ℓ, $\ell < k-1$, yields

$$\gamma_{k,\ell} = 0, \quad \ell = 0, 1, \dots, k-2. \tag{1.30}$$

Solving (1.27) for π_{k+1} then establishes (1.26), with α_k, β_k defined by (1.28) and (1.29), respectively. Clearly, $\beta_k > 0$.

The recursion (1.26) provides us with a practical scheme of generating orthogonal polynomials. Indeed, since $\pi_0 = 1$, we can compute α_0 by (1.28) with $k = 0$. This allows us to compute $\pi_1(t)$ for any t, using (1.26) with $k = 0$. Knowing π_0, π_1, we can go back to (1.28) and (1.29) and compute, respectively, α_1 and β_1. This gives us access to π_2 via (1.26) with $k = 1$. Proceeding in this fashion, using alternately (1.28), (1.29), and (1.26), we can generate as many of the orthogonal polynomials as are desired. This procedure — called *Stieltjes's procedure* — is particularly well suited for discrete orthogonal polynomials, since the inner product is then a finite sum, $(u, v) = \sum_{i=1}^{N} w_i u(t_i) v(t_i)$ (cf. (0.4′)), so that the computation of the α_k, β_k from (1.28) and (1.29) is straightforward. In the continuous case, the computation of the inner product requires integration, which complicates matters. Fortunately, for many important special measures $d\lambda(t) = w(t)dt$, the recursion coefficients are explicitly known (cf. Ch. 3, Table 3.2.1). In these cases, it is again straightforward to generate the orthogonal polynomials by (1.26).

The special case of *symmetry* (i.e., $d\lambda(t) = w(t)dt$ with $w(-t) = w(t)$ and $\text{supp}(d\lambda)$ symmetric with respect to the origin) deserves special mention. In this case, defining $p_k(t) = (-1)^k \pi_k(-t)$, one obtains by a simple change of variables that $(p_k, p_\ell) = (-1)^{k+\ell}(\pi_k, \pi_\ell) = 0$ if $k \neq \ell$. Since p_k is monic, it follows by uniqueness that $p_k(t) \equiv \pi_k(t)$; that is,

$$(-1)^k \pi_k(-t) \equiv \pi_k(t) \quad (d\lambda \text{ symmetric}). \tag{1.31}$$

Thus, if k is even, then π_k is an even polynomial, that is, a polynomial in t^2. Likewise, when k is odd, π_k contains only odd powers of t. As a consequence,

$$\alpha_k = 0 \quad \text{for all } k \geq 0 \quad (d\lambda \text{ symmetric}), \tag{1.32}$$

which also follows from (1.28), since the numerator on the right of this equation is an integral of an odd function over a symmetric set of points.

Example: Legendre polynomials

We may introduce the monic Legendre polynomials by

$$\pi_k(t) = (-1)^k \frac{k!}{(2k!)} \frac{d^k}{dt^k}(1-t^2)^k, \quad k = 0, 1, 2, \dots , \tag{1.33}$$

which is known as the *Rodrigues formula.*

We first verify orthogonality on the interval $[-1, 1]$ relative to the measure $d\lambda(t) = dt$. For any ℓ with $0 \le \ell < k$, repeated integration by parts gives

$$\int_{-1}^{1} \frac{d^k}{dt^k}(1-t^2)^k \cdot t^\ell dt$$

$$= \sum_{m=0}^{\ell} (-1)^m \ell(\ell-1)\cdots(\ell-m+1)t^{\ell-m} \frac{d^{k-m-1}}{dt^{k-m-1}}(1-t^2)^k \Big|_{-1}^{1} = 0,$$

the last equation since $0 \le k - m - 1 < k$. Thus,

$$(\pi_k, p) = 0 \quad \text{for every } p \in \mathbb{P}_{k-1},$$

proving orthogonality. Writing (by symmetry)

$$\pi_k(t) = t^k + \mu_k t^{k-2} + \cdots , \quad k \ge 2,$$

and noting (again by symmetry) that the recurrence relation has the form

$$\pi_{k+1}(t) = t\pi_k(t) - \beta_k \pi_{k-1}(t),$$

we obtain

$$\beta_k = \frac{t\pi_k(t) - \pi_{k+1}(t)}{\pi_{k-1}(t)} ,$$

which is valid for all t. In particular, as $t \to \infty$,

$$\beta_k = \lim_{t\to\infty} \frac{t\pi_k(t) - \pi_{k+1}(t)}{\pi_{k-1}(t)} = \lim_{t\to\infty} \frac{(\mu_k - \mu_{k+1})t^{k-1} + \cdots}{t^{k-1} + \cdots} = \mu_k - \mu_{k+1}.$$

(If $k = 1$, set $\mu_1 = 0$.) From Rodrigues's formula, however, we find

$$\pi_k(t) = \frac{k!}{(2k)!} \frac{d^k}{dt^k}(t^{2k} - kt^{2k-2} + \cdots) = \frac{k!}{(2k)!} \Big(2k(2k-1)\cdots(k+1)t^k$$

$$- k \cdot (2k-2)(2k-3)\cdots(k-1)t^{k-2} + \cdots \Big)$$

$$= t^k - \frac{k(k-1)}{2(2k-1)} t^{k-2} + \cdots ,$$

so that

$$\mu_k = -\frac{k(k-1)}{2(2k-1)}, \quad k \geq 2.$$

Therefore,

$$\beta_k = \mu_k - \mu_{k+1} = -\frac{k(k-1)}{2(2k-1)} + \frac{(k+1)k}{2(2k+1)} = \frac{k}{2}\,\frac{2k}{(2k+1)(2k-1)};$$

that is, since $\mu_1 = 0$,

$$\beta_k = \frac{1}{4-k^{-2}}, \quad k \geq 1. \tag{1.34}$$

We conclude with two remarks concerning discrete measures $d\lambda$ with $\operatorname{supp} d\lambda = \{t_1, t_2, \ldots, t_N\}$. As before, the L_2 errors decrease monotonically, but the last one is now zero, since there is a polynomial of degree $\leq N-1$ that interpolates f at the N points $t_1, t_2, \ldots, t_N$ (cf. §2.1). Thus,

$$\|f - \hat{\varphi}_0\| \geq \|f - \hat{\varphi}_1\| \geq \cdots \geq \|f - \hat{\varphi}_{N-1}\| = 0, \tag{1.35}$$

where $\hat{\varphi}_n$ is the L_2 approximant of degree $\leq n$,

$$\hat{\varphi}_n(t) = \sum_{j=0}^{n} \hat{c}_j \pi_j(t; d\lambda), \quad \hat{c}_j = \frac{(\pi_j, f)}{(\pi_j, \pi_j)}. \tag{1.36}$$

We see that the polynomial $\hat{\varphi}_{N-1}$ solves the interpolation problem for $\mathbb{P}_{N-1}$. Using (1.36) with $n = N-1$ to obtain the interpolation polynomial, however, is a roundabout way of solving the interpolation problem. We learn of more direct ways in the next section.

§2. Polynomial Interpolation

We now wish to approximate functions by matching their values at given points. Using polynomials as approximants gives rise to the following problem: given $n+1$ distinct points $x_0, x_1, \ldots, x_n$ and values $f_i = f(x_i)$ of some function f at these points, find a polynomial $p \in \mathbb{P}_n$ such that

$$p(x_i) = f_i, \quad i = 0, 1, 2, \ldots, n.$$

Since we have to satisfy $n+1$ conditions, and have at our disposal $n+1$ degrees of freedom — the coefficients of p — we expect the problem to have a unique solution. Other questions of interest, in addition to existence and uniqueness, are different ways of representing and computing the polynomial p, what can be said about the error $e(x) = f(x) - p(x)$ when $x \neq x_i$, $i = 0, 1, \dots, n$, and the quality of approximation $f(x) \approx p(x)$ when the number of points, and hence the degree of p, is allowed to increase indefinitely. Although these questions are not of the utmost interest in themselves, the results discussed here are widely used in the development of approximate methods for more important practical tasks such as solving initial and boundary value problems for ordinary and partial differential equations. It is in view of these and other applications that we study polynomial interpolation.

The simplest example is *linear interpolation*, that is, the case $n = 1$. Here it is obvious from Figure 2.2.1 that the interpolation problem has a unique solution. It is also clear that the error $e(x)$ can be as large as one likes (or dislikes) if nothing is known about f other than its two values at x_0 and x_1.

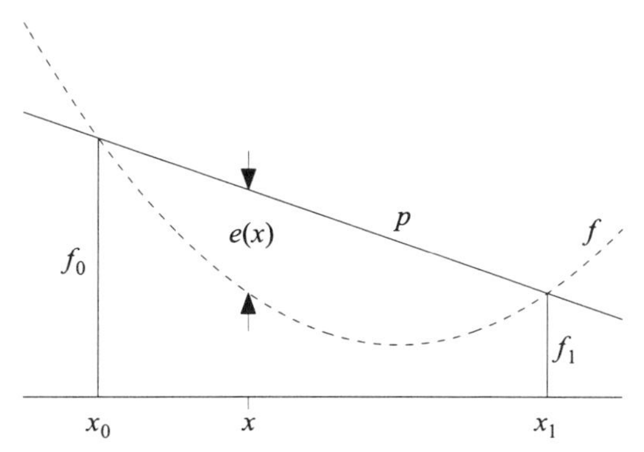

FIGURE 2.2.1. Linear interpolation

One way of writing down the linear interpolant p is as a weighted average of f_0 and f_1 (already taught in high school),

$$p(x) = \frac{x - x_1}{x_0 - x_1} f_0 + \frac{x - x_0}{x_1 - x_0} f_1.$$

This is the way Lagrange expressed p in the general case (cf. §2.1). However, we can write p also in Taylor's form, noting that its derivative at x_0 is equal to the "difference quotient,"

$$p(x) = f_0 + \frac{f_1 - f_0}{x_1 - x_0}(x - x_0).$$

This indeed is a prototype of Newton's form of the interpolation polynomial (cf. §2.6).

Interpolating to function values is referred to as *Lagrange interpolation.* More generally, we may wish to interpolate to function and consecutive derivative values of some function. This is called *Hermite interpolation.* It turns out that the latter can be solved as a limit case of the former (cf. §2.7).

§2.1. **Lagrange interpolation formula; interpolation operator.** We prove the existence of the interpolation polynomial by simply writing it down! It is clear, indeed, that

$$\ell_i(x) = \prod_{\substack{j=0 \\ j \neq i}}^{n} \frac{x - x_j}{x_i - x_j}, \quad i = 0, 1, \ldots, n, \tag{2.1}$$

is a polynomial of degree n that interpolates to 1 at $x = x_i$ and to 0 at all the other points. Multiplying it by f_i produces the correct value at x_i, and then adding up the resulting polynomials,

$$p(x) = \sum_{i=0}^{n} f_i \ell_i(x),$$

produces a polynomial, still of degree $\leq n$, that has the desired interpolation properties. To prove this formally, note that

$$\ell_i(x_k) = \delta_{ik} = \begin{cases} 1 & \text{if} \quad i = k, \\ 0 & \text{if} \quad i \neq k, \end{cases} \quad i, k = 0, 1, \ldots, n. \tag{2.2}$$

Therefore,

$$p(x_k) = \sum_{i=0}^{n} f_i \ell_i(x_k) = \sum_{i=0}^{n} f_i \delta_{ik} = f_k, \quad k = 0, 1, \ldots, n.$$

This establishes the *existence* of the interpolation polynomial. To prove *uniqueness*, assume that there are two polynomials of degree $\leq n$, say, p and p^*, both interpolating to f at x_i, $i = 0, 1, \ldots, n$. Then

$$d(x) = p(x) - p^*(x)$$

is a polynomial of degree $\leq n$ that satisfies

$$d(x_i) = f_i - f_i = 0, \quad i = 0, 1, \ldots, n.$$

In other words, d has $n+1$ distinct zeros x_i. There is only *one* polynomial in $\mathbb{P}_n$ with that many zeros, namely, $d(x) \equiv 0$. Therefore, $p^*(x) \equiv p(x)$.

We denote the unique polynomial $p \in \mathbb{P}_n$ interpolating f at the (distinct) points $x_0, x_1, \ldots, x_n$ by

$$p_n(f; x_0, x_1, \ldots, x_n; x) = p_n(f; x), \tag{2.3}$$

where we use the long form on the left if we want to place in evidence the points at which interpolation takes place, and the short form on the right if the choice of these points is clear from the context. We thus have what is called the *Lagrange*[1] *interpolation formula*

$$p_n(f; x) = \sum_{i=0}^{n} f(x_i)\ell_i(x), \tag{2.4}$$

with the $\ell_i(x)$ — the *elementary Lagrange interpolation polynomials* — defined in (2.1).

It is useful to look at Lagrange interpolation in terms of a (linear) operator P_n from (say) the space of continuous functions to the space of polynomials $\mathbb{P}_n$,

$$P_n : \quad C[a,b] \to \mathbb{P}_n, \quad p(\,\cdot\,) = p_n(f;\,\cdot\,). \tag{2.5}$$

The interval $[a, b]$ here is any interval containing all points x_i, $i = 0, 1, \ldots, n$. The operator P_n has the following properties.

(i) $P_n(\alpha f) = \alpha P_n f, \quad \alpha \in \mathbb{R}$ (homogeneity);

(ii) $P_n(f + g) = P_n f + P_n g$ (additivity).

[1] Joseph Louis Lagrange (1736–1813), born in Turin, became, through correspondence with Euler, his protégé. In 1766 he indeed succeeded Euler in Berlin. He returned to Paris in 1787. Clairaut wrote of the young Lagrange: " ... a young man, no less remarkable for his talents than for his modesty; his temperament is mild and melancholic; he knows no other pleasure than study." Lagrange made fundamental contributions to the calculus of variations and to number theory, but worked also on many problems in analysis. He is widely known for his representation of the remainder term in Taylor's formula. The interpolation formula appeared in 1794. His *Mécanique Analytique*, published in 1788, made him one of the founders of analytic mechanics.

Combining (i) and (ii) shows that P_n is a linear operator,

$$P_n(\alpha f + \beta g) = \alpha P_n f + \beta P_n g, \quad \alpha, \beta \in \mathbb{R}.$$

(iii) $P_n f = f$ for all $f \in \mathbb{P}_n$.

The last property — an immediate consequence of uniqueness of the interpolation polynomial — says that P_n leaves polynomials of degree $\le n$ unchanged, and hence is a *projection* operator.

A norm of the linear operator P_n can be defined (similarly as for matrices, cf. Ch. 1, (3.11)) by

$$\|P_n\| = \max_{f \in C[a,b]} \frac{\|P_n f\|}{\|f\|}, \tag{2.6}$$

where on the right one takes any convenient norm for functions. Taking the L_∞ norm (cf. Table 2.0.1), one obtains from Lagrange's formula (2.4)

$$\begin{aligned} \|p_n(f;\,\cdot\,)\|_\infty &= \max_{a \le x \le b} \left| \sum_{i=0}^{n} f(x_i)\ell_i(x) \right| \\ &\le \|f\|_\infty \max_{a \le x \le b} \sum_{i=0}^{n} |\ell_i(x)|. \end{aligned} \tag{2.7}$$

Indeed, equality holds for some continuous function f; cf. Ex. 27. Therefore,

$$\|P_n\|_\infty = \Lambda_n, \tag{2.8}$$

where

$$\Lambda_n = \|\lambda_n\|_\infty, \quad \lambda_n(x) = \sum_{i=0}^{n} |\ell_i(x)|. \tag{2.9}$$

The function $\lambda_n(x)$ and its maximum Λ_n are called, respectively, the *Lebesgue*[2] *function* and *Lebesgue constant* for Lagrange interpolation. They provide a first estimate for the interpolation error: let $\mathcal{E}_n(f)$ be the *best* (uniform) *approximation* of f by polynomials of degree $\le n$,

$$\mathcal{E}_n(f) = \min_{p \in \mathbb{P}_n} \|f - p\|_\infty = \|f - \hat{p}_n\|_\infty, \tag{2.10}$$

[2]Henri Lebesgue (1875–1941) was a French mathematician known for his fundamental work on the theory of real functions, notably the concepts of measure and integral that now bear his name.

where $\hat{p}_n$ is the nth-degree polynomial of best uniform approximation to f. Then, using the basic properties (i) through (iii) of P_n, in particular, the projection property (iii), and (2.7) and (2.9), one finds

$$\begin{aligned}\|f - p_n(f;\,\cdot\,)\|_\infty &= \|f - \hat{p}_n - p_n(f - \hat{p}_n;\,\cdot\,)\|_\infty \\ &\le \|f - \hat{p}_n\|_\infty + \Lambda_n\|f - \hat{p}_n\|_\infty;\end{aligned}$$

that is,

$$\|f - p_n(f;\,\cdot\,)\|_\infty \le (1 + \Lambda_n)\mathcal{E}_n(f). \tag{2.11}$$

Thus, the better f can be approximated by polynomials of degree $\le n$, the smaller the interpolation error. Unfortunately, Λ_n is not uniformly bounded: no matter how one chooses the nodes $x_i = x_i^{(n)}$, $i = 0, 1, \ldots, n$, one can show that always $\Lambda_n > O(\log n)$ as $n \to \infty$. It is not possible, therefore, to conclude from Weierstrass's approximation theorem (i.e., from $\mathcal{E}_n(f) \to 0$, $n \to \infty$) that Lagrange interpolation converges uniformly on $[a, b]$ for any continuous function, not even for judiciously selected nodes; indeed, one knows that it does not.

§2.2. **Interpolation error**. As noted earlier, we need to make some assumptions about the function f in order to be able to estimate the error of interpolation, $f(x) - p_n(f; x)$, for any $x \ne x_i$ in $[a, b]$. In (2.11) we made an assumption in terms of how well f can be approximated on $[a, b]$ by polynomials of degree $\le n$. Now we make an assumption on the magnitude of some appropriate derivative of f.

It is not difficult to guess how the formula for the error should look: since the error is zero at each x_i, $i = 0, 1, \ldots, n$, we ought to see a factor of the form $(x - x_0)\,(x - x_1) \cdots (x - x_n)$. On the other hand, by the projection property (iii) in §2.1, the error is also zero (even identically so) if $f \in \mathbb{P}_n$, which suggests another factor — the $(n+1)$st derivative of f. But evaluated where? Certainly not at x, since f would then have to satisfy a differential equation. So let us say that $f^{(n+1)}$ is evaluated at some point $\xi = \xi(x)$, which is unknown but must be expected to depend on x. Now if we test the formula so far conjectured on the simplest nontrivial polynomial, $f(x) = x^{n+1}$, we discover that a factor $1/(n + 1)!$ is missing. So, our final (educated) guess is the formula

$$f(x) - p_n(f; x) = \frac{f^{(n+1)}(\xi(x))}{(n + 1)!} \prod_{i=0}^{n} (x - x_i), \quad x \in [a, b]. \tag{2.12}$$

Here $\xi(x)$ is some number in the open interval (a, b), but otherwise unspecified,

$$a < \xi(x) < b. \tag{2.13}$$

The statement (2.12) and (2.13) is, in fact, correct if we assume that $f \in C^{n+1}[a, b]$. An elegant proof of it, due to Cauchy,[3] goes as follows. We can assume $x \neq x_i$ for $i = 0, 1, \ldots, n$, since otherwise (2.12) would be trivially true for any $\xi(x)$. So, fix $x \in [a, b]$ in this manner, and define a function F of the new variable t as follows,

$$F(t) = f(t) - p_n(f; t) - \frac{f(x) - p_n(f; x)}{\prod_{i=0}^{n}(x - x_i)} \prod_{i=0}^{n}(t - x_i). \tag{2.14}$$

Clearly, $F \in C^{n+1}[a, b]$. Furthermore,

$$F(x_i) = 0, \ i = 0, 1, \ldots, n; \quad F(x) = 0.$$

Thus, F has $n + 2$ distinct zeros in $[a, b]$. Applying repeatedly Rolle's Theorem, we conclude that

F'	has	at	least	$n+1$	distinct	zeros	in	(a, b)
F''	″	″	″	n	″	″	″	″
F'''	″	″	″	$n-1$	″	″	″	″
...	...	...	...	...	...	...	...	...
$F^{(n+1)}$	″	″	″	1		zero	″	″

since $F^{(n+1)}$ is still continuous on $[a, b]$. Denote by $\xi(x)$ a zero of $F^{(n+1)}$ whose existence we just established. It certainly satisfies (2.13) and, of

[3] Augustin Cauchy (1789–1857), active in Paris, is truly the father of modern analysis. He provided a firm foundation for analysis by basing it on a rigorous concept of limit. He is also the creator of complex analysis, of which "Cauchy's formula" (cf. (2.22)) is a centerpiece. In addition, Cauchy's name is attached to pioneering contributions to the theory of ordinary and partial differential equations, in particular, regarding questions of existence and uniqueness. As with many great mathematicians of the 18th and 19th centuries, his work also encompasses geometry, algebra, number theory, and mechanics, as well as theoretical physics.

course, will depend on x. Now differentiating F in (2.14) $n+1$ times with respect to t, and then setting $t = \xi(x)$, we get

$$0 = f^{(n+1)}(\xi(x)) - \frac{f(x) - p_n(f;x)}{\prod_{i=0}^{n}(x - x_i)} \cdot (n+1)!,$$

which, when solved for $f(x) - p_n(f;x)$, gives precisely (2.12). Actually, what we have shown is that $\xi(x)$ is contained in the span of $x_0, x_1, \dots, x_n, x$, that is, in the interior of the smallest closed interval containing $x_0, x_1, \dots, x_n$ and x.

Examples.

(1) Linear interpolation ($n = 1$). Assume that $x_0 \le x \le x_1$; that is, $[a,b] = [x_0, x_1]$, and let $h = x_1 - x_0$. Then by (2.12) and (2.13),

$$f(x) - p_1(f;x) = (x - x_0)(x - x_1)\frac{f''(\xi)}{2}, \quad x_0 < \xi < x_1,$$

and an easy computation gives

$$\|f - p_1(f;\cdot)\|_\infty \le \frac{M_2}{8} h^2, \quad M_2 = \|f''\|_\infty. \tag{2.15}$$

Here the ∞-norm refers to the interval $[x_0, x_1]$. Thus, on small intervals of length h, the error for linear interpolation is $O(h^2)$.

(2) Quadratic interpolation ($n = 2$) on *equally spaced* points x_0, $x_1 = x_0 + h$, $x_2 = x_0 + 2h$. We now have, for $x \in [x_0, x_2]$,

$$f(x) - p_2(f;x) = (x - x_0)(x - x_1)(x - x_2)\frac{f'''(\xi)}{6}, \quad x_0 < \xi < x_2,$$

and (cf. Ex. 40(a))

$$\|f - p_2(f;\cdot)\|_\infty \le \frac{M_3}{9\sqrt{3}} h^3, \quad M_3 = \|f'''\|_\infty,$$

giving an error of $O(h^3)$.

(3) nth-degree interpolation on *equally spaced* points $x_i = x_0 + ih$, $i = 0, 1, \dots, n$. When h is small, and $x_0 \le x \le x_n$, then $\xi(x)$ in (2.12) is constrained to a relatively small interval and $f^{(n+1)}(\xi(x))$ cannot vary a great deal. The behavior of the error, therefore, is mainly determined by

the product $\prod_{i=0}^{n}(x - x_i)$, the graph of which, for $n = 7$, is shown in Figure 2.2.2. We clearly have symmetry with respect to the midpoint $(x_0 + x_n)/2$. It can also be shown that the relative extrema decrease monotonically in modulus as one moves from the endpoints to the center (cf. Ex. 26).

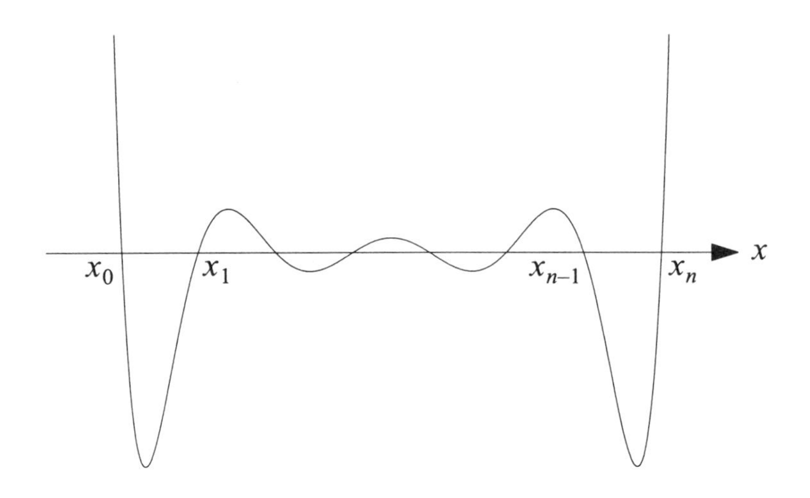

FIGURE 2.2.2. Interpolation error for eight equally spaced points

It is evident that the oscillations become more violent as n increases. In particular, the curve is extremely steep at the endpoints, and takes off to ∞ rapidly as x moves away from the interval $[x_0, x_n]$. Although it is true that the curve representing the interpolation error is scaled by a factor of $O(h^{n+1})$, it is also clear that one ought to interpolate near the center zone of the interval $[x_0, x_n]$, if at all possible, and should avoid interpolation near the end zones, or even *extra*polation outside the interval. The highly oscillatory nature of the error curve, when n is large, also casts some legitimate doubts about convergence of the interpolation process as $n \to \infty$. This is studied in the next section.

§2.3. **Convergence.** We first must define what we mean by "convergence." We assume that we are given a triangular array of interpolation nodes $x_i = x_i^{(n)}$, exactly $n+1$ distinct nodes for each $n = 0, 1, 2, \ldots$:

$$\begin{array}{lllll} x_0^{(0)} & & & & \\ x_0^{(1)} & x_1^{(1)} & & & \\ x_0^{(2)} & x_1^{(2)} & x_2^{(2)} & & \\ \cdots & \cdots & \cdots & & \\ x_0^{(n)} & x_1^{(n)} & x_2^{(n)} & \cdots & x_n^{(n)} \\ \cdots & \cdots & \cdots & \cdots & \cdots \end{array} \tag{2.16}$$

We further assume that all nodes $x_i^{(n)}$ are contained in some finite interval $[a, b]$. Then, for each n, we define

$$p_n(x) = p_n(f; x_0^{(n)}, x_1^{(n)}, \ldots, x_n^{(n)}; x), \quad x \in [a, b]. \tag{2.17}$$

We say that Lagrange interpolation based on the triangular array of nodes (2.16) *converges* if

$$p_n(x) \to f(x) \quad \text{as} \quad n \to \infty, \tag{2.18}$$

uniformly for $x \in [a, b]$.

Convergence clearly depends on the behavior of the kth derivative $f^{(k)}$ of f as $k \to \infty$. We assume that $f \in C^\infty[a, b]$, and that

$$|f^{(k)}(x)| \leq M_k \quad \text{for} \quad a \leq x \leq b, \quad k = 0, 1, 2, \ldots . \tag{2.19}$$

Since $|x - x_i^{(n)}| \leq b - a$ whenever $x \in [a, b]$ and $x_i^{(n)} \in [a, b]$, we have

$$|(x - x_0^{(n)})(x - x_1^{(n)}) \cdots (x - x_n^{(n)})| \leq (b - a)^{n+1}, \tag{2.20}$$

so that by (2.12)

$$|f(x) - p_n(x)| \leq (b - a)^{n+1} \frac{M_{n+1}}{(n+1)!}, \quad x \in [a, b].$$

We therefore have convergence if

$$\lim_{k \to \infty} \frac{(b - a)^k}{k!} M_k = 0. \tag{2.21}$$

We now show that (2.21) is true if f is analytic in a sufficiently large region in the complex plane containing the interval $[a, b]$. Specifically, let C_r be the circular (closed) disk with center at the midpoint of $[a, b]$ and radius r, and assume, for the time being, that $r > \frac{1}{2}(b - a)$, so that $[a, b] \subset C_r$. Assume f analytic in C_r. Then we can estimate the derivative in (2.19) by Cauchy's Formula,

$$f^{(k)}(x) = \frac{k!}{2\pi i} \oint_{\partial C_r} \frac{f(z)}{(z-x)^{k+1}}\, dz, \quad x \in [a, b]. \tag{2.22}$$

Noting that $|z - x| \geq r - \frac{1}{2}(b - a)$ (cf. Figure 2.2.3),

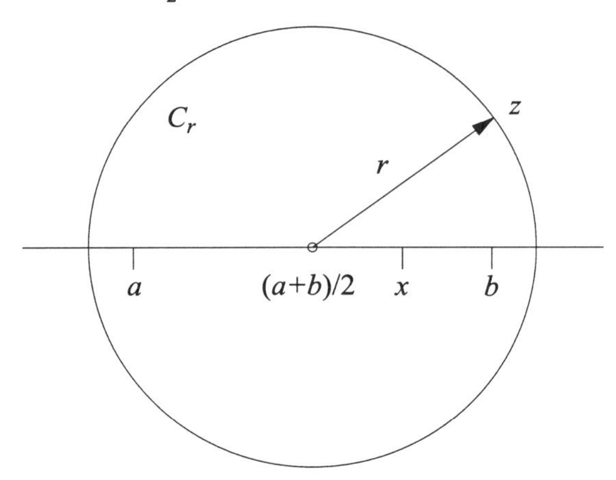

FIGURE 2.2.3. The circular disk C_r

we obtain

$$|f^{(k)}(x)| \leq \frac{k!}{2\pi} \frac{\max\limits_{z \in \partial C_r} |f(z)|}{[r - \frac{1}{2}(b-a)]^{k+1}} \cdot 2\pi r.$$

Therefore, we can take for M_k in (2.19)

$$M_k = \frac{r}{r - \frac{1}{2}(b-a)} \max_{z \in \partial C_r} |f(z)| \cdot \frac{k!}{[r - \frac{1}{2}(b-a)]^k}, \tag{2.23}$$

and (2.21) holds if

$$\left(\frac{b-a}{r - \frac{1}{2}(b-a)} \right)^k \to 0 \quad \text{as} \quad k \to \infty,$$

that is, if $b - a < r - \frac{1}{2}(b - a)$, or, equivalently,

$$r > \tfrac{3}{2}(b - a). \tag{2.24}$$

We have shown that *Lagrange interpolation converges* (uniformly on $[a, b]$) *for an arbitrary triangular set of nodes* (2.16) (all contained in $[a, b]$) *if f is analytic in the circular disk C_r centered at $(a + b)/2$ and having radius r sufficiently large so that* (2.24) *holds.*

Since our derivation of this result used rather crude estimates (see, in particular, (2.20)), the required domain of analyticity for f that we found is certainly not sharp. Using more refined methods, one can prove the following. Let $d\mu(t)$ be the "limit distribution" of the interpolation nodes, that is,

$$\int_a^x d\mu(t), \qquad a < x \le b,$$

the ratio of the number of nodes $x_i^{(n)}$ in $[a, x]$ to the total number, $n + 1$, of nodes, asymptotically as $n \to \infty$. (When the nodes are uniformly distributed over the interval $[a, b]$, then $d\mu(t) = dt/(b - a)$.) A curve of *constant logarithmic potential* is the locus of all complex $z \in \mathbb{C}$ such that

$$u(z) = \gamma, \quad u(z) = \int_a^b \ln \frac{1}{|z - t|}\, d\mu(t),$$

where γ is a constant. For large negative γ, these curves look like circles with large radii and center at $(a + b)/2$. As γ increases, the curves "shrink" towards the interval $[a, b]$. Let

$$\Gamma = \sup \gamma,$$

where the supremum is taken over all curves $u(z) = \gamma$ containing $[a, b]$ in their interior. The important domain (replacing C_r) is then the domain

$$C_\Gamma = \{z \in \mathbb{C} : \ u(z) \ge \Gamma\}, \tag{2.25}$$

in the sense that *if f is analytic in any domain C containing C_Γ in its interior* (no matter how closely C covers C_Γ), *then*

$$|f(z) - p_n(f; z)| \to 0 \ \ as \ \ n \to \infty \tag{2.26}$$

uniformly for $z \in C_\Gamma$.

Examples.

(1) Equally distributed nodes: $d\mu(t) = dt/(b-a)$, $a \le t \le b$. In this case, C_Γ is a lens-shaped domain as shown in Figure 2.2.4.

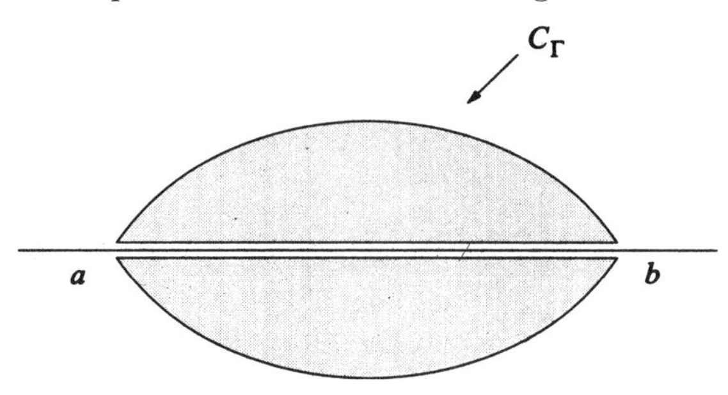

FIGURE 2.2.4. The domain C_Γ for uniformly distributed nodes

Thus, we have uniform convergence in C_Γ (not just on $[a, b]$, as before) provided f is analytic in a region slightly larger than C_Γ.

(2) Arc sine distribution on [–1,1]: $d\mu(t) = \dfrac{1}{\pi}\dfrac{dt}{\sqrt{1-t^2}}$. Here the nodes are more densely distributed near the endpoints of the interval [–1,1]. It turns out that in this case C_Γ = [–1,1], so that Lagrange interpolation converges uniformly on [–1,1] if f is "analytic on [1,1]," that is, analytic in any region, no matter how thin, that contains the interval [–1,1] in its interior.

(3) Runge's[4] example:

$$\begin{aligned} f(x) &= \frac{1}{1+x^2}\,, \quad -5 \le x \le 5, \\ x_k^{(n)} &= -5 + k\,\frac{10}{n}\,, \quad k = 0, 1, 2, \ldots, n. \end{aligned} \tag{2.27}$$

Here the nodes are equally spaced, hence asymptotically equally distributed. Note that $f(z)$ has poles at $z = \pm\, i$. These poles lie definitely inside the region C_Γ in Figure 2.2.4 for the interval [–5,5], so that f is *not* analytic

[4]Carl David Tolmé Runge (1856–1927) was active in the famous Göttingen school of mathematics and is one of the early pioneers of numerical mathematics. He is best known for the Runge-Kutta formula in ordinary differential equations, for which he provided the basic idea. He made also notable contributions to approximation theory in the complex plane.

in C_Γ. For this reason, we can no longer expect convergence on the whole interval [–5,5]. It has been shown, indeed, that

$$\lim_{n\to\infty} |f(x) - p_n(f;x)| = \begin{cases} 0 & \text{if} \quad |x| < 3.633\ldots\,, \\ \infty & \text{if} \quad |x| > 3.633\ldots\,. \end{cases} \tag{2.28}$$

We have convergence in the central zone of the interval [–5,5], but divergence in the lateral zones. With Figure 2.2.2 kept in mind, this is perhaps not all that surprising (cf. Mach. Ass. 7(b)).

(4) Bernstein's[5] example:

$$\begin{gathered} f(x) = |x|, \quad -1 \le x \le 1, \\ x_k^{(n)} = -1 + \frac{2k}{n}\,, \quad k = 0, 1, 2, \ldots, n. \end{gathered} \tag{2.29}$$

Here analyticity of f is completely gone, f being not even differentiable at $x = 0$. Accordingly one finds that

$$\begin{gathered} \lim_{n\to\infty} |f(x) - p_n(f;x)| = \infty \quad \text{for} \quad \textit{every} \quad x \in [-1,1], \\ \text{except} \quad x = -1, \; x = 0, \quad \text{and} \quad x = 1. \end{gathered} \tag{2.30}$$

The fact that $x = \pm\, 1$ are exceptional points is trivial, since they are interpolation nodes, where the error is zero. The same is true for $x = 0$ when n is even, but not if n is odd.

The failure of convergence in the last two examples can only in part be blamed on insufficient regularity of f. Another culprit is the equidistribution of the nodes. There are indeed better distributions, for example, the arc sine distribution of Example (2). An instance of the latter is discussed in the next section.

We add one more example, which involves *complex* nodes, and for which the preceding theory, therefore, no longer applies. We prove convergence directly.

---

[5]Sergei Natanovič Bernštein (1880–1968) made major contributions to polynomial approximation, continuing in the tradition of his countryman Chebyshev. He is also known for his work on partial differential equations and probability theory.

(5) Interpolation at the roots of unity (Fejér[6]): $z_k = \exp(k2\pi i/n)$, $k = 1, 2, \ldots, n$. We show that

$$p_{n-1}(f; z) \to f(z), \quad n \to \infty, \quad \text{for any } |z| < 1, \tag{2.31}$$

uniformly in any disk $|z| \le \rho < 1$, provided f is analytic in $|z| < 1$ and continuous on $|z| \le 1$.

We have

$$\omega_n(z) := \prod_{k=1}^{n}(z - z_k) = z^n - 1, \quad \omega_n'(z_k) = n z_k^{n-1} = \frac{n}{z_k},$$

so that the elementary Lagrange polynomials are

$$\begin{aligned}\ell_k(z) &= \frac{\omega_n(z)}{\omega_n'(z_k)(z - z_k)} = \frac{z^n - 1}{\frac{n}{z_k}(z - z_k)} \\ &= \frac{z_k}{n}\frac{1}{z_k - z} + z^n \frac{z_k}{(z - z_k)n}.\end{aligned}$$

Therefore,

$$p_{n-1}(f; z) = \sum_{k=1}^{n} \frac{f(z_k)}{z_k - z}\frac{z_k}{n} + z^n \sum_{k=1}^{n} \frac{f(z_k)}{z - z_k}\frac{z_k}{n}. \tag{2.32}$$

We interpret the first sum as a Riemann sum of an integral extended over the unit circle:

$$\begin{aligned}\sum_{k=1}^{n} \frac{f(z_k)}{z_k - z}\frac{z_k}{n} &= \frac{1}{2\pi i}\sum_{k=1}^{n} \frac{f(e^{ik2\pi/n})}{e^{ik2\pi/n} - z}\, i e^{ik2\pi/n} \cdot \frac{2\pi}{n} \\ \to \frac{1}{2\pi i}\int_0^{2\pi} \frac{f(e^{i\theta})}{e^{i\theta} - z}\, i e^{i\theta} d\theta &= \frac{1}{2\pi i}\oint_{|\zeta|=1} \frac{f(\zeta) d\zeta}{\zeta - z} \quad \text{as } n \to \infty.\end{aligned}$$

The last expression, by Cauchy's Formula, however, is precisely $f(z)$. The second term in (2.32), being just $-z^n$ times the first, converges to zero, uniformly in $|z| \le \rho < 1$.

---

[6]Leopold Fejér (1880–1959) was a leading Hungarian mathematician of the 20th century. Interestingly, Fejér had great difficulties in mathematics at the elementary and lower secondary school level, and even required private tutoring. It was an inspiring teacher in the upper-level secondary school who awoke Fejér's interest and passion for mathematics. He went on to discover — still a university student — an important result on the summability of Fourier series, which made him famous overnight. He continued to make further contributions to the theory of Fourier series, but also occupied himself with problems of approximation and interpolation in the real as well as complex domain. He in turn was an inspiring teacher to the next generation of Hungarian mathematicians.

§2.4. **Chebyshev polynomials and nodes.** The choice of nodes, as we saw in the previous section, distinctly influences the convergence character of the interpolation process. We now discuss a choice of points — the *Chebyshev points* — which leads to very favorable convergence properties. These points are useful, not only for interpolation, but also for other purposes (integration, collocation, etc.). We consider them on the canonical interval [–1,1], but they can be defined on any finite interval $[a, b]$ by means of a linear transformation of variables that maps [–1,1] onto $[a, b]$.

We begin with developing the Chebyshev polynomials. They arise from the fact that the cosine of a multiple argument is a polynomial in the cosine of the simple argument; more precisely,

$$\cos n\theta = T_n(\cos\theta), \quad T_n \in \mathbb{P}_n. \tag{2.33}$$

This is a consequence of the well-known trigonometric identity

$$\cos(k+1)\theta + \cos(k-1)\theta = 2\cos\theta\cos k\theta,$$

which, when solved for the first term, gives

$$\cos(k+1)\theta = 2\cos\theta\cos k\theta - \cos(k-1)\theta. \tag{2.34}$$

Therefore, if $\cos m\theta$ is a polynomial of degree m in $\cos\theta$ for all $m \le k$, then the same is true for $m = k+1$. Mathematical induction then proves (2.33). At the same time, it follows from (2.34) that

$$\begin{gathered} T_{k+1}(x) = 2xT_k(x) - T_{k-1}(x), \quad k = 1, 2, 3, \ldots\,, \\ T_0(x) = 1, \quad T_1(x) = x. \end{gathered} \tag{2.35}$$

The polynomials T_m so defined are called the *Chebyshev polynomials* (of the first kind). Thus, for example,

$$T_2(x) = 2x^2 - 1,$$

$$T_3(x) = 4x^3 - 3x,$$

$$T_4(x) = 8x^4 - 8x^2 + 1,$$

and so on.

Clearly, these polynomials are defined not only for x in [-1,1], but for arbitrary real or complex x. It is only that on the interval [-1,1] they satisfy the identity (2.33) (where θ is real).

It is evident from (2.35) that the leading coefficient of T_n is 2^{n-1} (if $n \geq 1$); the *monic Chebyshev polynomial* of degree n, therefore, is

$$\mathring{T}_n(x) = \frac{1}{2^{n-1}} T_n(x), \ n \geq 1; \quad \mathring{T}_0 = T_0. \tag{2.36}$$

The basic identity (2.33) allows us to immediately obtain the zeros $x_k = x_k^{(n)}$ of T_n: indeed, $\cos n\theta = 0$ if $n\theta = (2k-1)\pi/2$, so that

$$x_k^{(n)} = \cos \theta_k^{(n)}, \quad \theta_k^{(n)} = \frac{2k-1}{2n}\pi, \quad k = 1, 2, \ldots, n. \tag{2.37}$$

All zeros of T_n are thus real, distinct, and contained in the open interval (-1,1). They are the projections onto the real line of equally spaced points on the unit circle; cf. Figure 2.2.5 for the case $n = 4$.

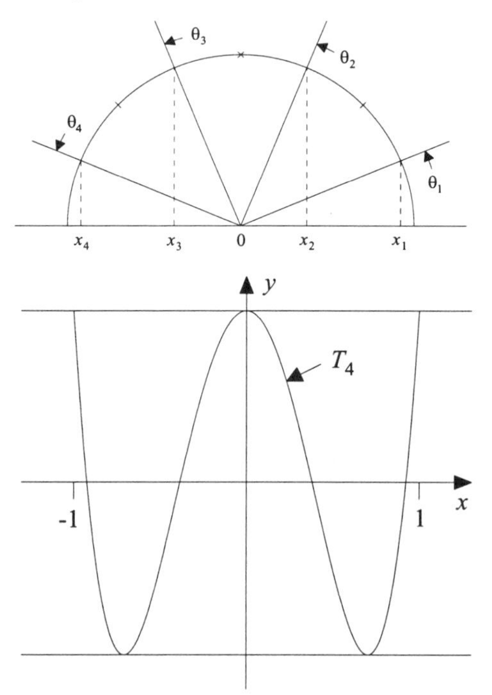

FIGURE 2.2.5. The Chebyshev polynomial $y = T_4(x)$

In terms of the zeros $x_k^{(n)}$ of T_n, we can write the monic polynomial in factored form as

$$\overset{\circ}{T}_n(x) = \prod_{k=1}^{n} (x - x_k^{(n)}). \qquad (2.38)$$

As we let θ increase from 0 to π, hence $x = \cos\theta$ decrease from +1 to −1, Eq. (2.33) shows that $T_n(x)$ oscillates between +1 and −1, attaining these extreme values at

$$y_k^{(n)} = \cos\eta_k^{(n)}, \quad \eta_k^{(n)} = k\frac{\pi}{n}, \quad k = 0, 1, 2, \ldots, n. \qquad (2.39)$$

In summary, then,

$$T_n(x_k^{(n)}) = 0 \;\text{ for }\; x_k^{(n)} = \cos\frac{2k-1}{2n}\pi, \quad k = 1, 2, \ldots, n; \qquad (2.40)$$

$$T_n(y_k^{(n)}) = (-1)^k \;\text{ for }\; y_k^{(n)} = \cos\frac{k}{n}\pi, \quad k = 0, 1, 2, \ldots, n. \qquad (2.41)$$

Chebyshev polynomials owe their importance and usefulness to the following theorem, due to Chebyshev.[7]

Theorem 2.2.1. *For an arbitrary monic polynomial $\overset{\circ}{p}_n$ of degree n, there holds*

$$\max_{-1\le x\le 1} |\overset{\circ}{p}_n(x)| \ge \max_{-1\le x\le 1} |\overset{\circ}{T}_n(x)| = \frac{1}{2^{n-1}}, \quad n \ge 1, \qquad (2.42)$$

where $\overset{\circ}{T}_n$ is the monic Chebyshev polynomial (2.36) *of degree n.*

Proof (by contradiction). Assume, contrary to (2.42), that

$$\max_{-1\le x\le 1} |\overset{\circ}{p}_n(x)| < \frac{1}{2^{n-1}}. \qquad (2.43)$$

Then the polynomial $d_n(x) = \overset{\circ}{T}_n(x) - \overset{\circ}{p}_n(x)$ (a polynomial of degree $\le n-1$), satisfies

$$d_n(y_0^{(n)}) > 0,\; d_n(y_1^{(n)}) < 0,\; d_n(y_2^{(n)}) > 0, \ldots,\; (-1)^n d_n(y_n^{(n)}) > 0. \qquad (2.44)$$

[7] Pafnuti Levovich Chebyshev (1821–1894) was the most prominent member of the St. Petersburg school of mathematics. He made pioneering contributions to number theory, probability theory, and approximation theory. He is regarded as the founder of constructive function theory, but also worked in mechanics, notably the theory of mechanisms, and in ballistics.

Thus d_n changes sign at least n times, and hence has at least n distinct real zeros. But having degree $\leq n-1$, it must vanish identically, $d_n(x) \equiv 0$. This contradicts (2.44); thus (2.43) cannot be true. $\square$

The result (2.42) can be given the following interesting interpretation: *the best uniform approximation* (on the interval [−1,1]) *to* $f(x) = x^n$ *from polynomials in* $\mathbb{P}_{n-1}$ *is given by* $x^n - \mathring{T}_n(x)$, that is, by the aggregate of terms of degree $\leq n-1$ in $\mathring{T}_n$ taken with the minus sign. From the theory of uniform polynomial approximation it is known that the best approximant is unique. Therefore, equality in (2.42) can only hold if $\mathring{p}_n = \mathring{T}_n$.

What is the significance of Chebyshev polynomials for interpolation? Recall (cf. (2.12)) that the interpolation error (on [−1,1], for a function $f \in C^{n+1}[-1,1]$) is given by

$$f(x) - p_n(f;x) = \frac{f^{(n+1)}(\xi(x))}{(n+1)!} \cdot \prod_{i=0}^{n}(x - x_i), \quad x \in [-1,1]. \tag{2.45}$$

The first factor is essentially independent of the choice of the nodes x_i. It is true that $\xi(x)$ *does* depend on the x_i, but we usually estimate $f^{(n+1)}$ by $\|f^{(n+1)}\|_\infty$, which removes this dependence. On the other hand, the product in the second factor, including its norm

$$\|\prod_{i=0}^{n}(\cdot - x_i)\|_\infty, \tag{2.46}$$

depends strongly on the x_i. It makes sense, therefore, to try to minimize (2.46) over all $x_i \in [-1,1]$. Since the product in (2.46) is a monic polynomial of degree $n+1$, it follows from Theorem 2.2.1 that *the optimal nodes* $x_i = \hat{x}_i^{(n)}$ in (2.45) *are precisely the zeros of* T_{n+1}; that is,

$$\hat{x}_i^{(n)} = \cos\frac{2i+1}{2n+2}\pi, \quad i = 0,1,2,\ldots,n. \tag{2.47}$$

For these nodes, we then have (cf. (2.42))

$$\|f(\cdot) - p_n(f;\cdot)\|_\infty \leq \frac{\|f^{(n+1)}\|_\infty}{(n+1)!} \cdot \frac{1}{2^n}\,. \tag{2.48}$$

One ought to compare the last factor in (2.48) with the much cruder bound given in (2.20), which, in the case of the interval [−1,1], is 2^{n+1}.

Since by (2.45) the error curve $y = f - p_n(f;\cdot)$ for Chebyshev points (2.47) is essentially *equilibrated* (modulo the variation in the factor $f^{(n+1)}$), and thus free of the violent oscillations we saw for equally spaced points, we would expect more favorable convergence properties for the triangular array (2.16) consisting of Chebyshev nodes. Indeed, one can prove, for example, that

$$p_n(f; \hat{x}_0^{(n)}, \hat{x}_1^{(n)}, \ldots, \hat{x}_n^{(n)}; x) \to f(x) \quad \text{as} \quad n \to \infty, \tag{2.49}$$

uniformly on [-1,1], provided only that $f \in C^1[-1,1]$. Thus we do not need analyticity of f for (2.49) to hold.

We finally remark — as already suggested by the recurrence relation (2.35) — that Chebyshev polynomials are a special case of orthogonal polynomials. Indeed, the measure in question is precisely (up to an unimportant constant factor) the arc sine measure

$$d\lambda(x) = \frac{dx}{\sqrt{1-x^2}} \quad \text{on} \quad [-1,1] \tag{2.50}$$

already mentioned in Example (2) of §2.3. This is easily verified from (2.33) and the orthogonality of the cosines (cf. §1.4, Eq. (1.22cc)):

$$\int_{-1}^{1} T_k(x)T_\ell(x)\frac{dx}{\sqrt{1-x^2}} = \int_0^\pi T_k(\cos\theta)T_\ell(\cos\theta)d\theta$$

$$= \int_0^\pi \cos k\theta \cos \ell\theta d\theta = \begin{cases} 0 & \text{if} \quad k \neq \ell, \\ \pi & \text{if} \quad k = \ell = 0, \\ \frac{1}{2}\pi & \text{if} \quad k = \ell > 0. \end{cases} \tag{2.51}$$

The Fourier expansion in Chebyshev polynomials (essentially the Fourier-cosine expansion) is therefore given by

$$f(x) = {\sum_{j=1}^{\infty}}' c_j T_j(x) = \tfrac{1}{2}c_0 + \sum_{j=1}^{\infty} c_j T_j(x), \tag{2.52}$$

where

$$c_j = \frac{2}{\pi}\int_{-1}^{1} f(x)T_j(x)\frac{dx}{\sqrt{1-x^2}}, \quad j = 0, 1, 2, \ldots\ . \tag{2.53}$$

Truncating (2.52) with the term of degree n gives a useful polynomial approximation of degree n,

$$\tau_n(x) = {\sum_{j=0}^{n}}' c_j T_j(x),$$

having an error

$$f(x) - \tau_n(x) = \sum_{j=n+1}^{\infty} c_j T_j(x) \approx c_{n+1} T_{n+1}(x). \tag{2.54}$$

The approximation on the far right is better the faster the Fourier coefficients c_j tend to zero. The error (2.54), therefore, essentially oscillates between $+c_{n+1}$ and $-c_{n+1}$ as x varies on the interval [–1,1], and thus is of "uniform" size. This is in stark contrast to Taylor's expansion at $x = 0$, where the nth-degree partial sum has an error proportional to x^{n+1} on [–1,1].

§2.5. **Barycentric formula**. Lagrange's formula (2.4) is attractive more for theoretical purposes than for practical computational work. It can be rewritten, however, in a form that makes it efficient computationally, and that also allows additional interpolation nodes to be added with ease. Having the latter feature in mind, we now assume a sequential set x_0, x_1, $x_2, \dots$ of interpolation nodes and denote by $p_n(f;\,\cdot)$ the polynomial of degree $\leq n$ interpolating to f at the first $n+1$ of them. We do not assume that the x_i are in any particular order, as long as they are mutually distinct.

We introduce a triangular array of auxiliary quantities defined by

$$\lambda_0^{(0)} = 1, \;\; \lambda_i^{(n)} = \prod_{\substack{j=0 \\ j \neq i}}^{n} \frac{1}{x_i - x_j}, \quad i = 0, 1, \dots, n; \;\; n = 1, 2, 3, \dots\,. \tag{2.55}$$

The elementary Lagrange interpolation polynomials of degree n, (2.1), can then be written in the form

$$\ell_i(x) = \frac{\lambda_i^{(n)}}{x - x_i}\, \omega_n(x), \;\; i = 0, 1, \dots, n; \;\; \omega_n(x) = \prod_{j=0}^{n} (x - x_j). \tag{2.56}$$

Dividing Lagrange's formula through by $1 \equiv \sum_{i=0}^{n} \ell_i(x)$, one finds

$$p_n(f;x) = \sum_{i=0}^{n} f_i \ell_i(x) = \frac{\sum_{i=0}^{n} f_i \ell_i(x)}{\sum_{i=0}^{n} \ell_i(x)} = \frac{\sum_{i=0}^{n} f_i \frac{\lambda_i^{(n)}}{x - x_i} \omega_n(x)}{\sum_{i=0}^{n} \frac{\lambda_i^{(n)}}{x - x_i} \omega_n(x)};$$

that is,

$$p_n(f;x) = \frac{\displaystyle\sum_{i=0}^{n} \frac{\lambda_i^{(n)}}{x - x_i} f_i}{\displaystyle\sum_{i=0}^{n} \frac{\lambda_i^{(n)}}{x - x_i}}, \qquad x \neq x_i \text{ for } i = 0, 1, \ldots, n. \tag{2.57}$$

This expresses the interpolation polynomial as a weighted average of the function values $f_i = f(x_i)$ and is, therefore, called the *barycentric formula* — a slight misnomer, since the weights are not necessarily all positive. The auxiliary quantities $\lambda_i^{(n)}$ involved in (2.57) are those in the row numbered n of the triangular array (2.55). Once they have been calculated, the evaluation of $p_n(f;x)$ by (2.57), for any fixed x, is straightforward and cheap.

Comparison with (2.4) shows that

$$\ell_i(x) = \frac{\dfrac{\lambda_i^{(n)}}{x - x_i}}{\displaystyle\sum_{i=0}^{n} \frac{\lambda_i^{(n)}}{x - x_i}}, \qquad i = 0, 1, \ldots, n. \tag{2.1$'$}$$

In order to arrive at an efficient algorithm for computing the required quantities $\lambda_i^{(n)}$, we first note that, for $k \geq 1$,

$$\lambda_i^{(k)} = \frac{\lambda_i^{(k-1)}}{x_i - x_k}, \qquad i = 0, 1, \ldots, k-1. \tag{2.58}$$

Furthermore, Lagrange's formula for $p_k(f;\,\cdot\,)$, with $f(\,\cdot\,) \equiv 1$, gives

$$\sum_{i=0}^{k} \lambda_i^{(k)} \prod_{\substack{j=0 \\ j \neq i}}^{k} (x - x_j) \equiv 1.$$

Comparing the coefficients of x^k on either side (assuming $k \geq 1$) yields

$$\sum_{i=0}^{k} \lambda_i^{(k)} = 0.$$

This gives us the means to compute the last quantity $\lambda_k^{(k)}$ missing in (2.58). Altogether, we arrive at the following algorithm.

$$\lambda_0^{(0)} = 1,$$
$$\text{for } k = 1, 2, \ldots, n \text{ do}$$
$$\left[\begin{array}{ll} \lambda_i^{(k)} = \dfrac{\lambda_i^{(k-1)}}{x_i - x_k}, & i = 0, 1, \ldots, k-1, \\ \lambda_k^{(k)} = -\displaystyle\sum_{i=0}^{k-1} \lambda_i^{(k)}. & \end{array}\right. \tag{2.59}$$

This requires exactly n^2 additions (including the subtractions in $x_i - x_k$) and $\frac{1}{2}n(n+1)$ divisions for computing the $n+1$ quantities $\lambda_0^{(n)}, \lambda_1^{(n)}, \ldots, \lambda_n^{(n)}$ in (2.57). If we decide to incorporate the next data point (x_{n+1}, f_{n+1}), all we need to do is extend the k-loop in (2.59) through $n+1$, that is, generate the next row of auxiliary quantities $\lambda_0^{(n+1)}, \lambda_1^{(n+1)}, \ldots, \lambda_{n+1}^{(n+1)}$. We are then ready to compute $p_{n+1}(f; x)$ from (2.57) with n replaced by $n+1$.

Although (2.1′) in combination with (2.59) is more efficient than (2.1) (which requires $O(n^3)$ operations to evaluate), it is also more exposed to the detrimental effects of rounding errors. The weak spot, indeed, is the summation in the last step of (2.59), which is subject to significant cancellation errors whenever $\max\limits_{0 \le i \le k-1} |\lambda_i^{(k)}|$ is much larger than $|\lambda_k^{(k)}|$. Unfortunately, this is more often the case than not. The extent of cancellation, however, may depend on the order in which the nodes x_i are arranged. It is recommended, for given n, that they be arranged in the order of decreasing distance from their midpoint. In contrast, the formula (2.1) is devoid of numerical difficulties since only benign operations — multiplication and division — are involved (disregarding the formation of differences such as $x - x_i$, which occur in both formulae).

§2.6. **Newton's[8] formula**. This is another way of organizing the work in §2.5. Although the computational effort remains essentially the same, it becomes easier to treat "confluent" interpolation points, that is, multiple points in which not only the function values, but also consecutive derivative values, are given (cf. §2.7).

[8]Sir Isaac Newton (1643–1727) was an eminent figure of 17th century mathematics and physics. Not only did he lay the foundations of modern physics, but he was also one of the coinventors of differential calculus. Another was Leibniz, with whom he became entangled in a bitter and life-long priority dispute. His most influential work was the *Principia*, which not only contains his ideas on interpolation, but also his suggestion to use the interpolating polynomial for purposes of integration (cf. Ch. 3, §2.2).

Using the same setup as in §2.5, we denote

$$p_n(x) = p_n(f; x_0, x_1, \dots, x_n; x), \quad n = 0, 1, 2, \dots \ . \tag{2.60}$$

We clearly have

$$p_0(x) = a_0,$$

$$\begin{aligned} p_n(x) = p_{n-1}(x) + a_n(x - x_0)(x - x_1) \cdots (x - x_{n-1}), \\ n = 1, 2, 3, \dots \ , \end{aligned} \tag{2.61}$$

for some constants a_0, a_1, $a_2, \dots$. This gives rise to a new form of the interpolation polynomial,

$$\begin{aligned} p_n(f; x) = a_0 + a_1(x - x_0) + a_2(x - x_0)(x - x_1) \\ + \dots + a_n(x - x_0)(x - x_1) \cdots (x - x_{n-1}), \end{aligned} \tag{2.62}$$

which is called *Newton's form.* The constants involved can be determined, in principle, by the interpolation conditions

$$f_0 = a_0,$$

$$f_1 = a_0 + a_1(x_1 - x_0),$$

$$f_2 = a_0 + a_1(x_2 - x_0) + a_2(x_2 - x_0)(x_2 - x_1),$$

and so on, which represent a triangular, nonsingular (why?) system of linear algebraic equations. This uniquely determines the constants; for example,

$$a_0 = f_0,$$

$$a_1 = \frac{f_1 - f_0}{x_1 - x_0},$$

$$a_2 = \frac{f_2 - a_0 - a_1(x_2 - x_0)}{(x_2 - x_0)(x_2 - x_1)},$$

and so on. Evidently, a_n is a linear combination of f_0, $f_1, \dots, f_n$, with coefficients that depend on x_0, $x_1, \dots, x_n$. We use the notation

$$a_n = [x_0, x_1, \dots, x_n]f, \quad n = 0, 1, 2, \dots \ , \tag{2.63}$$

for this linear combination, and call the right-hand side the nth *divided difference* of f relative to the nodes $x_0, x_1, \ldots, x_n$. Considered as a function of these $n+1$ variables, the divided difference is a *symmetric function*; that is, permuting the variables in any way does not affect the value of the function. This is a direct consequence of the fact that a_n in (2.62) is the *leading coefficient of* $p_n(f;x)$: the interpolation polynomial $p_n(f;\cdot)$ surely does not depend on the order in which we write down the interpolation conditions.

The name "divided difference" comes from the useful property

$$[x_0, x_1, x_2, \ldots, x_k]f = \frac{[x_1, x_2, \ldots, x_k]f - [x_0, x_1, \ldots, x_{k-1}]f}{x_k - x_0} \tag{2.64}$$

expressing the kth divided difference as a difference of $(k-1)$st divided differences, divided by a difference of the xs. Since we have symmetry, the order in which the variables are written down is immaterial; what is important is that the two divided differences (of the same order $k-1$) in the numerator have $k-1$ of the xs in common. The "extra" one in the first term, and the "extra" one in the second, are precisely the xs that appear in the denominator, in the same order.

To prove (2.64), let

$$r(x) = p_{k-1}(f; x_1, x_2, \ldots, x_k; x)$$

and

$$s(x) = p_{k-1}(f; x_0, x_1, \ldots, x_{k-1}; x).$$

Then

$$p_k(f; x_0, x_1, \ldots, x_k; x) = r(x) + \frac{x - x_k}{x_k - x_0}[r(x) - s(x)]. \tag{2.65}$$

Indeed, the polynomial on the right has clearly degree $\leq k$ and takes on the correct value f_i at x_i, $i = 0, 1, \ldots, k$. For example, if $i \neq 0$ and $i \neq k$,

$$r(x_i) + \frac{x_i - x_k}{x_k - x_0}[r(x_i) - s(x_i)] = f_i + \frac{x_i - x_k}{x_k - x_0}[f_i - f_i] = f_i,$$

and similarly for $i = 0$ and for $i = k$. By uniqueness of the interpolation polynomial, this implies (2.65). Now equating the leading coefficients on both sides of (2.65) immediately gives (2.64).

The formula (2.64) can be used to generate the *table of divided differences*:

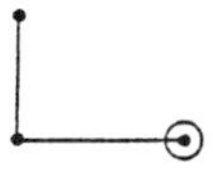

$$
\begin{array}{cc|llll}
x & f & & & & \\
\hline
x_0 & f_0 & & & & \\
x_1 & f_1 & [x_0, x_1]f & & & \\
x_2 & f_2 & [x_1, x_2]f & [x_0, x_1, x_2]f & & \\
x_3 & f_3 & [x_2, x_3]f & [x_1, x_2, x_3]f & [x_0, x_1, x_2, x_3]f & \\
\cdot & \cdot & \cdots & \cdots & \cdots &
\end{array}
\tag{2.66}
$$

The divided differences are here arranged in such a manner that their computation proceeds according to one single rule: *each entry is the difference of the entry immediately to the left and the one above it, divided by the difference of the x-value horizontally to the left and the one opposite the f-value found by going diagonally up.* Each entry, therefore, is calculated from its two neighbors immediately to the left, which is expressed by the computing stencil in (2.66).

The divided differences $a_0, a_1, \ldots, a_n$ (cf. (2.63)) that occur in Newton's formula (2.62) are precisely the first $n + 1$ *diagonal entries* in the table of divided differences. Their computation requires $n(n + 1)$ additions and $\frac{1}{2}n(n + 1)$ divisions, essentially the same effort that was required in computing the auxiliary quantities $\lambda_i^{(n)}$ in the barycentric formula. (Actually, the latter requires only n^2 additions, but then, when it comes to evaluating the formula itself, the barycentric formula is a little bit more expensive than Newton's formula, if the latter is evaluated efficiently; cf. Ex. 55). Adding another data point (x_{n+1}, f_{n+1}) requires the generation of the next line of divided differences. The last entry of this line is a_{n+1}, and we can update $p_n(f; x)$ by adding to it the term $a_{n+1}(x - x_0)(x - x_1) \cdots (x - x_n)$ to get p_{n+1} (cf. (2.61)).

Example.

x	f			
0	3			
1	4	$(4–3)/(1–0) = 1$		
2	7	$(7–4)/(2–1) = 3$	$(3–1)/(2–0) = 1$	
4	19	$(19–7)/(4–2) = 6$	$(6–3)/(4–1) = 1$	$(1–1)/(4–0) = 0$

The cubic interpolation polynomial is

$$\begin{aligned} p_3(f;x) &= 3 + 1\cdot(x-0) + 1\cdot(x-0)(x-1) + 0\cdot(x-0)(x-1)(x-2) \\ &= 3 + x + x(x-1) = 3 + x^2, \end{aligned}$$

which indeed is the function tabulated. Note that the leading coefficient of $p_3(f;\,\cdot\,)$ is zero, which is why the last divided difference turned out to be 0.

Newton's formula also yields a new representation for the error term in Lagrange interpolation. Let t temporarily denote an arbitrary "node" not equal to any of the $x_0, x_1, \ldots, x_n$. Then we have

$$\begin{aligned} &p_{n+1}(f;x_0,x_1,\ldots,x_n,t;x) \\ &\quad = p_n(f;x) + [x_0,x_1,\ldots,x_n,t]f\cdot\prod_{i=0}^{n}(x-x_i). \end{aligned}$$

Now put $x = t$; since the polynomial on the left interpolates to f at t, we get

$$f(t) = p_n(f;t) + [x_0,x_1,\ldots,x_n,t]f\cdot\prod_{i=0}^{n}(t-x_i).$$

Writing again x for t (which was arbitrary, after all), we find

$$f(x) - p_n(f;x) = [x_0,x_1,\ldots,x_n,x]f\cdot\prod_{i=0}^{n}(x-x_i). \tag{2.67}$$

This is the new formula for the interpolation error. Note that it involves no derivative of f, only function values. The trouble is, that $f(x)$ is one of them! Indeed, (2.67) is basically a tautology since, when everything is written out explicitly, the formula evaporates to $0 = 0$, which is correct, but not overly exciting.

In spite of this seeming emptiness of (2.67), we can draw from it an interesting and very useful conclusion. (For another application, see Ch. 3, Ex. 2.) Indeed, compare it with the earlier formula (2.12); one obtains

$$[x_0, x_1, \ldots, x_n, x]f = \frac{f^{(n+1)}(\xi(x))}{(n+1)!},$$

where $x_0, x_1, \ldots, x_n, x$ are arbitrary distinct points in $[a, b]$ and $f \in C^{n+1}[a, b]$. Moreover, $\xi(x)$ is strictly between the smallest and largest of these points (cf. the proof of (2.12)). We can now write $x = x_{n+1}$, and then replace $n+1$ by n to get

$$[x_0, x_1, \ldots, x_n]f = \frac{1}{n!} f^{(n)}(\xi). \tag{2.68}$$

Thus, for any $n+1$ distinct points in $[a, b]$ and any $f \in C^n[a, b]$, *the divided difference of f of order n is the* nth *scaled derivative of f at some* (unknown) *intermediate point.* If we now let all x_i, $i \geq 1$, tend to x_0, then ξ, being trapped between them, must also tend to x_0, and, since $f^{(n)}$ is continuous at x_0, we obtain

$$\underbrace{[x_0, x_0, \ldots, x_0]}_{n+1 \text{ times}} f = \frac{1}{n!} f^{(n)}(x_0). \tag{2.69}$$

This suggests that *the* nth *divided difference at $n+1$ "confluent"* (i.e., identical) *points be defined to be the* nth *derivative at this point divided by $n!$*. This allows us, in the next section, to solve the Hermite interpolation problem.

§2.7. **Hermite interpolation**. The general Hermite interpolation problem consists of the following: given $K + 1$ distinct points $x_0, x_1, \ldots, x_K$ in $[a, b]$ and corresponding integers $m_k \geq 1$, and given a function $f \in C^{M-1}[a, b]$, with $M = \max_k m_k$, find a polynomial p of lowest degree such that, for $k = 0, 1, \ldots, K$,

$$p^{(\mu)}(x_k) = f_k^{(\mu)}, \quad \mu = 0, 1, \ldots, m_k - 1, \tag{2.70}$$

where $f_k^{(\mu)} = f^{(\mu)}(x_k)$ is the μth derivative of f at x_k.

The problem can be thought of as a limiting case of Lagrange interpolation if we consider x_k to be a *point of multiplicity* m_k, that is, obtained by a confluence of m_k distinct points into a single point x_k. We can imagine setting up the table of divided differences, and Newton's interpolation formula, just before the confluence takes place, and then simply "go to the limit." To do this in practice requires that each point x_k be entered exactly m_k times in the first column of the table of divided differences. The formula (2.69) then allows us to *initialize* the divided differences for these points. For example, if $m_k = 4$, then

$$\begin{array}{ccccc} x & f & & & \\ \hline \cdot & \cdot & \cdot & \cdot & \cdot \\ x_k & f_k & & & \\ x_k & f_k & f'_k & & \\ x_k & f_k & f'_k & \frac{1}{2}f''_k & \\ x_k & f_k & f'_k & \frac{1}{2}f''_k & \frac{1}{6}f'''_k \\ \cdot & \cdot & \cdot & \cdot & \cdot \end{array} \tag{2.71}$$

Doing this initialization for each k, we are then ready to complete the table of divided differences in the usual way. (There will be no zero divisors; they have been taken care of during the initialization.) We obtain a table with $m_0 + m_1 + \cdots + m_K$ entries in the first column, and hence an interpolation polynomial of degree $\leq n = m_0 + m_1 + \cdots + m_K - 1$, which, as in the Lagrange case, is unique. The $n+1$ diagonal entries in the table give us the coefficients in Newton's formula, as before, except that in the product terms of the formula, some of the factors are repeated. Also the error term of interpolation remains in force, with the repetition of factors properly accounted for.

We illustrate the procedure with two simple examples.

(1) Find $p \in \mathbb{P}_3$ such that

$$p(x_0) = f_0,\ p'(x_0) = f'_0,\ p''(x_0) = f''_0,\ p'''(x_0) = f'''_0.$$

Here $K = 0$, $m_0 = 4$, that is, we have a single quadruple point. The table of divided differences is precisely the one in (2.71) (with $k = 0$); hence Newton's formula becomes

$$p(x) = f_0 + (x - x_0)f'_0 + \tfrac{1}{2}(x - x_0)^2 f''_0 + \tfrac{1}{6}(x - x_0)^3 f'''_0,$$

which is nothing but the Taylor polynomial of degree 3. Thus Taylor's polynomial is a special case of a Hermite interpolation polynomial. The error term of interpolation, furthermore, gives us

$$f(x) - p(x) = \tfrac{1}{24}(x - x_0)^4 f^{(4)}(\xi), \quad \xi \text{ between } x_0 \text{ and } x,$$

which is Lagrange's form of the remainder term in Taylor's formula.

(2) Find $p \in \mathbb{P}_3$ such that

$$p(x_0) = f_0, \quad p(x_1) = f_1, \quad p'(x_1) = f_1', \quad p(x_2) = f_2,$$

where $x_0 < x_1 < x_2$ (cf. Figure 2.2.6).

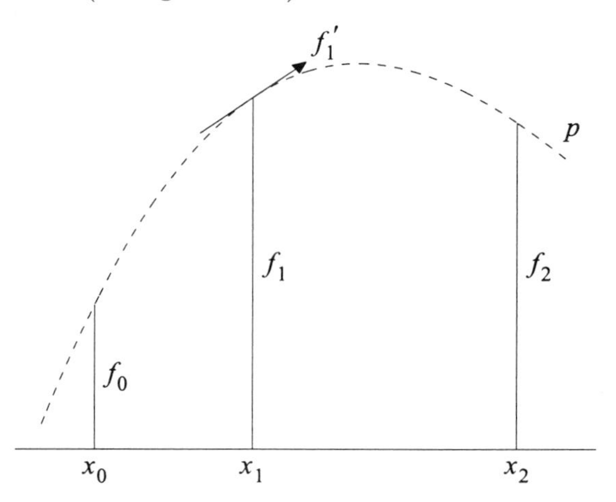

FIGURE 2.2.6. A Hermite interpolation problem

The table of divided differences now has the form

x	f			
x_0	f_0			
x_1	f_1	$[x_0, x_1]f$		
x_1	f_1	f_1'	$[x_0, x_1, x_1]f$	
x_2	f_2	$[x_1, x_2]f$	$[x_1, x_1, x_2]f$	$[x_0, x_1, x_1, x_2]f$.

If we denote the diagonal entries, as before, by a_0, a_1, a_2, a_3, Newton's formula takes the form

$$p(x) = a_0 + a_1(x - x_0) + a_2(x - x_0)(x - x_1) + a_3(x - x_0)(x - x_1)^2,$$

and the error formula becomes

$$f(x) - p(x) = (x - x_0)(x - x_1)^2(x - x_2)\frac{f^{(4)}(\xi)}{4!}, \quad x_0 < \xi < x_2.$$

For equally spaced points, say, $x_0 = x_1 - h$, $x_2 = x_1 + h$, we have, if $x = x_1 + th$, $-1 \le t \le 1$,

$$|(x - x_0)(x - x_1)^2(x - x_2)| = |(t^2 - 1)t^2 \cdot h^4| \le \tfrac{1}{4}h^4,$$

and so

$$\|f - p\|_\infty \le \frac{1}{4}h^4\frac{\|f^{(4)}\|_\infty}{24} = \frac{h^4}{96}\|f^{(4)}\|_\infty,$$

with the ∞-norm referring to the interval $[x_0, x_2]$.

§2.8. **Inverse interpolation**. An interesting application of interpolation — and, in particular, of Newton's formula — is to the solution of a nonlinear equation,

$$f(x) = 0. \tag{2.72}$$

Here f is a given (nonlinear) function, and we are interested in a root α of the equation for which we already have two approximations,

$$x_0 \approx \alpha, \quad x_1 \approx \alpha.$$

We assume further that near the root α, the function f is monotone, so that

$$y = f(x) \text{ has an inverse } x = f^{-1}(y).$$

Denote, for short,

$$g(y) = f^{-1}(y).$$

Since $\alpha = g(0)$, our problem is to evaluate $g(0)$. From our two approximations, we can compute $y_0 = f(x_0)$ and $y_1 = f(x_1)$, giving $x_0 = g(y_0)$, $x_1 = g(y_1)$. Hence, we can start a table of divided differences for the inverse function g:

y	g	
y_0	x_0	
y_1	x_1	$[y_0, y_1]g$.

Wanting to compute $g(0)$, we can get a first improved approximation by linear interpolation,

$$x_2 = x_0 + (0 - y_0)[y_0, y_1]g = x_0 - y_0[y_0, y_1]g.$$

Now evaluating $y_2 = f(x_2)$, we get $x_2 = g(y_2)$. Hence, the table of divided differences can be updated and becomes

y	g		
y_0	x_0		
y_1	x_1	$[y_0, y_1]g$	
y_2	x_2	$[y_1, y_2]g$	$[y_0, y_1, y_2]g$.

This allows us to use quadratic interpolation to get, again with Newton's formula,

$$x_3 = x_2 + (0 - y_0)(0 - y_1)[y_0, y_1, y_2]g = x_2 + y_0 y_1 [y_0, y_1, y_2]g$$

and then

$$y_3 = f(x_3), \quad \text{and} \quad x_3 = g(y_3).$$

Since y_0, y_1 are small, the product $y_0 y_1$ is even smaller, making the correction term added to the linear interpolant x_2 quite small. If necessary, we can continue updating the difference table,

y	g			
y_0	x_0			
y_1	x_1	$[y_0, y_1]g$		
y_2	x_2	$[y_1, y_2]g$	$[y_0, y_1, y_2]g$	
y_3	x_3	$[y_2, y_3]g$	$[y_1, y_2, y_3]g$	$[y_0, y_1, y_2, y_3]g$

and computing

$$x_4 = x_3 - y_0 y_1 y_2 [y_0, y_1, y_2, y_3]g, \quad y_4 = f(x_4), \quad x_4 = g(y_4),$$

giving us another data point to generate the next row of divided differences, and so on. In general, the process will converge rapidly: $x_k \to \alpha$ as $k \to \infty$.

The precise analysis of convergence, however, is not simple because of the complicated structure of the successive derivatives of the inverse function $g = f^{-1}$.

§3. Approximation and Interpolation by Spline Functions

Our concern in §2 was with approximation of functions by a single polynomial over a finite interval $[a,b]$. When more accuracy was wanted, we simply increased the degree of the polynomial, and under suitable assumptions the approximation indeed can be made as accurate as one wishes by choosing the degree of the approximating polynomial sufficiently large.

However, there are other ways to control accuracy. One is to impose a subdivision Δ upon the interval $[a,b]$,

$$\Delta: \quad a = x_1 < x_2 < x_3 < \cdots < x_{n-1} < x_n = b, \tag{3.1}$$

and use *low-degree* polynomials on each subinterval $[x_i, x_{i+1}]$ ($i = 1, 2, \ldots, n-1$) to approximate the given function. The rationale behind this is the recognition that on a sufficiently small interval, functions can be approximated arbitrarily well by polynomials of low degree, even degree 1, or zero, for that matter. Thus, measuring the "fineness" of the subdivision Δ by

$$|\Delta| = \max_{1 \le i \le n-1} \Delta x_i, \quad \Delta x_i = x_{i+1} - x_i, \tag{3.2}$$

we try to control (increase) the accuracy by varying (decreasing) $|\Delta|$, keeping the degrees of the polynomial pieces uniformly low.

To discuss these approximation processes, we make use of the class of functions (cf. Example (2) at the beginning of Chapter 2)

$$\mathbb{S}_m^k(\Delta) = \{s: \; s \in C^k[a,b], \; s\Big|_{[x_i,x_{i+1}]} \in \mathbb{P}_m, \quad i = 1, 2, \ldots, n-1\}, \tag{3.3}$$

where $m \ge 0$, $k \ge 0$ are given nonnegative integers. We refer to $\mathbb{S}_m^k(\Delta)$ as the *spline functions of degree m and smoothness class k* relative to the subdivision Δ. (If the subdivision is understood from the context, we omit Δ in the notation on the left of (3.3).) The point in the continuity assumption of (3.3), of course, is that the kth derivative of s is to be continuous *everywhere* on $[a,b]$, in particular, also at the subdivision points x_i ($i = 2, \ldots, n-1$) of Δ. One extreme case is $k = m$, in which case $s \in \mathbb{S}_m^m$ necessarily consists of just one single polynomial of degree m on the whole interval $[a,b]$; that is, $\mathbb{S}_m^m = \mathbb{P}_m$ (see Ex. 62). Since we want to get away from $\mathbb{P}_m$, we assume $k < m$. The other extreme is the case where no continuity at all (at the subdivision points x_i) is required; we then put $k = -1$. Thus $\mathbb{S}_m^{-1}(\Delta)$ is the class of piecewise polynomials of degree $\leq m$, where the polynomial pieces can be completely disjoint (see Figure 2.3.1).

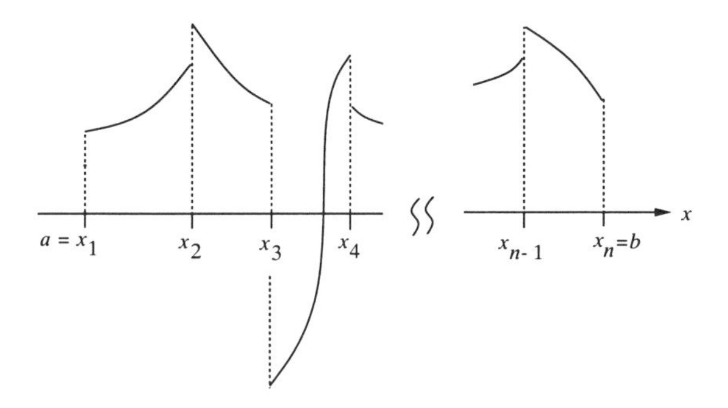

FIGURE 2.3.1. A function $s \in \mathbb{S}_m^{-1}$

We begin with the simplest case — piecewise linear approximation — that is, the case $m = 1$ (hence $k = 0$).

§3.1. **Interpolation by piecewise linear functions**. The problem here is to find an $s \in \mathbb{S}_1^0(\Delta)$ such that, for a given function f defined on $[a,b]$, we have

$$s(x_i) = f_i \quad \text{where} \quad f_i = f(x_i), \quad i = 1, 2, \ldots, n. \tag{3.4}$$

We conveniently let the interpolation nodes coincide with the points x_i of the subdivision Δ in (3.1). This simplifies matters, but is not necessary (cf. Ex. 68). The solution then indeed is trivial; see Figure 2.3.2. If we denote the (obviously unique) interpolant by $s(\cdot) = s_1(f; \cdot)$, then the formula of linear interpolation gives

$$s_1(f; x) = f_i + (x - x_i)[x_i, x_{i+1}]f \quad \text{for} \quad x_i \leq x \leq x_{i+1}, \quad i = 1, 2, \ldots, n-1. \tag{3.5}$$

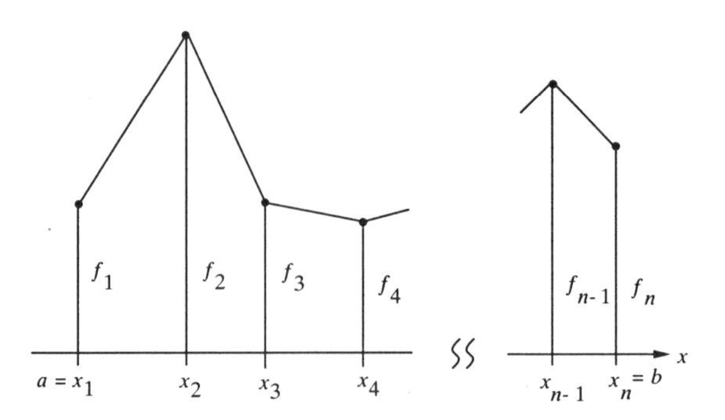

FIGURE 2.3.2. Piecewise linear interpolation

A bit more interesting is the analysis of the error. This, too, however, is quite straightforward, once we note that $s_1(f;\,\cdot\,)$ on $[x_i, x_{i+1}]$ is simply the linear interpolant to f. Thus, from the theory of (linear) interpolation,

$$f(x) - s_1(f;x) = (x - x_i)(x - x_{i+1})[x_i, x_{i+1}, x]f \quad \text{for} \quad x \in [x_i, x_{i+1}];$$

hence, if $f \in C^2[a,b]$,

$$|f(x) - s_1(f;x)| \leq \frac{(\Delta x_i)^2}{8} \max_{[x_i, x_{i+1}]} |f''|, \quad x \in [x_i, x_{i+1}].$$

It then follows immediately that

$$\|f(\,\cdot\,) - s_1(f;\,\cdot\,)\|_\infty \leq \tfrac{1}{8}\, |\Delta|^2 \|f''\|_\infty, \tag{3.6}$$

where the maximum norms are those on $[a,b]$; that is, $\|g\|_\infty = \max\limits_{[a,b]} |g|$. This shows that the error indeed can be made arbitrarily small, uniformly on $[a,b]$, by taking $|\Delta|$ sufficiently small. Making $|\Delta|$ smaller, of course, increases the number of polynomial pieces, and with it, the volume of data.

It is easy to show (see Ex. 73) that

$$\operatorname{dist}_\infty(f, \mathbb{S}_1^0) \leq \|f(\,\cdot\,) - s_1(f;\,\cdot\,)\|_\infty \leq 2 \operatorname{dist}_\infty(f, \mathbb{S}_1^0), \tag{3.7}$$

where, for any set of functions $\mathbb{S}$,

$$\operatorname{dist}_\infty(f, \mathbb{S}) := \inf_{s \in \mathbb{S}} \|f - s\|_\infty.$$

In other words, the piecewise linear interpolant $s_1(f;\cdot)$ is a nearly optimal approximation, its error differing from the error of the best approximant to f from $\mathbb{S}_1^0$ by at most a factor of 2.

§3.2. **A basis for $\mathbb{S}_1^0(\Delta)$.** What is the dimension of the space $\mathbb{S}_1^0(\Delta)$? In other words, how many degrees of freedom do we have? If, for the moment, we ignore the continuity requirement (i.e., if we look at $\mathbb{S}_1^{-1}(\Delta)$), then each linear piece has 2 degrees of freedom, and there are $n-1$ pieces; so $\dim \mathbb{S}_1^{-1}(\Delta) = 2n-2$. Each continuity requirement imposes one equation, and hence reduces the degree of freedom by 1. Since continuity must be enforced only at the interior subdivision points x_i, $i = 2,\ldots,n-1$, we find that $\dim \mathbb{S}_1^0(\Delta) = 2n-2-(n-2) = n$. So we expect that a basis of $\mathbb{S}_1^0(\Delta)$ must consist of exactly n basis functions.

We now define n such functions. For notational convenience, we let $x_0 = x_1$ and $x_{n+1} = x_n$; then, for $i = 1,2,\ldots,n$, we define

$$B_i(x) = \begin{cases} \dfrac{x - x_{i-1}}{x_i - x_{i-1}} & \text{if } x_{i-1} \le x \le x_i, \\[2ex] \dfrac{x_{i+1} - x}{x_{i+1} - x_i} & \text{if } x_i \le x \le x_{i+1}, \\[2ex] 0 & \text{otherwise.} \end{cases} \tag{3.8}$$

Note that the first equation, when $i = 1$, and the second, when $i = n$, are to be ignored, since x in both cases is restricted to a single point and the

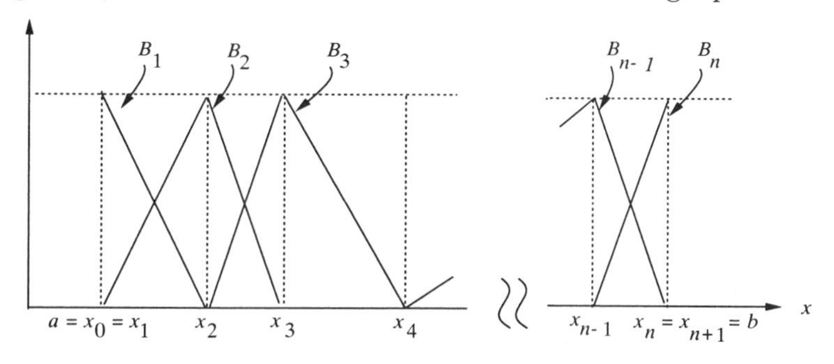

FIGURE 2.3.3. The functions B_i

ratio in question has the meaningless form 0/0. (It is the other ratio that provides the necessary information in these cases.) The functions B_i may be referred to as "hat functions" (Chinese hats), but note that the first and last hat is cut in half. The functions B_i are depicted in Figure 2.3.3. We expect these functions to form a basis of $\mathbb{S}_1^0(\Delta)$. To prove this, we must show:

(i) the functions $\{B_i\}_{i=1}^n$ are linearly independent; and

(ii) they span the space $\mathbb{S}_1^0(\Delta)$.

Both these properties follow from the basic fact that

$$B_i(x_j) = \delta_{ij} = \begin{cases} 1 & \text{if} \quad i = j, \\ 0 & \text{if} \quad i \neq j, \end{cases} \tag{3.9}$$

which one easily reads from Figure 2.3.3. To show (i), assume there is a linear combination of the B_i that vanishes identically on $[a,b]$,

$$s(x) = \sum_{i=1}^n c_i B_i(x), \quad s(x) \equiv 0 \quad \text{on} \quad [a, b]. \tag{3.10}$$

Putting $x = x_j$ in (3.10) and using (3.9) then gives $c_j = 0$. Since this holds for each $j = 1, 2, \ldots, n$, we see that only the trivial linear combination (with all $c_i = 0$) can vanish identically. To prove (ii), let $s \in \mathbb{S}_1^0(\Delta)$ be given arbitrarily. We must show that s can be represented as a linear combination of the B_i. We claim that, indeed,

$$s(x) = \sum_{i=1}^n s(x_i) B_i(x). \tag{3.11}$$

This is so, because the function on the right has the same values as s at each x_j, and therefore, being in $\mathbb{S}_1^0(\Delta)$, must coincide with s.

The formula (3.11), which holds for every $s \in \mathbb{S}_1^0(\Delta)$, may be thought of as the analogue of the Lagrange interpolation formula for polynomials. The role of the elementary Lagrange polynomials ℓ_i is now played by the B_i.

§3.3. **Least squares approximation**. As an application of the basis B_i we consider the problem of least squares approximation on $[a,b]$ by functions in $\mathbb{S}_1^0(\Delta)$. The discrete L_2 approximation problem with data given at the points x_i ($i = 1, 2, \ldots, n$), of course, has the trivial solution $s_1(f; \cdot)$, which drives the error to zero at each data point. We therefore consider only the continuous problem: given $f \in C[a,b]$, find $\hat{s}_1(f;\cdot) \in \mathbb{S}_1^0(\Delta)$ such that

$$\int_a^b [f(x) - \hat{s}_1(f;x)]^2 dx \le \int_a^b [f(x) - s(x)]^2 dx \quad \text{for all} \quad s \in \mathbb{S}_1^0(\Delta). \tag{3.12}$$

Writing

$$\hat{s}_1(f;x) = \sum_{i=1}^{n} \hat{c}_i B_i(x), \tag{3.13}$$

we know from the general theory of §1 that the coefficients $\hat{c}_i$ must satisfy the normal equations

$$\sum_{j=1}^{n} \left[\int_a^b B_i(x) B_j(x) dx \right] \hat{c}_j = \int_a^b B_i(x) f(x) dx, \quad i = 1, 2, \ldots, n. \tag{3.14}$$

Now the fact that B_i is nonzero only on (x_{i-1}, x_{i+1}) implies that $\int_a^b B_i(x) \cdot B_j(x) dx = 0$ if $|i-j| > 1$; that is, the system (3.14) is tridiagonal. An easy computation (cf. Ex. 70) indeed yields

$$\tfrac{1}{6}\Delta x_{i-1} \cdot \hat{c}_{i-1} + \tfrac{1}{3}(\Delta x_{i-1} + \Delta x_i)\hat{c}_i + \tfrac{1}{6}\Delta x_i \cdot \hat{c}_{i+1} = b_i, \; i = 1, 2, \ldots, n, \tag{3.14$'$}$$

where $b_i = \int_a^b B_i(x) f(x) dx = \int_{x_{i-1}}^{x_{i+1}} B_i(x) f(x) dx$. Note, by our convention, that $\Delta x_0 = 0$ and $\Delta x_n = 0$, so that (3.14$'$) is in fact a tridiagonal system for the unknowns $\hat{c}_1, \hat{c}_2, \ldots, \hat{c}_n$. Its matrix is given by

$$\begin{bmatrix} \frac{1}{3}\Delta x_1 & \frac{1}{6}\Delta x_1 & & & 0 \\ \frac{1}{6}\Delta x_1 & \frac{1}{3}(\Delta x_1 + \Delta x_2) & \frac{1}{6}\Delta x_2 & & \\ & \frac{1}{6}\Delta x_2 & & \ddots & \\ & & \ddots & \ddots & \frac{1}{6}\Delta x_{n-1} \\ 0 & & & \frac{1}{6}\Delta x_{n-1} & \frac{1}{3}\Delta x_{n-1} \end{bmatrix}.$$

As it must be, by the general theory of §1, the matrix is symmetric and positive definite, but it is also diagonally dominant, each diagonal element exceeding the sum of the (positive) off-diagonal elements in the same row by a factor of 2. The system (3.14′) can therefore be solved easily, rapidly, and accurately by the Gauss elimination procedure, and there is no need for pivoting.

Like the interpolant $s_1(f;\,\cdot\,)$, the least squares approximant $\hat{s}_1(f;\,\cdot\,)$, too, can be shown to be nearly optimal, in that

$$\text{dist}_\infty(f, \mathbb{S}_1^0) \le \|f(\,\cdot\,) - \hat{s}_1(f;\,\cdot\,)\|_\infty \le 4\ \text{dist}_\infty(f, \mathbb{S}_1^0). \tag{3.15}$$

The spread is now by a factor of 4, rather than 2, as in (3.7).

§3.4. **Interpolation by cubic splines**. The most widely used splines are cubic splines, in particular, cubic spline interpolants. We first discuss the interpolation problem for splines $s \in \mathbb{S}_3^1(\Delta)$. Continuity of the first derivative of any cubic spline interpolant $s_3(f;\,\cdot\,)$ can be enforced by prescribing the values of the first derivative at each point x_i, $i = 1, 2, \ldots, n$. Thus let m_1, $m_2, \ldots, m_n$ be arbitrary given numbers, and denote

$$s_3(f;\,\cdot\,)|_{[x_i, x_{i+1}]} = p_i(x), \quad i = 1, 2, \ldots, n-1. \tag{3.16}$$

Then we enforce $s_3'(f; x_i) = m_i$, $i = 1, 2, \ldots, n$, by selecting each piece p_i of $s_3(f;\,\cdot\,)$ to be the (unique) solution of a Hermite interpolation problem, namely,

$$\begin{aligned} p_i(x_i) &= f_i, \quad p_i(x_{i+1}) = f_{i+1}, \\ p_i'(x_i) &= m_i, \quad p_i'(x_{i+1}) = m_{i+1}, \end{aligned} \qquad i = 1, 2, \ldots, n-1. \tag{3.17}$$

We solve (3.17) by Newton's interpolation formula. The required divided differences are

$$\begin{array}{lllll}
x_i & f_i & & & \\
x_i & f_i & m_i & & \\
x_{i+1} & f_{i+1} & [x_i, x_{i+1}]f & \dfrac{[x_i, x_{i+1}]f - m_i}{\Delta x_i} & \\
x_{i+1} & f_{i+1} & m_{i+1} & \dfrac{m_{i+1} - [x_i, x_{i+1}]f}{\Delta x_i} & \dfrac{m_{i+1} + m_i - 2[x_i, x_{i+1}]f}{(\Delta x_i)^2}
\end{array}$$

and the interpolation polynomial (in Newton's form) is

$$p_i(x) = f_i + (x - x_i)m_i + (x - x_i)^2 \frac{[x_i, x_{i+1}]f - m_i}{\Delta x_i} + (x - x_i)^2(x - x_{i+1}) \frac{m_{i+1} + m_i - 2[x_i, x_{i+1}]f}{(\Delta x_i)^2} .$$

Alternatively, in Taylor's form, we can write

$$p_i(x) = c_{i,0} + c_{i,1}(x - x_i) + c_{i,2}(x - x_i)^2 + c_{i,3}(x - x_i)^3, \quad x_i \le x \le x_{i+1}, \tag{3.18}$$

where, by noting that $x - x_{i+1} = x - x_i - \Delta x_i$,

$$\begin{aligned} c_{i,0} &= f_i, \\ c_{i,1} &= m_i, \\ c_{i,2} &= \frac{[x_i, x_{i+1}]f - m_i}{\Delta x_i} - c_{i,3} \cdot \Delta x_i, \\ c_{i,3} &= \frac{m_{i+1} + m_i - 2[x_i, x_{i+1}]f}{(\Delta x_i)^2} . \end{aligned} \tag{3.19}$$

Thus to compute $s_3(f; x)$ for any given $x \in [a, b]$ that is not an interpolation node, one first locates the interval $[x_i, x_{i+1}]$ containing x and then computes the corresponding piece (3.16) by (3.18) and (3.19).

We now discuss some possible choices of the parameters $m_1, m_2, \dots, m_n$.

(i) *Piecewise cubic Hermite interpolation.* Here one selects $m_i = f'(x_i)$, assuming that these derivative values are known. This gives rise to a strictly *local* scheme, in that each piece p_i can be determined independently from the others. Furthermore, the error of interpolation is easily estimated, since from the theory of interpolation,

$$f(x) - p_i(x) = (x - x_i)^2(x - x_{i+1})^2[x_i, x_i, x_{i+1}, x_{i+1}, x]f, \quad x_i \le x \le x_{i+1};$$

hence, if $f \in C^4[a, b]$,

$$|f(x) - p_i(x)| \le \left(\frac{1}{2}\Delta x_i\right)^4 \max_{[x_i, x_{i+1}]} \frac{|f^{(4)}|}{4!}, \quad x_i \le x \le x_{i+1}.$$

There follows

$$\|f(\cdot) - s_3(f;\,\cdot)\|_\infty \le \tfrac{1}{384}\,|\Delta|^4\,\|f^{(4)}\|_\infty. \tag{3.20}$$

In the case of equally spaced points x_i, one has $|\Delta| = (b-a)/(n-1)$ and, therefore,

$$\|f(\cdot) - s_3(f;\,\cdot)\|_\infty = O(n^{-4}) \quad \text{as} \quad n \to \infty. \tag{3.20$'$}$$

This is quite satisfactory, but note that the derivative of f must be known at each point x_i, and the interpolant is only in $C^1[a,b]$.

As to the derivative values, one could approximate them by the derivatives of $p_2(f; x_{i-1}, x_i, x_{i+1}; x)$ at $x = x_i$, which requires only function values of f, except at the endpoints, where again the derivatives of f are involved, the points $a = x_0 = x_1$ and $b = x_n = x_{n+1}$ being double points (cf. Ex. 71). It can be shown that this degrades the accuracy to $O(|\Delta|^3)$ in the case of unequally spaced nodes, but maintains an $O(n^{-4})$ error in the case of equally spaced points.

(ii) *Cubic spline interpolation.* Here we require $s_3(f;\,\cdot) \in \mathbb{S}_3^2(\Delta)$, that is, continuity of the second derivative. In terms of the pieces (3.16) of $s_3(f;\,\cdot)$, this means that

$$p''_{i-1}(x_i) = p''_i(x_i), \quad i = 2, 3, \dots, n-1, \tag{3.21}$$

and translates into a condition for the Taylor coefficients in (3.18), namely,

$$2\,c_{i-1,2} + 6\,c_{i-1,3} \cdot \Delta x_{i-1} = 2\,c_{i,2}, \quad i = 2, 3, \dots, n-1.$$

Plugging in the explicit values (3.19) for these coefficients, we arrive at the linear system

$$(\Delta x_i) m_{i-1} + 2(\Delta x_{i-1} + \Delta x_i) m_i + (\Delta x_{i-1}) m_{i+1} = b_i, \quad i = 2, 3, \dots, n-1, \tag{3.22}$$

where

$$b_i = 3\{(\Delta x_i)[x_{i-1}, x_i]f + (\Delta x_{i-1})[x_i, x_{i+1}]f\}. \tag{3.23}$$

These are $n-2$ linear equations in the n unknowns $m_1, m_2, \dots, m_n$. Once m_1 and m_n have been chosen in some way, the system again becomes tridiagonal in the remaining unknowns and hence is readily solved by Gauss elimination. Here are some possible choices of m_1 and m_n.

(ii.1) *Complete splines:* $m_1 = f'(a)$, $m_n = f'(b)$. It is known that for this spline

$$\|f^{(r)}(\,\cdot\,) - s^{(r)}(f;\,\cdot\,)\|_\infty \le c_r|\Delta|^{4-r}\|f^{(4)}\|_\infty, r = 0, 1, 2, 3, \quad \text{if} \quad f \in C^4[a, b], \tag{3.24}$$

where $c_0 = \frac{5}{384}$, $c_1 = \frac{1}{24}$, $c_2 = \frac{3}{8}$, and c_3 is a constant depending on the mesh ratio $\frac{|\Delta|}{\min_i \Delta x_i}$. Rather remarkably, the bound for $r = 0$ is only five times larger than the bound (3.20) for the piecewise cubic Hermite interpolant, which requires derivative values of f at all interpolation nodes x_i, not just at the endpoints a and b.

(ii.2) *Matching of the second derivatives at the endpoints:* $s_3''(f; a) = f''(a)$, $s_3''(f; b) = f''(b)$. Each of these conditions gives rise to an additional equation, namely,

$$\begin{aligned} 2m_1 + m_2 &= 3[x_1, x_2]f - \tfrac{1}{2}f''(a)\Delta x_1, \\ m_{n-1} + 2m_n &= 3[x_{n-1}, x_n]f + \tfrac{1}{2}f''(b)\Delta x_{n-1}. \end{aligned} \tag{3.25}$$

One conveniently adjoins the first equation to the top of the system (3.22), and the second to the bottom, thereby preserving the tridiagonal structure of the system.

(ii.3) *Natural cubic spline:* $s''(f; a) = s''(f; b) = 0$. This again produces two additional equations, which can be obtained from (3.25) by putting there $f''(a) = f''(b) = 0$. They are adjoined to the system (3.22) as described in (ii.2). The nice thing about this spline is that it requires only function values of f — no derivatives! — but the price one pays is a degradation of the accuracy to $O(|\Delta|^2)$ near the endpoints (unless indeed $f''(a) = f''(b) = 0$).

(ii.4) *"Not-a-knot spline"* (C. de Boor): Here we require $p_1(x) \equiv p_2(x)$ and $p_{n-2}(x) \equiv p_{n-1}(x)$; that is, the first two pieces of the spline should be the same polynomial, and similarly for the last two pieces. In effect, this means that the first interior knot x_2, and the last one x_{n-1}, both are inactive (hence the name). This again gives rise to two supplementary equations expressing continuity of $s_3'''(f; x)$ at $x = x_2$ and $x = x_{n-1}$ (cf. Ex. 72).

§3.5. **Minimality properties of cubic spline interpolants.** The complete and natural splines defined in (ii.1) and (ii.3) of the preceding section have interesting optimality properties. To formulate them, it is convenient to consider not only the subdivision Δ in (3.1), but also the subdivision

$$\Delta' : \quad a = x_0 = x_1 < x_2 < x_3 < \cdots < x_{n-1} < x_n = x_{n+1} = b, \tag{3.26}$$

in which the endpoints are double knots. This means that whenever we interpolate on Δ', we interpolate to function values at all interior points, but to the function as well as first derivative values at the endpoints.

The first of the two theorems relates to the complete cubic spline interpolant, $s_{\text{compl}}(f; \cdot)$.

Theorem 2.3.1. *For any function $g \in C^2[a,b]$ that interpolates f on Δ', there holds*

$$\int_a^b [g''(x)]^2 dx \geq \int_a^b [s''_{\text{compl}}(f; x)]^2 dx, \tag{3.27}$$

with equality if and only if $g(\cdot) = s_{\text{compl}}(f; \cdot)$.

Note that $s_{\text{compl}}(f; \cdot)$ in Theorem 2.3.1 also interpolates f on Δ', and among all such interpolants its second derivative has the smallest L_2 norm.

Proof of Theorem 2.3.1. We write (for short) $s_{\text{compl}} = s$. The theorem follows, once we have shown that

$$\int_a^b [g''(x)]^2 dx = \int_a^b [g''(x) - s''(x)]^2 dx + \int_a^b [s''(x)]^2 dx. \tag{3.28}$$

Indeed, this immediately implies (3.27), and equality in (3.27) holds if and only if $g''(x) - s''(x) \equiv 0$, which, integrating twice from a to x and using the interpolation properties of s and g at $x = a$ gives $g(x) \equiv s(x)$.

To complete the proof, note that (3.28) is equivalent to

$$\int_a^b s''(x)[g''(x) - s''(x)]dx = 0. \tag{3.29}$$

Integrating by parts, we get

$$\begin{aligned} &\int_a^b s''(x)[g''(x) - s''(x)]dx \\ &= s''(x)[g'(x) - s'(x)]\big|_a^b - \int_a^b s'''(x)[g'(x) - s'(x)]dx \\ &= -\int_a^b s'''(x)[g'(x) - s'(x)]dx, \end{aligned} \tag{3.30}$$

since $s'(b) = g'(b) = f'(b)$, and similarly at $x = a$. But s''' is piecewise constant, so

$$\begin{aligned}
&\int_a^b s'''(x)[g'(x) - s'(x)]dx \\
&\quad = \sum_{\nu=1}^{n-1} s'''(x_\nu + 0) \int_{x_\nu}^{x_{\nu+1}} [g'(x) - s'(x)]dx \\
&= \sum_{\nu=1}^{n-1} s'''(x_\nu + 0)[g(x_{\nu+1}) - s(x_{\nu+1}) - (g(x_\nu) - s(x_\nu))] = 0,
\end{aligned}$$

since both s and g interpolate to f on Δ. This proves (3.29) and hence the theorem. □

For interpolation on Δ, the distinction of being optimal goes to the natural cubic spline interpolant $s_{\text{nat}}(f;\,\cdot\,)$. This is the content of the second theorem.

Theorem 2.3.2. *For any function $g \in C^2[a, b]$ that interpolates f on Δ* (not Δ'), *there holds*

$$\int_a^b [g''(x)]^2 dx \geq \int_a^b [s''_{\text{nat}}(f; x)]^2 dx, \tag{3.31}$$

with equality if and only if $g(\,\cdot\,) = s_{\text{nat}}(f;\,\cdot\,)$.

The proof of Theorem 2.3.2 is virtually the same as that of Theorem 2.3.1, since (3.30) holds again, this time because $s''(b) = s''(a) = 0$.

Putting $g(\,\cdot\,) = s_{\text{compl}}(f;\,\cdot\,)$ in Theorem 2.3.2 immediately gives

$$\int_a^b [s''_{\text{compl}}(f; x)]^2 dx \geq \int_a^b [s''_{\text{nat}}(f; x)]^2 dx. \tag{3.32}$$

Therefore, in a sense, the natural cubic spline is the "smoothest" interpolant.

The property expressed in Theorem 2.3.2 is the origin of the name "spline." A spline is a flexible strip of wood used in drawing curves. If its shape is given by the equation $y = g(x)$, $a \leq x \leq b$, and if the spline is constrained to pass through the points (x_i, g_i), then it assumes a form that minimizes the bending energy

$$\int_a^b \frac{[g''(x)]^2 dx}{(1 + [g'(x)]^2)^3}$$

over all functions g similarly constrained. For slowly varying g ($\|g'\|_\infty \ll 1$), this is nearly the same as the minimum property of Theorem 2.3.2.

NOTES TO CHAPTER 2

There are many excellent texts on the general problem of best approximation as exemplified by (0.1). One that emphasizes uniform approximation by polynomials is Feinerman and Newman [1974]; apart from the basic theory of best polynomial approximation, it also contains no fewer than four proofs of the fundamental theorem of Weierstrass. For approximation in the L_∞ and L_1 norm, which is related to linear programming, a number of constructive methods, notably the Remez algorithms and exchange algorithms, are known, both for polynomial and rational approximation. Early, but still very readable, expositions are given in Cheney [1966] and Rivlin [1981], and more recent accounts in Watson [1980] and Powell [1981]. Nearly best polynomial and rational approximations are widely used in computer routines for special functions; for a survey of work in this area, up to about 1975, see Gautschi [1975a], and for subsequent work, van der Laan and Temme [1984] and Németh [1992]. Much relevant material is also contained in the books by Luke [1975] and [1977]. The numerical approximation and software for special functions is exhaustively documented in Lozier and Olver [1994]; a recent package for some of the more esoteric functions is described in MacLeod [1996]. For an extensive (and mathematically demanding) treatment of rational approximation, the reader is referred to Petrushev and Popov [1987], and for L_1 approximation, to Pinkus [1989]. Methods of nonlinear approximation, including approximation by exponential sums, are studied in Braess [1986]. Other basic texts on approximation and interpolation are Natanson [1964,1965,1965] and Davis [1975] from the 1960s, and the more recent books by DeVore and Lorentz [1993] and its sequel, Lorentz, Golitschek, and Makovoz [1996]. A large variety of problems of interpolation and approximation by rational functions (including polynomials) in the complex plane is studied in Walsh [1969]. An example of a linear space Φ containing a denumerable set of nonrational basis functions are the sinc functions — scaled translates of $\frac{\sin \pi t}{\pi t}$. They are of importance in the Shannon sampling and interpolation theory (see, e.g., Zayed [1993]) but are also useful for approximation on infinite or semi-infinite domains in the complex plane; see Stenger [1993] for an extensive discussion of this. A reader interested in issues of current interest related to multivariate approximation can get a good start by consulting Cheney [1986].

Rich and valuable sources on polynomials and their numerous properties of interest in applied analysis are Milovanović, Mitrinović, and Rassias [1994] and Borwein and Erdélyi [1995]. Spline functions — in name and as a basic tool of approximation — were introduced in 1946 by Schoenberg [1946]; also see Schoenberg [1973]. They have generated enormous interest, owing both to their interesting mathematical theory and practical usefulness. There are now many texts available, treating splines from various points of view. A selected list is Ahlberg, Nilson, and Walsh [1967], Nürnberger [1989], and Schumaker [1993] for the basic theory, de Boor [1978] and Späth [1995] for more practical aspects including algorithms, Atteia [1992] for an abstract treatment based on Hilbert kernels, Bartels, Beatty, and Barsky [1987] and Dierckx [1993] for applications to computer graphics and geometric modeling, and Chui [1988], de Boor, Höllig, and Riemenschneider [1993], and Bojanov, Hakopian, and Sahakian [1993] for multivariate splines. The standard text on trigonometric series still is Zygmund [1988].

§1. Historically, the least squares principle evolved in the context of discrete linear approximation. The principle was first enunciated by Legendre in 1805 in a treatise on celestial mechanics (Legendre [1805]), although Gauss used it earlier in 1795, but published the method only in 1809 (in a paper also on celestial mechanics). For Gauss's subsequent treatises, published in 1821–1826, see the English translation in Gauss [1995]. The statistical justification of least squares as a minimum variance (unbiased) estimator is due to Gauss. If one were to disregard probabilistic arguments, then, as Gauss already remarked (Goldstine [1977, p. 212]), one could try to minimize the sum of any even (positive) power of the errors, and even let this power go to infinity, in which case one would minimize the maximum error. But by these principles " ... we should be led into the most complicated calculations." Interestingly, Laplace at about the same time also proposed discrete L_1 approximation (under the side condition that all errors add up to zero). A reader interested in the history of least squares may wish to consult the article by Sheynin [1993].

The choice of weights w_i in the discrete L_2 norm $\|\cdot\|_{2,w}$ can be motivated on statistical grounds if one assumes that the errors in the data $f(x_i)$ are uncorrelated and have zero mean and variances σ_i^2; an appropriate choice then is $w_i = \sigma_i^{-2}$.

The discrete problem of minimizing $\|f - \varphi\|_{2,w}$ over functions φ in Φ as given by (0.2) can be rephrased in terms of an overdetermined system of linear equations, $Pc = f$, where $P = [\pi_j(x_i)]$ is a rectangular matrix of size $N \times n$, and $f = [f(x_i)]$ the data vector of dimension N. If $r = f - Pc$, $r = [r_i]$ denotes the residual vector, one tries to find the coefficient vector $c \in \mathbb{R}^n$ such that $\sum_i w_i r_i^2$ is as small as possible. There is a vast literature dealing with overdetermined systems involving more general (full or sparse) matrices and their solution by the method of least squares. A large arsenal of modern techniques of matrix computation can be brought to bear on this problem; see, for example, Björck [1996] for an extensive discussion. In the special case considered here, the method of (discrete) orthogonal polynomials, however, is more efficient. It has its origin in the work of Chebyshev [1859]; a contemporary exposition, including computational and statistical issues, is given in Forsythe [1957].

There are interesting variations on the theme of polynomial least squares approximation. One is to minimize $\|f - p\|_{2,d\lambda}$ among all polynomials in $\mathbb{P}_n$ subject to interpolatory constraints at $m+1$ given points, where $m < n$. It turns out that this can be reduced to an unconstrained least squares problem, but for a different measure $d\lambda$ and a different function f; cf. Gautschi [1996, §2.1]. Something similar is true for approximation by rational functions with a prescribed denominator polynomial. A more substantial variation obtains if one wants to approximate simultaneously a function f and its first s derivatives. In the most general setting, this would require the minimization of $\int_{\mathbb{R}} \sum_{\sigma=0}^{s} [f^{(\sigma)}(t) - p^{(\sigma)}(t)]^2 d\lambda_\sigma(t)$ among all polynomials $p \in \mathbb{P}_n$, where $d\lambda_\sigma$ are given (continuous or discrete) positive measures. The problem can be solved, as in §1.2, by orthogonal polynomials, but they are now orthogonal with respect to the inner product $(u,v)_{H_s} = \sum_{\sigma=0}^{s} \int_{\mathbb{R}} u^{(\sigma)}(t)v^{(\sigma)}(t)d\lambda_\sigma(t)$ — a so-called Sobolev inner product. This gives rise to Sobolev orthogonal polynomials; see Gautschi [1996, §2.2] for some history on this problem and relevant literature.

§1.2. The alternative form (1.15′) of computing the coefficients $\hat{c}_j$ was suggested in the 1972 edition of Conte and de Boor [1980] and is further discussed by Shampine [1975]. The Gram-Schmidt procedure described at the end of this section is now called the classical Gram-Schmidt procedure. There are other, modified, versions of Gram-Schmidt that are computationally more effective; see, for example, Björck [1996, pp. 61ff].

§1.4. The standard text on Fourier series, as already mentioned, is Zygmund [1988], and on orthogonal polynomials, Szegö [1975]. Not only is it true that orthogonal polynomials satisfy a three-term recurrence relation (1.26), but the converse is also true: any system $\{\pi_k\}$ of monic polynomials satisfying (1.26) for all $k \geq 0$, with real coefficients α_k and $\beta_k > 0$, is necessarily orthogonal with respect to some (in general unknown) positive measure. This is known as Favard's Theorem (cf., e.g., Natanson [1965, Vol. 2, Ch. 8, §6]. The computation of orthogonal polynomials, when the recursion coefficients are not known explicitly, is not an easy task; a number of methods are surveyed in Gautschi [1996]. Orthogonal systems in $L_2(\mathbb{R})$ that have become prominent in recent years are wavelets, which are functions of the form $\psi_{j,k}(t) = 2^{j/2}\psi(2^j t - k)$, $j,k = 0, \pm 1, \pm 2, \ldots$, with ψ a "mother wavelet" — square integrable on $\mathbb{R}$ and (usually) satisfying $\int_{\mathbb{R}} \psi(t)dt = 0$. Among the growing textbook and monograph literature on this subject, we mention Chui [1992], Daubechies [1992], Walter [1994], Wickerhauser [1994], and Resnikoff and Wells [1997].

§2. Although interpolation by polynomials and spline functions is most common, it is sometimes appropriate to use other systems of approximants for interpolation, for example, trigonometric polynomials or rational functions. Trigonometric interpolation at equally spaced points is closely related to discrete Fourier analysis and hence accessible to the Fast Fourier Transform (FFT). For this, and also for rational interpolation algorithms, see, for example, Stoer and Bulirsch [1993, §§2.2 and 2.3]. For the fast Fourier transform and some of its important applications, see Henrici [1979a] and Van Loan [1992].

Besides Lagrange and Hermite interpolation, other types of interpolation processes have been studied in the literature. Among these are Fejér-Hermite interpolation, where one interpolates to given function values and zero values of the derivative, and Birkhoff (also called lacunary) interpolation, which is similar to Hermite interpolation, but derivatives of only preselected orders are being interpolated. Remarkably, Fejér-Hermite interpolation at the Chebyshev points (defined in §2.4) converges for every continuous function $f \in C[-1,1]$. The convergence theory of

Lagrange and Fejér-Hermite interpolation is the subject of a monograph by Szabados and Vértesi [1990]. The most comprehensive work on Birkhoff interpolation is the book by G. G. Lorentz, Jetter, and Riemenschneider [1983]. A more recent monograph by R. A. Lorentz [1992] deals with multivariate Birkhoff interpolation.

§2.1. The growth of the Lebesgue constants Λ_n is at least $O(\log n)$ as $n \to \infty$; specifically, $\Lambda_n > \frac{2}{\pi} \log n + c$ for any triangular array of interpolation nodes (cf. §2.3), where the constant c can be expressed in terms of Euler's constant γ (cf. Ch. 1., Mach. Ass. 4) by $c = \frac{2}{\pi}\left(\log \frac{8}{\pi} + \gamma\right) = .9625228\ldots$; see Rivlin [1990, Thm. 1.2]. The Chebyshev points achieve the optimal order $O(\log n)$; for them, $\Lambda_n \leq \frac{2}{\pi} \log n + 1$ (Rivlin [1990, Thm. 1.2]). Equally spaced nodes, on the other hand, lead to exponential growth of the Lebesgue constants inasmuch as $\Lambda_n \sim 2^{n+1}/(en \log n)$ for $n \to \infty$; see Trefethen and Weideman [1991] for some history on this result and Brutman [1997a] for a recent survey on Lebesgue constants. The very last statement of §2.1 is the content of Faber's Theorem (see, e.g., Natanson [1965, Vol. 3, Ch. 2, Thm. 2]), which says that, no matter how one chooses the triangular array of nodes (2.16) in $[a, b]$, there is always a continuous function $f \in C[a, b]$ for which the Lagrange interpolation process does not converge uniformly to f. Indeed, there is an $f \in C[a, b]$ for which Lagrange interpolation diverges almost everywhere in $[a, b]$; see Erdös and Vértesi [1980]. Compare this with Fejér-Hermite interpolation.

§2.3. A more complete discussion of how the convergence domain of Lagrange interpolation in the complex plane depends on the limit distribution of the interpolation nodes can be found in Krylov [1962, Ch. 12, §2].

Runge's example is further elucidated in Epperson [1987]. For an analysis of Bernstein's example, we refer to Natanson [1965, Vol. 3, Ch. 2, §2]. The same divergence phenomenon, incidentally, is exhibited also for a large class of nonequally spaced nodes; see Brutman and Passow [1995]. The proof of Example (5) follows Fejér [1918].

§2.4. The Chebyshev polynomial arguably is one of the most interesting polynomials from the point of view not only of approximation theory, but also of algebra and number theory. In Rivlin's words, it " ... is like a fine jewel that reveals different characteristics under illumination from various positions." In his text, Rivlin [1990] gives ample testimony in support of this view. Another text, unfortunately available only in Russian (or Polish), is Paszkowski [1983], which has an exhaustive account of analytic properties of Chebyshev polynomials as well as numerical applications.

The convergence result stated in (2.49) follows from (2.11) and the logarithmic growth of Λ_n, since $\mathcal{E}_n(f) \log n \to 0$ for $f \in C^1[-1, 1]$ by Jackson's theorems (cf. Cheney [1966, p. 147]). A more precise estimate for the error in (2.54) is $\mathcal{E}_n(f) \leq \|\tau_n - f\|_\infty \leq \left(4 + \frac{4}{\pi^2} \log n\right) \mathcal{E}_n(f)$ (Rivlin [1990, Thm. 3.3]), where the infinity norm refers to the interval $[-1, 1]$ and $\mathcal{E}_n(f)$ is the best uniform approximation of f by polynomials of degree n.

§2.5. The algorithm (2.59) for calculating the auxiliary quantities $\lambda_i^{(n)}$ in the barycentric formula is due to Werner [1984]. Scaling and translating the nodes will not affect the numerical properties of the algorithm, since the quantities $\frac{\lambda_i^{(k)}}{\lambda_k^{(k)}}$, which control cancellation, are invariant with respect to any linear transformation of the nodes. Ordering of the nodes, however, does, as was noted experimentally by Werner and also confirmed in Gautschi [1997, §3.1] in connection with the calculation of Cotes numbers (for Newton-Cotes integration).

Barycentric formulae have been developed also for trigonometric interpolation (see Henrici [1979b] for uniform, and Salzer [1949] and Berrut [1984] for nonuniform, distributions of the nodes), and for cardinal (sinc-) interpolation (Berrut [1989]).

§2.7. There are explicit formulae, analogous to Lagrange's formula, for Hermite interpolation in the most general case; see, for example, Stoer and Bulirsch [1993, §2.1.5]. For the important special case $m_k = 2$, see also Ch. 3, Ex. 33.

§2.8. To estimate the error for inverse interpolation, using an appropriate version of (2.12), one needs the derivatives of the inverse function f^{-1}. A general expression for the nth derivative of f^{-1} in terms of the first n derivatives of f is derived in Ostrowski [1973, Appendix C].

§3. The definition of the class of spline functions $\mathbf{S}_m^k(\Delta)$ can be refined to $\mathbf{S}_m^{\mathbf{k}}(\Delta)$, where $\mathbf{k}^T = [k_2, k_3, \ldots, k_{n-1}]$ is a vector with integer components $k_i \geq -1$ specifying the degree of smoothness at the interior knots x_i; that is, $s^{(j)}(x_i + 0) - s^{(j)}(x_i - 0) = 0$ for $j = 0, 1, \ldots, k_i$. Then $\mathbf{S}_m^k(\Delta)$ as defined in (3.3) becomes $\mathbf{S}_m^{\mathbf{k}}(\Delta)$ with $\mathbf{k} = [k, k, \ldots, k]$.

§3.1. As simple as the procedure of piecewise linear interpolation may seem, it can be applied to advantage in numerical Fourier analysis, for example. In trying to compute the (complex) Fourier coefficients $c_n(f) = \frac{1}{2\pi}\int_0^{2\pi} f(x)e^{-inx}dx$ of a 2π-periodic function f, one often approximates them by the "discrete Fourier transform" $\hat{c}_n(f) = \frac{1}{N}\sum_{k=0}^{N-1} f(x_k)e^{-inx_k}$, where $x_k = k\frac{2\pi}{N}$. This can be computed efficiently (for large N) by the Fast Fourier Transform. Note, however, that $\hat{c}_n(f)$ is periodic in n with period N, whereas the true Fourier coefficients $c_n(f)$ tend to zero as $n \to \infty$. To remove this deficiency, one can approximate f by some (simple) function φ and thereby approximate $c_n(f)$ by $c_n(\varphi)$. Then $c_n(\varphi)$ will indeed tend to zero as $n \to \infty$. The simplest choice for φ is precisely the piecewise linear interpolant $\varphi = s_1(f;\,\cdot\,)$ (relative to the uniform partition of $[0, 2\pi]$ into N subintervals). One then finds, rather remarkably (see Ch. 3, Ex. 13), that $c_n(\varphi)$ is a multiple of the discrete Fourier transform, namely, $c_n(f) = \tau_n\hat{c}_n(f)$, where $\tau_n = \left(\frac{\sin(n\pi/N)}{n\pi/N}\right)^2$; this still allows the application of the FFT but corrects the behavior of $\hat{c}_n(f)$ at infinity. The same modification of the discrete Fourier transform by an "attenuation factor" τ_n occurs for many other approximation processes $f \approx \varphi$; see Gautschi [1971/1972] for a general theory (and history) of attenuation factors.

The near optimality of the piecewise linear interpolant $s_1(f;\,\cdot\,)$, as expressed by the inequalities in (3.7), is noted by de Boor [1978, pp. 40–41].

§3.2. The basis (3.8) for $\mathbf{S}_1^0(\Delta)$ is a special case of a B-spline basis that can be defined for any space of spline functions $\mathbf{S}_m^{\mathbf{k}}(\Delta)$ previously introduced (cf. de Boor [1978, Thm. 9.1]. The B-splines are formed by means of divided differences of order $m+1$ applied to the truncated power $(t-x)_+^m$ (considered as a function of t). Like the basis in (3.8), each basis function of a B-spline basis is supported on at most $m+1$ consecutive intervals of Δ and is positive on the interior of the support.

§3.3. A proof of the near optimality of the piecewise linear least squares approximant $\hat{s}_1(f;\cdot)$, as expressed by the inequalities (3.15), can be found in de Boor [1978, pp. 41–44]. For smoothing and least squares approximation procedures involving cubic splines, see, for example, de Boor [1978, Ch. 14].

§3.4. (i) For the remark in the last paragraph of (i), see de Boor [1978, Ch. 4, Problems 3 and 4].

(ii.1) The error bounds in (3.24), which for $r=0$ and $r=1$ are asymptotically sharp, are due to Hall and Meyer [1976].

(ii.2) The cubic spline interpolant matching second derivatives at the endpoints satisfies the same error bounds as in (3.24) for $r = 0, 1, 2$, with constants $c_0 = \frac{3}{64}$, $c_1 = \frac{3}{16}$ and $c_2 = \frac{3}{8}$; see Kershaw [1971, Thm. 2]. The same is shown also for periodic spline interpolants s, satisfying $s^{(r)}(a) = s^{(r)}(b)$ for $r = 0, 1, 2$.

(ii.3) Even though the natural spline interpolant, in general, converges only with order $|\Delta|^2$ (e.g., for uniform partitions Δ), it has been shown by Atkinson [1968] that the order of convergence is $|\Delta|^4$ on any compact interval contained in the open interval (a,b), and by Kershaw [1971] even on intervals extending (in a sense made precise) to $[a,b]$ as $|\Delta| \to 0$. On such intervals, in fact, the natural spline interpolant s provides approximations to any $f \in C^4[a,b]$ with errors satisfying $\|f^{(r)} - s^{(r)}\|_\infty \le 8c_r K|\Delta|^{4-r}$, where $K = 2 + \frac{3}{8}\|f^{(4)}\|_\infty$ and $c_0 = \frac{1}{8}$, $c_1 = \frac{1}{2}$ and $c_2 = 1$.

(ii.4) The error of the "not-a-knot" spline interpolant is of the same order as the error of the complete spline; it follows from Beatson [1986, Eq. (2.1)] that for functions $f \in C^4[a,b]$, one has $\|f^{(r)} - s^{(r)}\|_\infty \le c_r|\Delta|^{4-r}\|f^{(4)}\|_\infty$, $r = 0, 1, 2$ (at least when $n \ge 6$), where c_r are constants independent of f and Δ. The same bounds are valid for other schemes that depend only on function values, for example, the scheme with m_1 equal to the first (or second) derivative of $p_3(f; x_1, x_2, x_3, x_4; \cdot)$ at $x = a$, and similarly for m_n. The first of these schemes (using first-order derivatives of p_3) is in fact the one recommended by Beatson and Chacko [1989], [1992] for general-purpose interpolation. Numerical experiments in Beatson and Chacko [1989] suggest values of approximately 1 for the constants c_r in the preceding error estimates. In Beatson and Chacko [1992] further comparisons are made among many other cubic spline interpolation schemes.

§3.5. The minimum norm property of natural splines (Theorem 2.3.1) and its proof based on the identity (3.28), called "the first integral relation" in Ahlberg, Nilson, and Walsh [1967], is due to Holladay [1957], who derived it in the context of numerical quadrature. "Much of the present-day theory of splines began with this theorem" (Ahlberg, Nilson, and Walsh [1967, p. 3]). An elegant alternative proof of (3.29), and hence of the theorem, can be based (cf. de Boor [1978, pp. 64–66]) on the Peano representation (see Ch. 3, §2.6) of the second divided difference of $g - s$, that is, $[x_{i-1}, x_i, x_{i+1}](g - s) = \int_{\mathbb{R}} K(t)(g''(t) - s''(t))dt$, by noting that the Peano kernel K, up to a constant, is the B-spline B_i defined in (3.8). Since the left-hand side is zero by the interpolation properties of g and s, it follows from the preceding equation that $g'' - s''$ is orthogonal to the span of the B_i, hence to s'', which lies in this span.

EXERCISES AND MACHINE ASSIGNMENTS TO CHAPTER 2

EXERCISES

1. Suppose you want to approximate the function

$$f(t) = \begin{cases} -1 & \text{if} \quad -1 \le t < 0, \\ 0 & \text{if} \quad t = 0, \\ 1 & \text{if} \quad 0 < t \le 1, \end{cases}$$

by a constant function $\varphi(x) = c$:

(a) on [–1,1] in the continuous L_1 norm,

(b) on $\{t_1, t_2, \ldots, t_N\}$ in the discrete L_1 norm,

(c) on [–1,1] in the continuous L_2 norm,

(d) on $\{t_1, t_2, \ldots, t_N\}$ in the discrete L_2 norm,

(e) on [–1,1] in the ∞-norm,

(f) on $\{t_1, t_2, \ldots, t_N\}$ in the discrete ∞-norm.

The weighting in all norms is uniform (i.e., $w(t) \equiv 1$, $w_i = 1$) and $t_i = -1 + \frac{2(i-1)}{N-1}$, $i = 1, 2, \ldots, N$. Determine the best constant c (or constants c, if there is nonuniqueness) and the minimum error.

2. Consider the data

$$f(t_i) = 1, \quad i = 1, 2, \ldots, N-1; \qquad f(t_N) = y \gg 1.$$

(a) Determine the discrete L_∞ approximant to f by means of a constant c (polynomial of degree zero).

(b) Do the same for discrete (equally weighted) least square approximation.

(c) Compare and discuss the results, especially as $N \to \infty$.

3. Let $x_0, x_1, \ldots, x_n$ be pairwise distinct points in $[a, b]$, $-\infty < a < b < \infty$, and $f \in C^1[a, b]$. Show that, given any $\varepsilon > 0$, there exists a polynomial p such that $\|f - p\|_\infty < \varepsilon$ and, at the same time, $p(x_i) = f(x_i)$, $i = 0, 1, \ldots, n$. Here $\|u\|_\infty = \max_{a \le x \le b} |u(x)|$. {*Hint*: Write $p = p_n(f; \cdot) + \omega_n q$, where $p_n(f; \cdot)$ is the interpolation polynomial of degree n (cf. §2.1, Eq. (2.3)), $\omega_n(x) = \prod_{i=0}^{n}(x - x_i)$, $q \in \mathbb{P}$, and apply Weierstrass's approximation theorem.}

4. Consider the function $f(t) = t^\alpha$ on $0 \le t \le 1$, where $\alpha > 0$. Suppose we want to approximate f best in the L_p norm by a constant c, $0 < c < 1$, that is, minimize the L_p error

$$E_p(c) = \|t^\alpha - c\|_p = \left(\int_0^1 |t^\alpha - c|^p dt \right)^{1/p}$$

as a function of c. Find the optimal $c = c_p$ for $p = \infty$, $p = 2$, and $p = 1$, and determine $E_p(c_p)$ for each of these p-values.

5. Taylor expansion yields the simple approximation $e^x \approx 1 + x$, $0 \le x \le 1$. Suppose you want to improve this by seeking an approximation of the form $e^x \approx 1 + cx$, $0 \le x \le 1$, for some suitable c.

   (a) How must c be chosen if the approximation is to be optimal in the (continuous, equally weighted) least squares sense?

   (b) Sketch the error curves $e_1(x) := e^x - (1 + x)$ and $e_2(x) := e^x - (1 + cx)$ with c as obtained in (a) and determine $\max_{0 \le x \le 1} |e_1(x)|$ and $\max_{0 \le x \le 1} |e_2(x)|$.

   (c) Solve the analogous problem (and provide error curves) with three instead of two terms in the modified Taylor expansion: $e^x \approx 1 + c_1 x + c_2 x^2$.

6. Discuss uniqueness and nonuniqueness of the least squares approximant in the case of a discrete set $T = \{t_1, t_2\}$ (i.e., $N = 2$) and $\Phi_n = \mathbb{P}_{n-1}$ (polynomials of degree $\le n - 1$). In case of nonuniqueness, determine *all* solutions.

7. Determine the least squares approximation

$$\varphi(t) = \frac{c_1}{1+t} + \frac{c_2}{(1+t)^2}\,, \quad 0 \le t \le 1,$$

to the exponential function $f(t) = e^{-t}$, assuming $d\lambda(t) = dt$ on $[0,1]$. Determine the condition number $\text{cond}_\infty A = \|A\|_\infty \, \|A^{-1}\|_\infty$ of the coefficient matrix A of the normal equations. Calculate the error $f(t) - \varphi(t)$ at $t = 0$, $t = 1/2$, and $t = 1$. {*Point of information*: The integral $\int_1^\infty t^{-m} e^{-xt} dt = E_m(x)$ is known as the "mth exponential integral".}

8. Approximate the circular quarter arc γ given by the equation $y(t) = \sqrt{1 - t^2}$, $0 \le t \le 1$ (see figure) by a straight line ℓ in the least squares sense, using either the weight function $w(t) = (1 - t^2)^{-1/2}$, $0 \le t \le 1$, or $w(t) = 1$, $0 \le t \le 1$. Where does ℓ intersect the coordinate axes in these two cases? {*Points of information*: $\int_0^{\pi/2} \cos^2 \theta d\theta = \frac{\pi}{4}$, $\int_0^{\pi/2} \cos^3 \theta d\theta = \frac{2}{3}$.}

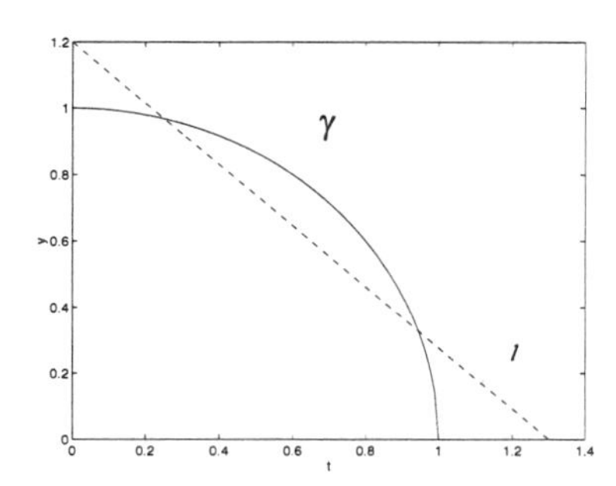

9. (a) Let the class Φ_n of approximating functions have the following properties. Each $\varphi \in \Phi_n$ is defined on an interval $[a, b]$ symmetric with respect to the origin (i.e., $a = -b$), and $\varphi(t) \in \Phi_n$ implies $\varphi(-t) \in \Phi_n$. Let $d\lambda(t) = \omega(t)dt$, with $\omega(t)$ an even function on $[a, b]$ (i.e., $\omega(-t) = \omega(t)$). Show: if f is an even function on $[a, b]$, then so is its least squares approximant, $\hat{\varphi}_n$, on $[a, b]$ from Φ_n.

   (b) Consider the "hat function" $f(t) = \begin{cases} 1 - t & \text{if } \ 0 \le t \le 1, \\ 1 + t & \text{if } \ -1 \le t \le 0. \end{cases}$

   Determine its least squares approximation on [−1,1] by a quadratic function. (Use $d\lambda(t) = dt$.) Simplify your calculation by using part (a). Determine where the error vanishes.

10. Suppose you want to approximate the step function

$$f(t) = \begin{cases} 1 & \text{if } 0 \le t \le 1, \\ 0 & \text{if } t > 1 \end{cases}$$

on the positive line $\mathbb{R}_+$ by a linear combination of exponentials $\pi_j(t) = e^{-jt}$, $j = 1, 2, \ldots, n$, in the (continuous, equally weighted) least squares sense.

   (a) Derive the normal equations. How is the matrix related to the Hilbert matrix?

   (b) If you have access to MATLAB or Linpack, solve the normal equations for $n = 1, 2, \ldots, 8$. Print n, the Euclidean condition number of the matrix, along with the solution. Plot the approximations vs. the exact function for $1 \le n \le 4$.

11. Let $\pi_j(t) = (t - a_j)^{-1}$, $j = 1, 2, \ldots, n$, where a_j are distinct real numbers with $|a_j| > 1$, $j = 1, 2, \ldots, n$. For $d\lambda(t) = dt$ on $-1 \le t \le 1$ and $d\lambda(t) = 0$, $t \notin [-1, 1]$, determine the matrix of the normal equations for the least squares problem $\int_{\mathbb{R}} (f - \varphi)^2 d\lambda(t) = \min$, $\varphi = \sum_{j=1}^{n} c_j \pi_j$. Can the sytem $\{\pi_j\}_{j=1}^{n}$, $n > 1$, be an orthogonal system for suitable choices of the constants a_j? Explain.

12. Define $\{\pi_j(t)\}_{j=0}^n$ on $0 \le t \le 1$ as follows:

$$\pi_0(t) = \begin{cases} 1 - nt & \text{if } 0 \le t \le \frac{1}{n}, \\ 0 & \text{otherwise}, \end{cases}$$

$$\pi_j(t) = \begin{cases} nt - (j-1) & \text{if } \frac{j-1}{n} \le t \le \frac{j}{n}, \\ j + 1 - nt & \text{if } \frac{j}{n} \le t \le \frac{j+1}{n}, \\ 0 & \text{otherwise}, \end{cases} \quad j = 1, \dots, n-1,$$

$$\pi_n(t) = \begin{cases} nt - n + 1 & \text{if } \frac{n-1}{n} \le t \le 1, \\ 0 & \text{otherwise}. \end{cases}$$

(a) Draw a picture of these functions. Describe in words the meaning of a linear combination $\pi(t) = \sum_{j=0}^n c_j \pi_j(t)$.

(b) Determine $\pi_j(k/n)$ for $j,\ k = 0, 1, \dots, n$.

(c) Show that the system $\{\pi_j(t)\}_{j=0}^n$ is linearly independent on the interval $0 \le t \le 1$. Is it also linearly independent on the set of points $0, \frac{1}{n}, \frac{2}{n}, \dots, \frac{n-1}{n}, 1$? Explain.

(d) Compute the matrix of the normal equations for $\{\pi_j\}$, assuming $d\lambda(t) = dt$ on [0,1]. That is, compute the $(n+1) \times (n+1)$ matrix $A = [a_{ij}]$, where $a_{ij} = \int_0^1 \pi_i(t)\pi_j(t)dt$.

13. Even though the function $f(t) = \ln(1/t)$ becomes infinite as $t \to 0$, it can be approximated on [0,1] arbitrarily well by polynomials of sufficiently high degree in the (continuous, equally weighted) least squares sense. Show this by proving

$$e_{n,2} := \min_{p \in \mathbb{P}_n} \|f - p\|_2 = \frac{1}{n+1}.$$

{*Hint*: Use the following known facts about the "shifted" Legendre polynomial $\pi_j(t)$ of degree j (orthogonal on [0,1] with respect to the weight function $w \equiv 1$ and normalized to satisfy $\pi_j(1) = 1$):

$$\int_0^1 \pi_j^2(t)dt = \frac{1}{2j+1}, \quad j \ge 0; \qquad \int_0^1 \pi_j(t)\ln(1/t)dt = \begin{cases} 1 & \text{if } j = 0, \\ \frac{(-1)^j}{j(j+1)} & \text{if } j > 0. \end{cases}$$

The first relation is well known from the theory of orthogonal polynomials; the second is due to J.L. BLUE, 1979.}

14. Let $d\lambda$ be a continuous (positive) measure on $[a,b]$ and $n \geq 1$ a given integer. Assume f continuous on $[a,b]$ and not a polynomial of degree $\leq n-1$. Let $\hat{p}_{n-1} \in \mathbb{P}_{n-1}$ be the least squares approximant to f on $[a,b]$ from polynomials of degree $\leq n-1$:

$$\int_a^b [\hat{p}_{n-1}(t) - f(t)]^2 d\lambda(t) \leq \int_a^b [p(t) - f(t)]^2 d\lambda(t), \qquad \text{all } p \in \mathbb{P}_{n-1}.$$

Prove: the error $e_n(t) = \hat{p}_{n-1}(t) - f(t)$ changes sign at least n times in $[a,b]$. {*Hint*: Assume the contrary and develop a contradiction.}

15. Let f be a given function on [0,1] satisfying $f(0) = 0,\ f(1) = 1$.

    (a) Reduce the problem of approximating f on [0,1] in the (continuous, equally weighted) least squares sense by a quadratic polynomial p satisfying $p(0) = 0,\ p(1) = 1$ to an unconstrained least squares problem (for a different function).

    (b) Apply the result of (a) to $f(t) = t^r,\ r > 2$. Plot the approximation against the exact function for $r = 3$.

16. Suppose you want to approximate $f(t)$ on $[a,b]$ by a function of the form $r(t) = \pi(t)/q(t)$ in the least squares sense with weight function w, where $\pi \in \mathbb{P}_n$ and q is a *given* function (e.g., a polynomial) such that $q(t) > 0$ on $[a,b]$. Formulate this problem as an ordinary polynomial least squares problem for an appropriate new function $\overline{f}$ and new weight function $\overline{w}$.

17. The Bernstein polynomials of degree n are defined by

$$B_j^n(t) = \binom{n}{j} t^j (1-t)^{n-j}, \qquad j = 0, 1, \ldots, n,$$

and are usually employed on the interval $0 \leq t \leq 1$.

    (a) Show that $B_0^n(0) = 1$, and for $j = 1, 2, \ldots, n$

$$\left.\frac{d^r}{dt^r} B_j^n(t)\right|_{t=0} = 0, \quad r = 0, 1, \ldots, j-1; \qquad \left.\frac{d^j}{dt^j} B_j^n(t)\right|_{t=0} \neq 0.$$

    (b) What are the analogous properties at $t = 1$, and how are they most easily derived?

    (c) Give a rough plot of the fourth-degree polynomials $B_j^4(t),\ j = 0, 1, \ldots, 4$, $0 \leq t \leq 1$.

(d) Use (a) to show that the system $\{B_j^n(t)\}_{j=0}^n$ is linearly independent on [0,1] and spans the space $\mathbb{P}_n$.

(e) Show that $\sum_{j=0}^n B_j^n(t) \equiv 1$. {*Hint*: Use the binomial theorem.}

18. Prove that, if $\{\pi_j\}_{j=1}^n$ is linearly dependent on the support of $d\lambda$, then the matrix $A = [a_{ij}]$, where $a_{ij} = (\pi_i, \pi_j)_{d\lambda} = \int_{\mathbb{R}} \pi_i(t)\pi_j(t)d\lambda(t)$, is singular.

19. Given the recursion relation $\pi_{k+1}(t) = (t-\alpha_k)\pi_k(t) - \beta_k\pi_{k-1}(t)$, $k = 0, 1, 2, \ldots,$ for the (monic) orthogonal polynomials $\{\pi_j(\,\cdot\,; d\lambda)\}$, and defining $\beta_0 = \int_{\mathbb{R}} d\lambda(t)$, show that $\|\pi_k\|^2 = \beta_0\beta_1 \cdots \beta_k$, $k = 0, 1, 2, \ldots$.

20. (a) Let $\pi_n(\,\cdot\,) = \pi_n(\,\cdot\,; d\lambda)$ be the (monic) orthogonal polynomial of degree n relative to the positive measure $d\lambda$ on $\mathbb{R}$. Show:

$$\int_{\mathbb{R}} \pi_n^2(t; d\lambda)d\lambda(t) \le \int_{\mathbb{R}} p^2(t)d\lambda(t), \qquad \text{all } p \in \mathbb{P}_n^0,$$

where $\mathbb{P}_n^0$ is the class of monic polynomials of degree n. Discuss the case of equality. {*Hint*: Represent p in terms of $\pi_j(\,\cdot\,; d\lambda)$, $j = 0, 1, \ldots, n$.}

(b) If $d\lambda(t) = d\lambda_N(t)$ is a discrete measure with exactly N support points $t_1, t_2, \ldots, t_N$, and $\pi_j(t) = \pi_j(\,\cdot\,; d\lambda_N)$, $j = 0, 1, \ldots, N-1$, are the corresponding (monic) orthogonal polynomials, let $\pi_N(t) = (t-\alpha_{N-1})\pi_{N-1}(t) - \beta_{N-1}\pi_{N-2}(t)$, with α_{N-1}, β_{N-1} defined as in §1.4(2). Show that $\pi_N(t_j) = 0$ for $j = 1, 2, \ldots, N$.

21. Let $(u, v) = \sum_{k=1}^N w_k u(t_k)v(t_k)$ be a discrete inner product on the interval [–1,1] with $-1 \le t_1 < t_2 < \cdots < t_N \le 1$, and let α_k, β_k be the recursion coefficients for the (monic) orthogonal polynomials $\{\pi_k(t)\}_{k=0}^{N-1}$ associated with (u, v):

$$\begin{cases} \pi_{k+1}(t) = (t-\alpha_k)\pi_k(t) - \beta_k\pi_{k-1}(t), \\ \qquad\qquad\qquad\qquad k = 0, 1, 2, \ldots, N-2, \\ \pi_0(t) = 1, \quad \pi_{-1}(t) = 0. \end{cases}$$

Let $x = \frac{b-a}{2}t + \frac{a+b}{2}$ map the interval [–1,1] to $[a, b]$, and the points $t_k \in [-1, 1]$ to $x_k \in [a, b]$. Define $(u, v)^* = \sum_{k=1}^N w_k u(x_k)v(x_k)$, and let $\{\pi_k^*(x)\}_{k=0}^{N-1}$ be the (monic) orthogonal polynomials associated with $(u, v)^*$. Express the recursion coefficients α_k^*, β_k^* for the $\{\pi_k^*\}$ in terms of those for $\{\pi_k\}$. {*Hint*: First show that $\pi_k^*(x) = (\frac{b-a}{2})^k \pi_k(\frac{2}{b-a}(x - \frac{a+b}{2}))$.}

22. Let

$$(\star)\qquad \begin{cases} \pi_{k+1}(t) = (t-\alpha_k)\pi_k(t) - \beta_k\pi_{k-1}(t), \\ \qquad\qquad\qquad\qquad k = 0, 1, 2, \ldots\ , \\ \pi_0(t) = 1, \quad \pi_{-1}(t) = 0 \end{cases}$$

and consider

$$p_n(t) = \sum_{j=0}^{n} c_j \pi_j(t).$$

Show that p_n can be computed by the following algorithm (*Clenshaw's algorithm*):

$$(\star\star)\qquad \begin{cases} u_n = c_n, \quad u_{n+1} = 0, \\ u_k = (t-\alpha_k)u_{k+1} - \beta_{k+1}u_{k+2} + c_k, \\ \qquad\qquad\qquad k = n-1, n-2, \ldots, 0, \\ p_n = u_0. \end{cases}$$

{*Hint*: Write $(\star)$ in matrix form in terms of the vector $\pi^T = [\pi_0, \pi_1, \ldots, \pi_n]$ and a unit triangular matrix. Do likewise for $(\star\star)$.}

23. Show that the elementary Lagrange interpolation polynomials $\ell_i(x)$ are invariant with respect to any linear transformation of the independent variable.

24. Prepare plots of the Lebesgue function for interpolation, $\lambda_n(x)$, $-1 \le x \le 1$, for $n = 5, 10, 20$, with the interpolation nodes x_i being given by

    (a) $x_i = -1 + \frac{2i}{n}$, $\quad i = 0, 1, 2, \ldots, n$;

    (b) $x_i = \cos\frac{2i+1}{2n+2}\pi$, $\quad i = 0, 1, 2, \ldots, n$.

    Compute $\lambda_n(x)$ on a grid obtained by dividing each interval $[x_{i-1}, x_i]$, $i = 1, 2, \ldots, n$, into 20 equal subintervals. Plot $\log_{10}\lambda_n(x)$ in case (a), and $\lambda_n(x)$ in case (b). Comment on the results.

25. Let $\omega_n(x) = \prod_{k=0}^{n}(x-k)$ and denote by x_n the location of the extremum of ω_n on [0,1], that is, the unique x in [0,1], where $\omega_n'(x) = 0$.

    (a) Prove or disprove that $x_n \to 0$ as $n \to \infty$.

    (b) Investigate the monotonicity of x_n as n increases.

26. Consider equidistant sampling points $x_k = k$ $(k = 0, 1, \ldots, n)$ and $\omega_n(x) = \prod_{k=0}^{n}(x-k)$.

(a) Show that $\omega_n(x) = (-1)^{n+1}\omega_n(n-x)$. What kind of symmetry does this imply?

(b) Show that $|\omega_n(x)| < |\omega_n(x+1)|$ for nonintegral $x > (n-1)/2$.

(c) Show that the relative maxima of $|\omega_n(x)|$ increase monotonically (from the center of $[0,n]$ outward).

27. Let

$$\lambda_n(x) = \sum_{i=0}^{n} |\ell_i(x)|$$

be the Lebesgue function for polynomial interpolation at the distinct points $x_i \in [a,b]$, $i = 0,1,\dots,n$, and $\Lambda_n = \|\lambda_n\|_\infty = \max_{a\le x\le b} |\lambda_n(x)|$ the Lebesgue constant. Let $p_n(f;\,\cdot\,)$ be the polynomial of degree $\le n$ interpolating f at the nodes x_i. Show that in the inequality

$$\|p_n(f;\,\cdot\,)\|_\infty \le \Lambda_n \|f\|_\infty, \quad \text{all } f \in C[a,b],$$

equality can be attained for some $f = \varphi \in C[a,b]$. {*Hint*: Let $\|\lambda_n\|_\infty = \lambda_n(x_\infty)$; take $\varphi \in C[a,b]$ piecewise linear and such that $\varphi(x_i) = \operatorname{sgn} \ell_i(x_\infty)$, $i = 0,1,\dots,n$.}

28. (a) Let $x_0, x_1, \dots, x_n$ be $n+1$ distinct points in $[a,b]$ and $f_i = f(x_i)$, $i = 0,1,\dots,n$, for some function f. Let $f_i^* = f_i + \varepsilon_i$, where $|\varepsilon_i| \le \varepsilon$. Use the Lagrange interpolation formula to show that $|p_n(f^*;x) - p_n(f;x)| \le \varepsilon\lambda_n(x)$, $a \le x \le b$, where $\lambda_n(x)$ is the Lebesgue function (cf. Ex. 27).

(b) Show: $\lambda_n(x_j) = 1$ for $j = 0,1,\dots,n$.

(c) For quadratic interpolation at three equally spaced points, show that $\lambda_2(x) \le 1.25$ for any x between the three points.

(d) Obtain $\lambda_2(x)$ for $x_0 = 0$, $x_1 = 1$, $x_2 = p$, where $p \gg 1$, and determine $\max_{1\le x\le p} \lambda_2(x)$. How fast does this maximum grow with p? {*Hint*: To simplify the algebra, note from (b) that $\lambda_2(x)$ on $1 \le x \le p$ must be of the form $\lambda_2(x) = 1 + c(x-1)(p-x)$ for some constant c.}

29. In a table of the Bessel function $J_0(x) = \frac{1}{\pi}\int_0^\pi \cos(x\sin\theta)d\theta$, where x is incremented in steps of size h, how small must h be chosen if the table is to be "linearly interpolable" with an error less that 10^{-6} in absolute value? {*Point of information*: $\int_0^{\pi/2} \sin^2\theta d\theta = \frac{\pi}{4}$.}

30. Suppose you have a table of the logarithm function $\ln x$ for positive integer values of x, and you compute $\ln 11.1$ by quadratic interpolation at $x_0 = 10$, $x_1 = 11$, $x_2 = 12$. Estimate the relative error incurred.

31. The "Airy function" $y(x) = \mathrm{Ai}(x)$ is a solution of the differential equation $y'' = xy(x)$ satisfying appropriate initial conditions. It is known that $\mathrm{Ai}(x)$

on $[0, \infty)$ is monotonically decreasing to zero and $\text{Ai}'(x)$ monotonically increasing to zero. Suppose you have a table of Ai and Ai′ (with tabular step h) and you want to interpolate

(a) linearly between x_0 and x_1,

(b) quadratically between x_0, x_1, and x_2,

where x_0, $x_1 = x_0 + h$, $x_2 = x_0 + 2h$ are (positive) tabular arguments. Determine close upper bounds for the respective errors in terms of quantities $y_k = y(x_k)$, $y'_k = y'(x_k)$, $k = 0, 1, 2$, contained in the table.

32. The error in linear interpolation of f at x_0, x_1 is known to be

$$f(x) - p_1(f; x) = (x - x_0)(x - x_1)\,\frac{f''(\xi(x))}{2}\,, \qquad x_0 < x < x_1,$$

if $f \in C^2[x_0, x_1]$. Determine $\xi(x)$ explicitly in the case $f(x) = \frac{1}{x}$, $x_0 = 1$, $x_1 = 2$, and find $\max_{1\le x\le 2} \xi(x)$ and $\min_{1\le x\le 2} \xi(x)$.

33. (a) Let $p_n(f; x)$ be the interpolation polynomial of degree $\le n$ interpolating $f(x) = e^x$ at the points $x_i = i/n$, $i = 0, 1, 2, \ldots, n$. Derive an upper bound for

$$\max_{0\le x\le 1} |e^x - p_n(f; x)|,$$

and determine the smallest n guaranteeing an error less than 10^{-6} on [0, 1]. {*Hint*: First show that for any integer i with $0 \le i \le n$ one has $\max_{0\le x\le 1} |(x - \frac{i}{n})(x - \frac{n-i}{n})| \le \frac{1}{4}$.}

(b) Solve the analogous problem for the nth-degree Taylor polynomial $t_n(x) = 1 + x + \frac{x^2}{2!} + \cdots + \frac{x^n}{n!}$, and compare the result with the one in (a).

34. Let $x_0 < x_1 < x_2 < \cdots < x_n$ and $H = \max_{0\le i\le n-1}(x_{i+1} - x_i)$. Defining $\omega_n(x) = \prod_{i=0}^{n}(x - x_i)$, find an upper bound for $\|\omega_n\|_\infty = \max_{x_0\le x\le x_n} |\omega_n(x)|$ in terms of H and n. {*Hint*: Assume $x_j \le x \le x_{j+1}$ for some $0 \le j < n$ and estimate $(x - x_j)(x - x_{j+1})$ and $\prod_{\substack{i\ne j\\ i\ne j+1}} (x - x_i)$ separately.}

35. Show that the power x^n on the interval $-1 \le x \le 1$ can be uniformly approximated by a linear combination of powers $1, x, x^2, \ldots, x^{n-1}$ with error $\le 2^{-(n-1)}$. In this sense, the powers of x become "less and less linearly independent" on [–1,1] as the exponents grow.

36. Determine

$$\min \max_{a\le x\le b} |a_0x^n + a_1x^{n-1} + \cdots + a_n|, \quad n \ge 1,$$

where the minimum is taken over all real $a_0, a_1, \ldots, a_n$ with $a_0 \ne 0$.

37. Let $a > 1$ and $\mathbb{P}_n^a = \{p \in \mathbb{P}_n : p(a) = 1\}$. Define $\hat{p}_n \in \mathbb{P}_n^a$ by $\hat{p}_n(x) = \frac{T_n(x)}{T_n(a)}$, where T_n is the Chebyshev polynomial of degree n, and let $\|\cdot\|_\infty$ denote the maximum norm on the interval $[-1, 1]$. Prove:

$$\|\hat{p}_n\|_\infty \le \|p\|_\infty \quad \text{for all} \quad p \in \mathbb{P}_n^a.$$

{*Hint*: Imitate the proof of Theorem 2.2.1.}

38. Let

$$f(x) = \int_5^\infty \frac{e^{-t}}{t - x} dt, \qquad -1 \le x \le 1,$$

and let $p_{n-1}(f; \cdot)$ be the polynomial of degree $\le n - 1$ interpolating f at the n Chebyshev points $x_\nu = \cos(\frac{2\nu-1}{2n}\pi)$, $\nu = 1, 2, \ldots, n$. Derive an upper bound for $\max_{-1\le x\le 1} |f(x) - p_{n-1}(f, x)|$.

39. Let f be a positive function defined on $[a, b]$ and assume

$$\min_{a\le x\le b} |f(x)| = m_0, \qquad \max_{a\le x\le b} |f^{(k)}(x)| = M_k, \quad k = 0, 1, 2, \ldots .$$

(a) Denote by $p_{n-1}(f; \cdot)$ the polynomial of degree $\le n - 1$ interpolating f at the n Chebyshev points (relative to the interval $[a, b]$). Estimate the maximum relative error $r_n = \max_{a\le x\le b} |(f(x) - p_{n-1}(f; x))/f(x)|$.

(b) Apply the result of (a) to $f(x) = \ln x$ on $I_r = \{e^r \le x \le e^{r+1}\}$, $r \ge 1$ an integer. In particular, show that $r_n \le \alpha(r, n)c^n$, where $0 < c < 1$ and α is slowly varying. Exhibit c.

(c) (This relates to the function $f(x) = \ln x$ of part (b).) How does one compute $f(\overline{x})$, $\overline{x} \in I_s$, from $f(x)$, $x \in I_r$?

40. (a) For quadratic interpolation on equally spaced points x_0, $x_1 = x_0 + h$, $x_2 = x_0 + 2h$, derive an upper bound for $\|f - p_2(f; \cdot)\|_\infty$ involving $\|f'''\|_\infty$ and h. (Here $\|u\|_\infty = \max_{x_0\le x\le x_2} |u(x)|$.)

(b) Compare the bound obtained in (a) with the analogous one for interpolation at the three Chebyshev points in $[x_0, x_2]$.

41. (a) Suppose the function $f(x) = \ln(2 + x)$, $-1 \le x \le 1$, is interpolated by a polynomial p_n of degree $\le n$ at the Chebyshev points $x_k = \cos\left(\frac{2k+1}{2n+2}\pi\right)$, $k = 0, 1, \ldots, n$. Derive a bound for the maximum error $\|f - p_n\|_\infty = \max_{-1\le x\le 1} |f(x) - p_n(x)|$.

(b) Compare the result of (a) with a bound for $\|f - t_n\|_\infty$, where $t_n(x) = f(0) + \frac{f'(0)}{1!}x + \cdots + \frac{f^{(n)}(0)}{n!}x^n$ is the nth-degree Taylor polynomial of f.

42. Consider $f(t) = \cos^{-1} t$, $-1 \le t \le 1$. Obtain the least squares approximation $\hat{\varphi} \in \mathbb{P}_n$ of f relative to the weight function $w(t) = (1-t)^{-\frac{1}{2}}$; that is, find the solution $\varphi = \hat{\varphi}$ of

$$\text{minimize} \left\{ \int_{-1}^{1} [f(t) - \varphi(t)]^2 \frac{dt}{\sqrt{1-t^2}} : \varphi \in \mathbb{P}_n \right\}.$$

Express φ in terms of Chebyshev polynomials $\pi_j(t) = T_j(t)$.

43. Compute $T_n'(0)$, where T_n is the Chebyshev polynomial of degree n.

44. Prove that the system of Chebyshev polynomials $\{T_k : 0 \le k < n\}$ is orthogonal with respect to the discrete inner product $(u, v) = \sum_{\nu=1}^{n} u(x_\nu)v(x_\nu)$, where x_ν are the Chebyshev points $x_\nu = \cos \frac{2\nu-1}{2n}\pi$.

45. Let $T_k(x)$ denote the Chebyshev polynomial of degree k. Clearly, $T_n(T_m(x))$ is a polynomial of degree $n \cdot m$. Identify it.

46. Let T_n denote the Chebyshev polynomial of degree $n \ge 2$. The equation

$$x = T_n(x)$$

is an algebraic equation of degree n and hence has exactly n roots. Identify them.

47. Let $f(x)$ be defined for all $x \in \mathbb{R}$ and infinitely often differentiable on $\mathbb{R}$. Assume further that

$$|f^{(m)}(x)| \le 1, \quad \text{all } x \in \mathbb{R}, \ m = 1, 2, 3, \ldots .$$

Let $h > 0$ and p_{2n-1} be the polynomial of degree $< 2n$ interpolating f at the $2n$ points $x = kh$, $k = \pm 1, \pm 2, \ldots, \pm n$. For what values of h is it true that

$$\lim_{n\to\infty} p_{2n-1}(0) = f(0)\,?$$

(Note that $x = 0$ is *not* an interpolation node.) Explain why the convergence theory discussed in §2.3 does not apply here. {*Point of information*: $n! \sim \sqrt{2\pi n}(n/e)^n$ as $n \to \infty$ (Stirling's formula).}

48. (a) Let $x_i^C = \cos\left(\frac{2i+1}{2n+2}\pi\right)$, $i = 0, 1, \ldots, n$, be the Chebyshev nodes on $[-1, 1]$. Obtain the analogous Chebyshev points t_i^C on $[a, b]$ (where $a < b$) and find an upper bound of $\prod_{i=0}^{n}(t - t_i^C)$ for $a \le t \le b$.

    (b) Consider $f(t) = \ln t$ on $[a, b]$, $0 < a < b$, and let $p_n(t) = p_n(f; t_0^{(n)}, t_1^{(n)}, \ldots, t_n^{(n)}; t)$. Given $a > 0$, how large can b be chosen such that $\lim_{n\to\infty} p_n(t) = f(t)$ for arbitrary nodes $t_i^{(n)} \in [a, b]$ and arbitrary $t \in [a, b]$?

(c) Repeat (b), but with $t_i^{(n)} = t_i^C$ (see (a)).

49. Let $\mathbb{P}_m^+$ be the set of all polynomials of degree $\leq m$ that are nonnegative on the real line,

$$\mathbb{P}_m^+ = \{p: \quad p \in \mathbb{P}_m, \ p(x) \geq 0 \text{ for all } x \in \mathbb{R}\}.$$

Consider the following interpolation problem: find $p \in \mathbb{P}_m^+$ such that $p(x_i) = f_i$, $i = 0, 1, \ldots, n$, where $f_i \geq 0$ and x_i are distinct points on $\mathbb{R}$.

(a) Show that, if $m = 2n$, the problem admits a solution for arbitrary $f_i \geq 0$.

(b) Prove: if a solution is to exist for arbitrary $f_i \geq 0$, then, necessarily, $m \geq 2n$. {*Hint*: Consider $f_0 = 1$, $f_1 = f_2 = \cdots = f_n = 0$.}

50. Defining forward differences by $\Delta f(x) = f(x+h) - f(x)$, $\Delta^2 f(x) = \Delta(\Delta f(x)) = f(x+2h) - 2f(x+h) + f(x)$, and so on, show that

$$\Delta^k f(x) = k! h^k [x_0, x_1, \ldots, x_k] f,$$

where $x_j = x + jh$, $j = 0, 1, 2, \ldots$. Prove an analogous formula for backward differences.

51. Let $f(x) = x^7$. Compute the fifth divided difference $[0,1,1,1,2,2]f$ of f. It is known that this divided difference is expressible in terms of the fifth derivative of f evaluated at some ξ, $0 < \xi < 2$. Determine ξ.

52. Show that

$$\frac{\partial}{\partial x_0}[x_0, x_1, \ldots, x_n]f = [x_0, x_0, x_1, \ldots, x_n]f,$$

assuming f is differentiable at x_0. What about the partial derivative with respect to one of the other variables?

53. (a) For $n+1$ distinct nodes x_ν, show that

$$[x_0, x_1, \ldots, x_n]f = \sum_{\nu=0}^{n} \frac{f(x_\nu)}{\prod_{\mu \neq \nu}(x_\nu - x_\mu)}.$$

(b) Show that

$$[x_0, x_1, \ldots, x_n](f g_j) = [x_0, x_1, \ldots, x_{j-1}, x_{j+1}, \ldots, x_n]f,$$

where $g_j(x) = x - x_j$.

54. (Mikeladze, 1941) Show that

$$[\underbrace{x_0, x_0, \ldots, x_0}_{m \text{ times}}, x_1, x_2, \ldots, x_n]f$$

$$= \frac{[\overbrace{x_0, \ldots, x_0}^{m \text{ times}}]f}{\prod_{\mu=1}^{n}(x_0 - x_\mu)} + \sum_{\nu=1}^{n} \frac{[\overbrace{x_0, \ldots, x_0}^{(m-1) \text{ times}}, x_\nu]f}{\prod_{\substack{\mu=0 \\ \mu \neq \nu}}^{n}(x_\nu - x_\mu)}.$$

{*Hint*: Use induction on m.}

55. Determine the number of additions and the number of multiplications/divisions required to

    (a) compute all divided differences for $n+1$ data points;

    (b) compute (efficiently) Newton's polynomial, once the divided differences are available.

    Compare these two counts with the analogous ones for the barycentric formula.

56. Consider the data $f(0) = 5$, $f(1) = 3$, $f(3) = 5$, $f(4) = 12$.

    (a) Use Newton's method to obtain the appropriate interpolation polynomial $p_3(f; x)$.

    (b) The data suggest that f has a minimum between $x = 1$ and $x = 3$. Find an approximate value for the location $x_{\min}$ of the minimum.

57. Let $f(x) = (1+a)^x$, $|a| < 1$. Show that $p_n(f; 0, 1, \ldots, n; x)$ is the truncation of the binomial series for f to $n+1$ terms. {*Hint*: Use Newton's form of the interpolation polynomial.}

58. Suppose f is a function on [0,3] for which one knows that

    $$f(0) = 1, \quad f(1) = 2, \quad f'(1) = -1, \quad f(3) = f'(3) = 0.$$

    (a) Estimate $f(2)$, using Hermite interpolation.

    (b) Estimate the maximum possible error of the answer given in (a) if one knows, in addition, that $f \in C^5[0,3]$ and $|f^{(5)}(x)| \leq M$ on [0,3]. Express the answer in terms of M.

59. (a) Use Hermite interpolation to find a polynomial of lowest degree satisfying $p(-1) = p'(-1) = 0$, $p(0) = 1$, $p(1) = p'(1) = 0$. Simplify your expression for p as much as possible.

    (b) Suppose the polynomial p of (a) is used to approximate the function $f(x) = [\cos(\pi x/2)]^2$ on $-1 \leq x \leq 1$.

    (b1) Express the error $e(x) = f(x) - p(x)$ (for some *fixed* x in $[-1, 1]$) in terms of an appropriate derivative of f.

(b2) Find an upper bound for $|e(x)|$ (still for a *fixed* $x \in [-1,1]$).
(b3) Estimate $\max_{-1 \le x \le 1} |e(x)|$.

60. Consider the problem of finding a polynomial $p \in \mathbb{P}_n$ such that

$$p(x_0) = f_0, \quad p'(x_i) = f_i', \quad i = 1, 2, \ldots, n,$$

where x_i, $i = 1, 2, \ldots, n$, are distinct nodes. (It is not excluded that $x_1 = x_0$.) This is neither a Lagrange nor a Hermite interpolation problem (why not?). Nevertheless, show that the problem has a unique solution and describe how it can be obtained.

61. Let

$$f(t) = \begin{cases} 0 & \text{if} \quad 0 \le t \le \frac{1}{2}, \\ 1 & \text{if} \quad \frac{1}{2} \le t \le 1. \end{cases}$$

(a) Find the linear least squares approximant $\hat{p}_1$ to f on $[0,1]$, that is, the polynomial $p_1 \in \mathbb{P}_1$ for which

$$\int_0^1 [p_1(t) - f(t)]^2 dt = \min.$$

Use the normal equations with $\pi_0(t) = 1, \ \pi_1(t) = t$.

(b) Can you do better with continuous piecewise linear functions (relative to the partition $[0,1] = [0, \frac{1}{2}] \cup [\frac{1}{2}, 1]$)? Use the normal equations for the B-spline basis B_0, B_1, B_2.

62. Show that $\mathbb{S}_m^m(\Delta) = \mathbb{P}_m$.

63. Let Δ be the subdivision

$$\Delta = [0,1] \cup [1,2] \cup [2,3]$$

of the interval [0,3]. Define the function s by

$$s(x) = \begin{cases} 2 - x(3 - 3x + x^2) & \text{if} \quad 0 \le x \le 1, \\ 1 & \text{if} \quad 1 \le x \le 2, \\ \frac{1}{4}\, x^2(3 - x) & \text{if} \quad 2 \le x \le 3. \end{cases}$$

To which class $\mathbb{S}_m^k(\Delta)$ does s belong?

64. Determine $p \in \mathbb{P}_3$ in

$$s(x) = \begin{cases} p(x) & \text{if} \quad 0 \le x \le 1, \\ (2 - x)^3 & \text{if} \quad 1 \le x \le 2 \end{cases}$$

such that $s(0) = 0$ and s is a cubic spline in $\mathbb{S}_3^2(\Delta)$ on the subdivision $\Delta = [0,1] \cup [1,2]$ of the interval [0,2]. Do you get a natural spline?

65. Let Δ: $a = x_1 < x_2 < x_3 < \cdots < x_N = b$ be a subdivision of $[a,b]$ into $N-1$ subintervals. What is the dimension of the space $\mathbb{S}_m^k = \{s \in C^k[a,b]\colon s|_{[x_i,x_{i+1}]} \in \mathbb{P}_m,\ i = 1,2,\dots,N-1\}$?

66. Given the subdivision $\Delta:\ a = x_1 < x_2 < \cdots < x_n = b$ of $[a,b]$, determine a basis of "hat functions" for the space $S = \{s \in \mathbb{S}_1^0:\ s(a) = s(b) = 0\}$.

67. Let the subdivision Δ of $[a,b]$ be given by

$$\Delta:\quad a = x_1 < x_2 < x_3 < \cdots < x_{n-1} < x_n = b, \quad n \geq 2,$$

and let $f_i = f(x_i)$, $i = 1,2,\dots,n$, for some function f. Suppose you want to interpolate this data by a quintic spline $s_5(f;\,\cdot\,)$ (a piecewise fifth-degree polynomial of smoothness class $C^4[a,b]$). By counting the number of parameters at your disposal and the number of conditions imposed, state how many additional conditions (if any) you expect are needed to make $s_5(f;\,\cdot\,)$ unique.

68. Let

$$\Delta:\qquad a = x_1 < x_2 < x_3 < \cdots < x_{n-1} < x_n = b.$$

Consider the following problem: given $n-1$ numbers f_ν and $n-1$ points ξ_ν with $x_\nu < \xi_\nu < x_{\nu+1}$ $(\nu = 1,2,\dots,n-1)$, find a piecewise linear function $s \in \mathbb{S}_1^0(\Delta)$ such that

$$s(\xi_\nu) = f_\nu \quad (\nu = 1,2,\dots,n-1), \quad s(x_1) = s(x_n).$$

Representing s in terms of the basis $B_1, B_2, \dots, B_n$ of "hat functions," determine the structure of the linear system of equations that you obtain for the coefficients c_j in $s(x) = \sum_{j=1}^n c_j B_j(x)$. Describe how you would solve the system.

69. Let $s_1(x) = 1 + c(x+1)^3$, $-1 \leq x \leq 0$, where c is a (real) parameter. Determine $s_2(x)$ on $0 \leq x \leq 1$ so that

$$s(x) := \begin{cases} s_1(x) & \text{if} \quad -1 \leq x \leq 0, \\ s_2(x) & \text{if} \quad 0 \leq x \leq 1 \end{cases}$$

is a natural cubic spline on [1,1] with knots at -1, 0, 1. How must c be chosen if one wants $s(1) = -1$?

70. Derive (3.14′).

71. Determine the quantities m_i in the variant of piecewise cubic Hermite interpolation mentioned at the end of §3.4(i).

72. (a) Derive the two extra equations for $m_1, m_2, \ldots, m_n$ that result from the "not-a-knot" condition imposed on the cubic spline interpolant $s \in \mathbb{S}_3^2(\Delta)$ (with Δ as in Ex. 68).

(b) Adjoin the first of these equations to the top and the second to the bottom of the system of $n-2$ equations derived in §3.4(ii). Then apply elementary row operations to produce a tridiagonal system. Display the new matrix elements in the first and last equations, simplified as much as possible.

(c) Is the tridiagonal system so obtained diagonally dominant?

73. Let $\mathbf{S}_1^0(\Delta)$ be the class of piecewise linear functions relative to the subdivision $a = x_1 < x_2 < \cdots < x_n = b$. Let $\|g\|_\infty = \max_{a \le x \le b} |g(x)|$, and denote by $s_1(g;\,\cdot\,)$ the piecewise linear interpolant (from $\mathbf{S}_1^0(\Delta)$) to g.

(a) Show: $\|s_1(g;\,\cdot\,)\|_\infty \le \|g\|_\infty$ for any $g \in C[a,b]$.

(b) Show: $\|f - s_1(f;\,\cdot\,)\|_\infty \le 2\|f - s\|_\infty$ for any $s \in \mathbf{S}_1^0$, $f \in C[a,b]$. {*Hint*: Use additivity of $s_1(f;\,\cdot\,)$ with respect to f.}

(c) Interpret the result in (b) when s is the best uniform spline approximant to f.

MACHINE ASSIGNMENTS

1. (a) A simple-minded approach to best uniform approximation of a function $f(x)$ on [0,1] by a linear function $ax+b$ is to first discretize the problem and then, for various (appropriate) trial values of a, solve the problem of (discrete) uniform approximation of $f(x) - ax$ by a constant b (which admits an easy solution). Write a program to implement this idea.

(b) Run your program for $f(x) = e^x$, $f(x) = 1/(1+x)$, $f(x) = \sin \frac{\pi}{2}x$, $f(x) = x^\alpha$ ($\alpha = 2,3,4,5$). Print the respective optimal values of a and b and the associated minimum error. What do you find particularly interesting in the results (if anything)?

(c) Give a heuristic explanation (and hence exact values) for the results, using the known fact that the error curve for the optimal linear approximation attains its maximum modulus at three consecutive points $0 \le x_0 < x_1 < x_2 \le 1$ with alternating signs (*Principle of Alternation*).

2. (a) Determine the $(n+1) \times (n+1)$ matrix $A = [a_{ij}]$, $a_{ij} = (B_i^n, B_j^n)$, of the normal equations relative to the Bernstein basis

$$B_j^n(t) = \binom{n}{j} t^j (1-t)^{n-j}, \qquad j = 0, 1, \ldots, n,$$

and weight function $w(t) \equiv 1$ on [0,1].

{*Point of information:* $\int_0^1 t^k(1-t)^\ell dt = k!\ell!/(k+\ell+1)!$}.

(b) Use Linpack or MATLAB to solve the normal equations of (a) (in single precision), for $n = 3(3)12$, when the function to be approximated is $f(t) \equiv 1$. What should the exact answer be? Print, for each n, an estimate for the condition number, the computed coefficient vector, and the associated error (absolute value of computed minus exact value). Comment on your results.

3. Compute discrete least squares approximations to the function $f(t) = \sin\left(\frac{\pi}{2}t\right)$ on $0 \le t \le 1$ by approximants of the form

$$\varphi_n(t) = t + t(1-t)\sum_{j=1}^{n} c_j t^{j-1}, \quad n = 1(1)5,$$

using N abscissae $t_k = k/(N+1)$, $k = 1, 2, \ldots, N$. Note that $\varphi_n(0) = 0$, $\varphi_n(1) = 1$ are the exact values of f at $t = 0$ and $t = 1$, respectively. {*Hint*: Approximate $f(t) - t$ by a linear combination of $\pi_j(t) = t(1-t)t^{j-1}$; $j = 1, 2, \ldots, n$.} Write a program that does the computation in both single and double precision, using Linpack to solve the normal equations $Ac = b$, $A = [(\pi_i, \pi_j)]$, $b = [(\pi_i, f - t)]$, $c = [c_j]$.

Output (for each $n = 1, 2, \ldots, 5$):

- The condition number of the system.
- The maximum relative error in the coefficients, $\max_{1\le j\le n} |(c_j^s - c_j^d)/c_j^d|$, where c_j^s are the single-precision values of c_j and c_j^d the double-precision values.
- The minimum and maximum error (in double precision):

$$e_{\min} = \min_{1\le k\le N} |\varphi_n(t_k) - f(t_k)|, \quad e_{\max} = \max_{1\le k\le N} |\varphi_n(t_k) - f(t_k)|.$$

Make two runs:

(a) $N = 5, 10, 20$

(b) $N = 4$

Comment on the results.

4. Write a program for discrete polynomial least squares approximation of a function f defined on [–1,1], using the inner product

$$(u, v) = \frac{2}{N+1}\sum_{i=0}^{N} u(t_i)v(t_i), \quad t_i = -1 + \frac{2i}{N}.$$

Follow these steps.

(a) The recurrence coefficients for the appropriate (monic) orthogonal polynomials $\{\pi_k(t)\}$ are known explicitly:

$$\alpha_k = 0, \quad k = 0, 1, \ldots, N; \qquad \beta_0 = 2,$$

$$\beta_k = \left(1 + \frac{1}{N}\right)^2 \left(1 - \left(\frac{k}{N+1}\right)^2\right) \left(4 - \frac{1}{k^2}\right)^{-1}, \quad k = 1, 2, \ldots, N.$$

(You do not have to prove this.) Define $\gamma_k = \|\pi_k\|^2 = (\pi_k, \pi_k)$.

(b) Using the recurrence formula with coefficients α_k, β_k given in (a), generate an array π of dimension $(N+2, N+1)$ containing $\pi_k(t_\ell)$, $k = 0, 1, \ldots, N+1$; $\ell = 0, 1, \ldots, N$. (Here k is the row index and ℓ the column index.) Define $\mu_k = \max_{0 \le \ell \le N} |\pi_k(t_\ell)|$, $k = 1, 2, \ldots, N+1$. Use both single and double precision.

Output ($N = 10$):

k	β_k	γ_k	μ_{k+1}
0	______	______	______
1	______	______	______
2	______	______	______
.	.	.	.
.	.	.	.
.	.	.	.
$N-1$	______	______	______
N	______	______	______

in `e15.7` resp. `d18.10` format

(c) With $\hat{p}_n = \sum_{k=0}^{n} \hat{c}_k \pi_k(t)$, $n = 0, 1, \ldots, N$, denoting the least squares approximation of degree $\le n$ to the function f on [–1,1], define

$$\|e_n\|_2 = \|\hat{p}_n - f\| = (\hat{p}_n - f, \hat{p}_n - f)^{1/2},$$

$$\|e_n\|_\infty = \max_{0 \le i \le N} |\hat{p}_n(t_i) - f(t_i)|.$$

Using the array π generated in part (b), compute $\hat{c}_n$, $\|e_n\|_2$, $\|e_n\|_\infty$, $n = 0, 1, \ldots, N$, for the following four functions:

$$f(t) = e^{-t}, \quad f(t) = \ln(2+t), \quad f(t) = \sqrt{1+t}, \quad f(t) = |t|.$$

Do the computation both in single and double precision. Be sure you compute $\|e_n\|_2$ as accurately as possible. Comment on your results.

Output ($N = 10$), for each f:

n	$\hat{c}_n$	$\|e_n\|_2$	$\|e_n\|_\infty$
0	______	______	______
1	______	______	______
2	______	______	______
.	.	.	.
.	.	.	.
.	.	.	.
$N-1$	______	______	______
N	______	______	______

in `e15.7` resp. `d18.10` format

5. (a) A Sobolev-type least squares approximation problem results if the inner product is defined by

$$(u,v)=\int_{\mathbb{R}}u(t)v(t)d\lambda_0(t)+\int_{\mathbb{R}}u'(t)v'(t)d\lambda_1(t),$$

where $d\lambda_0$, $d\lambda_1$ are positive measures. What does this type of approximation try to accomplish?

(b) Letting $d\lambda_0(t)=dt$, $d\lambda_1(t)=\lambda dt$ on [0,2], where $\lambda>0$ is a parameter, set up the normal equations for the Sobolev-type approximation in (a) of the function $f(t)=e^{-t^2}$ on [0,2] by means of a polynomial of degree $n-1$. Use the basis $\pi_j(t)=t^{j-1}$, $j=1,2,\dots,n$. {*Hint*: Express the components b_i of the right-hand vector of the normal equations in terms of the "incomplete gamma function" $\gamma(a,x)=\int_0^x t^{a-1}e^{-t}dt$ with $x=4$, $a=i/2$.}

(c) Use MATLAB or Linpack (in double precision) to solve the normal equations for $n=2,3,4,5$ and $\lambda=0,.5,1,2$. Print

$$\|\hat{\varphi}_n-f\|_\infty \quad \text{and} \quad \|\hat{\varphi}_n'-f'\|_\infty, \qquad n=2,3,4,5$$

(or a suitable approximation thereof) along with the condition numbers of the normal equations. {Use the following values for the incomplete gamma function: $\gamma(\frac{1}{2},4)=1.76416278152484$, $\gamma(1,4)=.981684361111266$, $\gamma(\frac{3}{2},4)=.845450112984953$, $\gamma(2,4)=.908421805556329$, $\gamma(\frac{5}{2},4)=1.12165005836756$.} Comment on the results.

6. With $\omega_n(x)=\prod_{k=0}^n(x-k)$, let M_n be the largest, and m_n the smallest, relative maximum of $|\omega_n(x)|$. Calculate M_n/m_n for $n=5,10,15,\dots,30$, to 3 decimal places.

7. (a) Write a subroutine that produces the value of the interpolation polymomial $p_n(f;x_0,x_1,\dots,x_n;t)$ at any real t, where $n\geq 0$ is a given integer, x_i are $n+1$ distinct nodes, and f is any function available in

the form of a function subroutine. Use Newton's interpolation formula and exercise frugality in the use of memory space when generating the divided differences. It is possible, indeed, to generate them "in place" in a single array of dimension $n+1$ that originally contains the values $f(x_i)$, $i = 0, 1, \ldots, n$. {*Hint*: Generate the divided differences from the bottom up.}

(b) Run your routine on the function $f(t) = \frac{1}{1+t^2}$, $-5 \le t \le 5$, using $x_i = -5 + 10\frac{i}{n}$, $i = 0, 1, \ldots, n$, and $n = 2(2)8$ (Runge's example). Plot the polynomials against the exact function.

8. (a) Write a program in single and double precision implementing the algorithm in §2.5, Eq. (2.59), for computing the auxiliary quantities $\lambda_i^{(n)}$ in the barycentric formula. Compute the maximum relative error (over $i = 0, 1, \ldots, n$) in the $\lambda_i^{(n)}$ by comparing single-precision and double-precision results.

(b) Run your program for equally spaced nodes

$$x_i = -1 + 2\,\frac{i}{n}, \qquad i = 0, 1, \ldots, n,$$

in this order, using $n = 5, 10, 15, 20$. Comment on the results. Rearrange the nodes so that they move inward from both ends, by letting, say,

$$x_i = \begin{cases} -1 + \dfrac{i}{n}, & i \text{ even}, \\ 1 - \dfrac{i-1}{n}, & i \text{ odd}, \end{cases} \qquad i = 0, 1, \ldots, n.$$

What do the results suggest? Do you expect scaling or shifting of the nodes to affect the accuracy of the results? Explain.

9. (a) Write a subroutine tri (n, a, b, c, u, v) for solving a tridiagonal (nonsymmetric) system

$$\begin{bmatrix} a_1 & c_1 & & & 0 \\ b_1 & a_2 & c_2 & & \\ & b_2 & a_3 & c_3 & \\ & & \ddots & \ddots \ \ddots & \\ & & & & c_{n-1} \\ 0 & & & b_{n-1} & a_n \end{bmatrix} \begin{bmatrix} u_1 \\ u_2 \\ u_3 \\ \vdots \\ u_{n-1} \\ u_n \end{bmatrix} = \begin{bmatrix} v_1 \\ v_2 \\ v_3 \\ \vdots \\ v_{n-1} \\ v_n \end{bmatrix}$$

by Gauss elimination without pivoting. Keep the program short.

(b) Write a program for computing the natural spline interpolant $s_{\text{nat}}(f;\,\cdot\,)$ on an arbitrary partition $a = x_1 < x_2 < x_3 < \cdots < x_{n-1} < x_n = b$ of $[a, b]$.

Print $\{i, \text{errmax}(i);\ i = 1, 2, \ldots, n-1\}$, where

$$\text{errmax}(i) = \max_{1 \le j \le 51} |s_{\text{nat}}(f; x_{i,j}) - f(x_{i,j})|, \quad x_{i,j} = x_i + \frac{j-1}{50}\,\Delta x_i.$$

(You will need the subroutine tri). Test the program for cases in which the error is zero (what are these, and why?).

(c) Write a second program for computing the complete cubic spline interpolant $s_{\text{compl}}(f;\,\cdot\,)$ by modifying the program in (b) with a minimum of changes. Highlight the changes in the program listing. Apply (and justify) a test similar to that of (b).

(d) Run the programs in (b) and (c) for $[a, b] = [0, 1]$, $n = 11$, and

(i) $x_i = \frac{i-1}{n-1}$, $i = 1, 2, \ldots, n$; $f(x) = e^{-x}$ and $f(x) = x^{5/2}$.

(ii) $x_i = \left(\frac{i-1}{n-1}\right)^2$, $i = 1, 2, \ldots, n$; $f(x) = x^{5/2}$.

Comment on the results.

CHAPTER 3

NUMERICAL DIFFERENTIATION AND INTEGRATION

Differentiation and integration are infinitary concepts of calculus; that is, they are defined by means of a limit process — the limit of the difference quotient in the first instance, the limit of Riemann sums in the second. Since limit processes cannot be carried out on the computer, we must replace them by finite processes. The tools to do so come from the theory of polynomial interpolation (Chapter 2, §2). They not only provide us with approximate formulae for the limits in question, but also permit us to estimate the errors committed and discuss convergence.

§1. Numerical Differentiation

For simplicity, we consider only the first derivative; analogous techniques apply to higher-order derivatives.

The problem can be formulated as follows: for a given differentiable function f, approximate the derivative $f'(x_0)$ in terms of the values of f at x_0 and at nearby points $x_1, x_2, \ldots, x_n$ (not necessarily equally spaced or in natural order). Estimate the error of the approximation obtained.

In §1.1 we solve this problem by means of interpolation. Examples are given in §1.2, and the problematic nature of numerical differentiation in the presence of rounding errors is briefly discussed in §1.3.

§1.1. **A general differentiation formula for unequally spaced points.** The idea is simply to differentiate not $f(\,\cdot\,)$, but its interpolation polynomial $p_n(f; x_0, \ldots, x_n;\,\cdot\,)$. By carrying along the error term of interpolation, we can analyze the error committed.

Thus recall from Chapter 2, §2, that, given the $n+1$ distinct points x_0, $x_1, \ldots, x_n$, we have

$$f(x) = p_n(f; x) + r_n(x), \tag{1.1}$$

where the interpolation polynomial can be written in Newton's form

$$\begin{aligned} p_n(f;x) = f_0 &+ (x-x_0)[x_0,x_1]f + (x-x_0)(x-x_1)[x_0,x_1,x_2]f + \cdots \\ &+ (x-x_0)(x-x_1)\cdots(x-x_{n-1})[x_0,x_1,\ldots,x_n]f, \end{aligned} \tag{1.2}$$

and the error term in the form

$$r_n(x) = (x - x_0)(x - x_1)\cdots(x - x_n)\,\frac{f^{(n+1)}(\xi(x))}{(n+1)!}\,, \tag{1.3}$$

assuming (as we do) that f has a continuous $(n+1)$st derivative in an interval that contains all x_i and x. Differentiating (1.2) with respect to x and then putting $x = x_0$ gives

$$\begin{aligned} p_n'(f;x_0) &= [x_0,x_1]f + (x_0 - x_1)[x_0,x_1,x_2]f + \cdots \\ &+ (x_0 - x_1)(x_0 - x_2)\cdots(x_0 - x_{n-1})[x_0,x_1,\ldots,x_n]f. \end{aligned} \tag{1.4}$$

Similarly, from (1.3) (assuming that f has in fact $n+2$ continuous derivatives in an appropriate interval) we get

$$r_n'(x_0) = (x_0 - x_1)(x_0 - x_2)\cdots(x_0 - x_n)\,\frac{f^{(n+1)}(\xi(x_0))}{(n+1)!}\,. \tag{1.5}$$

Therefore, differentiating (1.1), we find

$$f'(x_0) = p_n'(f;x_0) + e_n, \tag{1.6}$$

where the first term on the right, given by (1.4), represents the desired approximation, and the second,

$$e_n = r_n'(x_0), \tag{1.7}$$

given by (1.5), the respective error. If $H = \max_i |x_0 - x_i|$, we clearly obtain from (1.5) that

$$e_n = O(H^n) \quad \text{as} \quad H \to 0. \tag{1.8}$$

We can thus get approximation formulae of arbitrarily high order, but those with large n are of limited practical use; cf. §1.3.

§1.2 **Examples**. The most important uses of differentiation formulae are made in the discretization of differential equations — ordinary or partial. In these applications, the spacing of the points is usually uniform, but unequally distributed points arise when partial differential operators are to be discretized near the boundary of the domain of interest.

Example 1: $n = 1$, $x_1 = x_0 + h$. Here

$$p_1'(f; x_0) = [x_0, x_1]f = \frac{f_1 - f_0}{h} \,,$$

and (1.6) in conjunction with (1.7) and (1.5) gives

$$f'(x_0) = \frac{f_1 - f_0}{h} + e_1, \quad e_1 = -h\,\frac{f''(\xi)}{2} \,, \tag{1.9}$$

provided $f \in C^3[x_0, x_1]$. (Taylor's formula, actually, shows that $f \in C^2[x_0, x_1]$ suffices.) Thus the error is of $O(h)$ as $h \to 0$.

Example 2: $n = 2$, $x_1 = x_0 + h$, $x_2 = x_0 - h$. We also use the suggestive notation $x_2 = x_{-1}$, $f_2 = f_{-1}$. Here

$$p_2'(f; x_0) = [x_0, x_1]f + (x_0 - x_1)[x_0, x_1, x_2]f. \tag{1.10}$$

The table of divided differences is:

$$\begin{array}{llll} x_{-1} & f_{-1} & & \\ x_0 & f_0 & \dfrac{f_0 - f_{-1}}{h} & \\ x_1 & f_1 & \dfrac{f_1 - f_0}{h} & \dfrac{f_1 - 2f_0 + f_{-1}}{2h^2} \end{array} .$$

Therefore,

$$p_2'(f; x_0) = \frac{f_1 - f_0}{h} - h\,\frac{f_1 - 2f_0 + f_{-1}}{2h^2} = \frac{f_1 - f_{-1}}{2h} \,,$$

and (1.6), (1.7), and (1.5) give, if $f \in C^3[x_{-1}, x_1]$,

$$f'(x_0) = \frac{f_1 - f_{-1}}{2h} + e_2, \quad e_2 = -h^2\,\frac{f'''(\xi)}{6} \,. \tag{1.11}$$

Both approximations (1.9) and (1.11) are difference quotients; the former, however, is "one-sided" whereas the latter is "symmetric." As can be seen, the symmetric difference quotient is one order more accurate than the one-sided difference quotient.

Example 3: $n = 2$, $x_1 = x_0 + h$, $x_2 = x_0 + 2h$. In this case, we have the following table of divided differences,

$$\begin{array}{llll} x_0 & f_0 & & \\ x_1 & f_1 & \dfrac{f_1 - f_0}{h} & \\ x_2 & f_2 & \dfrac{f_2 - f_1}{h} & \dfrac{f_2 - 2f_1 + f_0}{2h^2} \end{array}$$

and (1.10) now gives

$$p_2'(f; x_0) = \frac{f_1 - f_0}{h} - h\,\frac{f_2 - 2f_1 + f_0}{2h^2} = \frac{-f_2 + 4f_1 - 3f_0}{2h} ;$$

hence, by (1.7) and (1.5),

$$f'(x_0) = \frac{-f_2 + 4f_1 - 3f_0}{2h} + e_2, \quad e_2 = h^2\,\frac{f'''(\xi)}{3} . \tag{1.12}$$

Compared to (1.11), this formula also is accurate to $O(h^2)$, but the error is now about twice as large, in modulus, than before. One always pays for destroying symmetry!

Example 4: For a function $u = u(x, y)$ of two variables, approximate $\partial u/\partial x$ "near the boundary."

Consider the points $P_0(x_0, y_0)$, $P_1(x_0 + h, y_0)$, $P_B(x_0 - \beta h, y_0)$, $0 < \beta < 1$ (see Figure 3.1.1); the problem is to approximate $(\partial u/\partial x)(P_0)$ in terms of $u_0 = u(P_0)$, $u_1 = u(P_1)$, $u_B = u(P_B)$.

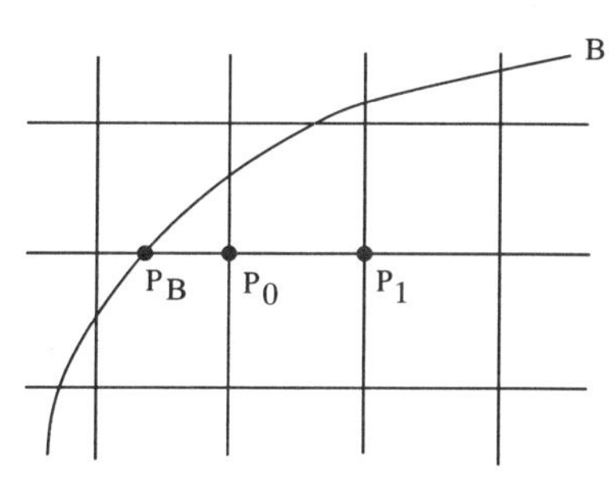

FIGURE 3.1.1. Partial derivative near the boundary

The relevant table of divided differences is:

$$
\begin{array}{llll}
x_B & u_B & & \\
x_0 & u_0 & \dfrac{u_0 - u_B}{\beta h} & \\
x_1 & u_1 & \dfrac{u_1 - u_0}{h} & \dfrac{\beta(u_1 - u_0) - (u_0 - u_B)}{\beta h(1+\beta)h}
\end{array} .
$$

Thus

$$p_2'(u; P_0) = \frac{u_1 - u_0}{h} - h\,\frac{\beta(u_1 - u_0) - (u_0 - u_B)}{\beta h(1+\beta)h} = \frac{\beta^2 u_1 + (1-\beta^2)u_0 - u_B}{\beta(1+\beta)h},$$

and the error is given by

$$e_2 = -\frac{\beta}{6}\,h^2\,\frac{\partial^3 u}{\partial x^3}(\xi, y_0).$$

§1.3. **Numerical differentiation with perturbed data.** Formulae for numerical differentiation become more accurate as the spacing h between evaluation points is made smaller, provided the function to be differentiated is sufficiently smooth. This, however, is true only in theory, since in practice the data are usually inaccurate, if for no other reason than rounding, and the problem of cancellation becomes more acute as h gets smaller. There will be a point of diminishing returns, beyond which the errors increase rather than decrease.

To give a simple analysis of this, take the symmetric differentiation formula (1.11),

$$f'(x_0) = \frac{f_1 - f_{-1}}{2h} + e_2, \quad e_2 = -h^2\,\frac{f'''(\xi)}{6}. \tag{1.13}$$

Suppose now that what are known are not the exact values $f_{\pm 1} = f(x_0 \pm h)$, but slight perturbations of them, say,

$$f_1^* = f_1 + \varepsilon_1, \; f_{-1}^* = f_{-1} + \varepsilon_{-1}, \quad |\varepsilon_{\pm 1}| \le \varepsilon. \tag{1.14}$$

Then our formula (1.13) becomes

$$f'(x_0) = \frac{f_1^* - f_{-1}^*}{2h} - \frac{\varepsilon_1 - \varepsilon_{-1}}{2h} + e_2. \tag{1.15}$$

Here the first term on the right is what we actually compute (assuming, for simplicity, that h is machine-representable and roundoff errors in forming the difference quotient are neglected). The corresponding error, therefore, is

$$E_2 = f'(x_0) - \frac{f_1^* - f_{-1}^*}{2h} = -\frac{\varepsilon_1 - \varepsilon_{-1}}{2h} + e_2$$

and can be estimated by

$$|E_2| \le \frac{\varepsilon}{h} + \frac{M_3}{6}\, h^2, \quad M_3 = \max_{[x_{-1}, x_1]} |f'''|. \tag{1.16}$$

The bound on the right is best possible. It consists of two parts, the term ε/h, which is due to noise in the data, and $\frac{1}{6}M_3h^2$, which is the truncation error introduced by replacing the derivative by a finite difference expression. Their behavior is shown in Figure 3.1.2.

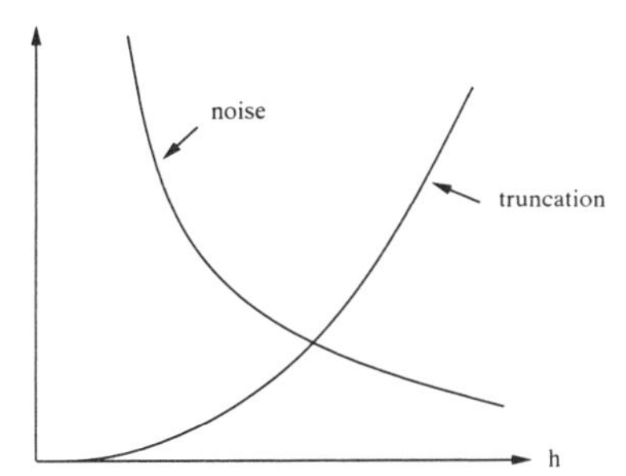

FIGURE 3.1.2. Truncation and noise error in numerical differentiation

If we denote the bound in (1.16) by $E(h)$,

$$E(h) = \frac{\varepsilon}{h} + \frac{M_3}{6}\, h^2, \tag{1.17}$$

then by determining its minimum, one finds

$$E(h) \ge E(h_0), \quad h_0 = \left(\frac{3\varepsilon}{M_3}\right)^{1/3},$$

and

$$E(h_0) = \frac{3}{2}\left(\frac{M_3}{3}\right)^{1/3} \varepsilon^{2/3}. \tag{1.18}$$

This shows that even in the best of circumstances, the error is $O(\varepsilon^{2/3})$, not $O(\varepsilon)$, as one would hope. This represents a significant loss of accuracy.

The same problem persists, indeed is more severe, in higher-order formulae. The only way one can escape from this dilemma is to use not *difference* formulae, but *summation* formulae, that is, integration. But to do this, one has to go into the complex plane and assume that the definition of f can be extended into a domain of the complex plane containing x_0. Then one can use Cauchy's theorem,

$$f'(x_0) = \frac{1}{2\pi i} \oint_C \frac{f(z)}{(z - x_0)^2} dz = \frac{1}{2\pi r} \int_0^{2\pi} e^{-i\theta} f(x_0 + re^{i\theta}) d\theta, \qquad (1.19)$$

in combination with numerical integration (cf. §2). Here C was taken to be a circular contour about x_0 with radius r, with r chosen such that $z = x_0 + re^{i\theta}$ remains in the domain of analyticity of f. Since the result is real, one can replace the integrand by its real part.

§2. Numerical Integration

The basic problem is to calculate the definite integral of a given function f, extended over a finite interval $[a,b]$. If f is well behaved, this is a routine problem for which the simplest integration rules, such as the composite trapezoidal or Simpson's rule (§2.1) will be quite adequate, the former having an edge over the latter if f is periodic with period $b - a$. Complications arise if f has an integrable singularity, or the interval of integration extends to infinity (which is just another manifestation of singular behavior). By breaking up the integral, if necessary, into several pieces, it can be assumed that the singularity, if its location is known, is at one (or both) ends of the interval $[a, b]$. Such "improper" integrals can usually be treated by weighted quadrature; that is, one incorporates the singularity into a weight function, which then becomes one factor of the integrand, leaving the other factor well behaved. The most important example of this is Gaussian quadrature relative to such a weight function (§§2.2 through 2.4). Finally, it is possible to accelerate the convergence of quadrature schemes by suitable recombinations. The best-known example of this is Romberg integration (§2.7).

§2.1. **The composite trapezoidal and Simpson's rules**. These may be regarded as the workhorses of numerical integration. They will do the job when the interval is finite and the integrand unproblematic. The trapezoidal rule is sometimes surprisingly effective even on infinite intervals.

Both rules are obtained by applying the simplest kind of interpolation on subintervals of the decomposition

$$a = x_0 < x_1 < x_2 < \cdots < x_{n-1} < x_n = b, \; x_k = a + kh, \quad h = \frac{b-a}{n} \tag{2.1}$$

of the interval $[a,b]$. In the trapezoidal rule, one interpolates linearly on each subinterval $[x_k, x_{k+1}]$, and obtains

$$\int_{x_k}^{x_{k+1}} f(x)dx = \int_{x_k}^{x_{k+1}} p_1(f;x)dx + \int_{x_k}^{x_{k+1}} R_1(x)dx, \tag{2.2}$$

where

$$p_1(f;x) = f_k + (x - x_k)[x_k, x_{k+1}]f, \; R_1(x) = (x - x_k)(x - x_{k+1}) \frac{f''(\xi(x))}{2}. \tag{2.3}$$

Here $f_k = f(x_k)$, and we assumed that $f \in C^2[a,b]$. The first integral on the right of (2.2) is easily obtained as the area of a trapezoid with "bases" f_k, f_{k+1} and "height" h, or else by direct integration of $p_1(f;\,\cdot\,)$ in (2.3). To the second integral we can apply the Mean Value Theorem of integration, since $(x - x_k)(x - x_{k+1})$ has constant (negative) sign on $[x_k, x_{k+1}]$. The result is

$$\int_{x_k}^{x_{k+1}} f(x)dx = \frac{h}{2}(f_k + f_{k+1}) - \frac{1}{12} h^3 f''(\xi_k), \tag{2.4}$$

where $x_k < \xi_k < x_{k+1}$. This is the *elementary trapezoidal rule.* Summing over all subintervals gives the *composite trapezoidal rule*

$$\int_a^b f(x)dx = h\left(\tfrac{1}{2} f_0 + f_1 + \cdots + f_{n-1} + \tfrac{1}{2} f_n\right) + E_n^T(f), \tag{2.5}$$

with error term

$$E_n^T(f) = -\frac{1}{12} h^3 \sum_{k=0}^{n-1} f''(\xi_k).$$

This is not a particularly elegant expression for the error. We can simplify it by writing

$$E_n^T(f) = -\frac{1}{12} h^2 \cdot (b-a) \left[\frac{1}{n} \sum_{k=0}^{n-1} f''(\xi_k) \right]$$

and noting that the expression in brackets is a mean value of second-derivative values, hence certainly contained between the algebraically smallest and largest value of the second derivative f'' on $[a,b]$. Since the function f'' was assumed continuous on $[a,b]$, it takes on every value between its smallest and largest, in particular, the bracketed value in question, at some interior point, say, ξ, of $[a,b]$. Consequently,

$$E_n^T(f) = -\tfrac{1}{12}(b-a)\,h^2 f''(\xi), \quad a < \xi < b. \tag{2.6}$$

Since f'' is bounded in absolute value on $[a,b]$, this shows that $E_n^T(f) = O(h^2)$ as $h \to 0$. In particular, the composite trapezoidal rule converges as $h \to 0$ (or, equivalently, $n \to \infty$) in (2.5), provided $f \in C^2[a,b]$.

It should be noted that (2.6) holds only for real-valued functions f and cannot be applied to complex-valued functions; cf. (2.10).

One expects an improvement if instead of linear interpolation one uses quadratic interpolation over two consecutive subintervals. This gives rise to the composite Simpson's formula.[1] Its "elementary" version, analogous to (2.4), is

$$\int_{x_k}^{x_{k+2}} f(x)dx = \frac{h}{3}(f_k + 4f_{k+1} + f_{k+2}) - \frac{1}{90}\,h^5 f^{(4)}(\xi_k), \; x_k < \xi_k < x_{k+2}, \tag{2.7}$$

where it has been assumed that $f \in C^4[a,b]$. The remainder term shown in (2.7) does not come about as easily as before in (2.4), since the Mean Value Theorem is no longer applicable, the factor $(x - x_k)\,(x - x_{k+1})\,(x - x_{k+2})$ changing sign at the midpoint of $[x_k, x_{k+2}]$. However, an alternative derivation of (2.7), using Hermite interpolation (cf. Ex. 8), not only produces the desired error term, but also explains its unexpectedly large order $O(h^5)$. If n is even, we can sum up all $n/2$ contributions in (2.7) and obtain the *composite Simpson's rule*,

$$\int_a^b f(x)dx = \frac{h}{3}\,(f_0 + 4f_1 + 2f_2 + 4f_3 + 2f_4 + \cdots + 4f_{n-1} + f_n) + E_n^S(f),$$
$$E_n^S(f) = -\tfrac{1}{180}(b-a)\,h^4 f^{(4)}(\xi), \quad a < \xi < b. \tag{2.8}$$

The error term in (2.8) is the result of a simplification similar to the one previously carried out for the trapezoidal rule (cf. Ex. 8). Comparing it with the one in (2.6), we see that we gained *two* orders of accuracy without any

[1]Thomas Simpson (1710–1761) was an English mathematician, self-educated, and author of many textbooks popular at the time. Simpson published his formula in 1743, but it was already known to Cavalieri (1639), Gregory (1668), and Cotes (1722), among others.

appreciable increase in work (same number of function evaluations). This is the reason why Simpson's rule has long been, and continues to be, one of the most popular general-purpose integration methods.

The composite trapezoidal rule, nevertheless, has its own advantages. Although it integrates exactly polynomials of degree 1 only, it does much better with *trigonometric* polynomials. Suppose, indeed (for simplicity), that the interval $[a,b]$ is $[0,2\pi]$, and denote by $T_m[0,2\pi]$ the class of trigonometric polynomials of degree m,

$$T_m[0,2\pi] = \{t(x):\ t(x) = a_0 + a_1\cos x + a_2\cos 2x + \cdots + a_m\cos mx \\ + b_1\sin x + b_2\sin 2x + \cdots + b_m\sin mx\}.$$

Then

$$E_n^T(f) = 0 \quad \text{for all} \quad f \in T_{n-1}[0,2\pi]. \tag{2.9}$$

This is most easily verified by taking for f the complex exponential $e_\nu(x) = e^{i\nu x}$ $(= \cos\nu x + i\sin\nu x)$, $\nu = 0,1,2,\ldots$:

$$\begin{aligned} E_n^T(e_\nu) &= \int_0^{2\pi} e_\nu(x)dx - \frac{2\pi}{n}\left[\frac{1}{2}e_\nu(0) + \sum_{k=1}^{n-1} e_\nu(k\cdot 2\pi/n) + \frac{1}{2}e_\nu(2\pi)\right] \\ &= \int_0^{2\pi} e^{i\nu x}dx - \frac{2\pi}{n}\sum_{k=0}^{n-1} e^{i\nu k\cdot 2\pi/n}. \end{aligned}$$

When $\nu = 0$, this is clearly zero, and otherwise, since $\int_0^{2\pi} e^{i\nu x}dx = (i\nu)^{-1} \cdot e^{i\nu x}\big|_0^{2\pi} = 0$,

$$E_n^T(e_\nu) = \begin{cases} -2\pi & \text{if } \nu = 0 \pmod n, \quad \nu > 0, \\ -\dfrac{2\pi}{n}\dfrac{1-e^{i\nu n\cdot 2\pi/n}}{1-e^{i\nu\cdot 2\pi/n}} = 0 & \text{if } \nu \neq 0 \pmod n. \end{cases} \tag{2.10}$$

In particular, $E_n^T(e_\nu) = 0$ for $\nu = 0,1,\ldots,n-1$, which proves (2.9). Taking real and imaginary parts in (2.10) gives

$$E_n^T(\cos\nu\,\cdot) = \left\{\begin{matrix} -2\pi, & \nu = 0 \pmod n, \quad \nu \neq 0, \\ 0 & \text{otherwise}, \end{matrix}\right\}, \qquad E_n^T(\sin\nu\,\cdot) = 0.$$

Therefore, if f is 2π-periodic and has a uniformly convergent Fourier expansion

$$f(x) = \sum_{\nu=0}^{\infty} [a_\nu(f) \cos \nu x + b_\nu(f) \sin \nu x], \tag{2.11}$$

where $a_\nu(f)$, $b_\nu(f)$ are the "Fourier coefficients" of f, then

$$\begin{aligned} E_n^T(f) &= \sum_{\nu=0}^{\infty} [a_\nu(f) E_n^T(\cos \nu \cdot) + b_\nu(f) E_n^T(\sin \nu \cdot)] \\ &= -2\pi \sum_{\ell=1}^{\infty} a_{\ell \cdot n}(f). \end{aligned} \tag{2.12}$$

From the theory of Fourier series it is known that the Fourier coefficients of f go to zero faster the smoother f is. More precisely, if $f \in C^r[\mathbb{R}]$, then $a_\nu(f) = O(\nu^{-r})$ as $\nu \to \infty$ (and similarly for $b_\nu(f)$). Since by (2.12), $E_n^T(f) \approx -2\pi a_n(f)$, it follows that

$$E_n^T(f) = O(n^{-r}) \quad \text{as} \quad n \to \infty \quad (f \in C^r[\mathbb{R}], \quad 2\pi\text{-periodic}), \tag{2.13}$$

which, if $r > 2$, is better than $E_n^T(f) = O(n^{-2})$, valid for nonperiodic functions f. In particular, if $r = \infty$, then the trapezoidal rule converges faster than any power of n^{-1}. It should be noted, however, that f must be smooth on the whole real line $\mathbb{R}$. (See (1.19) for an example.) Starting with a function $f \in C^r[0, 2\pi]$ and extending it periodically to $\mathbb{R}$, will *not* in general produce a function $f \in C^r[\mathbb{R}]$.

Another instance in which the composite trapezoidal rule excels is for functions f defined on $\mathbb{R}$ and having the following properties for some $r \geq 1$,

$$\begin{aligned} &f \in C^{2r+1}[\mathbb{R}], \quad \int_{\mathbb{R}} |f^{(2r+1)}(x)| dx < \infty, \\ &\lim_{x \to -\infty} f^{(2\rho-1)}(x) = \lim_{x \to \infty} f^{(2\rho-1)}(x) = 0, \quad \rho = 1, 2, \ldots, r. \end{aligned} \tag{2.14}$$

In this case, it can be shown that

$$\int_{\mathbb{R}} f(x) dx = h \sum_{k=-\infty}^{\infty} f(kh) + E(f; h) \tag{2.15}$$

has an error $E(f; h)$ satisfying $E(f; h) = O(h^{2r+1})$, $h \to 0$. Therefore, again, if (2.14) holds for all $r = 1, 2, 3, \ldots$, then the error goes to zero faster than any power of h.

§2.2. **(Weighted) Newton-Cotes and Gauss formulae.** A weighted quadrature formula is a formula of the type

$$\int_a^b f(t)w(t)dt = \sum_{k=1}^{n} w_k f(t_k) + E_n(f), \tag{2.16}$$

where w is a positive (or at least nonnegative) "weight function," assumed integrable over (a,b). The interval (a,b) may now be finite or infinite. If it is infinite, we must make sure that the integral in (2.16) is well defined, at least when f is a polynomial. We achieve this by requiring that all moments of the weight function,

$$\mu_s = \int_a^b t^s w(t)dt, \quad s = 0, 1, 2, \dots , \tag{2.17}$$

exist and be finite.

We say that the quadrature formula (2.16) has (polynomial) *degree of exactness* d if

$$E_n(f) = 0 \quad \text{for all} \quad f \in \mathbb{P}_d; \tag{2.18}$$

that is, the formula has zero error whenever f is a polynomial of degree $\leq d$. We call (2.16) *interpolatory*, if it has degree of exactness $d = n - 1$. Interpolatory formulae are precisely those "obtained by interpolation," that is, for which

$$\sum_{k=1}^{n} w_k f(t_k) = \int_a^b p_{n-1}(f; t_1, \dots, t_n; t)w(t)dt, \tag{2.19}$$

or, equivalently,

$$w_k = \int_a^b \ell_k(t)w(t)dt, \quad k = 1, 2, \dots, n, \tag{2.20}$$

where

$$\ell_k(t) = \prod_{\substack{\ell=1 \\ \ell \neq k}}^{n} \frac{t - t_\ell}{t_k - t_\ell} \tag{2.21}$$

are the elementary Lagrange interpolation polynomials associated with the nodes $t_1, t_2, \dots, t_n$. The fact that (2.16) with w_k given by (2.20) has degree of exactness $d = n-1$ is evident, since for any $f \in \mathbb{P}_{n-1}$ we have $p_{n-1}(f; \cdot) \equiv f(\cdot)$ in (2.19). Conversely, if (2.16) has degree of exactness $d = n - 1$, then putting $f(t) = \ell_r(t)$ in (2.16) gives $\int_a^b \ell_r(t)w(t)dt = \sum_{k=1}^n w_k \ell_r(t_k) = w_r$, $r = 1, 2, \ldots, n$, that is, (2.20).

We see, therefore, that given any n distinct nodes $t_1, t_2, \ldots, t_n$, it is always possible to construct a formula of type (2.16) which is exact for all polynomials of degree $\leq n-1$. In the case $w(t) \equiv 1$ on $[-1,1]$, and t_k equally spaced on $[-1,1]$, the feasibility of such a construction was already alluded to by Newton in 1687 and implemented in detail by Cotes[2] around 1712. By extension, we call the formula (2.16), with the t_k prescribed and the w_k given by (2.20), a *Newton-Cotes formula.*

The question naturally arises whether we can do better, that is, whether we can achieve $d > n-1$ by a judicious choice of the nodes t_k (the weights w_k being necessarily given by (2.20)). The answer is surprisingly simple and direct. To formulate it, we introduce the node polynomial

$$\omega_n(t) = \prod_{k=1}^n (t - t_k). \tag{2.22}$$

Theorem 3.2.1. *Given an integer k with $0 \leq k \leq n$, the quadrature formula* (2.16) *has degree of exactness $d = n - 1 + k$ if and only if both of the following conditions are satisfied.*

(a) *The formula* (2.16) *is interpolatory.*

(b) *The node polynomial ω_n in* (2.22) *satisfies $\int_a^b \omega_n(t)p(t)w(t)dt = 0$ for all $p \in \mathbb{P}_{k-1}$.*

The condition in (b) imposes k conditions on the nodes $t_1, t_2, \ldots, t_n$ of (2.16). (If $k = 0$, there is no restriction since, as we know, we can always get $d = n - 1$.) In effect, ω_n must be orthogonal to $\mathbb{P}_{k-1}$ relative to the weight function w. Since $w(t) \geq 0$, we have necessarily $k \leq n$; otherwise, ω_n would have to be orthogonal to $\mathbb{P}_n$, in particular, orthogonal to itself, which is impossible. Thus $k = n$ is optimal, giving rise to a quadrature rule of

[2]Roger Cotes (1682–1716), precocious son of an English country pastor, was entrusted with the preparation of the second edition of Newton's *Principia.* He worked out in detail Newton's idea of numerical integration and published the coefficients — now known as Cotes numbers — of the n-point formula for all $n < 11$. Upon his death at the early age of 33, Newton said of him: "If he had lived, we might have known something."

maximum degree of exactness $d_{\max} = 2n - 1$. Condition (b) then amounts to orthogonality of ω_n to all polynomials of lower degree; that is, $\omega_n(\,\cdot\,) = \pi_n(\,\cdot\,; w)$ is precisely the nth-degree orthogonal polynomial belonging to the weight function w (cf. Ch. 2, §1.4(2)). This optimal formula is called the *Gaussian quadrature formula* associated with the weight function w. Its nodes, therefore, are the zeros of $\pi_n(\,\cdot\,; w)$, and the weights w_k are given as in (2.20); thus,

$$\begin{aligned} &\pi_n(t_k; w) = 0, \\ &w_k = \int_a^b \frac{\pi_n(t; w)}{(t - t_k)\pi_n'(t_k; w)}\, w(t)dt, \end{aligned} \qquad k = 1, 2, \ldots, n. \tag{2.23}$$

The formula was developed in 1814 by Gauss[3] for the special case $w(t) \equiv 1$ on [–1,1], and extended to more general weight functions by Christoffel[4] in 1877. It is, therefore, also referred to as the *Gauss-Christoffel quadrature formula.*

Proof of Theorem 3.2.1. We first prove the *necessity* of (a) and (b). Since, by assumption, the degree of exactness is $d = n-1+k \geq n-1$, condition (a) is trivial. Condition (b) also follows immediately, since, for any $p \in \mathbb{P}_{k-1}$, the product $\omega_n p$ is in $\mathbb{P}_{n-1+k}$; hence

$$\int_a^b \omega_n(t)p(t)w(t)dt = \sum_{k=1}^{n} w_k \omega_n(t_k)p(t_k),$$

[3] Carl Friedrich Gauss (1777–1855) was one of the greatest mathematicians of the 19th century — and perhaps of all time. He spent almost his entire life in Göttingen, where he was Director of the Observatory for some 40 years. Already as a student in Göttingen, Gauss discovered that the regular 17-gon can be constructed by compass and ruler, thereby settling a problem that had been open since antiquity. His dissertation gave the first proof of the Fundamental Theorem of Algebra (that an algebraic equation of degree n has exactly n roots). He went on to make fundamental contributions to number theory, differential and non-Euclidean geometry, elliptic and hypergeometric functions, celestial mechanics, geodesy, and various branches of physics, notably magnetism and optics. His computational efforts in celestial mechanics and geodesy, based on the principle of least squares, required the solution (by hand) of large systems of linear equations, for which he used what today are known as Gauss elimination and relaxation methods. Gauss's work on quadrature builds upon the earlier work of Newton and Cotes.

[4] Elvin Bruno Christoffel (1829–1900) was active for short periods of time in Berlin and Zurich and, for the rest of his life, in Strasbourg. He is best known for his work in geometry, in particular, tensor analysis, which became important in Einstein's theory of relativity.

which vanishes, since $\omega_n(t_k) = 0$ for $k = 1, 2, \ldots, n$.

To prove the *sufficiency* of (a), (b), we must show that for any $p \in \mathbb{P}_{n-1+k}$ we have $E_n(p) = 0$ in (2.16). Given any such p, divide it by ω_n, so that

$$p = q\omega_n + r, \quad q \in \mathbb{P}_{k-1}, \quad r \in \mathbb{P}_{n-1},$$

where q is the quotient and r the remainder. There follows

$$\int_a^b p(t)w(t)dt = \int_a^b q(t)\omega_n(t)w(t)dt + \int_a^b r(t)w(t)dt.$$

The first integral on the right, by (b), is zero, since $q \in \mathbb{P}_{k-1}$, whereas the second, by (a), since $r \in \mathbb{P}_{n-1}$, equals

$$\sum_{k=1}^{n} w_k r(t_k) = \sum_{k=1}^{n} w_k [p(t_k) - q(t_k)\omega_n(t_k)] = \sum_{k=1}^{n} w_k p(t_k),$$

the last equality following again from $\omega_n(t_k) = 0$, $k = 1, 2, \ldots, n$. This completes the proof. □

The case $k = n$ of Theorem 3.2.1 (i.e., the Gauss quadrature rule) is discussed further in §2.3. Here we still mention two special cases with $k < n$, which are of some practical interest. The first is the *Gauss-Radau* quadrature formula in which one endpoint, say, a, is finite and serves as a quadrature node, say, $t_1 = a$. The maximum degree of exactness attainable then is $d = 2n - 2$ and corresponds to $k = n - 1$ in Theorem 3.2.1. Part (b) of that theorem tells us that the remaining nodes $t_2, \ldots, t_n$ must be the zeros of $\pi_{n-1}(\,\cdot\,; w_a)$, where $w_a(t) = (t - a)w(t)$. Similarly, in the *Gauss-Lobatto* formula, both endpoints are finite and serve as nodes, say, $t_1 = a$, $t_n = b$, and the remaining nodes $t_2, \ldots, t_{n-1}$ are taken to be the zeros of $\pi_{n-2}(\,\cdot\,; w_{a,b})$, $w_{a,b}(t) = (t - a)(b - t)$, thus achieving maximum degree of exactness $d = 2n - 3$.

Example: Two-point Newton-Cotes vs. two-point Gauss

We compare the Newton-Cotes with the Gauss formula in the case $n = 2$ and for the weight function $w(t) = t^{-1/2}$ on [0,1]. The two prescribed nodes in the Newton-Cotes formula are taken to be the endpoints; thus

$$\int_0^1 t^{-1/2} f(t)dt \approx \begin{cases} w_1^{NC} f(0) + w_2^{NC} f(1) & \text{(Newton-Cotes)}, \\[2ex] w_1^{G} f(t_1) + w_2^{G} f(t_2) & \text{(Gauss)}. \end{cases}$$

To get the coefficients in the Newton-Cotes formula, we use (2.20), where

$$\ell_1(t) = \frac{t-1}{0-1} = 1-t, \quad \ell_2(t) = \frac{t-0}{1-0} = t.$$

This gives

$$w_1^{NC} = \int_0^1 t^{-1/2}\ell_1(t)dt = \int_0^1 (t^{-1/2} - t^{1/2})dt = \left(2t^{1/2} - \tfrac{2}{3}\,t^{3/2}\right)\Big|_0^1 = \tfrac{4}{3}\,,$$
$$w_2^{NC} = \int_0^1 t^{-1/2}\ell_2(t)dt = \int_0^1 t^{1/2}dt = \tfrac{2}{3}t^{3/2}\Big|_0^1 = \tfrac{2}{3}\,.$$

Thus,

$$\int_0^1 t^{-1/2} f(t)dt = \tfrac{2}{3}\,(2f(0) + f(1)) + E_2^{NC}(f). \tag{2.24}$$

Note how the square root singularity at the origin causes the value $f(0)$ to receive a weight twice as large as that of $f(1)$.

To develop the Gauss formula, we first construct the required orthogonal polynomial

$$\pi_2(t) = t^2 - p_1 t + p_2.$$

Since it is orthogonal to the constant 1, and to t, we get

$$\begin{aligned} 0 &= \int_0^1 t^{-1/2}\pi_2(t)dt = \int_0^1 (t^{3/2} - p_1 t^{1/2} + p_2 t^{-1/2})dt \\ &= \left(\tfrac{2}{5}\,t^{5/2} - \tfrac{2}{3}\,p_1 t^{3/2} + 2p_2 t^{1/2}\right)\Big|_0^1 = \tfrac{2}{5} - \tfrac{2}{3}\,p_1 + 2p_2, \\ 0 &= \int_0^1 t^{-1/2}\cdot t\pi_2(t)dt = \int_0^1 (t^{5/2} - p_1 t^{3/2} + p_2 t^{1/2})dt \\ &= \left(\tfrac{2}{7}\,t^{7/2} - \tfrac{2}{5}\,p_1 t^{5/2} + \tfrac{2}{3}\,p_2 t^{3/2}\right)\Big|_0^1 = \tfrac{2}{7} - \tfrac{2}{5}\,p_1 + \tfrac{2}{3}\,p_2, \end{aligned}$$

that is, the linear system

$$\tfrac{1}{3}\,p_1 - p_2 = \tfrac{1}{5}\,,$$

$$\tfrac{1}{5}\,p_1 - \tfrac{1}{3}\,p_2 = \tfrac{1}{7}\,.$$

The solution is $p_1 = \frac{6}{7}$, $p_2 = \frac{3}{35}$; thus

$$\pi_2(t) = t^2 - \tfrac{6}{7}\,t + \tfrac{3}{35}\,.$$

The Gauss nodes — the zeros of π_2 — are therefore

$$t_1 = \tfrac{1}{7}\left(3 - 2\sqrt{\tfrac{6}{5}}\right) = .1155871100, \quad t_2 = \tfrac{1}{7}\left(3 + 2\sqrt{\tfrac{6}{5}}\right) = .7415557471.$$

For the weights w_1^G, w_2^G, we could use again (2.20), but it is simpler to set up a linear system of equations which expresses the fact that the formula is exact for $f(t) \equiv 1$ and $f(t) \equiv t$:

$$w_1^G + w_2^G = \int_0^1 t^{-1/2} dt = 2,$$

$$t_1 w_1^G + t_2 w_2^G = \int_0^1 t^{-1/2} \cdot t dt = \tfrac{2}{3}.$$

This yields

$$w_1^G = \frac{-2t_2 + \frac{2}{3}}{t_1 - t_2}, \quad w_2^G = \frac{2t_1 - \frac{2}{3}}{t_1 - t_2},$$

or, with the values of t_1, t_2 substituted from the preceding,

$$w_1^G = 1 + \tfrac{1}{3}\sqrt{\tfrac{5}{6}} = 1.304290310, \quad w_2^G = 1 - \tfrac{1}{3}\sqrt{\tfrac{5}{6}} = .6957096903.$$

Again, w_1^G is larger than w_2^G, this time by a factor $1.874\ldots$.

Summarizing, we obtain the Gauss formula

$$\int_0^1 t^{-1/2} f(t) dt = \left(1 + \tfrac{1}{3}\sqrt{\tfrac{5}{6}}\right) f\left(\tfrac{1}{7}\left(3 - 2\sqrt{\tfrac{6}{5}}\right)\right) + \left(1 - \tfrac{1}{3}\sqrt{\tfrac{5}{6}}\right)$$
$$\cdot f\left(\tfrac{1}{7}\left(3 + 2\sqrt{\tfrac{6}{5}}\right)\right) + E_2^G(f). \tag{2.25}$$

To illustrate, consider $f(t) = \cos(\frac{1}{2}\pi t)$; that is,

$$I = \int_0^1 t^{-1/2} \cos(\tfrac{1}{2}\pi t) dt = 2C(1) = 1.5597865.$$

($C(x)$ is the Fresnel integral $\int_0^x \cos(\frac{1}{2}\pi t^2) dt$.) Newton-Cotes and Gauss give the following approximations,

$$I^{NC} = \tfrac{4}{3} = 1.333\ldots\,,$$

$$I^G = 1.282851067 + .274738493 = 1.55758956.$$

The respective errors are

$$E_2^{NC} = .226, \quad E_2^G = .00220,$$

demonstrating the superiority of the Gauss formula even for $n = 2$.

§2.3. **Properties of Gaussian quadrature rules.** The Gaussian quadrature rule (2.16) and (2.23), in addition to being optimal, has some interesting and useful properties. The more important ones are now listed, with most of the proofs relegated to the exercises.

(i) All nodes t_k are real, distinct, and contained in the open interval (a,b). This is a well-known property satisfied by the zeros of orthogonal polynomials (cf. Ex. 31).

(ii) All weights w_k are positive. The formula (2.20) for the weights gives no clue as to their signs; however, an ingenious observation of Stieltjes[5] proves it almost immediately. Indeed,

$$0 < \int_a^b \ell_j^2(t)w(t)dt = \sum_{k=1}^{n} w_k \ell_j^2(t_k) = w_j, \quad j = 1, 2, \dots, n,$$

the first equality following since $\ell_j^2 \in \mathbb{P}_{2n-2}$ and the degree of exactness is $d = 2n - 1$.

(iii) If $[a,b]$ is a finite interval, then the Gauss formula converges for any continuous function; that is, $E_n(f) \to 0$ as $n \to \infty$ whenever $f \in C[a,b]$. This is basically a consequence of the Weierstrass Approximation Theorem, which implies that, if $\hat{p}_{2n-1}(f;\,\cdot\,)$ denotes the polynomial of degree $2n-1$ that approximates f best on $[a,b]$ in the uniform norm, then

$$\lim_{n\to\infty} \|f(\,\cdot\,) - \hat{p}_{2n-1}(f;\,\cdot\,)\|_\infty = 0.$$

[5]Thomas Jan Stieltjes (1856–1894), born in the Netherlands, studied at the Technical Institute of Delft, but never finished to get his degree because of a deep-seated aversion to examinations. He nevertheless got a job at the Observatory of Leiden as a "computer assistant for astronomical calculations." His early publications caught the attention of Hermite, who was able to eventually secure a university position for Stieltjes in Toulouse. A life-long friendship evolved between these two great men, of which two volumes of their correspondence (see Hermite and Stieltjes [1905]) gives vivid testimony (and still makes fascinating reading). Stieltjes is best known for his work on continued fractions and the moment problem, which, among other things, led him to invent a new concept of integral which now bears his name. He died very young of tuberculosis at the age of 38.

Since $E_n(\hat{p}_{2n-1}) = 0$ (polynomial degree of exactness $d = 2n - 1$), it follows that

$$
\begin{aligned}
|E_n(f)| &= |E_n(f - \hat{p}_{2n-1})| \\
&= \left| \int_a^b [f(t) - \hat{p}_{2n-1}(f;t)]w(t)dt - \sum_{k=1}^{n} w_k[f(t_k) - \hat{p}_{2n-1}(f;t_k)] \right| \\
&\le \int_a^b |f(t) - \hat{p}_{2n-1}(f;t)|w(t)dt + \sum_{k=1}^{n} w_k|f(t_k) - \hat{p}_{2n-1}(f;t_k)| \\
&\le \|f(\cdot) - \hat{p}_{2n-1}(f;\cdot)\|_\infty \left[\int_a^b w(t)dt + \sum_{k=1}^{n} w_k \right].
\end{aligned}
$$

Here the positivity of the weight w_k has been used crucially. Noting that

$$
\sum_{k=1}^{n} w_k = \int_a^b w(t)dt = \mu_0 \quad \text{(cf. (2.17))},
$$

we thus conclude

$$
|E_n(f)| \le 2\mu_0 \|f - \hat{p}_{2n-1}\|_\infty \to 0 \ \text{ as } \ n \to \infty.
$$

(iv) The nodes in the n-point Gauss formula separate those of the $(n+1)$-point formula. (Cf. Ex. 32).

The next property forms the basis of an efficient algorithm for computing Gaussian quadrature formulae.

(v) Let $\alpha_k = \alpha_k(w)$, $\beta_k = \beta_k(w)$ be the recursion coefficients for the orthogonal polynomials $\pi_k(\,\cdot\,) = \pi_k(\,\cdot\,; w)$; that is (cf. Ch. 2, §1.4(2)),

$$
\begin{gathered}
\pi_{k+1}(t) = (t - \alpha_k)\pi_k(t) - \beta_k \pi_{k-1}(t), \quad k = 0, 1, 2 \dots\,, \\
\pi_0(t) = 1, \quad \pi_{-1}(t) = 0,
\end{gathered}
\tag{2.26}
$$

with β_0 (as is customary) defined by $\beta_0 = \int_a^b w(t)dt \; (= \mu_0)$. The nth-order *Jacobi matrix* for the weight function w is a tridiagonal symmetric matrix defined by

$$J_n = J_n(w) = \begin{bmatrix} \alpha_0 & \sqrt{\beta_1} & & & 0 \\ \sqrt{\beta_1} & \alpha_1 & \sqrt{\beta_2} & & \\ & \sqrt{\beta_2} & & \ddots & \\ & & \ddots & \ddots & \sqrt{\beta_{n-1}} \\ 0 & & & \sqrt{\beta_{n-1}} & \alpha_{n-1} \end{bmatrix}.$$

Then the nodes t_k are the eigenvalues of J_n (cf. Ex. 40),

$$J_n v_k = t_k v_k, \quad v_k^T v_k = 1, \quad k = 1, 2, \dots, n, \tag{2.27}$$

and the weights w_k expressible in terms of the first component $v_{k,1}$ of the corresponding (normalized) eigenvectors by

$$w_k = \beta_0 v_{k,1}^2, \quad k = 1, 2, \dots, n. \tag{2.28}$$

Thus, to compute the Gauss formula, we must solve an eigenvalue/eigenvector problem for a symmetric tridiagonal matrix. This is a routine problem in numerical linear algebra, and very efficient methods (the QR algorithm, e.g.) are known for solving it.

(vi) Markov[6] observed in 1885 that the Gauss quadrature formula can also be obtained by Hermite interpolation on the nodes t_k, each counting as a double node, if one requires that after integration all coefficients of the derivative terms should be zero (cf. Ex. 33). An interesting consequence of this new interpretation is the following expression for the remainder term, which follows directly from the error term of Hermite interpolation (cf. Ch. 2, §2.7) and the Mean Value Theorem of integration,

$$E_n(f) = \frac{f^{(2n)}(\xi)}{(2n)!} \int_a^b [\pi_n(t; w)]^2 w(t) dt, \quad a < \xi < b. \tag{2.29}$$

Here $\pi_n(\,\cdot\,; w)$ is the orthogonal polynomial, with leading coefficient 1, relative to the weight function w. It is assumed, of course, that $f \in C^{2n}[a, b]$.

---

[6] Andrei Andrejevich Markov (1856–1922) was a Russian mathematician active in St. Petersburg who made important contributions to probability theory, number theory, and constructive approximation theory.

We conclude this section with a table of some classical weight functions, their corresponding orthogonal polynomials, and the recursion coefficients α_k, β_k for generating the orthogonal polynomials as in (2.26). We

$w(t)$	$[a,b]$	Orth. Pol.	Notation	α_k	β_k
1	$[-1,1]$	Legendre[7]	P_n	0	$2\ (k=0)$ $(4-k^{-2})^{-1}\ (k>0)$
$(1-t^2)^{-\frac{1}{2}}$	$[-1,1]$	Chebyshev #1	T_n	0	$\pi\ (k=0)$ $\frac{1}{2}\ (k=1)$ $\frac{1}{4}\ (k>1)$
$(1-t^2)^{\frac{1}{2}}$	$[-1,1]$	Chebyshev #2	U_n	0	$\frac{1}{2}\pi\ (k=0)$ $\frac{1}{4}\ (k>0)$
$(1-t)^{\alpha}\ (1+t)^{\beta}$ $\alpha>-1,\ \beta>-1$	$[-1,1]$	Jacobi[8]	$P_n^{(\alpha,\beta)}$	known	known
$t^{\alpha}e^{-t},\ \alpha>-1$	$[0,\infty]$	Laguerre[9]	$L_n^{(\alpha)}$	$2k+\alpha+1$	$\Gamma(1+\alpha)\ (k=0)$ $k(k+\alpha)\ (k>0)$
e^{-t^2}	$[-\infty,\infty]$	Hermite[10]	H_n	0	$\sqrt{\pi}\ (k=0)$ $\frac{1}{2}k\ (k>0)$

TABLE 3.2.1. Classical orthogonal polynomials

also include the standard notations for these polynomials (these usually do

[7]Adrien Marie Legendre (1752–1833) was a French mathematician active in Paris, best known for his treatise on elliptic integrals, but also famous for his work in number theory and geometry. He is considered the originator (in 1805) of the method of least squares, although Gauss had already used it in 1794, but published it only in 1809.

[8]Carl Gustav Jakob Jacobi (1804–1851) was a contemporary of Gauss, and with him one of the most important 19th-century mathematicians in Germany. His name is connected with elliptic functions, partial differential equations of dynamics, calculus of variations, celestial mechanics; functional determinants also bear his name. In his work on celestial mechanics he invented what is now called the Jacobi method for solving linear algebraic systems.

[9]Edmond Laguerre (1834–1886) was a French mathematician active in Paris, who made essential contributions to geometry, algebra, and analysis.

[10]Charles Hermite (1822–1901) was a leading French mathematician, Academician in Paris, known for his extensive work in number theory, algebra, and analysis. He is famous for his proof in 1873 of the transcendental nature of the number e.

not refer to the monic polynomials). Note that the recursion coefficients α_k are all zero for even weight functions on intervals symmetric with respect to the origin (cf. Ch. 2, §1.4(2)). For Jacobi polynomials, the recursion coefficients are explicitly known, but the formulae are a bit lengthy and are not given here.

§2.4. **Some applications of the Gauss quadrature rule**. In many applications the integrals to be computed have the weight function w already built in. In others, one has to figure out for oneself what the most appropriate weight function should be. Several examples of this are given in this section. We begin, however, with the easy exercise of transforming the Gauss-Jacobi quadrature rule from an *arbitrary* finite interval to the canonical interval [–1,1].

(i) *The Gauss-Jacobi formula for the interval* $[a,b]$. We assume $[a,b]$ a finite interval. What is essential about the weight function in the Jacobi case is the fact that it has an algebraic singularity (with exponent α) at the right endpoint, and an algebraic singularity (with exponent β) at the left endpoint. The integral in question, therefore, is

$$\int_a^b (b-x)^\alpha (x-a)^\beta g(x)dx.$$

A linear transformation of variables

$$x = \tfrac{1}{2}(b-a)t + \tfrac{1}{2}(b+a), \qquad dx = \tfrac{1}{2}(b-a)dt,$$

maps the x-interval $[a,b]$ onto the t-interval $[-1,1]$, and the integral becomes

$$\int_{-1}^1 \left(\tfrac{1}{2}(b-a)(1-t)\right)^\alpha \left(\tfrac{1}{2}(b-a)(1+t)\right)^\beta g\left(\tfrac{1}{2}(b-a)t+\tfrac{1}{2}(b+a)\right)\tfrac{1}{2}(b-a)dt$$
$$= \left(\tfrac{1}{2}(b-a)\right)^{\alpha+\beta+1}\int_{-1}^1 (1-t)^\alpha(1+t)^\beta f(t)dt,$$

where we have set

$$f(t) = g\left(\tfrac{1}{2}(b-a)t+\tfrac{1}{2}(b+a)\right), \quad -1\le t\le 1.$$

The last integral is now in standard form for the application of the Gauss-Jacobi quadrature formula. One thus finds

$$\begin{aligned}&\int_a^b (b-x)^\alpha (x-a)^\beta g(x)dx \\ &= \left(\tfrac{1}{2}(b-a)\right)^{\alpha+\beta+1}\sum_{k=1}^n w_k^J g\left(\tfrac{1}{2}(b-a)t_k^J+\tfrac{1}{2}(b+a)\right) + E_n(g),\end{aligned} \tag{2.30}$$

where t_k^J, w_k^J are the (standard) Gauss-Jacobi nodes and weights. Since with g, also f is a polynomial, and both have the same degree, the formula (2.30) is exact for all $g \in \mathbb{P}_{2n-1}$.

(ii) *Iterated integrals.* Let I denote the integral operator

$$(Ig)(t) := \int_0^t g(\tau)d\tau,$$

and I^p its pth power (the identity operator if $p = 0$). Then

$$(I^{p+1}g)(1) = \int_0^1 (I^p g)(t)dt$$

is the pth *iterated integral* of g. It is a well-known fact from calculus that an iterated integral can be written as a simple integral, namely,

$$(I^{p+1}g)(1) = \int_0^1 \frac{(1-t)^p}{p!} g(t)dt. \tag{2.31}$$

Thus, to the integral on the right of (2.31) we could apply any of the standard integration procedures, such as a composite trapezoidal or Simpson's rule. If g is smooth, this works well for p relatively small. As p gets larger, the factor $(1-t)^p$ becomes rather unpleasant, since as $p \to \infty$ it approaches the discontinuous function equal to 1 at $t = 0$, and 0 elsewhere on (0,1]. This adversely affects the performance of any standard quadrature scheme. However, noting that $(1-t)^p$ is a Jacobi weight on the interval [0,1], with parameters $\alpha = p$, $\beta = 0$, we can apply (2.30); that is,

$$(I^{p+1}g)(1) = \frac{1}{2^{p+1}p!} \sum_{k=1}^{n} w_k^J g\left(\tfrac{1}{2}(1+t_k^J)\right) + E_n(g), \tag{2.32}$$

and get accurate results with moderate values of n, even when p is quite large (cf. Mach. Ass. 4).

(iii) *Integration over* $\mathbb{R}$. Compute $\int_{-\infty}^{\infty} F(x)dx$, assuming that, for some $a > 0$,

$$F(x) \sim e^{-ax^2}, \quad x \to \pm\infty. \tag{2.33}$$

Instead of a weight function we are given here information about the asymptotic behavior of the integrand. It is natural, then, to introduce the new function

$$f(t) = e^{t^2} F\left(\frac{t}{\sqrt{a}}\right), \tag{2.34}$$

which tends to 1 as $t \to \pm\infty$. The change of variables $x = t/\sqrt{a}$ then gives

$$\int_{-\infty}^{\infty} F(x)dx = \frac{1}{\sqrt{a}} \int_{-\infty}^{\infty} F\left(\frac{t}{\sqrt{a}}\right) dt = \frac{1}{\sqrt{a}} \int_{-\infty}^{\infty} e^{-t^2} f(t)dt.$$

The last integral is now in a form suitable for the application of the Gauss-Hermite formula. There results

$$\int_{-\infty}^{\infty} F(x)dx = \frac{1}{\sqrt{a}} \sum_{k=1}^{n} w_k^H e^{(t_k^H)^2} F\left(\frac{t_k^H}{\sqrt{a}}\right) + E_n(f), \tag{2.35}$$

where t_k^H, w_k^H are the Gauss-Hermite nodes and weights. The remainder $E_n(f)$ vanishes whenever f is a polynomial of degree $\leq 2n-1$; that is,

$$F(x) = e^{-ax^2} p(x), \quad p \in \mathbb{P}_{2n-1}.$$

Since the coefficients in (2.35) involve the products $w_k^H \cdot \exp((t_k^H)^2)$, some tables of Gauss-Hermite formulae also provide these products, in addition to the nodes and weights.

(iv) *Integration over* $\mathbb{R}_+$. Compute $\int_0^{\infty} F(x)dx$, assuming that

$$F(x) \sim \begin{cases} x^p & \text{as } x \downarrow 0 \quad (p > -1), \\ x^{-q} & \text{as } x \to \infty \quad (q > 1). \end{cases} \tag{2.36}$$

Similarly as in (iii), we now define f by

$$F(x) = \frac{x^p}{(1+x)^{p+q}} f(x), \quad x \in \mathbb{R}_+ , \tag{2.37}$$

so as to again have $f(x) \to 1$ as $x \downarrow 0$ and $x \to \infty$. The change of variables

$$x = \frac{1+t}{1-t}, \quad dx = \frac{2dt}{(1-t)^2}$$

then yields

$$\int_0^{\infty} F(x)dx = \int_{-1}^{1} \left(\frac{1+t}{1-t}\right)^p \left(\frac{2}{1-t}\right)^{-(p+q)} f\left(\frac{1+t}{1-t}\right) \frac{2dt}{(1-t)^2}$$

$$= \frac{1}{2^{p+q-1}} \int_{-1}^{1} (1-t)^{q-2}(1+t)^p g(t)dt,$$

where

$$g(t) = f\left(\frac{1+t}{1-t}\right).$$

This calls for Gauss-Jacobi quadrature with parameters $\alpha = q-2$, $\beta = p$:

$$\int_0^\infty F(x)dx = \frac{1}{2^{p+q-1}}\left(\sum_{k=1}^{n} w_k^J g(t_k^J) + E_n(g)\right).$$

It remains to re-express g in terms of f, and f in terms of F, to obtain the final formula

$$\int_0^\infty F(x)dx = 2\sum_{k=1}^{n} w_k^J (1+t_k^J)^{-p}(1-t_k^J)^{-q} F\left(\frac{1+t_k^J}{1-t_k^J}\right) + 2^{-p-q+1}E_n(g). \tag{2.38}$$

This is exact whenever $g(t)$ is a polynomial of degree $\leq 2n-1$, for example a polynomial of that degree in the variable $1-t$; hence, since $f(x) = g\left(\frac{x-1}{x+1}\right)$, for any $F(x)$ of the form

$$F(x) = \frac{x^p}{(1+x)^{p+q+\lambda}}, \quad \lambda = 0, 1, 2, \ldots, 2n-1.$$

§2.5. **Approximation of linear functionals: method of interpolation vs. method of undetermined coefficients**. Up until now, we heavily relied on interpolation to obtain approximations to derivatives, integrals, and the like. This is not the only possible approach, however, to construct approximation formulae, and indeed not usually the simplest one. Another is the method of undetermined coefficients. Both approaches can be described in a vastly more general setting, in which a given linear functional is to be approximated by other linear functionals so as to have exactness on a suitable finite-dimensional function space. Although both approaches yield the same formulae, the mechanics involved are quite different.

Before stating the general approximation problem, we illustrate the two methods on some simple examples.

Example 1: Obtain an approximation formula of the type

$$\int_0^1 f(x)dx \approx a_1 f(0) + a_2 f(1). \tag{2.39}$$

Method of interpolation. Instead of integrating f, we integrate the (linear) polynomial interpolating f at $x = 0$ and $x = 1$. This gives

$$\begin{aligned}\int_0^1 f(x)dx &\approx \int_0^1 p_1(f; 0, 1; x)dx \\ &\approx \int_0^1 [(1-x)f(0) + xf(1)]dx = \tfrac{1}{2}\,[f(0) + f(1)],\end{aligned}$$

the trapezoidal rule, to nobody's surprise.

Method of undetermined coefficients. We simply require (2.39) to be exact for all linear functions. This is the same as requiring equality in (2.39) when $f(x) = 1$ and $f(x) = x$. (Then, by linearity, one gets equality also for $f(x) = c_0 + c_1 x$ for arbitrary constants c_0, c_1.) This immediately produces the linear system

$$\begin{aligned} 1 &= a_1 + a_2, \\ \tfrac{1}{2} &= \phantom{a_1 + {}} a_2, \end{aligned}$$

hence $a_1 = a_2 = \frac{1}{2}$, as before.

Example 2: Find a formula of the type

$$\int_0^1 \sqrt{x} f(x)dx \approx a_1 f(0) + a_2 \int_0^1 f(x)dx. \tag{2.40}$$

Such a formula may come in handy if we already know the integral on the right and want to use this information, together with the value of f at $x = 0$, to approximate the weighted integral on the left.

Method of interpolation. "Interpolation" here means interpolation to the given data — the value of f at $x = 0$ and the integral of f from 0 to 1. We are thus seeking a polynomial $p \in \mathbb{P}_1$ such that

$$p(0) = f(0), \quad \int_0^1 p(x)dx = \int_0^1 f(x)dx, \tag{2.41}$$

which we then substitute in place of f in the left-hand integral of (2.40). If we let $p(x) = c_0 + c_1 x$, the interpolation conditions (2.41) give

$$\begin{aligned} c_0 \qquad &= f(0), \\ c_0 + \tfrac{1}{2} c_1 &= \int_0^1 f(x)dx; \end{aligned}$$

hence

$$c_0 = f(0), \quad c_1 = 2\left\{\int_0^1 f(x)dx - f(0)\right\}.$$

Therefore,

$$\begin{aligned} &\int_0^1 \sqrt{x} f(x)dx \approx \int_0^1 \sqrt{x} p(x)dx = \int_0^1 \sqrt{x}(c_0 + c_1 x)dx \\ &= \tfrac{2}{3} c_0 + \tfrac{2}{5} c_1 = \tfrac{2}{3} f(0) + \tfrac{2}{5} \cdot 2\left\{\int_0^1 f(x)dx - f(0)\right\}; \end{aligned}$$

that is,

$$\int_0^1 \sqrt{x} f(x)dx \approx -\tfrac{2}{15} f(0) + \tfrac{4}{5} \int_0^1 f(x)dx. \tag{2.42}$$

Method of undetermined coefficients. Equality in (2.40) for $f(x) = 1$ and $f(x) = x$ immediately yields

$$\begin{aligned} \tfrac{2}{3} &= a_1 + a_2, \\ \tfrac{2}{5} &= \qquad \tfrac{1}{2} a_2; \end{aligned}$$

hence $a_1 = -\frac{2}{15}$, $a_2 = \frac{4}{5}$, the same result as in (2.42), but produced incomparably faster.

In both examples we insisted on exactness for polynomials (of degree 1). In place of polynomials, we could have chosen other classes of functions, as long as we make sure that their dimension matches the number of "free parameters."

The essence of the two examples consists of showing how a certain linear functional can be approximated in terms of other (presumably simpler) linear functionals by forming a suitable linear combination of the latter. We recall that a *linear functional* L on a function space $\mathbb{F}$ is a map

$$L: \ \mathbb{F} \to \mathbb{R} \tag{2.43}$$

which satisfies the conditions of additivity and homogeneity:

$$L(f+g) = Lf + Lg, \quad \text{all} \quad f, g \in \mathbb{F}, \tag{2.44$_1$}$$

and

$$L(cf) = cLf, \quad \text{all} \quad c \in \mathbb{R}, \ f \in \mathbb{F}. \tag{2.44$_2$}$$

The function class $\mathbb{F}$, of course, must be a *linear space*, that is, closed under addition and multiplication by a scalar: $f, g \in \mathbb{F}$ implies $f + g \in \mathbb{F}$, and $f \in \mathbb{F}$ implies $cf \in \mathbb{F}$ for any $c \in \mathbb{R}$.

Here are some examples of linear functionals and appropriate spaces $\mathbb{F}$ on which they live:

(i) $Lf = f(0)$; $\mathbb{F} = \{f : f \text{ is defined at } x = 0\}$.

(ii) $Lf = f''(\frac{1}{2})$; $\mathbb{F} = \{f : f \text{ has a second derivative at } x = \frac{1}{2}\}$.

(iii) $Lf = \int_0^1 f(x)dx$; $\mathbb{F} = C[0,1]$ (or, more generally, f is Riemann integrable, or Lebesgue integrable).

(iv) $Lf = \int_0^1 f(x)w(x)dx$, where w is a given (integrable) "weight function;" $\mathbb{F} = C[0,1]$.

(v) Any linear combination (with constant coefficients) of the preceding linear functionals.

Examples of *nonlinear functionals* are

(i′) $Kf = |f(0)|$.

(ii′) $Kf = \int_0^1 [f(x)]^2 dx$, and so on.

We are now ready to formulate the *general approximation problem*: given a linear functional L on $\mathbb{F}$ (to be approximated), n special linear functionals $L_1, L_2, \ldots, L_n$ on $\mathbb{F}$ and their values (the "data") $\ell_i = L_i f$, $i = 1, 2, \ldots, n$, applied to some function f, and given a linear subspace $\Phi \subset \mathbb{F}$ with $\dim \Phi = n$, we want to find an approximation formula of the type

$$Lf \approx \sum_{i=1}^{n} a_i L_i f \tag{2.45}$$

that is exact (i.e., holds with equality) whenever $f \in \Phi$.

It is natural (since we want to "interpolate") to make the following *Assumption*: the "interpolation problem"

$$\text{find } \varphi \in \Phi \text{ such that } L_i\varphi = s_i, \quad i = 1, 2, \ldots, n, \tag{2.46}$$

has a unique solution, $\varphi(\,\cdot\,) = \varphi(s;\,\cdot\,)$, for arbitrary $s = [s_1, s_2, \ldots, s_n]^T$.

We can express our assumption more explicitly in terms of a given *basis* $\varphi_1, \varphi_2, \ldots, \varphi_n$ of Φ and the associated "Gram[11] matrix"

$$G = [L_i\varphi_j] = \begin{bmatrix} L_1\varphi_1 & L_1\varphi_2 & \cdots & L_1\varphi_n \\ L_2\varphi_1 & L_2\varphi_2 & \cdots & L_2\varphi_n \\ \cdots & \cdots & \cdots & \cdots \\ L_n\varphi_1 & L_n\varphi_2 & \cdots & L_n\varphi_n \end{bmatrix} \in \mathbb{R}^{n\times n}. \tag{2.47}$$

What we require is that

$$\det G \neq 0. \tag{2.46$'$}$$

It is easily seen (cf. Ex. 44) that this condition is independent of the particular choice of basis. To show that unique solvability of (2.46) and (2.46′) are equivalent, we express φ in (2.46) as a linear combination of the basis functions,

$$\varphi = \sum_{j=1}^{n} c_j\varphi_j \tag{2.48}$$

and note that the interpolation conditions

$$L_i\left(\sum_{j=1}^{n} c_j\varphi_j\right) = s_i, \quad i = 1, 2, \ldots, n,$$

by the linearity of the functionals L_i, can be written in the form

$$\sum_{j=1}^{n} c_j L_i\varphi_j = s_i, \quad i = 1, 2, \ldots, n;$$

[11] Jórgen Pedersen Gram (1850–1916) was a farmer's son who studied at the University of Copenhagen. After graduation, he entered an insurance company as computer assistant and, moving up the ranks, eventually became its director. He was interested in series expansions of special functions and also contributed to Chebyshev and least squares approximation. The "Gram determinant" was introduced by him in connection with his study of linear independence.

that is,

$$Gc = s, \quad c = [c_1, c_2, \dots, c_n]^T, \quad s = [s_1, s_2, \dots, s_n]^T. \tag{2.49}$$

This has a unique solution for arbitrary s if and only if (2.46′) holds.

Method of interpolation. We solve the general approximation problem "by interpolation,"

$$Lf \approx L\varphi(\ell;\, \cdot\,), \quad \ell = [\ell_1, \ell_2, \dots, \ell_n]^T, \quad \ell_i = L_i f. \tag{2.50}$$

In other words, we apply L not to f, but to $\varphi(\ell;\, \cdot\,)$ — the solution of the interpolation problem (2.46) in which $s = \ell$, the given "data." Our assumption guarantees that $\varphi(\ell;\, \cdot\,)$ is uniquely determined. In particular, if $f \in \Phi$, then (2.50) holds with equality, since trivially $\varphi(\ell;\, \cdot\,) \equiv f(\,\cdot\,)$ in this case. Thus, our approximation (2.50) already satisfies the exactness condition required for (2.45). It remains only to show that (2.50) indeed produces an approximation of the form (2.45). To do so, observe that the interpolant in (2.50) is

$$\varphi(\ell;\, \cdot\,) = \sum_{j=1}^{n} c_j \varphi_j(\,\cdot\,),$$

where the vector $c = [c_1, c_2, \dots, c_n]^T$ satisfies (2.49) with $s = \ell$,

$$Gc = \ell, \quad \ell = [L_1 f, L_2 f, \dots, L_n f]^T.$$

Writing

$$\lambda_j = L\varphi_j, \quad j = 1, 2, \dots, n; \quad \lambda = [\lambda_1, \lambda_2, \dots, \lambda_n]^T, \tag{2.51}$$

we have by the linearity of L,

$$L\varphi(\ell;\, \cdot\,) = \sum_{j=1}^{n} c_j L\varphi_j = \lambda^T c = \lambda^T G^{-1} \ell = [(G^T)^{-1}\lambda]^T \ell;$$

that is,

$$L\varphi(\ell;\, \cdot\,) = \sum_{i=1}^{n} a_i L_i f, \quad a = [a_1, a_2, \dots, a_n]^T = (G^T)^{-1}\lambda. \tag{2.52}$$

Method of undetermined coefficients. Here we determine the coefficients a_i in (2.45) such that equality holds for all $f \in \Phi$, which, by the linearity of both L and L_i is equivalent to equality for $f = \varphi_1$, $f = \varphi_2, \ldots, f = \varphi_n$; that is,

$$\left(\sum_{j=1}^{n} a_j L_j\right) \varphi_i = L\varphi_i, \quad i = 1, 2, \ldots, n,$$

or, by (2.51),

$$\sum_{j=1}^{n} a_j L_j \varphi_i = \lambda_i, \quad i = 1, 2, \ldots, n.$$

Evidently, the matrix of this system is G^T, so that

$$a = [a_1, a_2, \ldots, a_n]^T = (G^T)^{-1}\lambda,$$

in agreement with (2.52). Thus, *the method of interpolation and the method of undetermined coefficients are mathematically equivalent* — they produce exactly the same approximation.

§2.6. **Peano representation of linear functionals.** It may be argued that the method of interpolation, at least in the case of polynomials (i.e., $\Phi = \mathbb{P}_d$), is more powerful than the method of undetermined coefficients because it also yields an expression for the error term (if we carry along the remainder term of interpolation). The method of undetermined coefficients, in contrast, generates only the coefficients in the approximation and gives no clue as to the approximation error.

There is, however, a device due to Peano[12] that allows us to discuss the error after the approximation has been found. The point is that the error,

$$Ef = Lf - \sum_{i=1}^{n} a_i L_i f, \tag{2.53}$$

is itself a linear functional, one that annihilates all polynomials, say, of degree d,

$$Ep = 0, \quad \text{all } \ p \in \mathbb{P}_d. \tag{2.54}$$

[12] Giuseppe Peano (1858–1932), an Italian mathematician active in Turin, made fundamental contributions to mathematical logic, set theory, and the foundations of mathematics. General existence theorems in ordinary differential equations also bear his name. He created his own mathematical language, using symbols of the algebra of logic, and even promoted (and used) a simplified Latin (his "latino") as a world language for scientific publication.

Now suppose that $\mathbb{F}$ consists of all functions f having a continuous $(d+1)$st derivative on the finite interval $[a,b]$, $\mathbb{F} = C^{d+1}[a,b]$. Then, by Taylor's theorem with the remainder in integral form, we have

$$\begin{aligned} f(x) = f(a) + (x-a)\,\frac{f'(a)}{1!} + \cdots + (x-a)^d\,\frac{f^{(d)}(a)}{d!} \\ + \frac{1}{d!}\int_a^x (x-t)^d f^{(d+1)}(t)dt. \end{aligned} \tag{2.55}$$

The last integral can be extended to $t = b$ if we replace $x - t$ by 0 when $t > x$:

$$(x-t)_+ = \begin{cases} x-t & \text{if} \quad x-t \geq 0, \\ 0 & \text{if} \quad x-t < 0. \end{cases}$$

Thus,

$$\int_a^x (x-t)^d f^{(d+1)}(t)dt = \int_a^b (x-t)_+^d f^{(d+1)}(t)dt. \tag{2.56}$$

Now by applying the linear functional E to both sides of (2.55), with the integral written as in (2.56), yields by linearity of E and (2.54)

$$\begin{aligned} Ef &= \frac{1}{d!}\,E\left\{\int_a^b (x-t)_+^d f^{(d+1)}(t)dt\right\} \\ &= \frac{1}{d!}\int_a^b [E_{(x)}(x-t)_+^d] f^{(d+1)}(t)dt, \end{aligned}$$

provided the interchange of E with the integral is legitimate. (For most functionals it is.) The subscript x in $E_{(x)}$ is to indicate that E acts on the variable x (not t). Defining

$$K_d(t) = \frac{1}{d!}\,E_{(x)}(x-t)_+^d, \quad t \in \mathbb{R}, \tag{2.57}$$

we thus have the following representation for the error E,

$$Ef = \int_a^b K_d(t) f^{(d+1)}(t)dt. \tag{2.58}$$

This is called the *Peano representation* of the functional E, and K_d the dth *Peano kernel* for E.

If the functional E makes reference only to values of x in $[a,b]$ (e.g., Ef may involve values of f or of a derivative of f at some points in $[a,b]$, or integration over $[a,b]$), then it follows from (2.57) that $K_d(t) = 0$ for $t \notin [a,b]$ (cf. Ex. 45). In this case, the integral in (2.58) can be extended over the whole real line.

The functional E is called *definite of order* d if its Peano kernel K_d does not change sign. (We then also say that K_d is definite.) For such functionals E we can use the Mean Value Theorem of integration to write (2.58) in the form

$$Ef = f^{(d+1)}(\tau)\int_a^b K_d(t)dt, \quad a < \tau < b \quad (E \text{ definite of order } d).$$

The integral on the right (which is nonzero by assumption) is easily evaluated by putting $f(t) = t^{d+1}/(d+1)!$ in (2.58). This gives

$$Ef = e_{d+1} f^{(d+1)}(\tau), \quad e_{d+1} = E\,\frac{t^{d+1}}{(d+1)!} \quad (E \text{ definite of order } d), \tag{2.59}$$

and hence the estimate

$$|Ef| \le |e_{d+1}|\,\|f^{(d+1)}\|_\infty. \tag{2.60}$$

Conversely, a functional E satisfying (2.59) with $e_{d+1} \neq 0$ is necessarily definite of order d (see Ex. 46). For nondefinite functionals E, we must estimate by

$$|Ef| \le \|f^{(d+1)}\|_\infty \int_a^b |K_d(t)|dt, \tag{2.60$'$}$$

which, in view of the form (2.57) of K_d, can be rather laborious.

As an example, consider the formula obtained in (2.42); here

$$Ef = \int_0^1 \sqrt{x} f(x)dx + \tfrac{2}{15} f(0) - \tfrac{4}{5}\int_0^1 f(x)dx,$$

and $Ef = 0$ for all $f \in \mathbb{P}_1$ $(d = 1)$. Assuming that $f \in C^2[0,1]$, we thus have by (2.58),

$$Ef = \int_0^1 K_1(t) f''(t)dt, \quad K_1(t) = E_{(x)}(x-t)_+ \ .$$

Furthermore,

$$
\begin{aligned}
K_1(t) &= \int_0^1 \sqrt{x}(x-t)_+ dx + \tfrac{2}{15}\cdot 0 - \tfrac{4}{5}\int_0^1 (x-t)_+ dx \\
&= \int_t^1 \sqrt{x}(x-t)dx - \tfrac{4}{5}\int_t^1 (x-t)dx \\
&= \tfrac{2}{5}\, x^{\frac{5}{2}}\Big|_t^1 - t\cdot \tfrac{2}{3}\, x^{\frac{3}{2}}\Big|_t^1 - \tfrac{4}{5}\left(\tfrac{1}{2}\, x^2\Big|_t^1 - t(1-t)\right) \\
&= \tfrac{2}{5}(1-t^{\frac{5}{2}}) - \tfrac{2}{3}\, t(1-t^{\frac{3}{2}}) - \tfrac{2}{5}(1-t^2) + \tfrac{4}{5}\, t(1-t) \\
&= \tfrac{4}{15}\, t^{\frac{5}{2}} - \tfrac{2}{5}\, t^2 + \tfrac{2}{15}\, t \\
&= \tfrac{2}{15}\, t(2t^{\frac{3}{2}} - 3t + 1).
\end{aligned}
$$

Now the function in parentheses, say, $q(t)$, satisfies $q(0) = 1$, $q(1) = 0$, $q'(t) = -3(1-t^{1/2}) < 0$ for $0 < t < 1$. There follows $q(t) \geq 0$ on [0,1], and the kernel K_1 is (positive) definite. Furthermore,

$$
\begin{aligned}
e_2 = E\left(\frac{t^2}{2!}\right) &= \int_0^1 \sqrt{x}\,\frac{x^2}{2} dx + \tfrac{2}{15}\cdot 0 - \tfrac{4}{5}\int_0^1 \frac{x^2}{2} dx \\
&= \tfrac{1}{2}\,\tfrac{2}{7}\, x^{\frac{7}{2}}\Big|_0^1 - \tfrac{2}{5}\,\tfrac{1}{3}\, x^3\Big|_0^1 \\
&= \tfrac{1}{7} - \tfrac{2}{15} = \tfrac{1}{105},
\end{aligned}
$$

so that finally, by (2.59),

$$
Ef = \tfrac{1}{105}\, f''(\tau), \quad 0 < \tau < 1.
$$

§2.7. **Extrapolation methods**. Many methods of approximation depend on a positive parameter, say, h, which controls the accuracy of the method. As $h \downarrow 0$, the approximations typically converge to the exact solution. An example of this is the composite trapezoidal rule of integration, where h is the spacing of the quadrature nodes. Other important examples are finite difference methods in ordinary and partial differential equations. In practice, one usually computes several approximations to a solution, corresponding to different values of the parameter h. It is then natural to try "extrapolating to the limit $h = 0$," that is, constructing a linear combination of these approximations that is more accurate than either of them. This is the basic idea behind extrapolation methods. We apply it here to the composite trapezoidal rule, which gives rise to an interesting and powerful integration technique known as Romberg integration. To develop the general principle, suppose the approximation in question is $A(h)$, a scalar-valued function of h, and thus $A(0)$ the exact solution. The approximation may be defined only for a set of discrete values of h, which, however, have $h = 0$ as a limit point (e.g., $h = (b-a)/n$, $n = 1, 2, 3, \dots$, in the case of the composite trapezoidal rule). We call these *admissible* values of h. When in the following we write $h \to 0$, we mean that h goes to zero over these admissible values. About the approximation $A(h)$, we first assume that there exist constants a_1, a_2 independent of h, and two positive numbers p, p' with $p' > p$, such that

$$A(h) = a_0 + a_1 h^p + O(h^{p'}), \quad h \to 0. \tag{2.61}$$

The order term here has the usual meaning of a quantity bounded (for all sufficiently small h) by a constant times $h^{p'}$, where the constant does not depend on h. We only assume the existence of such constants a_0, a_1; their values are usually not known. Indeed, $a_0 = A(0)$, for example, is the exact solution. The value p, on the other hand, is assumed to be known.

Now let $q < 1$ be a fixed positive number, and $q^{-1}h$ an admissible parameter. Then we have

$$\left.\begin{aligned} A(h) &= a_0 + a_1 h^p + O(h^{p'}), \\ A(q^{-1}h) &= a_0 + a_1 q^{-p} h^p + O(h^{p'}), \end{aligned}\right. \quad h \to 0.$$

Eliminating the middle terms on the right, we find

$$a_0 = A(0) = A(h) + \frac{A(h) - A(q^{-1}h)}{q^{-p} - 1} + O(h^{p'}), \quad h \to 0.$$

Thus, from two approximations, $A(h)$, $A(q^{-1}h)$, whose errors are both $O(h^p)$, we obtain an improved approximation,

$$A_{\text{impr}}(h) = A(h) + \frac{A(h) - A(q^{-1}h)}{q^{-p} - 1}, \tag{2.62}$$

with a smaller error of $O(h^{p'})$. The passage from A to A_{impr} is called *Richardson*[13] *extrapolation.*

If we want to repeat this process, we have to know more about the approximation $A(h)$; there must be more terms to be eliminated! This leads to the idea of an asymptotic expansion: we say that $A(h)$ admits an *asymptotic expansion*

$$A(h) = a_0 + a_1 h^{p_1} + a_2 h^{p_2} + \cdots, \quad 0 < p_1 < p_2 < \cdots, \quad h \to 0 \tag{2.63}$$

(with coefficients a_i independent of h), if for each $k = 1, 2, \ldots$

$$A(h) - (a_0 + a_1 h^{p_1} + \cdots + a_k h^{p_k}) = O(h^{p_{k+1}}), \quad h \to 0. \tag{2.64}$$

We emphasize that (2.63) need not (and in fact usually does not) converge for any fixed $h > 0$; all that is required is (2.64). If (2.64) holds only for finitely many k, say, $k = 1, 2, \ldots, K$, then the expansion (2.63) is finite and is referred to as an *asymptotic approximation* to $K + 1$ terms.

It is now clear that if $A(h)$ admits an asymptotic expansion (or approximation), we can successively eliminate the terms of the expansion exactly as we did for a 2-term approximation, thereby obtaining a (finite or infinite) sequence of successively improved approximations. We formulate this in the form of a theorem.

Theorem 3.2.2 (Repeated Richardson extrapolation). *Let $A(h)$ admit the asymptotic expansion* (2.63) *and define, for some fixed positive $q < 1$,*

$$\begin{aligned} A_1(h) &= A(h), \\ A_{k+1}(h) &= A_k(h) + \frac{A_k(h) - A_k(q^{-1}h)}{q^{-p_k} - 1}, \quad k = 1, 2, \ldots\ . \end{aligned} \tag{2.65}$$

Then, for each $n = 1, 2, 3, \ldots$, $A_n(h)$ admits an asymptotic expansion

$$A_n(h) = a_0 + a_n^{(n)} h^{p_n} + a_{n+1}^{(n)} h^{p_{n+1}} + \cdots\ , \quad h \to 0, \tag{2.66}$$

[13]Lewis Fry Richardson (1881–1953), born, educated, and active in England, did pioneering work in numerical weather prediction, proposing to solve the hydrodynamical and thermodynamical equations of meteorology by finite difference methods. Although this was the precomputer age, Richardson envisaged that the job could be done "with tier upon tier of human computers fitted into an Albert Hall structure" (P.S. Sheppard in *Nature*, vol. 172, 1953, p. 1127). He also did a penetrating study of atmospheric turbulence, where a nondimensional quantity introduced by him is now called "Richardson's number." At the age of 50 he earned a degree in psychology and began to develop a scientific theory of international relations. He was elected a Fellow of the Royal Society in 1926.

with certain coefficients $a_n^{(n)}, a_{n+1}^{(n)}, \ldots$ *not depending on* h.

We remark that if (2.63) is only an approximation to $K+1$ terms, then the recursion (2.65) is applicable only for $k = 1, 2, \ldots, K$, and (2.66) holds for $n = 1, 2, \ldots, K$, whereas for $n = K+1$ one has $A_{K+1}(h) = a_0 + O(h^{p_{K+1}})$.

It is easily seen from (2.65) that $A_{k+1}(h)$ is a linear combination of $A(h)$, $A(q^{-1}h)$, $A(q^{-2}h), \ldots, A(q^{-k}h)$, where it was tacitly assumed that h, $q^{-1}h$, $q^{-2}h, \ldots$ are admissible values of the parameter.

We now rework Theorem 3.2.2 into a practical algorithm. To do so, we assume that we initially compute $A(h)$ for a succession of parameter values

$$h_0, \quad qh_0, \quad q^2h_0, \ldots \quad (q < 1),$$

all being admissible. Then we define

$$A_{m,k} = A_{k+1}(q^m h_0), \qquad m, k = 0, 1, 2, \ldots \; . \tag{2.67}$$

The idea behind (2.67) is to provide two mechanisms for improving the accuracy: one is to increase m, which reduces the parameter h, the other is increasing k, which engages a more accurate approximation. Ideally, one employs both mechanisms simultaneously, which suggests that the diagonal entries $A_{m,m}$ are the ones of most interest.

Putting $h = q^m h_0$ in (2.65) produces the *extrapolation algorithm*

$$\begin{aligned} A_{m,k} &= A_{m,k-1} + \frac{A_{m,k-1} - A_{m-1,k-1}}{q^{-p_k} - 1}, \quad m \geq k \geq 1, \\ A_{m,0} &= A(q^m h_0). \end{aligned} \tag{2.68}$$

This allows us to compute the triangular scheme of approximations shown in Figure 3.2.1.

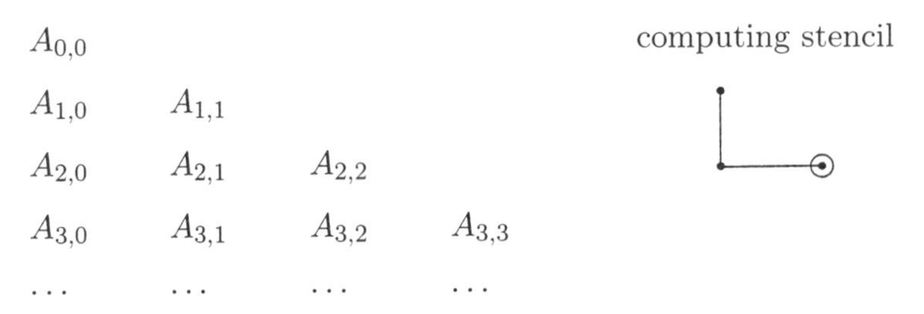

FIGURE 3.2.1. Extrapolation algorithm

Each entry is computed in terms of its neighbors horizontally and diagonally above to the left, as indicated in the computing stencil of Figure 3.2.1. The entries in the first column are the approximations initially computed. The generation of the triangular scheme, once the first column has been computed, is extremely cheap, yet can dramatically improve the accuracy, especially down the diagonal. If (2.63) is a finite asymptotic approximation to $K+1$ terms, then the array in Figure 3.2.1 has a trapezoidal shape, consisting of $K+1$ columns (including the one with $k=0$).

We now apply (2.68) and Theorem 3.2.2 to the case of the composite trapezoidal rule,

$$a_0 = \int_a^b f(x)dx, \tag{2.69}$$

$$A(h) = h\left\{\tfrac{1}{2}f(a) + \sum_{k=1}^{n-1} f(a+kh) + \tfrac{1}{2}f(b)\right\}, \quad h = \frac{b-a}{n}. \tag{2.70}$$

The development of an asymptotic expansion for $A(h)$ in (2.70) is far from trivial. In fact, the result is the content of a well-known formula due to Euler[14] and Maclaurin[15].

Before stating it, we need to define the *Bernoulli*[16] *numbers* B_k; these

[14]Leonhard Euler (1707–1783) was the son of a minister interested in mathematics who followed lectures of Jakob Bernoulli at the University of Basel. Euler himself was allowed to see Johann Bernoulli on Saturday afternoons for private tutoring. At the age of 20, after he was unsuccessful in obtaining a professorship in physics at the University of Basel, because of a lottery system then in use (Euler lost), he emigrated to St. Petersburg; later, he moved on to Berlin, and then back again to St. Petersburg. Euler unquestionably was the most prolific mathematician of the 18th century, working in virtually all branches of the differential and integral calculus and, in particular, being one of the founders of the calculus of variations. He also did pioneering work in the applied sciences, notably hydrodynamics, mechanics of deformable materials and rigid bodies, optics, astronomy, and the theory of the spinning top. Not even his blindness at the age of 59 managed to break his phenomenal productivity. Euler's collected works are still being edited, 71 volumes having already been published.

[15]Colin Maclaurin (1698–1764) was a Scottish mathematician who applied the new infinitesimal calculus to various problems in geometry. He is best known for his power series expansion, but also contributed to the theory of equations.

[16]Jakob Bernoulli (1654–1705), the elder brother of Johann Bernoulli, was active in Basel. He was one of the first to appreciate Leibniz's and Newton's differential and integral calculus and enriched it by many original contributions of his own, often in (not always amicable) competition with his younger brother. He is also known in probability theory for his "law of large numbers."

are the coefficients in the expansion

$$\frac{z}{e^z - 1} = \sum_{k=0}^{\infty} \frac{B_k}{k!} z^k, \quad |z| < 2\pi. \tag{2.71}$$

It is known that

$$\begin{aligned} &B_0 = 1, \quad B_2 = \tfrac{1}{6}, \quad B_4 = -\tfrac{1}{30}, \quad B_6 = \tfrac{1}{42}, \dots , \\ &B_1 = -\tfrac{1}{2}, \quad B_3 = B_5 = \cdots = 0. \end{aligned} \tag{2.72}$$

Furthermore,

$$|B_k| \sim 2\,\frac{k!}{(2\pi)^k} \quad \text{as } k \text{ (even)} \to \infty. \tag{2.73}$$

We now state without proof[17]

---

[17] A heuristic derivation of the formal expansion (2.74) and (2.75), very much in the spirit of Euler, may proceed as follows. We start from Taylor's expansion (where $x, x+h \in [a,b]$)

$$f(x+h) - f(x) = \sum_{k=1}^{\infty} \frac{h^k D^k}{k!} f(x) = (e^{hD} - 1) f(x), \quad D = \frac{d}{dx}.$$

Solving formally for $f(x)$, we get

$$f(x) = (e^{hD} - 1)^{-1}[f(x+h) - f(x)] = \sum_{r=0}^{\infty} \frac{B_r}{r!} (hD)^{r-1}[f(x+h) - f(x)];$$

that is, in view of (2.72),

$$f(x) = [(hD)^{-1} - \tfrac{1}{2} + \sum_{r=2}^{\infty} \frac{B_r}{r!}(hD)^{r-1}][f(x+h) - f(x)]$$

$$= (hD)^{-1}[f(x+h) - f(x)] - \tfrac{1}{2}[f(x+h) - f(x)] + \sum_{r=2}^{\infty} \frac{B_r}{r!} h^{r-1}[f^{(r-1)}(x+h) - f^{(r-1)}(x)]$$

$$= \frac{1}{h}\int_x^{x+h} f(t)dt - \tfrac{1}{2}[f(x+h) - f(x)] + \sum_{k=1}^{\infty} \frac{B_{2k}}{(2k)!} h^{2k-1}[f^{(2k-1)}(x+h) - f^{(2k-1)}(x)].$$

Therefore, bringing the second term to the left-hand side, and multiplying through by h,

$$\frac{h}{2}[f(x+h) + f(x)] = \int_x^{x+h} f(t)dt + \sum_{k=1}^{\infty} \frac{B_{2k}}{(2k)!} h^{2k}[f^{(2k-1)}(x+h) - f^{(2k-1)}(x)].$$

Theorem 3.2.3 (Euler-Maclaurin formula). *Let $A(h)$ be defined by* (2.70), *where $f \in C^{2K+1}[a,b]$ for some integer $K \geq 1$. Then*

$$A(h) = a_0 + a_1 h^2 + a_2 h^4 + \cdots + a_K h^{2K} + O(h^{2K+1}), \quad h \to 0, \tag{2.74}$$

where a_0 is given by (2.69) *and*

$$a_k = \frac{B_{2k}}{(2k)!}\,[f^{(2k-1)}(b) - f^{(2k-1)}(a)], \quad k = 1, 2, \ldots, K. \tag{2.75}$$

Thus, under the assumption of Theorem 3.2.3, we have in (2.74) an asymptotic approximation to $K+1$ terms for $A(h)$, with

$$p_k = 2k, \quad k = 1, 2, \ldots, K, \quad p_{K+1} = 2K + 1, \tag{2.76}$$

and an asymptotic expansion, if $K = \infty$ (i.e., if f has continuous derivatives of any order). Choosing $h_0 = b - a$, $q = \frac{1}{2}$ in (2.67), which is certainly permissible, the scheme (2.68) becomes

$$A_{m,k} = A_{m,k-1} + \frac{A_{m,k-1} - A_{m-1,k-1}}{4^k - 1}, \quad m \geq k \geq 1,$$

and is known as the *Romberg integration* scheme. The choice $q = \frac{1}{2}$ is particularly convenient here, since in the generation of $A_{m,0}$ we can reuse function values already computed (cf. Mach. Ass. 5).

There is an important instance in which the application of Romberg integration would be pointless, namely, when all coefficients $a_1, a_2, \ldots$ in (2.74) are zero. This is the case when f is periodic with period $b - a$, and smooth on $\mathbb{R}$. Indeed, we already know that the composite trapezoidal rule is then exceptionally accurate (cf. §2.1, (2.13)).

Now letting $x = a + ih$ and summing over i from 0 to $n-1$ gives

$$A(h) = \int_a^b f(t)dt + \sum_{k=1}^{\infty} \frac{B_{2k}}{(2k)!} h^{2k} [f^{(2k-1)}(b) - f^{(2k-1)}(a)].$$

NOTES TO CHAPTER 3

§1. Here we are dealing strictly with numerical differentiation, that is, with the problem of obtaining approximations to derivatives that can be used for numerical evaluation. The problem of symbolic differentiation, where the goal is to obtain analytic expressions for derivatives of functions given in analytic form, is handled by most computer algebra systems such as Mathematica and MACSYMA, and we refer to texts in this area cited in §0.3 under *Computer Algebra*. Another important approach to differentiation is what is referred to as automatic differentiation. Here the objective is to create a program (i.e., a piece of software) for computing the derivatives of a function which in turn is given in the form of a program or algorithm. Notable applications are to optimization (calculation of Jacobian and Hessian matrices) and to the solution of ordinary differential equations by Taylor expansion. For an early paper on this subject, see Kedem [1980], for a good cross-section of current activity, Griewank and Corliss [1991], and for a software package Griewank, Juedes, and Utke [1997].

§1.1. The interpolatory formulae for differentiation derived here are analogous to the Newton-Cotes formulae for integration (cf. §2.2) in the sense that one fixes $n+1$ distinct nodes $x_0, x_1, \ldots, x_n$ and interpolates on them by a polynomial of degree n to ensure polynomial degree of exactness n. In analogy to Gauss quadrature formulae, one might well ask whether by a suitable choice of nodes one could substantially increase the degree of precision. If x_0 is the point at which f' is to be approximated, there is actually no compelling reason for including it among the points where f is evaluated. One may thus consider approximating $f'(x_0)$ by $L(f; x_0, h) = h^{-1} \sum_{i=0}^{n} w_i f(x_0 + t_i h)$ and choosing the (real) numbers w_i, t_i so that the integer d for which $Ef = f'(x_0) - L(f; x_0, h) = 0$, all $f \in \mathbb{P}_d$, is as large as possible. This, however, is where the analogy with Gauss ends. For one thing, the t_i need to be normalized somehow, since multiplying all t_i by a constant c and dividing w_i by c yields essentially the same formula. One way of normalization is to require that $|t_i - t_j| \geq 1$ for all $i \neq j$. More important, the possible improvement over $d = n$ (which is achievable by interpolation) is disappointing: the best we can do is obtain $d = n+1$. This can be shown by a simple matrix argument; see Ash, Janson, and Jones [1984]. Among the formulae with $d = n+1$ (which are not unique), one may define an optimal one that minimizes the absolute value of the coefficient in the leading term of the truncation error Ef. These have been derived for each n in Ash, Janson, and Jones [1984], not only for the first, but also for the second derivative. They seem to be optimal also in the presence of noise (for $n = 2$, see Ash and Jones [1981, Thm. 3]), but are still subject to the magnification of noise as exemplified in (1.18). To strike a balance between errors due to truncation and those due to noise, an appropriate step h may be found adaptively; see, for example, Stepleman and Winarsky [1979] and Oliver [1980].

For the sth derivative and its approximation by a formula L as in the preceding paragraph, where h^{-1} is to be replaced by h^{-s}, one may alternatively wish to minimize the "condition mumber" $\sum_{i=0}^{n} |w_i|$. Interestingly, if n and s have the same parity, the optimum is achieved by the extreme points of the nth-degree Chebyshev polynomial T_n; see Rivlin [1975] and Miel and Mooney [1985].

One can do better, especially for high-order derivatives, if one allows the t_i and w_i to be complex and assumes analyticity for f. For the sth derivative, it is then possible (cf. Lyness [1968]) to achieve degree of exactness $n + s$ by choosing the t_i to be the $n + 1$ roots of unity; specifically, one applies the trapezoidal rule to Cauchy's integral for the sth derivative (see (1.19) for $s = 1$). A more sophisticated use of these trapezoidal sums is made in Lyness and Moler [1967]. For practical implementations of these ideas, and algorithms, see Lyness and Sande [1971] and Fornberg [1981].

Considering the derivative of a function f on some interval, say, $[0, 1]$, as the solution on this interval of the (trivial) integral equation $\int_0^x u(t)dt = f(x)$, one can try to combat noise in the data by applying "Tikhonov regularization" to this operator equation; this approach is studied, for example, in King and Murio [1986].

§2. The standard text on the numerical evaluation of integrals — simple as well as multiple — is Davis and Rabinowitz [1984]. It contains a valuable bibliography of over 1500 items. Other useful texts are Krylov [1962], Brass [1977], Engels [1980], and Evans [1993], the last being more practical and application-oriented than the others and containing a detailed discussion of oscillatory integrals. Most quadrature rules are designed to integrate exactly polynomials up to some degree d, that is, all solutions of the differential equation $u^{(d+1)} = 0$. One can generalize and require the rules to be exact for all solutions of a linear homogeneous differential equation of order $d + 1$. This is the approach taken in the book by Ghizzetti and Ossicini [1970]. There are classes of quadrature rules, perhaps more of theoretical than practical interest, that are not covered in our text. One such concerns rules, first studied by Chebyshev, whose weights are all equal (which minimizes the effects of small random errors in the function values), or whose weights have small variance. Surveys on these are given in Gautschi [1976a], Förster [1993], and Brutman [1997b]. Another class includes optimal quadrature formulae, which minimize, for prescribed or variable nodes, the supremum of the modulus of the remainder term taken over some suitable class of functions. For these, and their close relationship to "monosplines," we refer to Nikol'skiĭ [1988] or Levin and Girshovich [1979].

Symbolic integration is inherently more difficult than symbolic differentiation since integrals are often not expressible in analytic form, even if the integrand is an elementary function. A great amount of attention, however, has been given to the problem of automatic integration. Here the user is required to specify the limits of integration, to provide a routine for evaluating the integrand function, and to indicate an error tolerance (absolute, relative, or mixed) and an upper bound for the number of function evaluations to be performed. The automatic integrator is then expected to either return an answer satisfying the user's criteria, or to confess that the error criterion could not be satisfied within the desired volume of computation. In the latter event, a user-friendly integrator will offer a best-possible estimate for the value of the integral along with an estimate of the error. A popular collection of automatic integrators is Quadpack (see Piessens, de Doncker-Kapenga, Überhuber, and Kahaner [1983]), and a good description of the internal workings of automatic integration routines can be found in Davis and Rabinowitz [1984, Ch. 6].

For the important (and difficult) problem of multiple integration and related computational tools, we must refer to special texts, for example, Stroud [1971], Mysovskikh [1981] (for readers familiar with Russian), and Sloan and Joe [1994]. An update to Stroud [1971] is available in Cools and Rabinowitz [1993]. Monte Carlo methods are widely used in statistical physics to compute high-dimensional integrals; recent texts discussing these methods are Kalos and Whitlock [1986], Niederreiter [1992], and Sobol' [1994].

In dealing with definite integrals, one should never lose sight of the many analytical tools available that may help in evaluating or approximating integrals. The reader will find the old, but still pertinent, essay of Abramowitz [1954] informative in this respect, and may also wish to consult Zwillinger [1992a].

§2.1. The result (2.14) and (2.15), in a slightly different form, is proved in Davis and Rabinowitz [1984, p. 209]. Their proof carries over to integrals of the form $\int_{\mathbb{R}} f(x)dx$, if one first lets $n \to \infty$, and then $a \to -\infty$, in their proof.

§2.2. The classical Newton-Cotes formulae (2.16) involve equally spaced nodes (on $[-1,1]$) and weight function $w \equiv 1$. They are useful only for relatively small values of n, since for large n the weights become large and oscillatory in sign. Choosing the Chebyshev points instead removes this obstacle and gives rise to the Fejér quadrature rule. Close relatives are the Filippi rule, which uses the local extreme points of T_{n+1}, and the Clenshaw-Curtis rule, which uses all extreme points (including ± 1) of T_{n-1} as nodes. All three quadrature rules have weights that can be explicitly expressed in terms of trigonometric functions, and they are all positive. The latter has been proved for the first two rules by Fejér [1933], and for the last by Imhof [1963]. Algorithms for accurately computing weighted Newton-Cotes formulae are discussed in Kautsky and Elhay [1982] and Gautschi [1997]; for a computer program, see Elhay and Kautsky [1987].

It is difficult to trace the origin of Theorem 3.2.1, but in essence, Jacobi already was aware of it in 1826, and the idea of the proof, using division by the node polynomial, is his (Jacobi [1826]). There are other noteworthy applications of Theorem 3.2.1. One is to quadrature rules with $2n+1$ points, where n of them are Gauss points and the remaining $n+1$ are to be selected, together with all the weights, so as to make the degree of exactness as large as possible. These are called Gauss-Kronrod formulae (cf. Ex. 18) and have found use in automatic integration routines. Interest has focused on the polynomial of degree $n+1$ — the Stieltjes polynomial — whose zeros are the $n+1$ nodes that were added to the Gauss nodes. In particular, this polynomial must be orthogonal to all polynomials of lower degree with respect to the "weight function" $\pi_n(t;w)w(t)$. The oscillatory character of this weight function poses intriguing questions regarding the location relative to the Gauss nodes, or even the reality, of the added nodes. For a discussion of these and related matters, see the surveys in Gautschi [1988] and Notaris [1994].

There is a theorem analogous to Theorem 3.2.1 that deals with quadrature rules having multiple nodes. The simplest one, first studied by Turán [1950], has constant multiplicity $2s+1$ ($s \geq 0$) for each node; that is, on the right of (2.16) there are also terms involving derivatives up to order $2s$ for each node t_k (cf. Ex. 19 for $s = 1$). If one applies Gauss's principle of maximum algebraic degree of exactness to them, one is led to define the t_k as the zeros of a polynomial of degree n whose $(2s+1)$st power is orthogonal to all lower-degree polynomials (cf. Gautschi [1981, §2.2.1]). This gives rise to what are called s-orthogonal polynomials and to generalizations thereof pertaining to multiplicities that vary from node to node; a good reference for this is Ghizzetti and Ossicini [1970, Ch. 3, §9]; see also Gautschi [1981, §2.2] and Ch. 4, §1.4.

Another class of Gauss-type formulae, where exactness is required not only for polynomials (if at all), but also for rational functions (with prescribed poles), has recently been developed by Van Assche and Vanherwegen [1993] and Gautschi [1993]. They are particularly useful if the integrand has poles near the interval of integration. One can, of course, require exactness for still other systems of functions; for literature on this, see Gautschi [1981, §2.3.3].

§2.3. (iii) The convergence theory for Gauss formulae on infinite intervals is more delicate. Some general results can be found in the book by Freud [1971, Ch.3, §1].

(v) The importance for Gauss quadrature rules of eigenvalues and eigenvectors of the Jacobi matrix and related computational algorithms was first elaborated by Golub and Welsch [1969], although the idea is older. Similar eigenvalue techniques apply to Gauss-Radau and Gauss-Lobatto formulae (Golub [1973]), and indeed to Gauss-Kronrod formulae as well (Laurie [1997]).

A list of published tables of Gauss formulae for various classical and nonclassical weight functions is contained in Gautschi [1981, §5.4], where one also finds a detailed history of Gauss-Christoffel quadrature rules and extensions thereof. A table of recursion coefficients, in particular, also for Jacobi weight functions, can be found in the Appendix to Chihara [1978]. For practical purposes it is important to be able to automatically generate Gauss formulae as needed, even if the Jacobi matrix for them is unknown (and must itself be computed). A major first step in this direction is the computer package in Gautschi [1994b] based on earlier work of the author in Gautschi [1982].

§2.4. Other applications of classical Gaussian quadrature rules, notably to product integration of multiple integrals, are described in the book by Stroud and Secrest [1966], which also contains extensive high-precision tables of Gauss formulae. Prominent use of Gaussian quadrature, especially with Jacobi weight functions, is made in the evaluation of Cauchy principal value integrals in connection with singular integral equations; for this, see, for example, Gautschi [1981, §3.2]. A number of problems in approximation theory that can be solved by nonclassical Gaussian quadrature rules are discussed in Gautschi [1996].

§2.7. A classical account of Romberg integration — one that made this procedure popular — is Bauer, Rutishauser, and Stiefel [1963]. The basic idea, however, can be traced back to 19th-century mathematics, and even beyond. An extensive survey not only of the history, but also of the applications and modifications of extrapolation methods, is given in Joyce [1971] and supplemented in Rabinowitz [1992]. See also Engels [1979] and Dutka [1984] for additional historical accounts.

Richardson extrapolation is just one of many techniques to accelerate the convergence of sequences. For others, we refer to the books by Wimp [1981] and Brezinski and Redivo-Zaglia [1991]. The latter also has computer programs.

EXERCISES AND MACHINE ASSIGNMENTS TO CHAPTER 3

EXERCISES

1. From (1.6) with $n = 3$ we know that

$$f'(x_0) = [x_0, x_1]f + (x_0 - x_1)[x_0, x_1, x_2]f + (x_0 - x_1)(x_0 - x_2)[x_0, x_1, x_2, x_3]f + e,$$

where $e = (x_0 - x_1)(x_0 - x_2)(x_0 - x_3)\frac{f^{(4)}(\xi)}{4!}$. Apply this to

$$x_0 = 0, \quad x_1 = \tfrac{1}{8}, \quad x_2 = \tfrac{1}{4}, \quad x_3 = \tfrac{1}{2}$$

and express $f'(0)$ as a linear combination of $f_k = f(x_k)$, $k = 0, 1, 2, 3$, and e. Also estimate the error e in terms of $M_4 = \max_{0 \le x \le \frac{1}{2}} |f^{(4)}(x)|$.

2. Derive a formula for the error term $r_n'(x)$ of numerical differentiation analogous to Eq. (1.5) but for $x \neq x_0$. {*Hint*: Use Ch. 2, Eq. (2.67) in combination with Ch. 2, Ex. 52.}

3. Let x_i, $i = 0, 1, \ldots, n$, be $n + 1$ distinct points with $H = \max_{1 \le i \le n} |x_i - x_0|$ small.

   (a) Show that for $k = 0, 1, \ldots, n$ one has

$$\frac{d^k}{dx^k} \prod_{i=1}^{n} (x - x_i) \Bigg|_{x = x_0} = O(H^{n-k}) \quad \text{as } H \to 0.$$

   (b) Prove that

$$f^{(n)}(x_0) = n!\, [x_0, x_1, \ldots, x_n]f + e,$$

   where

$$e = \begin{cases} O(H^2) & \text{if } x_0 = \frac{1}{n}\sum_{i=1}^{n} x_i, \\ O(H) & \text{otherwise,} \end{cases}$$

   assuming that f is sufficiently often (how often?) differentiable in the interval spanned by the x_i. {*Hint*: Use the Newton interpolation formula with remainder, in combination with Leibniz's rule of differentiation.}

   (c) Specialize the formula in (b) to equally spaced points x_i with spacing h and express the result in terms of either the nth forward difference $\Delta^n f_0$ or the nth backward difference $\nabla^n f_n$ of the values $f_i = f(x_i)$. {Here $\Delta f_0 = f_1 - f_0$, $\Delta^2 f_0 = \Delta(\Delta f_0) = \Delta f_1 - \Delta f_0 = f_2 - 2f_1 + f_0$, etc., and similarly for ∇f_1, $\nabla^2 f_2$, and so on.}

4. Approximate $\partial u/\partial x|_{P_1}$ in terms of $u_0 = u(P_0)$, $u_1 = u(P_1)$, $u_2 = u(P_2)$ (see figure, where the curve represents a quarter arc of the unit circle). Estimate the error.

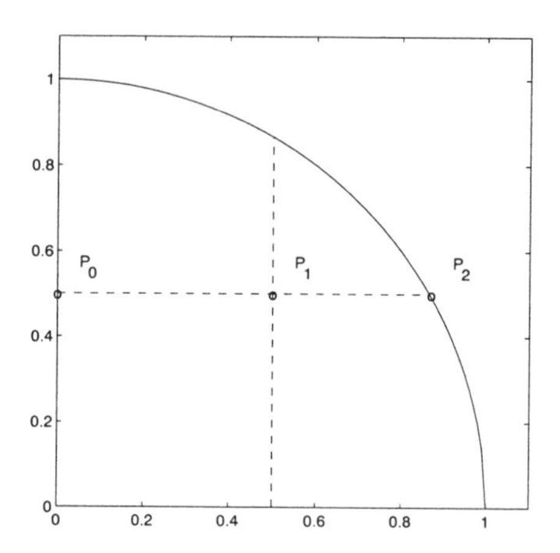

5. Consider the integral $I = \int_{-1}^{1} |x| dx$, whose exact value is evidently 1. Suppose I is approximated (as it stands) by the composite trapezoidal rule $T(h)$ with $h = 2/n$, $n = 1, 2, 3, \ldots$.

   (a) Show (without any computation) that $T(2/n) = 1$ if n is even.

   (b) Determine $T(2/n)$ for n odd and comment on the speed of convergence.

6. Let

$$I(h) = \int_0^h f(x)dx, \quad T(h) = \frac{h}{2}\,[f(0) + f(h)].$$

   (a) Evaluate $I(h)$, $T(h)$, and $E(h) = I(h) - T(h)$ explicitly for $f(x) = x^2 + x^{5/2}$.

   (b) Repeat for $f(x) = x^2 + x^{1/2}$. Explain the discrepancy that you will observe in the order of the error terms.

7. (a) Derive the "midpoint rule" of integration

$$\int_{x_k}^{x_k+h} f(x)dx = hf(x_k + \tfrac{1}{2}h) + \frac{1}{24}h^3 f''(\xi), \quad x_k < \xi < x_k + h.$$

   {*Hint*: Use Taylor's theorem centered at $x_k + \frac{1}{2}h$.}

   (b) Obtain the composite midpoint rule for $\int_a^b f(x)dx$, including the error term, subdividing $[a, b]$ into n subintervals of length $h = \frac{b-a}{n}$.

8. (a) Show that the elementary Simpson's rule can be obtained as follows:

$$\int_{-1}^{1} y(t)dt = \int_{-1}^{1} p_3(y; -1, 0, 0, 1; t)dt + E^S(y).$$

(b) Use (a) to obtain a formula for the remainder $E^S(y)$, assuming $y \in C^4[-1, 1]$.

(c) Using (a) and (b), derive the composite Simpson's rule for $\int_a^b f(x)dx$, including the remainder term.

9. Let $E_n^S(f)$ be the remainder term of the composite Simpson's rule for $\int_0^{2\pi} f(x)dx$ using n subintervals (n even). Evaluate $E_n^S(f)$ for $f(x) = e^{imx}$ ($m = 0, 1, \dots$). Hence determine for what values of d Simpson's rule integrates exactly (on $[0, 2\pi]$) trigonometric polynomials of degree d.

10. Estimate the number of subintervals required to obtain $\int_0^1 e^{-x^2} dx$ to 6 correct decimal places (absolute error $\leq \frac{1}{2} \times 10^{-6}$)

(a) by means of the composite trapezoidal rule,

(b) by means of the composite Simpson's rule.

11. Let f be an arbitrary (continuous) function on [0,1] satisfying $f(x) + f(1 - x) \equiv 1$ for $0 \leq x \leq 1$.

(a) Show that $\int_0^1 f(x)dx = \frac{1}{2}$.

(b) Show that the composite trapezoidal rule for computing $\int_0^1 f(x)dx$ is exact.

(c) Show, with as little computation as possible, that the composite Simpson's rule and more general symmetric rules are also exact.

12. (a) Construct a trapezoidal-like formula

$$\int_0^h f(x)dx = af(0) + bf(h) + E(f), \quad 0 < h < \pi,$$

which is exact for $f(x) = \cos x$ and $f(x) = \sin x$. Does this formula integrate constants exactly?

(b) Show that a similar formula holds for $\int_c^{c+h} g(t)dt$.

13. Given the subdivision Δ of $[0, 2\pi]$ into N equal subintervals, $0 = x_0 < x_1 < x_2 < \cdots < x_{N-1} < x_N = 2\pi$, $x_k = kh$, $h = 2\pi/N$, and a (2π)-periodic function f, construct a quadrature rule for the mth (complex) Fourier coefficients of f,

$$\frac{1}{2\pi} \int_0^{2\pi} f(x)e^{-imx} dx,$$

by approximating f by the spline interpolant $s_1(f;\,\cdot\,)$ from $\mathbb{S}_1^0(\Delta)$. Write the result in the form of a "modified" composite trapezoidal approximation. {*Hint*: Express $s_1(f;\,\cdot\,)$ in terms of the hat functions defined in Chapter 2, Eq. (3.8).}

14. The composite trapezoidal rule for computing $\int_0^1 f(x)dx$ can be generalized to subdivisions

$$\Delta: \quad 0 = x_0 < x_1 < x_2 < \cdots < x_{n-1} < x_n = 1$$

of the interval [0,1] in subintervals of arbitrary length $\Delta x_i = x_{i+1} - x_i$, $i = 0, 1, \ldots, n-1$, by approximating

$$\int_0^1 f(x)dx \approx \int_0^1 s_1(f;x)dx,$$

where $s_1(f;\,\cdot\,) \in \mathbb{S}_1^0(\Delta)$ is the piecewise linear continuous spline interpolating f at $x_0, x_1, \ldots, x_n$.

(a) Use the basis of hat functions $B_0, B_1, \ldots, B_n$ to represent $s_1(f;\,\cdot\,)$ and calculate $\int_0^1 s_1(f;x)dx$.

(b) Discuss the error $E(f) = \int_0^1 f(x)dx - \int_0^1 s_1(f;x)dx$. In particular, find a formula of the type $E(f) = \text{const} \cdot f''(\xi)$, $0 < \xi < 1$, where the constant depends only on Δ.

15. (a) Construct the weighted Newton-Cotes formula

$$\int_0^1 f(x)x^\alpha dx = a_0 f(0) + a_1 f(1) + E(f), \qquad \alpha > -1.$$

Explain why the formula obtained makes good sense.

(b) Derive an expression for the error term $E(f)$ in terms of an appropriate derivative of f.

(c) From the formulae in (a) and (b) derive an approximate integration formula for $\int_0^h g(t)t^\alpha dt$ ($h > 0$ small), including an expression for the error term.

16. (a) Construct the weighted Newton-Cotes formula

$$\int_0^1 f(x) \cdot x\ln(1/x)dx \approx a_0 f(0) + a_1 f(1).$$

{*Hint:* Use $\int_0^1 x^r \ln(1/x)dx = (r+1)^{-2}$, $r = 0, 1, 2, \ldots$.}

(b) Discuss how the formula in (a) can be used to approximate $\int_0^h g(t) \cdot t\ln(1/t)dt$ for small $h > 0$. {*Hint*: Make a change of variables.}

17. Let s be the function defined by

$$s(x) = \begin{cases} (x+1)^3 & \text{if } -1 \le x \le 0, \\ (1-x)^3 & \text{if } 0 \le x \le 1. \end{cases}$$

(a) With Δ denoting the subdivision of $[-1,1]$ into the two subintervals $[-1,0]$ and $[0,1]$, to what class $\mathbb{S}_m^k(\Delta)$ does the spline s belong?

(b) Estimate the error of the composite trapezoidal rule applied to $\int_{-1}^{1} s(x)dx$, when $[-1,1]$ is divided into n subintervals of equal length $h = 2/n$ and n is even.

(c) What is the error of the composite Simpson's rule applied to $\int_{-1}^{1} s(x)dx$, with the same subdivision of $[-1,1]$ as in (b)?

(d) What is the error resulting from applying the 2-point Gauss-Legendre rule to $\int_{-1}^{0} s(x)dx$ and $\int_{0}^{1} s(x)dx$ separately and summing?

18. (Gauss-Kronrod rule) Let $\pi_n(\,\cdot\,;w)$ be the (monic) orthogonal polynomial of degree n relative to a nonnegative weight function w on $[a,b]$, and $t_k^{(n)}$ its zeros. Use Theorem 3.2.1 to determine conditions on w_k, w_k^*, t_k^* for the quadrature rule

$$\int_a^b f(t)w(t)dt = \sum_{k=1}^{n} w_k f(t_k^{(n)}) + \sum_{k=1}^{n+1} w_k^* f(t_k^*) + E_n(f)$$

to have degree of exactness at least $3n+1$; that is, $E_n(f) = 0$ for all $f \in \mathbb{P}_{3n+1}$.

19. (Turán quadrature formula) Let w be a nonnegative weight function on $[a,b]$. Prove: the quadrature formula

$$\int_a^b f(t)w(t)dt = \sum_{k=1}^{n} [w_k f(t_k) + w_k' f'(t_k) + w_k'' f''(t_k)] + E_n(f)$$

has degree of exactness $d = 4n-1$ if and only if the following conditions are satisfied.

(a) The formula is (Hermite-) interpolatory; that is, $E_n(f) = 0$ if $f \in \mathbb{P}_{3n-1}$.

(b) The node polynomial $\omega_n(t) = \Pi_{k=1}^{n}(t - t_k)$ satisfies

$$\int_a^b [\omega_n(t)]^3 p(t)w(t)dt = 0 \qquad \text{for all } \ p \in \mathbb{P}_{n-1}.$$

{*Hint*: Simulate the proof of Theorem 3.2.1.}

20. Consider $s > 1$ weight functions $w_\sigma(t)$, $\sigma = 1, 2, \ldots, s$, integers m_σ such that $\sum_{\sigma=1}^{s} m_\sigma = n$, and s quadrature rules

$$Q_\sigma : \quad \int_a^b f(t) w_\sigma(t) dt = \sum_{k=1}^{n} w_{k,\sigma} f(t_k) + E_{n,\sigma}(f), \quad \sigma = 1, 2, \ldots, s,$$

which share n common nodes t_k but have individual weights $w_{k,\sigma}$. State necessary and sufficient conditions for Q_σ to have degree of exactness $n + m_\sigma - 1$, $\sigma = 1, 2, \ldots, s$, and explain why this is likely to be optimal.

21. Consider a quadrature formula of the form

$$\int_0^1 f(x) dx \approx a_0 f(0) + a_1 f'(0) + \sum_{k=1}^{n} w_k f(x_k) + b_0 f(1).$$

(a) Call the formula "Hermite-interpolatory" if the right-hand side is obtained by integrating on the left instead of f the (Hermite) interpolation polynomial p satisfying

$$p(0) = f(0), \quad p'(0) = f'(0), \quad p(1) = f(1),$$
$$p(x_k) = f(x_k), \quad k = 1, 2, \ldots, n.$$

What degree of exactness does the formula have in this case (regardless of how the nodes x_k are chosen, as long as they are mutually distinct and strictly inside the interval $[0, 1]$)?

(b) What is the maximum degree of exactness expected to be if all coefficients and nodes x_k are allowed to be freely chosen?

(c) Show that for the maximum degree of exactness to be achieved, it is necessary that $\{x_k\}$ are the zeros of the polynomial π_n of degree n which is orthogonal on $[0, 1]$ with respect to the weight function $w(x) = x^2(1 - x)$. Identify this polynomial in terms of one of the classical orthogonal polynomials.

(d) Show that the choice of the x_k in (c) together with the requirement of the quadrature formula to be Hermite-interpolatory, is sufficient for the maximum degree of exactness to be attained.

22. Show that the Gauss-Radau as well as the Gauss-Lobatto formulae are positive if the weight function w is nonnegative and not identically zero. {*Hint*: Modify the proof given for the Gauss formula in §2.3(ii).}

23. (FEJÉR, 1933). Let t_k, $k = 1, 2, \ldots, n$ be the zeros of

$$\omega_n(t) = P_n(t) + \alpha P_{n-1}(t) + \beta P_{n-2}(t), \quad n \geq 2,$$

where $\{P_k\}$ are the Legendre polynomials, and assume $\alpha \in \mathbb{R}$, $\beta \le 0$, and the zeros t_k real and pairwise distinct. Show that the Newton-Cotes formula

$$\int_{-1}^{1} f(t)dt = \sum_{k=1}^{n} w_k f(t_k) + E_n(f), \quad E_n(\mathbb{P}_{n-1}) = 0,$$

has all weights positive: $w_k > 0$ for $k = 1, 2, \ldots, n$. $\{$*Hint*: Define $\Delta_k(t) = [\ell_k(t)]^2 - \ell_k(t)$ and show that $\int_{-1}^{1} \Delta_k(t)dt \le 0.\}$

24. (a) Determine by Newton's interpolation formula the quadratic polynomial p interpolating f at $x = 0$ and $x = 1$ and f' at $x = 0$. Also express the error in terms of an appropriate derivative (assumed continuous on [0,1]).

    (b) Based on the result of (a), derive an integration formula of the type

$$\int_0^1 f(x)dx = a_0 f(0) + a_1 f(1) + b_0 f'(0) + E(f).$$

    Determine a_0, a_1, b_0 and an appropriate expression for $E(f)$.

    (c) Transform the result of (b) to obtain an integration rule, with remainder, for $\int_c^{c+h} y(t)dt$, where $h > 0$. $\{$Do not rederive this rule from scratch.$\}$

25. Imitate the procedures used in Ch. 2, §1.4, for monic Legendre polynomials to show orthogonality on $[0, \infty)$ relative to the Laguerre measure $d\lambda(t) = t^\alpha e^{-t}dt$ of the (monic) polynomials

$$\pi_k(t) = (-1)^k t^{-\alpha} e^t \frac{d^k}{dt^k}(t^{\alpha+k} e^{-t}), \quad k = 0, 1, 2, \ldots ,$$

and to derive explicit formulae for the recursion coefficients α_k, β_k. $\{$*Hint*: Express α_k and β_k in terms of the coefficients λ_k, μ_k in $\pi_k(t) = t^k + \lambda_k t^{k-1} + \mu_k t^{k-2} + \cdots .\}$

26. Show that

$$\pi_k(t) = \frac{(-1)^k}{2^k} e^{t^2} \frac{d^k}{dt^k}(e^{-t^2}), \quad k = 0, 1, 2, \ldots ,$$

are the monic orthogonal polynomials on $\mathbb{R}$ relative to the Hermite measure $d\lambda(t) = e^{-t^2}dt$. Use this "Rodrigues formula" directly to derive the recurrence relation for the (monic) Hermite polynomials.

27. (a) Construct the quadratic polynomial $\pi_2(\,\cdot\,; w)$ orthogonal on $(0, \infty)$ with respect to the weight function $w(t) = e^{-t}$. $\{$*Hint*: Use $\int_0^\infty t^m e^{-t}dt = m!.\}$

(b) Obtain the two-point Gauss-Laguerre quadrature formula,

$$\int_0^\infty f(t)e^{-t}dt = w_1 f(t_1) + w_2 f(t_2) + E_2(f),$$

including a representation for the remainder $E_2(f)$.

(c) Apply the formula in (b) to approximate $I = \int_0^\infty e^{-t}dt/(t+1)$. Use the remainder term $E_2(f)$ to estimate the error, and compare your estimate with the true error $\{I = .596347361\ldots\}$. Knowing the true error, identify the unknown quantity $\xi > 0$ contained in the error term $E_2(f)$.

28. Derive the 2-point Gauss-Hermite quadrature formula,

$$\int_{-\infty}^\infty f(t)e^{-t^2}dt = w_1 f(t_1) + w_2 f(t_2) + E_2(f),$$

including an expression for the remainder $E_2(f)$.

$\{$*Hint*: Use $\int_0^\infty t^{2n}e^{-t^2}dt = \frac{(2n)!}{n!2^{2n}}\frac{\sqrt{\pi}}{2}.\}$

29. Let $\pi_n(\,\cdot\,;w)$ be the nth-degree orthogonal polynomial with respect to the weight function w on $[a,b]$, $t_1, t_2, \ldots, t_n$ its n zeros, and $w_1, w_2, \ldots, w_n$ the n Gauss weights.

(a) Assuming $n > 1$, show that the n polynomials $\pi_0, \pi_1, \ldots, \pi_{n-1}$ are also orthogonal with respect to the *discrete* inner product $(u,v) = \sum_{\nu=1}^n w_\nu u(t_\nu)v(t_\nu)$.

(b) With $\ell_i(t) = \prod_{k\neq i}[(t-t_k)/(t_i-t_k)]$, $i = 1, 2, \ldots, n$, denoting the elementary Lagrange interpolation polynomials associated with the nodes $t_1, t_2, \ldots, t_n$, show that

$$\int_a^b \ell_i(t)\ell_k(t)w(t)dt = 0 \quad \text{if} \quad i \neq k.$$

30. Consider a quadrature formula of the type

$$\int_0^\infty e^{-x}f(x)dx = af(0) + bf(c) + E(f).$$

(a) Find a, b, c such that the formula has degree of exactness $d = 2$. Can you identify the formula so obtained? $\{$*Point of information*: $\int_0^\infty e^{-x}x^r dx = r!\}$

(b) Let $p_2(x) = p_2(f; 0, 2, 2; x)$ be the Hermite interpolation polynomial interpolating f at the (simple) point $x = 0$ and the double point $x = 2$. Determine $\int_0^\infty e^{-x}p_2(x)dx$ and compare with the result in (a).

(c) Obtain the remainder $E(f)$ in the form $E(f) = \text{const} \cdot f'''(\xi)$, $\xi > 0$.

31. In this problem, $\pi_j(\,\cdot\,; w)$ denotes the monic polynomial of degree j orthogonal on the interval $[a, b]$ relative to a weight function $w \geq 0$.

    (a) Show that $\pi_n(\,\cdot\,; w)$, $n > 0$, has at least one real zero in the interior of $[a, b]$ at which π_n changes sign.

    (b) Prove that all zeros of $\pi_n(\,\cdot\,; w)$ are real, simple, and contained in the interior of $[a, b]$. {*Hint*: Put $r_0 = \max\{r \geq 1$: $t_{k_1}^{(n)}, t_{k_2}^{(n)}, \ldots, t_{k_r}^{(n)}$ are distinct real zeros of π_n in (a, b) at each of which π_n changes sign$\}$. Show that $r_0 = n$.}

32. Prove that the zeros of $\pi_n(\,\cdot\,; w)$ interlace with those of $\pi_{n+1}(\,\cdot\,; w)$.

33. Consider the Hermite interpolation problem: Find $p \in \mathbb{P}_{2n-1}$ such that

$$(*) \qquad p(\tau_\nu) = f_\nu, \quad p'(\tau_\nu) = f'_\nu, \quad \nu = 1, 2, \ldots, n.$$

There are "elementary Hermite interpolation polynomials" h_ν, k_ν such that the solution of (*) can be expressed (in analogy to Lagrange's formula) in the form

$$p(t) = \sum_{\nu=1}^{n} [h_\nu(t) f_\nu + k_\nu(t) f'_\nu].$$

   (a) Seek h_ν and k_ν in the form

$$h_\nu(t) = (a_\nu + b_\nu t)\ell_\nu^2(t), \quad k_\nu(t) = (c_\nu + d_\nu t) l_\nu^2(t),$$

   where ℓ_ν are the elementary Lagrange polynomials. Determine the constants a_ν, b_ν, c_ν, d_ν.

   (b) Obtain the quadrature rule

$$\int_a^b f(t) w(t) dt = \sum_{\nu=1}^{n} [\lambda_\nu f(\tau_\nu) + \mu_\nu f'(\tau_\nu)] + E_n(f)$$

   with the property that $E_n(f) = 0$ for all $f \in \mathbb{P}_{2n-1}$.

   (c) What conditions on the node polynomial $\omega_n(t) = \prod_{\nu=1}^{n}(t - \tau_\nu)$ (or on the nodes τ_ν) must be imposed in order that $\mu_\nu = 0$ for $\nu = 1, 2, \ldots, n$?

34. Show that $\int_0^1 (1-t)^{-1/2} f(t) dt$, when f is smooth, can be computed accurately by Gauss-Legendre quadrature. {*Hint*: Substitute $1 - t = x^2$.}

35. The Gaussian quadrature rule for the (Chebyshev) weight function $w(t) = (1 - t^2)^{-1/2}$ is known to be

$$\int_{-1}^{1} f(t)(1 - t^2)^{-1/2} dt \approx \frac{\pi}{n} \sum_{k=1}^{n} f(t_k^C), \qquad t_k^C = \cos\left(\frac{2k - 1}{2n}\pi\right).$$

(The nodes t_k^C are the n Chebyshev points.) Use this fact to show that the unit disk has area π.

36. Assuming f is a well-behaved function, discuss how the following integrals can be approximated by *standard* Gauss-type rules (i.e., with canonical intervals and weight functions).

   (a) $\int_a^b f(x)dx \quad (a < b)$.

   (b) $\int_1^\infty e^{-ax} f(x)dx \quad (a > 0)$.

   (c) $\int_{-\infty}^\infty e^{-(ax^2+bx)} f(x)dx \quad (a > 0)$. {*Hint*: Complete the square.}

   (d) $\int_0^\infty \frac{e^{-xt}}{y+t} dt,\ x > 0,\ y > 0$. Is the approximation you get for the integral too small or too large? Explain.

37. Let f be a smooth function on $[0, \pi]$. Explain how best to evaluate

$$I_{\alpha,\beta}(f) = \int_0^\pi f(\theta)[\cos \tfrac{1}{2}\theta]^\alpha [\sin \tfrac{1}{2}\theta]^\beta d\theta, \quad \alpha > -1, \quad \beta > -1.$$

38. Given a nonnegative weight function w on $[-1, 1]$ and $x > 1$, let

$$(G) \qquad \int_{-1}^1 f(t)\frac{w(t)}{x^2 - t^2} dt = \sum_{k=1}^n w_k^G f(t_k^G) + E_n^G(f), \quad n \geq 2,$$

be the n-point Gaussian quadrature formula for the weight function $\frac{w(t)}{x^2-t^2}$. (Note that t_k^G, w_k^G both depend on n and x.) Consider the quadrature rule

$$\int_{-1}^1 g(t)w(t)dt = \sum_{k=1}^n w_k g(t_k) + E_n(g),$$

where $t_k = t_k^G$, $w_k = [x^2 - (t_k^G)^2]w_k^G$. Prove:

   (a) $E_n(g) = 0$ if $g(t) = \frac{1}{t \pm x}$.

   (b) $E_n(g) = 0$ whenever g is a polynomial of degree $\leq 2n - 3$.

39. Let ξ_ν, $\nu = 1, 2, \ldots, 2n$, be $2n$ preassigned distinct numbers satisfying $-1 < \xi_\nu < 1$, and let w be a positive weight function on [-1,1]. Define $\omega_{2n}(x) = \prod_{\nu=1}^{2n} (1 + \xi_\nu x)$. (Note that ω_{2n} is positive on [-1,1].) Let x_k^G, w_k^G be the nodes and weights of the n-point Gauss formula for the weight function $w^*(x) = \frac{w(x)}{\omega_{2n}(x)}$:

$$\int_{-1}^1 p(x)w^*(x)dx = \sum_{k=1}^n w_k^G p(x_k^G), \quad p \in \mathbb{P}_{2n-1}.$$

Define $x_k^* = x_k^G$, $w_k^* = w_k^G \omega_{2n}(x_k^G)$. Show that the quadrature formula

$$\int_{-1}^{1} f(x)w(x)dx = \sum_{k=1}^{n} w_k^* f(x_k^*) + E_n^*(f)$$

is exact for the $2n$ rational functions

$$f(x) = \frac{1}{1+\xi_\nu x}, \quad \nu = 1, 2, \ldots, 2n.$$

40. Prove (2.27).

41. (a) Use the method of undetermined coefficients to obtain an integration rule (having degree of exactness $d = 2$) of the form

$$\int_0^1 y(s)ds \approx ay(0) + by(1) - c[y'(1) - y'(0)].$$

(b) Transform the rule in (a) into one appropriate for approximating $\int_x^{x+h} f(t)dt$.

(c) Obtain a composite integration rule based on the formula in (b) for approximating $\int_a^b f(t)dt$. Interpret the result.

42. Determine the quadrature formula of the type

$$\int_{-1}^{1} f(t)dt = \alpha_{-1}\int_{-1}^{-1/2} f(t)dt + \alpha_0 f(0) + \alpha_1 \int_{1/2}^{1} f(t)dt + E(f)$$

having maximum degree of exactness d. What is the value of d?

43. (a) Determine the quadratic spline $s_2(x)$ on $[-1, 1]$ with a single knot at $x = 0$ and such that $s_2(x) \equiv 0$ on $[-1, 0]$ and $s_2(1) = 1$.

(b) Consider a function $s(x)$ of the form

$$s(x) = c_0 + c_1 x + c_2 x^2 + c_3 s_2(x), \quad c_i = \text{const},$$

where $s_2(x)$ is as defined in (a). What kind of function is s? Determine s such that

$$s(-1) = f_{-1}, \quad s(0) = f_0, \quad s'(0) = f_0', \quad s(1) = f_1,$$

where $f_{-1} = f(-1)$, $f_0 = f(0)$, $f_0' = f'(0)$, $f_1 = f(1)$ for some function f on $[-1, 1]$.

(c) What quadrature rule does one obtain if one approximates $\int_{-1}^{1} f(x)dx$ by $\int_{-1}^{1} s(x)dx$, with s as obtained in (b)?

44. Prove that the condition (2.46′) does not depend on the choice of the basis $\varphi_1, \varphi_2, \ldots, \varphi_n$.

45. Show that the Peano kernel $K_d(t)$ of a functional E vanishes for $t \notin [a,b]$, where $[a,b]$ is the interval of function values referenced by E.

46. Show that a linear functional E satisfying $Ef = e_{d+1} f^{(d+1)}(\bar{t})$, $e_{d+1} \neq 0$, for any $f \in C^{d+1}[a,b]$, is necessarily definite of order d if it has a continuous Peano kernel K_d.

47. Let $E(f) = \int_a^b f(x)dx - \sum_{k=1}^n w_k f(x_k)$ be the error of a quadrature rule having degree of exactness d. Show that none of the Peano kernels K_0, $K_1, \ldots, K_{d-1}$ can be definite.

48. Assume, in Simpson's rule

$$\int_{-1}^{1} f(x)dx = \frac{1}{3}[f(-1) + 4f(0) + f(1)] + E(f),$$

that f is only of class $C^2[-1,1]$ instead of class $C^4[-1,1]$ as normally assumed.

(a) Find an error estimate of the type

$$|E(f)| \leq \text{const} \cdot \|f''\|_\infty, \quad \|f''\|_\infty = \max_{-1 \leq x \leq 1} |f''(x)|.$$

{*Hint*: Apply Peano's theorem to $E(f)$.}

(b) Transform the result in (a) to obtain Simpson's formula, with remainder estimate, for the integral

$$\int_{c-h}^{c+h} g(t)dt, \quad g \in C^2[c-h, c+h], \quad h > 0.$$

49. Consider the trapezoidal formula "with mean values,"

$$\int_0^1 f(x)dx = \frac{1}{2}\left[\frac{1}{\varepsilon}\int_0^{\varepsilon} f(x)dx + \frac{1}{\varepsilon}\int_{1-\varepsilon}^{1} f(x)dx\right] + E(f), \quad 0 < \varepsilon < \frac{1}{2}.$$

(a) Determine the degree of exactness of this formula.

(b) Express the remainder $E(f)$ by means of the Peano theorem, in terms of f'', assuming $f \in C^2[0,1]$.

(c) Show that the first Peano kernel K_1 is definite, and thus express the remainder in the form $E(f) = e_2 f''(\tau)$, $0 < \tau < 1$.

(d) Study (and explain) the limit cases $\varepsilon \downarrow 0$ and $\varepsilon \to \frac{1}{2}$.

50. (a) Use the method of undetermined coefficients to construct a quadrature formula of the type

$$\int_0^1 f(x)dx = af(0) + bf(1) + cf''(\gamma) + E(f)$$

having maximum degree of exactness d, the variables being a, b, c, and γ.

(b) Show that the Peano kernel K_d for the formula obtained in (a) is definite, and hence express the remainder in the form $E(f) = e_{d+1}f^{(d+1)}(\xi)$, $0 < \xi < 1$.

51. (a) Use the method of undetermined coefficients to construct a quadrature formula of the type

$$\int_0^1 f(x)dx = -\alpha f'(0) + \beta f(\tfrac{1}{2}) + \alpha f'(1) + E(f)$$

that has maximum degree of exactness.

(b) What is the precise degree of exactness of the formula obtained in (a)?

(c) Use the Peano kernel of the error functional E to express $E(f)$ in terms of the appropriate derivative of f reflecting the result of (b).

(d) Transform the formula in (a) to one that is appropriate to evaluate $\int_c^{c+h} g(t)dt$, and then obtain the corresponding composite formula for $\int_a^b g(t)dt$, using n subintervals of equal length, and derive an error term. Interpret your result.

52. Consider a quadrature rule of the form

$$\int_0^1 x^\alpha f(x)dx \approx Af(0) + B\int_0^1 f(x)dx, \quad \alpha > -1, \quad \alpha \neq 0.$$

(a) Determine A and B such that the formula has degree of exactness $d = 1$.

(b) Let $E(f)$ be the error functional of the rule determined in (a). Show that the Peano kernel $K_1(t) = E_{(x)}((x-t)_+)$ of E is positive definite if $\alpha > 0$, and negative definite if $\alpha < 0$.

(c) Based on the result of (b), determine the constant e_2 in $E(f) = e_2 f''(\xi)$, $0 < \xi < 1$.

53. (a) Consider a quadrature formula of the type

$$(*) \qquad \int_0^1 f(x)dx = \alpha f(x_1) + \beta[f(1) - f(0)] + E(f)$$

and determine α, β, x_1 such that the degree of exactness is as large as possible. What is the maximum degree attainable?

(b) Use interpolation theory and the Peano theorem to obtain a bound on $|E(f)|$ in terms of $\|f^{(r)}\|_\infty = \max_{0\le x\le 1} |f^{(r)}(x)|$ for some suitable r.

(c) Adapt (*), including the bound on $|E(f)|$, to an integral of the form $\int_c^{c+h} f(t)dt$, where c is some constant and $h > 0$.

(d) Apply the result of (c) to develop a composite quadrature rule for $\int_a^b f(t)dt$ by subdividing $[a, b]$ into n subintervals of equal length $h = \frac{b-a}{n}$. Find a bound for the total error.

54. Construct a quadrature rule

$$\int_0^1 x^\alpha f(x)dx \approx a_1 \int_0^1 f(x)dx + a_2 \int_0^1 xf(x)dx, \quad 0 < \alpha < 1,$$

(a) which is exact for all polynomials p of degree ≤ 1;

(b) which is exact for all $f(x) = x^{1/2}p(x)$, $p \in \mathbb{P}_1$.

55. Let

$$a = x_0 < x_1 < x_2 < \cdots < x_{n-1} < x_n = b, \quad x_k = a + kh, \quad h = \frac{b-a}{n},$$

be a subdivision of $[a, b]$ into n equal subintervals.

(a) Derive an elementary quadrature formula for the integral $\int_{x_k}^{x_{k+1}} f(x)\,dx$, including a remainder term, by approximating f by the cubic Hermite interpolation polynomial $p_3(f;\, x_k,\, x_k,\, x_{k+1},\, x_{k+1};\, x)$ and then integrating over $[x_k,\, x_{k+1}]$. Interpret the result.

(b) Develop the formula obtained in (a) into a composite quadrature rule, with remainder term, for the integral $\int_a^b f(x)dx$.

56. (a) Given a function $g(x, y)$ on the unit square $0 \le x \le 1$, $0 \le y \le 1$, determine a "bilinear polynomial" $p(x, y) = a + bx + cy + dxy$ such that p has the same values as g at the four corners of the square.

(b) Use (a) to obtain a cubature formula for $\int_0^1\int_0^1 g(x, y)dxdy$ that involves the values of g at the four corners of the unit square. What rule does this reduce to if g is a function of x only (i.e., does not depend on y)?

(c) Use (b) to find a "composite cubature rule" for $\int_0^1\int_0^1 g(x, y)dxdy$ involving the values $g_{i,j} = g(ih, jh)$, $i, j = 0, 1, \ldots, n$, where $h = 1/n$.

57. (a) Let $d_1(h) = (f(h) - f(0))/h$, $h > 0$, be the difference quotient of f at the origin. Describe how the extrapolation method based on a suitable expansion of $d_1(h)$ can be used to approximate $f'(0)$ to successively higher accuracy.

(b) Develop a similar method for calculating $f''(0)$, based on $d_2(h) = [f(h) - 2f(0) + f(-h)]/h^2$.

MACHINE ASSIGNMENTS

1. Let $f(x) = \frac{1}{1-\pi x}$ and $f_i = f(ih)$, $i = -2, -1, 0, 1, 2$. Compute (in double precision)

$$e_n(h) = f'(0) - \frac{1}{h^n} \nabla^n f_n, \quad n = 1, 2, 3, 4,$$

and try to verify the speed of convergence as $h \to 0$ by printing, for $n = 1, \ldots, 4$,

$$e_n(h_k), \quad r_k := \frac{e_n(h_k)}{e_n(h_{k-1})}, \quad h_k = \tfrac{1}{4} \cdot 2^{-k}, \qquad \text{for } k = 1, 2, \ldots, 10.$$

Comment on the results.

2. Let

$$f(z) = \tan z, \qquad |z| < \tfrac{1}{2}\pi.$$

(a) Express

$$y_m(\theta) = \text{Re}\,\{e^{-im\theta} f(re^{i\theta})\}, \quad 0 < r < \tfrac{1}{2}\pi, \quad m = 1, 2, 3, \ldots$$

explicitly as a function of θ. {*Hint*: Use Euler's identities $\sin z = \frac{1}{2i}(e^{iz} - e^{-iz})$, $\cos z = \frac{1}{2}(e^{iz} + e^{-iz})$, valid for arbitrary complex z.}

(b) Obtain the analogue to (1.19) for the mth derivative and thus write $f^{(m)}(x_0)$ as a definite integral over $[0, 2\pi]$.

(c) Write a MATLAB script to compute $f^{(m)}(0)$ for $m = 1(1)5$ using the integral in (b) in conjunction with the composite trapezoidal rule (cf. §2.1) relative to a subdivision of $[0, 2\pi]$ into n subintervals. Use $r = \frac{1}{12}\rho\pi$, $\rho = 1, 2, 3, 4, 5$, and $n = 5(5)50$. For each m print a table whose columns contain r, n, the trapezoidal approximation $t_n^{(m)}$, and the (absolute) error, in this order. Comment on the results; in particular, try to explain the convergence behavior as r increases and the difference in behavior for n even and n odd. {*Hint*: Prepare plots of the integrand.}

(d) Do the same as (c), but for $f^{(m)}(\frac{7}{16}\pi)$ and $r = \frac{1}{32}\pi$.

(e) Write and run a script for approximating $f^{(m)}(0)$, $m = 1(1)5$, by central difference formulae with steps $h = 1, \frac{1}{5}, \frac{1}{25}, \frac{1}{125}, \frac{1}{625}$. Comment on the results.

(f) Do the same as (e), but for $f^{(m)}(\frac{7}{16}\pi)$ and $h = \frac{1}{32}\pi, \frac{1}{160}\pi, \frac{1}{800}\pi, \frac{1}{4000}\pi$.

3. Below are a number of suggestions as to how the following integrals may be computed,

$$I_c = \int_0^1 \frac{\cos x}{\sqrt{x}} dx, \quad I_s = \int_0^1 \frac{\sin x}{\sqrt{x}} dx.$$

   (a) Use the composite trapezoidal rule with n intervals of equal length $h = 1/n$, "ignoring" the singularity at $x = 0$ (i.e., arbitrarily using zero as the value of the integrand at $x = 0$).

   (b) Use the composite trapezoidal rule over the interval $[h, 1]$ with $n - 1$ intervals of length $h = 1/n$ in combination with a weighted Newton-Cotes rule with weight function $w(x) = x^{-1/2}$ on the interval $[0, h]$. {Adapt the formula (2.24) to the interval $[0, h]$.}

   (c) Make the change of variables $x = t^2$ and apply the composite trapezoidal rule to the resulting integrals.

   (d) Use Gauss-Legendre quadrature on the integrals obtained in (c).

   (e) Use Gauss-Jacobi quadrature with parameters $\alpha = 0$ and $\beta = -\frac{1}{2}$ directly on the integrals I_c and I_s.

   {As a *point of information*, $I_c = \sqrt{2\pi}\, C\left(\sqrt{\frac{2}{\pi}}\right) = 1.809045218947\ldots$, $I_s = \sqrt{2\pi}\, S\left(\sqrt{\frac{2}{\pi}}\right) = .620549071924\ldots$, where $C(x)$, $S(x)$ are the Fresnel integrals.}

   Implement and run the proposed methods for $n = 100(100)1000$ in (a) and (b), for $n = 20(20)200$ in (c), and for $n = 1(1)4$ in (d) and (e). Try to explain the results you obtain. {To get the required subroutines for Gaussian quadrature, use the commands

```
mail netlib@ornl.gov
send Algorithm 726 from toms .
```

   Extract the routines `r1mach`, `recur`, and `gauss`, and name them `r1mach.f`, `recur.f`, and `gauss.f`. In the first routine, highlight the data for "machine constants for ieee arithmetic machines" and comment out the data for "machine constants for cdc cyber 205 and eta-10".}

4. For a natural number p let

$$I_p = \int_0^1 (1 - t)^p f(t) dt$$

   be (except for the factor $1/p!$) the pth iterated integral of f; cf. Eq. (2.31). Compare the composite trapezoidal rule based on n subintervals with the

n-point Gauss-Jacobi rule with parameters $\alpha = p$ and $\beta = 0$. Take, for example, $f(t) = \tan t$ and $p = 5(5)20$, and let $n = 10(10)50$ in the case of the trapezoidal rule, and $n = 1(1)5$ for the Gauss rule. {See Mach. Ass. 3 for instructions on how to obtain routines for generating the Gaussian quadrature rules.}

5. (a) Let $M_h(f) = h\sum_{k=1}^{n} f(a + (k - \frac{1}{2})h)$, $h = \frac{b-a}{n}$, denote the composite midpoint rule for $\int_a^b f(x)dx$. Show that the first column of the Romberg array $\{T_{km}\}$ can be generated recursively as follows:

$$T_{0,0} = \frac{b-a}{2}\,[f(a) + f(b)],$$

$$T_{k+1,0} = \tfrac{1}{2}\,[T_{k,0} + M_{h_k}(f)], \quad k = 0, 1, 2, \ldots\,,$$

where $h_k = (b-a)/2^k$.

(b) Write a subroutine for computing $\int_a^b f(x)dx$ by the Romberg integration scheme, with $h_k = (b-a)/2^k$, $k = 0, 1, \ldots, n-1$.

Formal parameters: `a, b, n, f, t`, where `t` is the two-dimensional Romberg array, of dimension (n,n), `f` a function subroutine for $f(x)$, and `a, b` the limits of integration.

Order of computation: Generate `t` row by row; generate the trapezoidal sums recursively as in part (a).

Program size: Keep it down to about 20 lines of FORTRAN code.

Output: $\mathtt{t}_{k,0}$, $\mathtt{t}_{k,k}$, $k = 0, 1, \ldots, n-1$.

(c) Call your subroutine (with $n = 10$) to approximate the following integrals.

(i) $\int_1^2 \frac{e^x}{x}\,dx$ ("exponential integral")

(ii) $\int_0^1 \frac{\sin x}{x}\,dx$ ("sine integral")

(iii) $\frac{1}{\pi}\int_0^\pi \cos(yx)dx, \quad y = 1.7$

(iv) $\frac{1}{\pi}\int_0^\pi \cos(y \sin x)dx, \quad y = 1.7$

(v) $\int_0^1 \sqrt{1-x^2}\,dx$

(vi) $\int_0^2 f(x)dx, \quad f(x) = \begin{cases} x, & 0 \le x \le \sqrt{2}, \\ \dfrac{\sqrt{2}}{2-\sqrt{2}}\,(2-x), & \sqrt{2} \le x \le 2 \end{cases}$

(vii) $\displaystyle\int_0^2 f(x)dx, \quad f(x) = \begin{cases} x, & 0 \leq x \leq \frac{3}{4}, \\ \frac{3}{5}(2-x), & \frac{3}{4} \leq x \leq 2 \end{cases}$

(d) Comment on the behavior of the Romberg scheme in each of the seven cases in part (c).

CHAPTER 4

NONLINEAR EQUATIONS

The problems discussed in this chapter may be written generically in the form

$$f(x) = 0, \tag{0.1}$$

but allow different interpretations depending on the meaning of x and f. The simplest case is a *single equation* in a single unknown, in which case f is a given function of a real or complex variable, and we are trying to find values of this variable for which f vanishes. Such values are called *roots of the equation* (0.1), or *zeros of the function* f. If x in (0.1) is a vector, say, $x = [x_1, x_2, \ldots, x_d]^T \in \mathbb{R}^d$, and f is also a vector, each component of which is a function of d variables $x_1, x_2, \ldots, x_d$, then (0.1) represents a *system of equations.* It is said to be a *nonlinear system* if at least one component of f depends nonlinearly on at least one of the variables x_1, $x_2, \ldots, x_d$. If all components of f are linear functions of $x_1, x_2, \ldots, x_d$, then we call (0.1) a *system of linear algebraic equations*, which (if $d > 1$) is of considerable interest in itself, but is not discussed in this chapter. Still more generally, (0.1) could represent a *functional equation*, if x is an element in some function space and f a (linear or nonlinear) operator acting on this space. In each of these interpretations, the zero on the right of (0.1), of course, has a different meaning: the number zero in the first case, the zero vector in the second, and the function identically equal to zero in the last case.

Much of this chapter is devoted to single nonlinear equations. Such equations are often encountered in the analysis of vibrating systems, where the roots correspond to critical frequencies (resonance). The special case of *algebraic equations*, where f in (0.1) is a polynomial, is also of considerable importance and merits special treatment. Systems of nonlinear equations are briefly considered at the end of the chapter.

§1. Examples

§1.1. **A transcendental equation.** Nonalgebraic equations are referred to as "transcendental." An example is

$$\cos x \cosh x - 1 = 0 \tag{1.1}$$

and is typical for equations arising in problems of resonance. Before one starts computing roots, it is helpful to gather some qualitative properties about them: are there any symmetries among the roots? How many roots are there? Where approximately are they located? With regard to symmetry, one notes immediately from (1.1) that the roots are located symmetrically with respect to the origin: if α is a root, so is $-\alpha$. Also, $\alpha = 0$ is a trivial root (which is uninteresting in applications). It suffices therefore to consider positive roots.

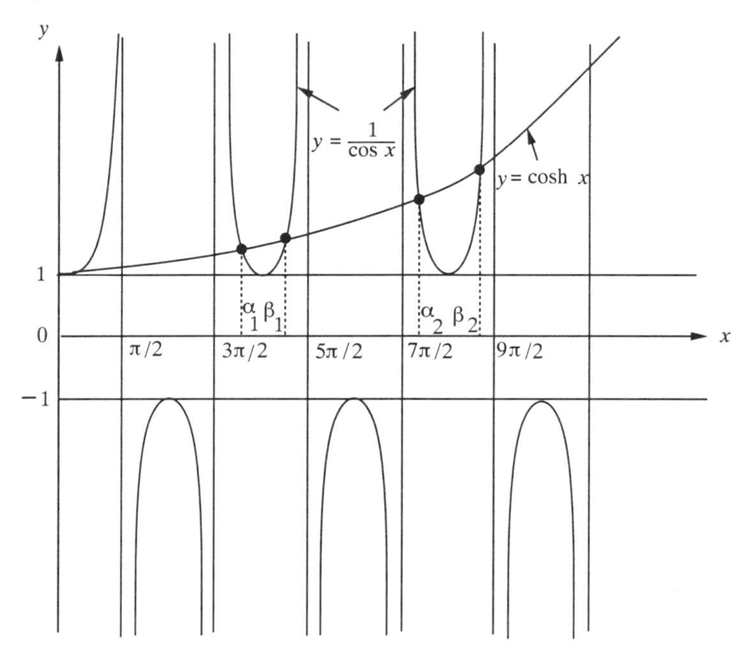

FIGURE 4.1.1. Graphical interpretation of (1.1′)

A quick way to get insight into the number and location of roots of (1.1) is to divide the equation by $\cos x$ and to rewrite it in the form

$$\cosh x = \frac{1}{\cos x} \ . \tag{1.1′}$$

No roots are being lost by this transformation, since clearly $\cos x \neq 0$ at any root $x = \alpha$. Now one graphs the function on the right and the function on the left and observes where the two graphs intersect. The respective abscissas of intersection are the desired (real) roots of the equation (1.1). This is illustrated in Figure 4.1.1 (not drawn to scale). It is evident from this figure that there are infinitely many positive roots. Indeed, each interval $[(2n - \frac{1}{2})\pi, (2n + \frac{1}{2})\pi]$, $n = 1, 2, 3, \ldots$, has exactly two roots, $\alpha_n < \beta_n$, with α_n rapidly approaching the left endpoint, and β_n the right endpoint, as n increases. These account for all positive roots and thus, by symmetry, for all real roots. In applications, it is likely that only the smallest positive root, α_1, will be of interest.

§1.2. **A two-point boundary value problem.** Here we are looking for a function $y \in C^2[0,1]$ satisfying the differential equation

$$y'' = g(x, y, y'), \qquad 0 \le x \le 1, \tag{1.2}$$

and the boundary conditions

$$y(0) = y_0, \qquad y(1) = y_1, \tag{1.3}$$

where g is a given (typically nonlinear) function on $[0,1] \times \mathbb{R} \times \mathbb{R}$, and y_0, y_1 are given numbers. At first sight, this does not look like a problem of the form (0.1), but it can be reduced to it if one introduces the associated initial value problem

$$\begin{aligned} &u'' = g(x, u, u'), \qquad 0 \le x \le 1, \\ &u(0) = y_0, \qquad u'(0) = s, \end{aligned} \tag{1.4}$$

where s (for "slope") is an unknown to be determined. Suppose, indeed, that for each s, (1.4) has a unique solution that exists on the whole interval [0,1]. Denote it by $u(x) = u(x; s)$. Then the problem (1.2) and (1.3) is equivalent to the problem

$$u(1; s) - y_1 = 0 \tag{1.5}$$

in the sense that to each solution of (1.5) there corresponds a solution of (1.2) and (1.3), and vice versa (cf. Ch. 7, §1.2). Thus, by defining

$$f(s) := u(1; s) - y_1, \tag{1.6}$$

we have precisely a problem of the form (0.1). It is to be noted, however, that $f(s)$ is not given explicitly as a function of s; rather, to evaluate $f(s)$ for any s, one has to solve the initial value problem (1.4) over the whole interval [0,1] to find the value of $u(x; s)$ at $x = 1$, and hence of $f(s)$ in (1.6).

A very natural way to go about solving (1.2) and (1.3) is to evaluate $f(s)$ for some initial guess s. If $f(s)$ is, say, positive, we lower the value of s until we find one for which $f(s)$ is negative. Then we have two slopes s, one that "overshoots" the target, and one that "undershoots" it. We now take as our next aim the average of these slopes and "shoot" again. Depending on whether we hit above the target or below, we discard the first or second initial slope and continue to repeat the same procedure. In the terminology of boundary value problems, this is called the *shooting method.* To shoot is tantamount to solving an initial value problem for a second-order differential equation, which in fact is the equation of the trajectory a bullet would traverse if it were fired from a gun.

§1.3. **A nonlinear integral equation**. Suppose we want to find a solution $y \in C[0,1]$ of the integral equation

$$y(x) - \int_0^1 K(x,t) f(t, y(t))dt = a(x), \qquad 0 \le x \le 1, \tag{1.7}$$

where K, the "kernel" of the equation, is a given (integrable) function on $[0,1] \times [0,1]$, f a given function on $[0,1] \times \mathbb{R}$, typically nonlinear in the second argument, and a also a given function on $[0,1]$. One way to approximately solve the equation (1.7) is to approximate the kernel by a degenerate kernel,

$$K(x,t) \approx k_n(x,t), \quad k_n(x,t) = \sum_{i=1}^{n} c_i(t) \pi_i(x). \tag{1.8}$$

We may think of the degenerate kernel as coming from truncating (to n terms) an infinite expansion of $K(x,t)$ in a system of basis functions $\{\pi_i(x)\}$, with coefficients c_i depending only on t. Replacing K in (1.7) by k_n then yields an approximate solution y_n, which is to satisfy

$$y_n(x) - \int_0^1 k_n(x,t) f(t, y_n(t))dt = a(x), \quad 0 \le x \le 1. \tag{1.9}$$

If we substitute (1.8) into (1.9) and define

$$\alpha_i = \int_0^1 c_i(t) f(t, y_n(t))dt, \qquad i = 1, 2, \ldots, n, \tag{1.10}$$

we can write y_n in the form

$$y_n(x) = a(x) + \sum_{i=1}^{n} \alpha_i \pi_i(x). \tag{1.11}$$

All that remains to be done is to compute the coefficients α_i in this representation. It is at this point where one is led to a system of nonlinear equations. Indeed, by (1.10), the α_i must satisfy

$$\alpha_i - \int_0^1 c_i(t) f(t, a(t) + \sum_{j=1}^{n} \alpha_j \pi_j(t)) dt = 0, \quad i = 1, 2, \ldots, n, \tag{1.12}$$

where the left-hand sides are functions f_i of $\alpha_1, \alpha_2, \ldots, \alpha_n$ which can be evaluated by numerical integration. It is seen how techniques of approximation (to obtain k_n) discussed in Ch. 2, and techniques of integration (to compute the integrals in (1.12)) discussed in Ch. 3, usefully combine to provide an approximate solution of (1.7).

§1.4. **s-Orthogonal polynomials**. In Ex. 19 of Ch. 3 we encountered an instance ($s = 1$) of "power orthogonality," that is, a (monic) polynomial π_n of degree n satisfying

$$\int_{\mathbb{R}} [\pi_n(t)]^{2s+1} p(t) d\lambda(t) = 0, \quad \text{all} \quad p \in \mathbb{P}_{n-1}. \tag{1.13}$$

This is called an s-orthogonal polynomial relative to the (positive) measure $d\lambda$. We can reinterpret power orthogonality as ordinary orthogonality

$$\int_{\mathbb{R}} \pi_n(t) p(t) \pi_n^{2s}(t) d\lambda(t) = 0,$$

but relative to the (positive) measure $d\lambda_n^s(t) = \pi_n^{2s}(t) d\lambda(t)$ depending on π_n. Thus, orthogonality is defined implicitly. The point, however, is that if we denote by $\{\pi_{k,n}\}_{k=0}^{n}$ the first $n+1$ orthogonal polynomials relative to $d\lambda_n^s$, we have $\pi_n = \pi_{n,n}$, and we can formally generate $\pi_{n,n}$ by a three-term recurrence relation,

$$\pi_{k+1,n}(t) = (t - \alpha_k)\pi_{k,n} - \beta_k \pi_{k-1,n}, \quad k = 0, 1, \ldots, n-1, \tag{1.14}$$

where $\pi_{-1,n}(t) = 0$, $\pi_{0,n}(t) = 1$. The coefficients $\alpha_0, \alpha_1, \ldots, \alpha_{n-1}$; β_0, β_1, $\ldots$, β_{n-1} are unknown and must be determined. Here is how a system of $2n$ nonlinear equations can be constructed for them.

From Ch. 2, Eqs. (1.28) and (1.29), one has

$$\alpha_k = \frac{(t\pi_{k,n}, \pi_{k,n})_{d\lambda_n^s}}{(\pi_{k,n}, \pi_{k,n})_{d\lambda_n^s}}, \qquad k = 0, 1, \dots, n-1;$$

$$\beta_0 = (1,1)_{d\lambda_n^s}, \quad \beta_k = \frac{(\pi_{k,n}, \pi_{k,n})_{d\lambda_n^s}}{(\pi_{k-1,n}, \pi_{k-1,n})_{d\lambda_n^s}}, \quad k = 1, \dots, n-1.$$

Consequently,

$$\begin{aligned}
f_0 &:= \beta_0 - \int_{\mathbb{R}} \pi_{n,n}^{2s}(t)d\lambda(t) = 0, \\
f_{2\nu+1} &:= \int_{\mathbb{R}} (\alpha_\nu - t)\pi_{\nu,n}^2(t)\pi_{n,n}^{2s}(t)d\lambda(t) = 0, \quad \nu = 0, 1, \dots, n-1, \\
f_{2\nu} &:= \int_{\mathbb{R}} [\beta_\nu \pi_{\nu-1,n}^2(t) - \pi_{\nu,n}^2(t)]\pi_{n,n}^{2s}(t)d\lambda(t) = 0, \quad \nu = 1, \dots, n-1.
\end{aligned} \tag{1.15}$$

Each of the $\pi_{\nu,n}$, $\nu = 1, 2, \dots, n$ depends on $\alpha_0, \dots, \alpha_{\nu-1}; \beta_1, \dots, \beta_{\nu-1}$ via the three-term recurrence relation (1.14). Therefore, we have $2n$ equations depending nonlinearly on the $2n$ unknowns $\alpha_0, \dots, \alpha_{n-1}; \beta_0, \dots, \beta_{n-1}$:

$$f(\rho) = 0, \quad \rho^T = [\alpha_0, \dots, \alpha_{n-1}; \beta_0, \dots, \beta_{n-1}].$$

Since the components of f are integrals of polynomials of degree at most $2(s+1)n - 1$, they can be computed exactly by an $(s+1)n$-point Gauss quadrature rule relative to the measure $d\lambda$ (cf. Mach. Ass. 8).

§2. Iteration, Convergence, and Efficiency

Even the simplest of nonlinear equations — for example, algebraic equations — are known to not admit solutions that are expressible rationally in terms of the data. It is therefore impossible, in general, to compute roots of nonlinear equations in a finite number of arithmetic operations. What is required is an *iterative method*, that is, a procedure that generates an infinite sequence of approximations, $\{x_n\}_{n=0}^\infty$, such that

$$\lim_{n\to\infty} x_n = \alpha \tag{2.1}$$

for some root α of the equation. In case of a system of equations, both x_n and α are vectors of appropriate dimension, and convergence is to be understood in the sense of componentwise convergence.

Although convergence of an iterative process is certainly desirable, it takes more than just convergence to make it practical. What one wants is fast convergence. A basic concept to measure the speed of convergence is the *order of convergence.*

Definition 2.1. *Linear convergence.* One says that x_n converges to α (at least) linearly if

$$|x_n - \alpha| \le \varepsilon_n, \tag{2.2}$$

where $\{\varepsilon_n\}$ is a positive sequence satisfying

$$\lim_{n\to\infty} \frac{\varepsilon_{n+1}}{\varepsilon_n} = c, \qquad 0 < c < 1. \tag{2.3}$$

If (2.2) and (2.3) hold with the inequality in (2.2) replaced by an equality, then c is called the *asymptotic error constant.*

The phrase "at least" in this definition relates to the fact that we have only inequality in (2.2), which in practice is all we can usually ascertain. So, strictly speaking, it is the *bounds* ε_n that converge linearly, meaning that eventually (e.g., for n large enough) each of these error bounds is approximately a constant fraction of the preceding one.

Definition 2.2. *Convergence of order p.* One says that x_n converges to α with (at least) order $p \ge 1$ if (2.2) holds with

$$\lim_{n\to\infty} \frac{\varepsilon_{n+1}}{\varepsilon_n^p} = c, \qquad c > 0. \tag{2.4}$$

(If $p = 1$, one must assume, in addition, $c < 1$.)

Thus, convergence of order 1 is the same as linear convergence, whereas convergence of order $p > 1$ is faster. Note that in this latter case there is no restriction on the constant c: once ε_n is small enough, it will be the exponent p that takes care of convergence. The constant c is again referred to as the asymptotic error constant if we have equality in (2.2).

The same definitions apply also to vector-valued sequences; one only needs to replace absolute values in (2.2) by (any) vector norm.

The classification of convergence with respect to order is still rather crude, as there are types of convergence that "fall between the cracks." Thus, a sequence $\{\varepsilon_n\}$ may converge to zero more slowly than linearly, for example, such that $c = 1$ in (2.3). We may call this type of convergence *sublinear.* Likewise, $c = 0$ in (2.3) gives rise to *superlinear* convergence, if (2.4) does not hold for any $p > 1$ (cf. also Ex. 4).

It is instructive to examine the behavior of ε_n if instead of the limit relations (2.3) and (2.4) we had strict equality from some n on, say,

$$\frac{\varepsilon_{n+1}}{\varepsilon_n^p} = c, \qquad n = n_0, n_0+1, n_0+2, \dots . \tag{2.5}$$

For n_0 large enough, this is almost true. A simple induction argument then shows that

$$\varepsilon_{n_0+k} = c^{\frac{p^k-1}{p-1}} \varepsilon_{n_0}^{p^k}, \qquad k = 0, 1, 2, \dots , \tag{2.6}$$

which certainly holds for $p > 1$, but also for $p = 1$ in the limit as $p \downarrow 1$:

$$\varepsilon_{n_0+k} = c^k \varepsilon_{n_0}, \qquad k = 0, 1, 2, \dots \quad (p = 1). \tag{2.7}$$

Assuming then ε_{n_0} sufficiently small so that the approximation x_{n_0} has several correct decimal digits, we write $\varepsilon_{n_0+k} = 10^{-\delta_k}\varepsilon_{n_0}$. Then δ_k, according to (2.2), approximately represents the number of additional correct decimal digits in the approximation x_{n_0+k} (as opposed to x_{n_0}). Taking logarithms in (2.7) and (2.6) gives

$$\delta_k = \begin{cases} k \log \dfrac{1}{c} & \text{if } \; p = 1, \\ p^k \left[\dfrac{1 - p^{-k}}{p - 1} \log \dfrac{1}{c} + (1 - p^{-k}) \log \dfrac{1}{\varepsilon_{n_0}} \right] & \text{if } \; p > 1; \end{cases}$$

hence, as $k \to \infty$,

$$\delta_k \sim C_1 \, k \;\; (p = 1); \qquad \delta_k \sim C_p p^k \;\; (p > 1), \tag{2.8}$$

where $C_1 = \log \frac{1}{c} > 0$ if $p = 1$, and $C_p = \frac{1}{p-1} \log \frac{1}{c} + \log \frac{1}{\varepsilon_{n_0}}$. (We assume here that n_0 is large enough, and hence ε_{n_0} small enough, to have $C_p > 0$.) This shows that the number of correct decimal digits increases linearly with k, when $p = 1$, but exponentially when $p > 1$. In the latter case, $\delta_{k+1}/\delta_k \sim p$, meaning that ultimately (for large k) the number of correct decimal digits increases, per iteration step, by a factor of p.

If each iteration requires m units of work (a "unit of work" typically is the work involved in computing a function value or a value of one of its derivatives), then the *efficiency index* of the iteration may be defined by $\lim_{k\to\infty} [\delta_{k+1}/\delta_k]^{1/m} = p^{1/m}$. It provides a common basis on which to compare different iterative methods with one another (cf. Ex. 6). Methods that converge linearly have efficiency index 1.

Practical computation requires the employment of a *stopping rule* that terminates the iteration once the desired accuracy is (or is believed to be) obtained. Ideally, one stops as soon as the absolute value (or norm) of the error $x_n - \alpha$ is smaller than a prescribed error tolerance. Since α is not known, one commonly replaces $x_n - \alpha$ by $x_n - x_{n-1}$ and requires

$$\|x_n - x_{n-1}\| \leq \text{tol}, \tag{2.9}$$

where

$$\text{tol} = \|x_n\|\epsilon_r + \epsilon_a \tag{2.10}$$

with ϵ_r, ϵ_a prescribed tolerances. As a safety measure, one might require (2.9) not just for one, but for a few consecutive values of n. Choosing $\epsilon_r = 0$ or $\epsilon_a = 0$ will make (2.10) an absolute (resp., relative) error tolerance. It is prudent, however, to use a "mixed error tolerance," say, $\epsilon_r = \epsilon_a = \epsilon$. Then, if $\|x_n\|$ is small or moderately large, one effectively controls the absolute error, whereas for $\|x_n\|$ very large, it is in effect the relative error that is controlled.

§3. The Methods of Bisection and Sturm Sequences

Both these methods generate a sequence of nested intervals $[a_n, b_n]$, $n = 0, 1, 2, \ldots$, whereby each interval is guaranteed to contain at least one root of the equation. As $n \to \infty$, the length of these intervals tends to zero, so that in the limit exactly one (isolated) root is captured. The first method applies to any equation (0.1) with f a continuous function, but has no built-in control of steering the iteration to any particular (real) root if there is more than one. The second method does have such a control mechanism, but applies only to a restricted class of equations, for example, the characteristic equation of a symmetric tridiagonal matrix.

§3.1. **Bisection method.** We assume that two numbers a, b with $a < b$ are known such that

$$f \in C[a, b], \qquad f(a) < 0, \qquad f(b) > 0. \tag{3.1}$$

What is essential here is that f has opposite signs at the endpoints of $[a, b]$; the particular sign combination in (3.1) is not essential as it can always be obtained, if necessary, by multiplying f by -1. The assumptions (3.1), in particular, guarantee that f has at least one zero in (a, b).

By repeatedly bisecting the interval and discarding endpoints in such a manner that the sign property in (3.1) remains preserved, it is possible to generate a sequence of nested intervals whose lengths are continuously halved and each of which contains a zero of f.

Specifically, the procedure is as follows. Define $a_1 = a$, $b_1 = b$. Then

for $n = 1, 2, 3, \ldots$ do

$$x_n = \tfrac{1}{2}(a_n + b_n)$$

if $f(x_n) < 0$ then $a_{n+1} = x_n, \quad b_{n+1} = b_n$ else

$$a_{n+1} = a_n, \quad b_{n+1} = x_n.$$

Since $b_n - a_n = 2^{-(n-1)}(b - a)$, $n = 1, 2, 3, \ldots$, and x_n is the midpoint of $[a_n, b_n]$, if α is the root eventually captured, we have

$$|x_n - \alpha| \leq \frac{1}{2}(b_n - a_n) = \frac{b - a}{2^n}\,. \tag{3.2}$$

Thus, (2.2) holds with $\varepsilon_n = 2^{-n}(b - a)$, and

$$\frac{\varepsilon_{n+1}}{\varepsilon_n} = \tfrac{1}{2}, \qquad \text{all } n. \tag{3.3}$$

This shows that the bisection method converges (at least) linearly with asymptotic error constant (for the bound ε_n) equal to $\frac{1}{2}$.

Given an (absolute) error tolerance tol > 0, the error in (3.2) will be less than or equal to tol if

$$\frac{b - a}{2^n} \leq \text{tol}.$$

Solved explicitly for n, this will be satisfied if

$$n = \left\lceil \frac{\log \frac{b-a}{\text{tol}}}{\log 2} \right\rceil, \tag{3.4}$$

where $\lceil x \rceil$ denotes the "ceiling" of x (i.e., the smallest integer $\geq x$). Thus, we know a priori how many steps are necessary to achieve a prescribed accuracy.

It should be noted that when $f(x_n)$ approaches the level of machine precision, its sign may be computed incorrectly, hence the wrong half of the current interval may be chosen. Normally, this should be of little concern since by this time, the interval has already become sufficiently small. In this sense, bisection is a method that is robust at the level of machine precision. Problems may occur when f is very flat near the root α, in which case, however, the root is numerically not well determined (cf. Mach. Ass. 1).

When implementing the procedure on a computer, it is clearly unnecessary to provide arrays to store a_n, b_n, x_n; one simply keeps overwriting. Assuming a and b have been initialized and tol assigned, one could use the following FORTRAN procedure with `ntol = alog((b-a)/tol)/alog(2.)+1`:

```
      subroutine bisec(a,b,ntol,x,f)
      do 10 n=1,ntol
        x=.5*(a+b)
        if(f(x).lt.0.) then
          a=x
        else
          b=x
        end if
   10 continue
      return
      end
```

As an example, we run the program to compute the smallest positive root of the equation (1.1), that is, of $f(x) = 0$ with $f(x) = \cos x \cosh x - 1$. By taking $a = \frac{3}{2}\pi$, $b = 2\pi$ (cf. Figure 4.1.1), we enclose exactly one root,

ntol	α_1
25	4.730041
52	4.73004074486270
112	4.730040744862704026018086528886672

TABLE 4.3.1. The bisection method applied to (1.1)

α_1, from the start, and bisection is guaranteed to converge to α_1. The results in single, double, and quadruple precision obtained on the Sparc 2 workstation corresponding, respectively, to tol $= \frac{1}{2} \times 10^{-7}$, tol $= \frac{1}{2} \times 10^{-15}$, and tol $= \frac{1}{2} \times 10^{-33}$, are shown in Table 4.3.1. As can be seen, bisection requires a fair amount of work (over a hundred iterations) to obtain α_1 to high precision. Had we run the program with initial interval $[a, b]$, $a = \frac{3}{2}\pi$, $b = \frac{9}{2}\pi$, which contains four roots, we would have converged to the third root α_2, with about the same amount of work.

§3.2. **Method of Sturm sequences.** There are situations in which it is desirable to be able to select one particular root among many and have the iterative scheme converge to it. This is the case, for example, in orthogonal polynomials, where we know that all zeros are real and distinct (cf. Ch. 3, §2.3(i)). It may well be that we are interested in the second-largest or third-largest zero, and should be able to compute it without computing any of the others. This is indeed possible if we combine bisection with the theorem of Sturm.[1]

Thus, consider

$$f(x) = \pi_d(x), \tag{3.5}$$

where π_d is a polynomial of degree d orthogonal with respect to some positive measure. We know (cf. Ch. 3, §2.3(v)) that π_d is the characteristic polynomial of a symmetric tridiagonal matrix and can be computed recursively by a three-term recurrence relation

$$\begin{gathered} \pi_0(x) = 1, \qquad \pi_1(x) = x - \alpha_0, \\ \pi_{k+1}(x) = (x - \alpha_k)\pi_k(x) - \beta_k \pi_{k-1}(x), \qquad k = 1, 2, \ldots, d-1, \end{gathered} \tag{3.6}$$

with all β_k positive. The recursion (3.6) is not only useful to compute $\pi_d(x)$ for any fixed x, but has also the following interesting property due to Sturm: *Let $\sigma(x)$ be the number of sign changes* (zeros do not count) *in the sequence of numbers*

$$\pi_d(x),\ \pi_{d-1}(x),\ \ldots\ ,\pi_1(x),\ \pi_0(x). \tag{3.7}$$

Then, for any two numbers a, b with $a < b$, the number of real zeros of π_d in the interval $a < x \le b$ is equal to $\sigma(a) - \sigma(b)$.

---

[1] Jacques Charles François Sturm (1803–1855), a Swiss analyst and theoretical physicist of Alsatian parentage, is best known for his theorem on Sturm sequences, discovered in 1829, and his theory of Sturm-Liouville differential equations, published in 1834, which earned him the Grand Prix des Sciences Mathématiques. He also contributed significantly to differential and projective geometry. A member of the French Academy of Sciences since 1836, he succeeded Poisson in the chair of mechanics at the École Polytechnique in Paris in 1839.

Since $\pi_k(x) = x^k + \cdots$, it is clear that $\sigma(-\infty) = d$, $\sigma(+\infty) = 0$, so that indeed the number of real zeros of π_d is $\sigma(-\infty) - \sigma(+\infty) = d$. Moreover, if $\xi_1 > \xi_2 > \cdots > \xi_d$ denote the zeros of π_d in decreasing order, we have the behavior of $\sigma(x)$ as shown in Figure 4.3.1.

It is now easy to see that

$$\sigma(x) \leq r-1 \quad \text{iff} \quad x \geq \xi_r. \tag{3.8}$$

Indeed, suppose that $x \geq \xi_r$. Then $\{\#\text{zeros} \leq x\} \geq d+1-r$; hence, by Sturm's theorem, $\sigma(-\infty) - \sigma(x) = d - \sigma(x) = \{\#\text{zeros} \leq x\} \geq d+1-r$; that is, $\sigma(x) \leq r-1$. Conversely, if $\sigma(x) \leq r-1$, then, again by Sturm's theorem, $\{\#\text{zeros} \leq x\} = d - \sigma(x) \geq d+1-r$, which implies $x \geq \xi_r$ (cf. Figure 4.3.1).

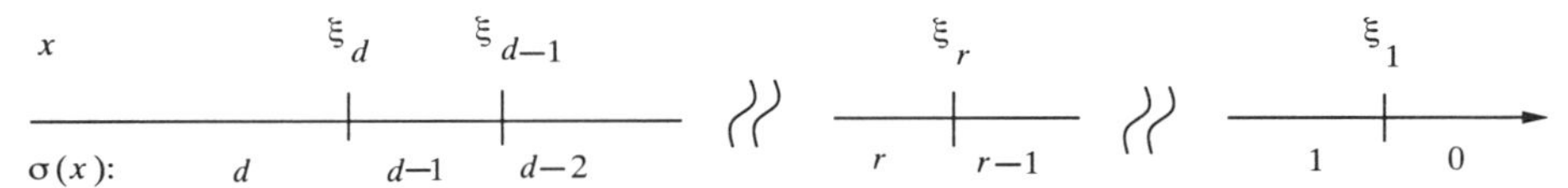

FIGURE 4.3.1. Sturm's theorem

The basic idea now is to control the bisection process not, as before, by checking the sign of $\pi_d(x)$, but rather, by checking the inequality (3.8) to see whether we are on the right or left side of the zero ξ_r. In order to initialize the procedure, we need two values $a_1 = a$, $b_1 = b$ such that $a < \xi_d$ and $b > \xi_1$. These are trivially obtained as the endpoints of the interval of orthogonality for π_d, if it is finite. More generally, one can apply Gershgorin's theorem[2] to the Jacobi matrix J_d associated with (3.6) (cf. Ch. 3, §2.3(v)) by recalling that the zeros of π_d are precisely the eigenvalues of J_d. In this way, a can be chosen to be the smallest and b the largest of the d numbers $\alpha_0 + \sqrt{\beta_1}$, $\alpha_1 + \sqrt{\beta_1} + \sqrt{\beta_2}, \ldots, \alpha_{d-2} + \sqrt{\beta_{d-2}} + \sqrt{\beta_{d-1}}$, $\alpha_{d-1} + \sqrt{\beta_{d-1}}$. The method of Sturm sequences then proceeds as follows, for any given r with $1 \leq r \leq d$.

for $n = 1, 2, 3, \ldots$ do

$\quad x_n = \frac{1}{2}(a_n + b_n)$

$\quad$ if $\sigma(x_n) > r - 1$ then $a_{n+1} = x_n$, $b_{n+1} = b_n$ else

$\qquad a_{n+1} = a_n$, $b_{n+1} = x_n$.

[2]Gershgorin's theorem states that the eigenvalues of a matrix $A = [a_{ij}]$ of order d are located in the union of the disks $\{z \in \mathbb{C} : \ |z - a_{ii}| \leq r_i\}$, $i = 1, 2, \ldots d$, where $r_i = \sum_{j \neq i} |a_{ij}|$.

Since initially $\sigma(a) = d > r-1$, $\sigma(b) = 0 \le r-1$, it follows by construction that

$$\sigma(a_n) > r-1, \quad \sigma(b_n) \le r-1, \quad \text{all} \quad n = 1, 2, 3, \ldots\ , \tag{3.9}$$

meaning that $\xi_r \in [a_n, b_n]$ for all $n = 1, 2, 3, \ldots$. Moreover, as in the bisection method, $b_n - a_n = 2^{-(n-1)}(b-a)$, so that $|x_n - \xi_r| \le \varepsilon_n$ with $\varepsilon_n = 2^{-n}(b-a)$. The method converges (at least) linearly to the root ξ_r. A computer implementation can be modeled after the one for the bisection method by modifying the `if-else` statement appropriately.

§4. Method of False Position

As in the method of bisection, we assume two numbers $a < b$ such that

$$f \in C[a,b], \qquad f(a)f(b) < 0, \tag{4.1}$$

and generate a sequence of nested intervals $[a_n, b_n]$, $n = 1, 2, 3, \ldots$, with $a_1 = a$, $b_1 = b$, such that $f(a_n)f(b_n) < 0$. Unlike the bisection method, however, we are not taking the midpoint of $[a_n, b_n]$ to determine the next interval, but rather the solution $x = x_n$ of the linear equation

$$p_1(f; a_n, b_n; x) = 0, \tag{4.2}$$

where $p_1(f; a_n, b_n; \cdot\,)$ is the linear interpolant of f at a_n and b_n. This would appear to be more flexible than bisection, as x_n will come to lie closer to the endpoint at which $|f|$ is smaller. Also, if f is a linear function, we obtain the root in one step rather than in an infinite number of steps. This explains the somewhat strange name given to this method (cf. Notes to §4).

More explicitly, the method proceeds as follows: define $a_1 = a$, $b_1 = b$. Then,

for $n = 1, 2, 3, \ldots$ do

$$x_n = a_n - \frac{a_n - b_n}{f(a_n) - f(b_n)} f(a_n)$$

if $f(x_n)f(a_n) > 0$ then $a_{n+1} = x_n$, $b_{n+1} = b_n$ else

$a_{n+1} = a_n$, $b_{n+1} = x_n$.

One may terminate the iteration as soon as $\min(x_n - a_n, b_n - x_n) \le \text{tol}$, where tol is a prescribed error tolerance, although this is not entirely foolproof (why not?).

As in the bisection method, when implementing the method on a computer, the as and bs can be overwritten. On the other hand, it is no longer known a priori how many iterations it takes to achieve the desired accuracy. It is prudent, then, to put a limit on the number of iterations. An implementation in FORTRAN may look as follows.

```
      subroutine false(a,b,tol,ntol,x,f)
      ntol=0
      fa=f(a)
      fb=f(b)
   10 ntol=ntol+1
      x=a-(a-b)*fa/(fa-fb)
      if(x-a.lt.tol .or. b-x.lt.tol .or. ntol.gt.100) then
        return
      else
        fx=f(x)
        if(fx*fa.gt.0.) then
          a=x
          fa=fx
        else
          b=x
          fb=fx
        end if
      end if
      go to 10
      end
```

The convergence behavior is most easily analyzed if we assume that f is convex or concave on $[a, b]$. To fix ideas, suppose f is convex, say,

$$f''(x) > 0 \quad \text{for} \quad a \le x \le b, \ \text{and} \ f(a) < 0, \ f(b) > 0. \tag{4.3}$$

Then f has exactly one zero, α, in $[a, b]$. Moreover, the secant connecting $f(a)$ and $f(b)$ lies entirely above the graph of $y = f(x)$, and hence intersects the real line to the left of α. This will be the case for all subsequent secants, which means that the point $x = b$ remains fixed while the other endpoint a

gets continuously updated, producing a monotonically increasing sequence of approximations defined by

$$x_{n+1} = x_n - \frac{x_n - b}{f(x_n) - f(b)} f(x_n), \qquad n = 1, 2, 3, \dots , \tag{4.4}$$

where $x_1 = a$. Any such sequence, being bounded from above by α, necessarily converges, and letting $n \to \infty$ in (4.4) shows immediately that $f(x_n) \to 0$, that is, $x_n \to \alpha$. To determine the speed of convergence, we subtract α from both sides of (4.4) and use the fact that $f(\alpha) = 0$:

$$x_{n+1} - \alpha = x_n - \alpha - \frac{x_n - b}{f(x_n) - f(b)} [f(x_n) - f(\alpha)].$$

Now divide by $x_n - \alpha$ to get

$$\frac{x_{n+1} - \alpha}{x_n - \alpha} = 1 - \frac{x_n - b}{f(x_n) - f(b)} \frac{f(x_n) - f(\alpha)}{x_n - \alpha} .$$

Letting here $n \to \infty$ and using the fact that $x_n \to \alpha$, we obtain

$$\lim_{n \to \infty} \frac{x_{n+1} - \alpha}{x_n - \alpha} = 1 - (b - \alpha) \frac{f'(\alpha)}{f(b)} . \tag{4.5}$$

Thus, we have linear convergence with asymptotic error constant equal to

$$c = 1 - (b - \alpha) \frac{f'(\alpha)}{f(b)} . \tag{4.6}$$

It is clear on geometric grounds that $0 < c < 1$ under the assumptions made. An analogous result will hold, with a constant $|c| < 1$, provided f is either convex or concave on $[a, b]$, and has opposite signs at the endpoints a and b. One of these endpoints then remains fixed while the other moves monotonically to the root α.

If f does not satisfy these convexity properties on the whole interval but is such that $f \in C^2[a, b]$ and $f''(\alpha) \neq 0$ at the root α eventually approached, then the convergence behavior described sets in for n large enough, since f'' has constant sign in a neighborhood of α and x_n will eventually come to lie in this neighborhood.

The fact that one of the "false positions" remains at a fixed point from some n on may speak against this method, especially if this occurs early in the iteration, or even from the beginning, as under the assumptions (4.3) made previously. Thus, in the case of the equation (1.1), for example, when $a = \frac{3}{2}\pi$ and $b = 2\pi$, we have $f''(x) = -2\sin x \sinh x > 0$ on $[a, b]$, and $f(a) = -1$, $f(b) = \cosh(2\pi) - 1 > 0$, so that we are precisely in a case where (4.3) holds. Accordingly, we have found that to compute the root α_1 to single, double, and quadruple precision, we now need 26, 75, and 177 iterations, as opposed to 25, 52, and 112 in the case of the bisection method.

Exceptionally slow convergence is likely to occur when f is very flat near α, the point a is nearby, and b further away. In this case, b will typically remain fixed while a is slowly creeping towards α (cf. Mach. Ass. 1).

§5. Secant Method

The secant method is a simple variant of the method of false position in which it is no longer required that the function f have opposite signs at the endpoints of each interval generated, not even the initial interval. In other words, one starts with two arbitrary initial approximations x_0, x_1 and continues with

$$x_{n+1} = x_n - \frac{x_n - x_{n-1}}{f(x_n) - f(x_{n-1})} f(x_n), \qquad n = 1, 2, 3, \ldots \tag{5.1}$$

This precludes the formation of a fixed false position, as in the method of false positions, and hence suggests potentially faster convergence. On the other hand, we can no longer be sure that each interval $[x_{n-1}, x_n]$ contains at least one root. It will turn out indeed that, if the method converges, then it does so with an order of convergence larger than 1 (but less than 2). However, it converges only "locally," that is, only if the initial approximations x_0, x_1 are sufficiently close to a root.

This can be seen by relating the three consecutive errors of x_{n+1}, x_n, and x_{n-1} as follows. Subtract α on both sides of (5.1), and use $f(\alpha) = 0$, to get

$$\begin{aligned}
x_{n+1} - \alpha &= x_n - \alpha - \frac{f(x_n)}{[x_{n-1}, x_n]f} \\
&= (x_n - \alpha)\left(1 - \frac{f(x_n) - f(\alpha)}{(x_n - \alpha)[x_{n-1}, x_n]f}\right) \\
&= (x_n - \alpha)\left(1 - \frac{[x_n, \alpha]f}{[x_{n-1}, x_n]f}\right) \\
&= (x_n - \alpha)\,\frac{[x_{n-1}, x_n]f - [x_n, \alpha]f}{[x_{n-1}, x_n]f}\;;
\end{aligned}$$

hence, by the definition of divided differences,

$$x_{n+1} - \alpha = (x_n - \alpha)(x_{n-1} - \alpha)\, \frac{[x_{n-1}, x_n, \alpha]f}{[x_{n-1}, x_n]f}\,, \qquad n = 1, 2, \ldots\,. \tag{5.2}$$

This is the fundamental relation holding between three consecutive errors.

From (5.2) it follows immediately that if α is a simple root,

$$f(\alpha) = 0, \qquad f'(\alpha) \neq 0, \tag{5.3}$$

and if $x_n \to \alpha$, then convergence is faster than linear, at least if $f \in C^2$ near α. Indeed,

$$\lim_{n\to\infty} \frac{x_{n+1} - \alpha}{x_n - \alpha} = 0,$$

since the divided differences in the numerator and denominator of (5.2) converge to $\frac{1}{2} f''(\alpha)$ and $f'(\alpha)$, respectively. But just how fast is convergence?

We can discover the order of convergence (assuming convergence) by a simple heuristic argument: we replace the ratio of divided differences in (5.2) by a constant, which is almost true if n is large. Letting then $e_k = |x_k - \alpha|$, we have

$$e_{n+1} = e_n e_{n-1} \cdot C, \qquad C > 0.$$

Multiplying both sides by C and defining $E_n = Ce_n$ gives

$$E_{n+1} = E_n E_{n-1}, \qquad E_n \to 0.$$

Taking logarithms on both sides, and defining $y_n = \log \frac{1}{E_n}$, we get

$$y_{n+1} = y_n + y_{n-1}, \tag{5.4}$$

the well-known difference equation for the Fibonacci sequence. Its characteristic equation is $t^2 - t - 1 = 0$, which has the two roots t_1, t_2 with

$$t_1 = \tfrac{1}{2}(1 + \sqrt{5}), \qquad t_2 = \tfrac{1}{2}(1 - \sqrt{5}),$$

and $t_1 > 1$, $|t_2| < 1$. The general solution of (5.4), therefore, is

$$y_n = c_1 t_1^n + c_2 t_2^n, \qquad c_1, c_2 \ \text{ constant}.$$

Since $y_n \to \infty$, we have $c_1 \neq 0$ and $y_n \sim c_1 t_1^n$ as $n \to \infty$, which translates into $\frac{1}{E_n} \sim e^{c_1 t_1^n}$, $\frac{1}{e_n} \sim Ce^{c_1 t_1^n}$, and thus

$$\frac{e_{n+1}}{e_n^{t_1}} \sim \frac{C^{t_1} e^{c_1 t_1^n \cdot t_1}}{Ce^{c_1 t_1^{n+1}}} = C^{t_1 - 1}, \qquad n \to \infty.$$

The order of convergence, therefore, is $t_1 = \frac{1}{2}(1+\sqrt{5}) = 1.61803\ldots$ (the golden mean).

We now give a rigorous proof of this, and begin with a proof of *local convergence.*

Theorem 4.5.1. *Let α be a simple zero of f. Let $I_\varepsilon = \{x \in \mathbb{R} : |x-\alpha| \le \varepsilon\}$ and assume $f \in C^2[I_\varepsilon]$. Define, for sufficiently small ε,*

$$M(\varepsilon) = \max_{\substack{s\in I_\varepsilon \\ t\in I_\varepsilon}} \left| \frac{f''(s)}{2f'(t)} \right| . \tag{5.5}$$

Assume ε so small that

$$\varepsilon M(\varepsilon) < 1. \tag{5.6}$$

Then the secant method converges to the unique root $\alpha \in I_\varepsilon$ for any starting values $x_0 \ne x_1$ with $x_0 \in I_\varepsilon$, $x_1 \in I_\varepsilon$.

Note that $\lim_{\varepsilon\to 0} M(\varepsilon) = \left|\frac{f''(\alpha)}{2f'(\alpha)}\right| < \infty$, so that (5.6) can certainly be satisfied for ε small enough. The local nature of convergence is thus quantified by the requirement $x_0, x_1 \in I_\varepsilon$.

Proof of Theorem 4.5.1. First of all, observe that α is the only zero of f in I_ε. This follows from Taylor's formula applied at $x = \alpha$:

$$f(x) = f(\alpha) + (x-\alpha)f'(\alpha) + \frac{(x-\alpha)^2}{2} f''(\xi),$$

where $f(\alpha) = 0$ and ξ is between x and α. Thus, if $x \in I_\varepsilon$, then so is ξ, and we have

$$f(x) = (x-\alpha)f'(\alpha)\left\{1 + \frac{x-\alpha}{2}\,\frac{f''(\xi)}{f'(\alpha)}\right\}, \qquad \xi \in I_\varepsilon .$$

Here, if $x \ne \alpha$, all three factors are different from zero, the last one since by assumption

$$\left|\frac{x-\alpha}{2}\,\frac{f''(\xi)}{f'(\alpha)}\right| \le \varepsilon M(\varepsilon) < 1.$$

Thus, f on I_ε can only vanish at $x = \alpha$.

Next we show that all $x_n \in I_\varepsilon$ and two consecutive iterates are distinct, unless $f(x_n) = 0$ for some n, in which case $x_n = \alpha$ and the method converges in a finite number of steps. We prove this by induction: assume that $x_{n-1} \in I_\varepsilon$, $x_n \in I_\varepsilon$ for some n and $x_n \ne x_{n-1}$. (By assumption, this is true for

$n = 1$.) Then, from known properties of divided differences, and by our assumption that $f \in C^2[I_\varepsilon]$, we have

$$[x_{n-1}, x_n]f = f'(\xi_1), \quad [x_{n-1}, x_n, \alpha]f = \tfrac{1}{2} f''(\xi_2), \quad \xi_i \in I_\varepsilon, \quad i = 1, 2.$$

Therefore, by (5.2),

$$|x_{n+1} - \alpha| \le \varepsilon^2 \left| \frac{f''(\xi_2)}{2f'(\xi_1)} \right| \le \varepsilon \cdot \varepsilon M(\varepsilon) < \varepsilon,$$

showing that $x_{n+1} \in I_\varepsilon$. Furthermore, by (5.1), $x_{n+1} \ne x_n$, unless $f(x_n) = 0$, hence $x_n = \alpha$.

Finally, again using (5.2), we have

$$|x_{n+1} - \alpha| \le |x_n - \alpha| \, \varepsilon M(\varepsilon), \qquad n = 1, 2, 3, \dots ,$$

which, applied repeatedly, yields

$$|x_n - \alpha| \le [\varepsilon M(\varepsilon)]^{n-1} |x_1 - \alpha|.$$

Since $\varepsilon M(\varepsilon) < 1$, it follows that $x_n \to \alpha$ as $n \to \infty$. □

We next prove that the order of convergence is indeed what we derived it to be heuristically.

Theorem 4.5.2. *The secant method is locally convergent with order of convergence (at least)* $p = \frac{1}{2}(1 + \sqrt{5}) = 1.61803\dots$.

Proof. Local convergence is the content of Theorem 4.5.1 and so needs no further proof. We assume that x_0, $x_1 \in I_\varepsilon$, where ε satisfies (5.6), and that all x_n are distinct. Then we know that $x_n \ne \alpha$ for all n, and $x_n \to \alpha$ as $n \to \infty$.

Now the number p in the theorem satisfies

$$p^2 = p + 1. \tag{5.7}$$

From (5.2), we have

$$|x_{n+1} - \alpha| \le |x_n - \alpha| \, |x_{n-1} - \alpha| \cdot M, \tag{5.8}$$

where we write simply M for $M(\varepsilon)$. Define

$$E_n = M |x_n - \alpha|. \tag{5.9}$$

Then, multiplying (5.8) by M, we get

$$E_{n+1} \le E_n E_{n-1}.$$

It follows easily by induction that

$$E_n \le E^{p^n}, \qquad E = \max(E_0, E_1^{1/p}). \tag{5.10}$$

Indeed, this is trivially true for $n = 0$ and $n = 1$. Suppose (5.10) holds for n as well as for $n-1$. Then

$$E_{n+1} \le E_n E_{n-1} \le E^{p^n} E^{p^{n-1}} = E^{p^{n-1}(p+1)} = E^{p^{n-1}\cdot p^2} = E^{p^{n+1}},$$

where (5.7) has been used. This proves (5.10) for $n+1$, and hence for all n. It now follows from (5.9) that

$$|x_n - \alpha| \le \varepsilon_n, \qquad \varepsilon_n = \frac{1}{M} E^{p^n}.$$

Since $E_0 = M|x_0 - \alpha| \le \varepsilon M(\varepsilon) < 1$, and the same holds for E_1, we have $E < 1$. Now it suffices to note that

$$\frac{\varepsilon_{n+1}}{\varepsilon_n^p} = M^{p-1} \frac{E^{p^{n+1}}}{E^{p^n \cdot p}} = M^{p-1}, \quad \text{all } n,$$

to establish the theorem. □

The method is easily programmed for a computer:

```
      subroutine secant(a,b,tol,ntol,x,f)
      ntol=0
      fa=f(a)
      x=b
   10 ntol=ntol+1
      b=a
      fb=fa
      a=x
      fa=f(x)
      x=a-(a-b)*fa/(fa-fb)
      if(abs(x-a).lt.tol .or. ntol.gt.100) then
        return
      else
        go to 10
      end if
      end
```

It is assumed here that $a = x_0$, $b = x_1$ and that the iteration (5.1) is terminated as soon as $|x_{n+1} - x_n| < \text{tol}$ or $n > 100$. The routine not only produces the approximation x to the root, but also the number of iterations required to obtain it to within an error of tol.

Since only one evaluation of f is required in each iteration step (the statement `fa = f(x)` in the preceding program), the secant method has efficiency index $p = 1.61803\ldots$ (cf. §2).

To illustrate the considerable gain in speed attainable by the secant method, we again apply the routine to the equation (1.1) with $x_0 = \frac{3}{2}\pi$, $x_1 = 2\pi$. In view of the convexity of f it is easily seen that after the first step of the method, all approximations remain in the interval $[\frac{3}{2}\pi, \frac{7}{4}\pi]$ of length $\frac{1}{4}\pi$, containing the unique root α_1. For s and t in that interval, we have

$$\left|\frac{f''(s)}{2f'(t)}\right| \leq \frac{\sinh\left(\frac{7}{4}\pi\right)}{2\cosh\left(\frac{3}{2}\pi\right)} = 1.0965\ldots\ ,$$

as follows by a simple calculation. Since this bound multiplied by $\frac{1}{4}\pi$ — the length of the interval — is $.86121\ldots < 1$, it follows from an argument analogous to the one in the proof of Theorem 4.5.1 that all x_n remain in $[\frac{3}{2}\pi, \frac{7}{4}\pi]$ and converge to α_1, the order of convergence being $p = \frac{1}{2}(1 + \sqrt{5})$ by Theorem 4.5.2. To find the root α_1 to single, double, and quadruple precision, we found that 6, 7, and 9 iterations, respectively, suffice. This should be contrasted with 25, 52, and 112 iterations (cf. §3.1) for the bisection method.

In contrast to bisection, the secant method is *not* robust at the level of machine precision and may even fail if `fa` happens to become equal to `fb`. The method is, therefore, rarely used on its own, but more often in combination with the method of bisection (cf. Mach. Ass. 3).

§6. Newton's Method

Newton's method can be thought of as a limit case of the secant method (5.1) if we let there x_{n-1} move into x_n. The result is

$$x_{n+1} = x_n - \frac{f(x_n)}{f'(x_n)}\ , \qquad n = 0, 1, 2, \ldots\ , \tag{6.1}$$

where x_0 is some appropriate initial approximation. Another, more fruitful interpretation is that of *linearization of the equation* $f(x) = 0$ *at* $x = x_n$.

In other words, we replace $f(x)$ for x near x_n by the linear approximation $f(x) \approx f_a(x) := f(x_n) + (x - x_n)f'(x_n)$ obtained by truncating the Taylor expansion of f centered at x_n after the linear term, and then solve the resulting linear equation, $f_a(x) = 0$, calling the solution x_{n+1}. This again leads to the equation (6.1). Viewed in this manner, Newton's method can be vastly generalized to nonlinear equations of all kinds, not only single equations as in (6.1), but also systems of nonlinear equations (cf. §9.2) and even functional equations, in which case the deviative $f'(x)$ is to be understood as a Fréchet derivative.

It is useful to begin with a few simple examples of single equations to get a feel for how Newton's method may behave.

Example 6.1: Square root $\alpha = \sqrt{a}$, $a > 0$. The equation here is $f(x) = 0$ with

$$f(x) = x^2 - a. \tag{6.2}$$

Equation (6.1) then immediately gives

$$x_{n+1} = \frac{1}{2}\left(x_n + \frac{a}{x_n}\right), \qquad n = 0, 1, 2, \ldots \tag{6.3}$$

(a method already used by the Babylonians long before Newton). Because of the convexity of f it is clear that the iteration (6.3) converges to the positive square root for each $x_0 > 0$ and is monotonically decreasing (except for the first step in the case $0 < x_0 < \alpha$). We have here an elementary example of *global convergence.*

Example 6.2: $f(x) = \sin x$, $|x| < \frac{1}{2}\pi$. There is exactly one root in this interval, the trivial root $\alpha = 0$. Newton's method becomes

$$x_{n+1} = x_n - \tan x_n, \qquad n = 0, 1, 2, \ldots . \tag{6.4}$$

It exhibits the following amusing phenomenon (cf. Figure 4.6.1). If $x_0 = x^*$, where

$$\tan x^* = 2x^*, \tag{6.5}$$

then $x_1 = -x^*$, $x_2 = x^*$; that is, after two steps of Newton's method we end up where we started! This is called a *cycle.*

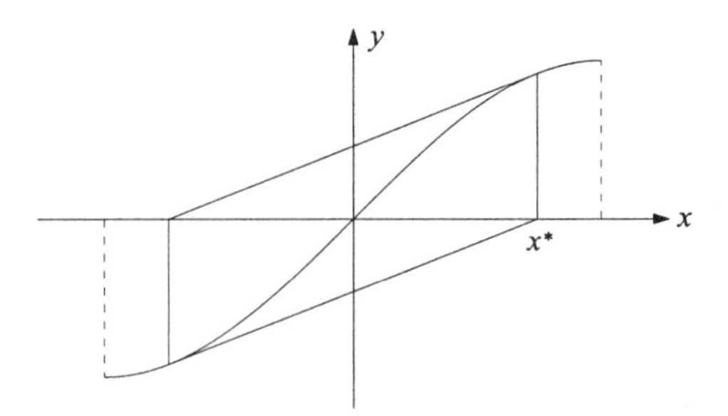

FIGURE 4.6.1. A cycle in Newton's method

For this starting value, Newton's method does not converge, let alone to $\alpha = 0$. It does converge, however, for any starting value x_0 with $|x_0| < x^*$, generating a sequence of alternately increasing and decreasing approximations x_n converging necessarily to $\alpha = 0$. The value of the critical number x^* can itself be computed by Newton's method applied to (6.5). The result is $x^* = 1.16556\ldots$. In a sense, we have here an example of *local convergence*, since convergence cannot hold for all $x_0 \in [-\frac{1}{2}\pi, \frac{1}{2}\pi]$. (If $x_0 = \frac{1}{2}\pi$, we even get thrown off to ∞.)

Example 6.3: $f(x) = x^{20} - 1$, $x > 0$. This has exactly one positive (simple) root $\alpha = 1$. Newton's method yields the iteration

$$x_{n+1} = \frac{19}{20}\, x_n + \frac{1}{20x_n^{19}}\,, \qquad n = 0, 1, 2, \ldots\,, \tag{6.6}$$

which provides a good example to illustrate one of the dangers in Newton's method: unless one starts sufficiently close to the desired root, it may take a long while to approach it. Thus, suppose we take $x_0 = \frac{1}{2}$; then $x_1 \approx \frac{2^{19}}{20} = 2.62144 \times 10^4$, a huge number. What is worse, it is going to be a slow ride back to the vicinity of $\alpha = 1$, since for x_n very large, one has

$$x_{n+1} \approx \tfrac{19}{20}\, x_n, \qquad x_n \gg 1.$$

At each step the approximation is reduced only by a fraction $\frac{19}{20} = .95$. It takes about 200 steps to get back to near the desired root. But once we come close to $\alpha = 1$, the iteration speeds up dramatically and converges to the root quadratically (see Theorem 4.6.1). Since f is again convex, we actually have global convergence on $\mathbb{R}_+$, but, as we have seen, this is of little comfort.

Example 6.4: Let $f \in C^2[a,b]$ be such that

$$\begin{cases} f \text{ is convex (or concave) on } [a,b]; \\ f(a)f(b) < 0; \\ \text{the tangents at the endpoints of } [a,b] \\ \quad \text{intersect the real line within } [a,b]. \end{cases} \tag{6.7}$$

In this case, it is clear on geometric grounds that Newton's method converges globally, that is, for any $x_0 \in [a,b]$. Note that the tangent condition in (6.7) is automatically satisfied at *one* of the endpoints.

The following is a subroutine implementing Newton's method and returning not only an approximation x to the root, but also the number of iterations required to obtain it. The initial approximation is input by `a`, and `tol` is the error tolerance. There are two function routines `f`, `fd` as parameters, which evaluate f and f' and must be declared `external` when the routine is called.

```
      subroutine newton(a,tol,ntol,x,f,fd)
      ntol=0
      x=a
   10 ntol=ntol+1
      a=x
      fa=f(x)
      fad=fd(x)
      x=a-fa/fad
      if(abs(x-a).lt.tol .or. ntol.gt.100) then
        return
      else
        goto 10
      end if
      end
```

Run on the example of equation (1.1), with $a = \frac{3}{2}\pi$, $b = 2\pi$, the routine yields α_1 in single, double, and quadruple precision with `ntol` $= 4$, 5, and 6, respectively. This is slightly, but not much, faster than the secant method.

We see shortly, however, that the efficiency index is smaller for Newton's method than for the secant method.

To study the error in Newton's iteration, subtract α — a presumed simple root of the equation — from both sides of (6.1) to get

$$\begin{aligned} x_{n+1} - \alpha &= x_n - \alpha - \frac{f(x_n)}{f'(x_n)} \\ &= (x_n - \alpha)\left(1 - \frac{f(x_n) - f(\alpha)}{(x_n - \alpha) f'(x_n)}\right) \\ &= (x_n - \alpha)\left(1 - \frac{[x_n, \alpha]f}{[x_n, x_n]f}\right) \\ &= (x_n - \alpha)^2 \, \frac{[x_n, x_n, \alpha]f}{[x_n, x_n]f} \, . \end{aligned} \tag{6.8}$$

Therefore, if $x_n \to \alpha$, then

$$\lim_{n \to \infty} \frac{x_{n+1} - \alpha}{(x_n - \alpha)^2} = \frac{f''(\alpha)}{2f'(\alpha)} \, , \tag{6.9}$$

that is, Newton's method converges quadratically if $f''(\alpha) \neq 0$. Since it requires at each step one function evaluation and one derivative evaluation, the efficiency index is $\sqrt{2} = 1.41421\ldots$, which is less than the one for the secant method. The formula (6.9) suggests writing the equation $f(x) = 0$ in alternative, but equivalent, ways so as to reduce the asymptotic error constant on the right of (6.9).

The proof of *local convergence* of Newton's method is virtually the same as the one for the secant method (cf. Theorem 4.5.1). We only state the result.

Theorem 4.6.1. *Let α be a simple root of the equation $f(x) = 0$ and let $I_\varepsilon = \{x \in \mathbb{R} : |x - \alpha| \leq \varepsilon\}$. Assume that $f \in C^2[I_\varepsilon]$. Define*

$$M(\varepsilon) = \max_{\substack{s \in I_\varepsilon \\ t \in I_\varepsilon}} \left| \frac{f''(s)}{2f'(t)} \right| \, . \tag{6.10}$$

If ε is so small that

$$2\varepsilon M(\varepsilon) < 1, \tag{6.11}$$

then for every $x_0 \in I_\varepsilon$, Newton's method is well defined and converges quadratically to the only root $\alpha \in I_\varepsilon$. (The extra factor 2 in (6.11) comes from the requirement that $f'(x) \neq 0$ for $x \in I_\varepsilon$.)

§7. Fixed Point Iteration

Often, in applications, a nonlinear equation presents itself in the form of a *fixed point problem*: find x such that

$$x = \varphi(x). \tag{7.1}$$

A number α satisfying this equation is called a *fixed point of* φ. Any equation $f(x) = 0$, in fact, can (in many different ways) be written equivalently in the form (7.1). For example, if $f'(x) \neq 0$ in the interval of interest, we can take

$$\varphi(x) = x - \frac{f(x)}{f'(x)}\,. \tag{7.2}$$

If x_0 is an initial approximation of a fixed point α of (7.1), the *fixed point iteration* generates a sequence of approximants by

$$x_{n+1} = \varphi(x_n), \qquad n = 0, 1, 2, \dots\ . \tag{7.3}$$

If it converges, it clearly converges to a fixed point of φ if φ is continuous. Note that (7.3) is precisely Newton's method for solving $f(x) = 0$ if φ is defined by (7.2). So Newton's method can be viewed as a fixed point iteration, but not the secant method (why not?).

For any iteration of the form (7.3), assuming that $x_n \to \alpha$ as $n \to \infty$, it is straightforward to determine the order of convergence. Suppose, indeed, that at the fixed point α we have

$$\varphi'(\alpha) = \varphi''(\alpha) = \cdots = \varphi^{(p-1)}(\alpha) = 0, \qquad \varphi^{(p)}(\alpha) \neq 0. \tag{7.4}$$

(We tacitly assume $\varphi \in C^p$ near α.) This defines the integer $p \geq 1$. We then have by Taylor's theorem

$$\begin{aligned}\varphi(x_n) &= \varphi(\alpha) + (x_n - \alpha)\varphi'(\alpha) + \cdots + \frac{(x_n-\alpha)^{p-1}}{(p-1)!}\,\varphi^{(p-1)}(\alpha)\\ &+ \frac{(x_n-\alpha)^p}{p!}\,\varphi^{(p)}(\xi_n) = \varphi(\alpha) + \frac{(x_n-\alpha)^p}{p!}\,\varphi^{(p)}(\xi_n),\end{aligned}$$

where ξ_n is between α and x_n. Since $\varphi(x_n) = x_{n+1}$ and $\varphi(\alpha) = \alpha$, we get

$$\frac{x_{n+1} - \alpha}{(x_n - \alpha)^p} = \frac{1}{p!}\,\varphi^{(p)}(\xi_n).$$

As $x_n \to \alpha$, since ξ_n is trapped between x_n and α, we conclude, by the continuity of $\varphi^{(p)}$ at α, that

$$\lim_{n\to\infty} \frac{x_{n+1}-\alpha}{(x_n-\alpha)^p} = \frac{1}{p!}\,\varphi^{(p)}(\alpha) \neq 0. \tag{7.5}$$

This shows that convergence is exactly of the order p, and

$$c = \frac{1}{p!}\,\varphi^{(p)}(\alpha) \tag{7.6}$$

is the asymptotic error constant. Combining this with the usual local convergence argument, we obtain the following result.

Theorem 4.7.1. *Let α be a fixed point of φ and $I_\varepsilon = \{x \in \mathbb{R}\colon |x-\alpha| \le \varepsilon\}$. Assume $\varphi \in C^p[I_\varepsilon]$ satisfies* (7.4). *If*

$$M(\varepsilon) := \max_{t\in I_\varepsilon} |\varphi'(t)| < 1, \tag{7.7}$$

then the fixed point iteration (7.3) *converges to α for any $x_0 \in I_\varepsilon$. The order of convergence is p and the asymptotic error constant given by* (7.6).

Applying the theorem to (7.2) recovers second-order convergence of Newton's method. Indeed, with φ given in (7.2), we have

$$\varphi'(x) = 1 - \frac{[f'(x)]^2 - f(x)f''(x)}{[f'(x)]^2} = f(x)\,\frac{f''(x)}{[f'(x)]^2}\ ;$$

hence $\varphi'(\alpha) = 0$ (if $f'(\alpha) \neq 0$), and

$$\varphi''(x) = f(x)\left(\frac{f''(x)}{[f'(x)]^2}\right)' + \frac{f''(x)}{f'(x)}\ ;$$

hence $\varphi''(\alpha) = \frac{f''(\alpha)}{f'(\alpha)} \neq 0$, unless $f''(\alpha) = 0$. In the exceptional case $f''(\alpha) = 0$, Newton's method converges cubically (at least).

§8. Algebraic Equations

There are many iterative methods specifically designed to solve algebraic equations. Here we only describe how Newton's method applies in this context, essentially confining ourselves to a discussion of an efficient way to evaluate simultaneously the value of a polynomial and its first derivative. In the special case where all zeros of the polynomial are known to be real and simple, we describe an improved variant of Newton's method.

§8.1. **Newton's method applied to an algebraic equation.** We consider an algebraic equation of degree d,

$$f(x) = 0, \qquad f(x) = x^d + a_{d-1}x^{d-1} + \cdots + a_0, \tag{8.1}$$

where the leading coefficient is assumed (without restricting generality) to be 1 and where we may also assume $a_0 \neq 0$ without loss of generality. For simplicity we assume all coefficients to be real.

To apply Newton's method to (8.1), one needs good methods for evaluating a polynomial and its derivative. Underlying such methods are *division algorithms* for polynomials. Let t be some parameter and suppose we want to divide $f(x)$ by $x - t$. We write

$$f(x) = (x - t)(x^{d-1} + b_{d-1}x^{d-2} + \cdots + b_1) + b_0 \tag{8.2}$$

and compare coefficients of powers of x on both sides. This leads immediately to the equations

$$\begin{aligned} b_d &= 1, \\ b_k &= tb_{k+1} + a_k, \qquad k = d-1, d-2, \ldots, 0. \end{aligned} \tag{8.3}$$

The coefficients b_k so determined of course depend on t; indeed, b_k is a polynomial in t of degree $d - k$. We now make three useful observations:

(a) From (8.2), we note that $b_0 = f(t)$. Thus, in (8.3), we have an algorithm to evaluate $f(t)$ for any given t. It is known as *Horner's scheme*, although Newton already knew it. It requires d multiplications and d additions, which is more efficient than the naive way of computing f, which would be to first form the successive powers of t and then multiply them into their coefficients. This would require twice as many multiplies as Horner's scheme, but the same number of additions. It is an interesting question of complexity whether the number of multiplications can be further reduced. (It is known by a theorem of Ostrowski that the number d of additions is optimal.) This indeed is possible, and schemes using less than d multiplications have been developed by Pan and others. Unfortunately, the reduction in complexity comes with a price of increased numerical instability. Horner's scheme, therefore, is still the most widely used technique for evaluating a polynomial.

(b) Suppose $t = \alpha$, where α is a zero of f. Then $b_0 = 0$, and (8.2) allows division by $x - \alpha$ without remainder:

$$x^{d-1} + b_{d-1}x^{d-2} + \cdots + b_1 = \frac{f(x)}{x - \alpha} \, . \tag{8.4}$$

This is the *deflated polynomial*, in which the zero α has been "removed" from f. To compute its coefficients, therefore, all we need to do is apply Horner's scheme with $t = \alpha$. This comes in very handy in Newton's method when f is evaluated by Horner's scheme: once the method has converged to a root α, the final evaluation of f at (or very near) α automatically provides us with the coefficients of the deflated polynomial, and we are ready to reapply Newton's method to this deflated polynomial to compute the remaining zeros of f.

(c) By differentiating (8.2) with respect to x, and then putting $x = t$, we obtain

$$f'(t) = t^{d-1} + b_{d-1}t^{d-2} + \cdots + b_1. \tag{8.5}$$

Thus, we can apply Horner's scheme again to this polynomial to evaluate $f'(t)$. Both applications of Horner's scheme are conveniently combined into the following *double Horner scheme*:

$$\begin{aligned} & b_d = 1, \quad c_d = 1 \\ & \left.\begin{aligned} b_k &= t b_{k+1} + a_k \\ c_k &= t c_{k+1} + b_k \end{aligned}\right\} \quad k = d-1, d-2, \ldots, 1, 0. \end{aligned} \tag{8.6}$$

Then

$$f(t) = b_0, \qquad f'(t) = c_1. \tag{8.7}$$

The last step in (8.6) for c_0 is actually redundant, but it does not seem worth the complication in programming to eliminate this extra step.

We now have a convenient way to compute $f(x_n)$ and $f'(x_n)$ in each Newton step $x_{n+1} = x_n - \frac{f(x_n)}{f'(x_n)}$: apply the algorithm (8.6) with $t = x_n$ and use (8.7). Once x_n has converged to α, the bs generated in (8.6) give us the coefficients of the deflated polynomial (8.4) and we are ready to reapply Newton's method to the deflated polynomial.

It is clear that any real initial approximation x_0 generates a sequence of *real* iterates x_n and, therefore, can only be applied to compute real zeros of f (if any). For complex zeros one must start with a complex x_0, and the whole computation proceeds in complex arithmetic. It is possible, however, to use division algorithms with quadratic divisors to compute quadratic factors of f entirely in real arithmetic (Bairstow's method). See Ex. 41 for details.

One word of caution is in order when one tries to compute all zeros of f successively by Newton's method. It is true that Newton's method combined with Horner's scheme has a built-in mechanism of deflation, but this mechanism is valid only if one assumes convergence to the exact roots. This of course is impossible, partly because of rounding errors and partly because of the stopping criterion used to terminate convergence prematurely. Thus, there is a build-up of errors in the successively deflated polynomials, which may well have a significant effect on the accuracy of the respective roots (cf. Ch. 1, §3.2(2)). It is therefore imperative, once all roots have been computed, to "purify" them by applying Newton's method one more time to the *original* polynomial f in (8.1), using the computed roots as initial approximations.

§8.2. **An accelerated Newton method for equations with real roots**. If the equation (8.1) has only real distinct roots,

$$f(\alpha_\nu) = 0, \qquad \alpha_1 > \alpha_2 > \cdots > \alpha_d, \tag{8.8}$$

one can try to speed up Newton's method by approaching each root from the right with double the Newton steps until it is overshot, at which time one switches back to the ordinary Newton iteration to finish off the root.

Underlying this method is the following interesting theorem.

Theorem 4.8.1. *Let f be a polynomial having only real zeros as in* (8.8), *and let α_1' be the largest zero of f'. Then for every $z > \alpha_1$, defining*

$$z' := z - \frac{f(z)}{f'(z)}, \quad y := z - 2\frac{f(z)}{f'(z)}, \quad y' := y - \frac{f(y)}{f'(y)}, \tag{8.9}$$

one has

$$\alpha_1' < y, \qquad \alpha_1 \le y' \le z'. \tag{8.10}$$

The theorem suggests the following algorithm: start with some $x_0 > \alpha_1$ and apply

$$x_{k+1} = x_k - 2\,\frac{f(x_k)}{f'(x_k)}\,, \qquad k = 0, 1, 2, \ldots \,. \tag{8.11}$$

Then there are the following possibilities.

(i) We have $x_0 > x_1 > x_2 > \cdots > \alpha_1$ and $x_k \downarrow \alpha_1$ as $k \to \infty$. Since we use double Newton steps in (8.11), convergence in this case is faster than for the ordinary Newton iteration.

(ii) There exists a first index $k = k_0$ such that

$$f(x_0)f(x_k) > 0 \;\text{ for }\; 0 \le k < k_0; \;\; f(x_0)f(x_{k_0}) < 0.$$

Then $y := x_{k_0}$ is to the left of α_1 (we overshot!) but, by (8.10), to the right of α_1'. Using now y as the starting value in the ordinary Newton iteration,

$$y_0 = y, \quad y_{k+1} = y_k - \frac{f(y_k)}{f'(y_k)}\,, \qquad k = 0, 1, 2 \ldots \,,$$

brings us back to the right of α_1 in the first step, and then monotonically down to α_1.

In either case, having obtained α_1, we apply the same procedure to the deflated polynomial $f_1(x) = \frac{f(x)}{x-\alpha_1}$ to compute the next smaller zero. As starting value we can take α_1, or better, if case (ii) has occurred, $y = x_{k_0}$. The procedure can obviously be continued until all roots have been computed.

§9. Systems of Nonlinear Equations

Although most of the methods discussed earlier allow extensions to systems of nonlinear equations,

$$f(x) = 0, \qquad f:\; \mathbb{R}^d \to \mathbb{R}^d, \tag{9.1}$$

we consider here only two such methods: the fixed point iteration and Newton's method.

§9.1. **Contraction mapping principle**. We write (9.1) in fixed point form, $f(x) = x - \varphi(x)$, and consider the fixed point iteration

$$x_{n+1} = \varphi(x_n), \quad n = 0, 1, 2, \ldots . \tag{9.2}$$

We say that $\varphi : \mathbb{R}^d \to \mathbb{R}^d$ is a *contraction map* (or is *contractive*) on a set $\mathcal{D} \subseteq \mathbb{R}^d$ if there exists a constant γ with $0 < \gamma < 1$ such that, in some appropriate vector norm,

$$\|\varphi(x) - \varphi(x^*)\| \leq \gamma\|x - x^*\| \quad \text{for all } x \in \mathcal{D},\ x^* \in \mathcal{D}. \tag{9.3}$$

Theorem 4.9.1 (Contraction Mapping Principle). *Let $\mathcal{D} \subseteq \mathbb{R}^d$ be a complete subset of $\mathbb{R}^d$* (i.e., either bounded and closed, or all of $\mathbb{R}^d$). *If $\varphi : \mathbb{R}^d \to \mathbb{R}^d$ is contractive in the sense of* (9.3) *and maps $\mathcal{D}$ into $\mathcal{D}$, then*

(i) *the iteration* (9.2) *is well defined and converges to a unique fixed point $\alpha \in \mathcal{D}$,*

$$\lim_{n\to\infty} x_n = \alpha; \tag{9.4}$$

(ii) *for $n = 1, 2, 3, \ldots$ there holds*

$$\|x_n - \alpha\| \leq \frac{\gamma^n}{1-\gamma}\|x_1 - x_0\| \tag{9.5}$$

and

$$\|x_n - \alpha\| \leq \gamma^n\|x_0 - \alpha\|. \tag{9.6}$$

Proof. (i) Since $\varphi(\mathcal{D}) \subseteq \mathcal{D}$, the iteration (9.2) is well defined. We have, for $n = 1, 2, 3, \ldots$,

$$\|x_{n+1} - x_n\| = \|\varphi(x_n) - \varphi(x_{n-1})\| \leq \gamma\|x_n - x_{n-1}\|.$$

Repeated application of this yields

$$\|x_{n+1} - x_n\| \leq \gamma^n\|x_1 - x_0\|,$$

and hence, since

$$x_{n+p} - x_n = (x_{n+p} - x_{n+p-1}) + (x_{n+p-1} - x_{n+p-2}) + \cdots + (x_{n+1} - x_n),$$

more generally,

$$\begin{aligned}\|x_{n+p} - x_n\| &\le \sum_{k=1}^{p} \|x_{n+k} - x_{n+k-1}\| \le \sum_{k=1}^{p} \gamma^{n+k-1}\|x_1 - x_0\| \\ &\le \gamma^n \sum_{k=1}^{\infty} \gamma^{k-1}\|x_1 - x_0\| = \frac{\gamma^n}{1-\gamma}\|x_1 - x_0\|.\end{aligned} \tag{9.7}$$

Since $\gamma^n \to 0$, it follows that $\{x_n\}$ is a Cauchy sequence in $\mathcal{D}$, and hence, since $\mathcal{D}$ is complete, converges to some $\alpha \in \mathcal{D}$,

$$\lim_{n\to\infty} x_n \to \alpha.$$

The limit α must be a fixed point of φ since

$$\|x_n - \varphi(\alpha)\| = \|\varphi(x_{n-1}) - \varphi(\alpha)\| \le \gamma\|x_{n-1} - \alpha\|; \tag{9.8}$$

hence $\alpha = \lim_{n\to\infty} x_n = \varphi(\alpha)$. Moreover, there can be only one fixed point in $\mathcal{D}$, since $\alpha = \varphi(\alpha)$, $\alpha^* = \varphi(\alpha^*)$, and $\alpha \in \mathcal{D}$, $\alpha^* \in \mathcal{D}$ imply $\|\alpha - \alpha^*\| = \|\varphi(\alpha) - \varphi(\alpha^*)\| \le \gamma\|\alpha - \alpha^*\|$; that is, $(1-\gamma)\|\alpha - \alpha^*\| \le 0$, and hence $\alpha = \alpha^*$, since $1 - \gamma > 0$.

(ii) Letting $p \to \infty$ in (9.7) yields the first inequality in Theorem 4.9.1(ii). The second follows by a repeated application of (9.8), since $\varphi(\alpha) = \alpha$. □

The inequality (9.6) shows that the fixed point iteration converges (at least) linearly, with an error bound having asymptotic error constant equal to γ.

§9.2. **Newton's method for systems of equations**. As we mentioned earlier, Newton's method can be easily adapted to deal with systems of nonlinear equations, reducing the nonlinear problem to an infinite sequence of linear problems, that is, systems of linear algebraic equations. The tool is linearization at the current approximation.

Thus, given the equation (9.1), written more explicitly as

$$f^i(x^1, x^2, \ldots, x^d) = 0, \qquad i = 1, 2, \ldots, d \tag{9.1$'$}$$

(where components are indexed by superscripts), and given an approximation x_0 to a solution $\alpha \in \mathbb{R}^d$, the ith equation in (9.1′) is linearized at $x = x_0$ by truncating the Taylor expansion of f^i at x_0 after the linear terms. This gives

$$f^i(x_0) + \sum_{j=1}^{d} \frac{\partial f^i}{\partial x^j}(x_0)(x^j - x_0^j) = 0, \qquad i = 1, 2, \ldots, d,$$

or, written in vector form,

$$f(x_0) + \frac{\partial f}{\partial x}(x_0)(x - x_0) = 0, \tag{9.9}$$

where

$$\frac{\partial f}{\partial x}(x) := \left[\frac{\partial f^i}{\partial x^j}(x)\right]_{i,j=1}^{d} \tag{9.10}$$

is the *Jacobian matrix* of f. This is the natural generalization of the first derivative of a single function to systems of functions. The solution x of (9.9) — a system of linear algebraic equations — will be taken to be the next approximation. Thus, in general, starting with an initial approximation x_0, Newton's method will generate a sequence of approximations $x_n \in \mathbb{R}^d$ by means of

$$\begin{aligned} \frac{\partial f}{\partial x}(x_n)\Delta_n &= -f(x_n), \\ x_{n+1} &= x_n + \Delta_n, \end{aligned} \qquad n = 0, 1, 2, \ldots, \tag{9.11}$$

where we assume that the matrix $(\partial f/\partial x)(x_n)$ in the first equation is nonsingular for each n. This will be the case if $(\partial f/\partial x)(\alpha)$ is nonsingular and x_0 is sufficiently close to α, in which case one can prove as in the one-dimensional case $d = 1$ that Newton's method converges quadratically to α; that is, $\|x_{n+1} - \alpha\| = O(\|x_n - \alpha\|^2)$ as $n \to \infty$.

Writing (9.11) in the form

$$x_{n+1} = x_n - \left[\frac{\partial f}{\partial x}(x_n)\right]^{-1} f(x_n), \qquad n = 0, 1, 2, \ldots, \tag{9.11$'$}$$

brings out the formal analogy with Newton's method (6.1) for a single equation. However, it is not necessary to compute the inverse of the Jacobian at each step; it is more efficient to solve the linear system directly as in (9.11).

There are many ways to modify the initial stages of Newton's method (for systems of nonlinear equations) to force the iteration to make good progress toward approaching a solution. These usually go under the name *quasi-Newton methods* and they all share the idea of employing a suitable approximate inverse of the Jacobian rather than the exact one. For these, and also for generalizations of the secant method and the method of false position to systems of nonlinear equations, we refer to specialized texts.

NOTES TO CHAPTER 4

Texts dealing largely with single nonlinear equations are Traub [1964], Householder [1970], and Brent [1973]. Traub's book provides a systematic treatment of iterative methods, both old and new. Although it deals also with questions of convergence, the emphasis is on the derivation, classification, and cataloguing of iterative methods. Householder's book is strong on algebraic and analytic tools (often rooted in 19th-century mathematics) underlying numerical methods and less so on the methods themselves. Brent's book, although mainly devoted to optimization, gives a detailed account, and a computer program, of an algorithm essentially due to Dekker, combining the secant method with bisection in a manner similar to, but more clever than, the procedure in Mach. Ass. 3. Other well-tested algorithms available in current software are, for example, the IMSL routine `zreal`, implementing Muller's method (Muller [1956]), and `zporc` for (real) algebraic equations, implementing the Jenkins-Traub three-stage algorithm (Jenkins and Traub [1970]). The routine `zplrc`, based on Laguerre's method (cf., e.g., Fröberg [1985, §11.5]), also finds complex roots.

For the solution of systems of nonlinear equations, the book by Ortega and Rheinboldt [1970] gives a well-organized, and perhaps the most comprehensive, description and mathematical analysis of nonlinear iterative methods. A less exhaustive, but mathematically penetrating, text, also dealing with operator equations in Banach space, is Ostrowski [1973]. At a lower level of mathematical sophistication, but richer in algorithmic details, is the book by Dennis and Schnabel [1996], which gives equal treatment to nonlinear equations and optimization. A recent book by Kelley [1995] dealing with iterative methods for linear and nonlinear systems of equations provides e-mail and Web site addresses for respective MATLAB codes. A useful guide on available software for problems in optimization (including nonlinear equations) is Moré and Wright [1993]. For older precomputer literature, the two-volume treatise by Durand [1960,1961] is still a valuable source.

§1. A sampling of typical systems of nonlinear equations occurring in applied analysis is given in Ortega and Rheinboldt [1970, Ch. 1]. Many nonlinear equation problems arise from minimization problems, where one tries to find a minimum of a function of several variables. If it occurs at an interior point, then indeed the gradient of the function must vanish.

§1.2. For more on shooting methods, see Ch. 7, §2.

§1.3. Another, even simpler, way of approximating the solution y of (1.7) by y_n is to apply an n-point quadrature rule to the integral on the left, thus writing $y_n(x) = a(x) + \sum_{k=1}^{n} w_k K(x, t_k) f(t_k, y_n(t_k))$, $0 \le x \le 1$, and determining $\{y_n(t_k)\}_{k=1}^{n}$ by putting $x = t_i$, $i = 1, 2, \ldots, n$, in this equation. Again, we are led to a system of nonlinear equations.

§1.4. This example is taken from Gautschi and Milovanović [submitted], where one also finds a discussion on how to compute the Turán-type quadrature rule associated with an s-orthogonal polynomial π_n. This is the quadrature rule of maximum degree of exactness that involves a function and its derivatives up to order $2s$ evaluated at the zeros of π_n.

§2. More refined measures for the speed of convergence are defined in Brent [1973, p. 21] and, especially, in Ortega and Rheinboldt [1970, Ch. 9], where the concepts of Q-order and R-order of convergence are introduced, the former relating to quotients of errors, as in (2.4), the latter to nth roots $\varepsilon_n^{1/n}$. Also see Potra [1989]. These are quantities that characterize the asymptotic behavior of the error as $n \to \infty$; they say nothing about the initial behavior of the error. If one wants to describe the overall behavior of the iteration, one needs a function, not a number, to define the rate of convergence, for example, a function (if one exists) that relates $\|x_{n+1} - x_n\|$ to $\|x_n - x_{n-1}\|$ (cf. Ex. 22(b)). This is the approach taken in Potra and Pták [1984] to describe the convergence behavior of iterative processes in complete metric spaces. For similar ideas in connection with the convergence behavior of continued fractions, also see Gautschi [1983].

The efficiency index was introduced by Ostrowski [1973, Ch. 3, §11], who also coined the word "horner" for a unit of work.

§3.2. For Sturm sequences and Sturm's theorem see, for example, Henrici [1988, p. 444ff], and for Gershgorin's theorem, Golub and Van Loan [1996, p. 320]. The bisection method based on Sturm sequences, in the context of eigenvalues of a symmetric tridiagonal matrix, is implemented in the Eispack routine `BISECT` (Smith et al. [1976, p. 211]).

§4. The method of false position is very old, originating in medieval Arabic mathematics, and even earlier in 5th-century Indian texts (Plofker [1996, p. 254]). Leonardo Pisano (better known as "Fibonacci"), in the 13th century, calls it "*regula duarum falsarum positionum*," which, in the 16th and 17th centuries became abbreviated to "*regula positionum*" or also "*regula falsi*." Peter Bienewitz (1527), obviously having a linear equation in mind, explains the method in these (old German) words: "*Vnd heisst nit darum falsi dass sie falsch vnd unrecht wehr, sunder, dass sie auss zweyen falschen vnd vnwahrhaftigen zalen, vnd zweyen lügen die wahrhaftige vnd begehrte zal finden lernt.*" (Cf. Maas [1985].)

In the form (4.4), the regula falsi can be thought of as a discretized Newton's method, if in the latter, (6.1), one replaces the derivative $f'(x_n)$ by the difference quotient $(f(x_n) - f(b))/(x_n - b)$. This suggests one (of many) possible extension to systems of equations in $\mathbb{R}^d$ (cf. Ortega and Rheinboldt [1970, p. 205]): define vector difference quotients $\Delta_{n,i} = (x_n^i - b^i)^{-1}[f(x_n) - f(x_n + (b^i - x_n^i)e_i)]$, $i = 1, 2, \ldots, d$, where e_i is the ith coordinate vector, $x_n^T = [x_n^1, \ldots, x_n^d]$, and $b^T = [b^1, \ldots, b^d]$ a fixed vector. (Note that the arguments in the two f-vectors are the same except for the ith one, which is x_n^i in the first, and b^i in the second.) If we let $\Delta_n = [\Delta_{n,1}, \ldots, \Delta_{n,d}]$, a "difference quotient matrix" of order d, the regula falsi becomes $x_{n+1} = x_n - \Delta_n^{-1} f(x_n)$. As in the one-dimensional case, the method converges no faster than linearly, if at all (Ortega and Rheinboldt [1970, p. 366]).

§5. The secant method is also rather old; it has been used, for example, in 15th-century Indian texts to compute the sine function (Plofker [1996]).

The heuristic motivation of Theorem 4.5.1 uses some simple facts from the theory of linear difference equations with constant coefficients. A reader not familiar with these may wish to consult Henrici [1988, §7.4 and Notes to §7.4 on p. 663].

Like the method of false position, the secant method, too, can be extended to $\mathbb{R}^d$ in many different ways (see, e.g., Ortega and Rheinboldt [1970, §7.2], Dennis and Schnabel [1996, Ch. 8]), one of which is to replace the preceding vector b by the vector x_{n-1}. Theorem 4.5.2 then remains in force (Ortega and Rheinboldt [1970, 11.2.9]).

§6. The history of Newton's method is somewhat shrouded in obscurity. Newton's original ideas on the subject, around 1669, were considerably more complicated and not even remotely similar to what is now conceived to be his method. Raphson in approximately 1690 gave a simplified version of Newton's algorithm, possibly without knowledge of Newton's work. In the English literature, the method is therefore often called the Newton-Raphson method. According to Kollerstrom [1992] and Ypma [1995], Newton's and Raphson's procedures are both purely algebraic without mention of derivatives (or fluxions, as it were). They credit Simpson with being the first to give a calculus description of Newton's method in 1740, without referring either to Newton or to Raphson. As noted by Ypma [loc. cit.], Simpson applied Newton's method even to a 2×2 system of nonlinear equations. The modern version of Newton's iteration seems to appear first in a paper by Fourier published posthumously in 1831. See also Alexander [1996] for further historical comments.

Global convergence results for Newton's method analogous to the one in Example 6.4 exist also in higher dimension; see, for example, Ortega and Rheinboldt [1970, 13.3.4 and 13.3.7]. Local convergence, including its quadratic order, is now well established; see Ortega and Rheinboldt [1970, §12.6 and NR 12.6-1] for precise statements and for a brief but informative history. Crucial in this development was the work of Kantorovich, who studied Newton's method not only in finite-dimensional, but also in infinite-dimensional spaces. A good reference for the latter is Kantorovich and Akilov [1982, Ch. 18]. For a modification of Newton's method which is cubically convergent but requires an extra evaluation of f, see Ortega and Rheinboldt [1970, p. 315].

§7. Ostrowski [1973, Ch. 4, §2] calls α a point of attraction for the iteration (7.3) if for any x_0 in a sufficiently close neighborhood of α one has $x_n \to \alpha$, and a point of repulsion otherwise. Theorem 4.7.1 tells us that α is a point of attraction if $|\varphi'(\alpha)| < 1$; it is clearly a point of repulsion if $|\varphi'(\alpha)| > 1$. An analogous situation holds in $\mathbb{R}^d$ (Ostrowski [loc. cit., Ch. 22]): if the Jacobian matrix $\partial\varphi/\partial x$ at $x = \alpha$ has spectral radius < 1, then α is a point of attraction and hence the fixed point iteration is locally convergent. If the spectral radius is > 1, then α is a point of repulsion.

§8. An unusually detailed treatment of algebraic equations and their numerical solution, especially by older methods, is given in the French work by Durand [1960]. Algebraic equations are special enough that detailed information can be had about the location of their roots, and a number of methods can be devised specifically tailored to them. Good accounts of localization theorems can be found in Householder [1970, Ch. 2] and Marden [1966]. Among the classical methods appropriate for algebraic equations, the best known is Graeffe's method, which basically attempts to separate the moduli of the roots by successive squaring. Another is Cauchy's method — a quadratic extension of Newton's method — which requires second derivatives and is thus applied more readily to polynomial equations than to general nonlinear equations. Combined with the more recent method of Muller [1956] — a quadratic extension of the secant method — it can be made the basis of a reliable rootfinder (Young and Gregory [1988, Vol. 1, §5.4]). Among contemporary methods, mention should be made of the Lehmer-Schur method (Lehmer [1961], Henrici [1988, §6.10]), which constructs a sequence of shrinking circular disks in the complex plane eventually capturing a root, and of Rutishauser's QD algorithm (Rutishauser [1957], Henrici [1988, §7.6]), which under appropriate separation assumptions allows all zeros of a polynomial to be computed simultaneously. The same global character is shared by the Durand-Kerner method, which is basically Newton's method applied to the system of equations $a_i = \sigma_{d-i}(\alpha_1, \ldots, \alpha_d)$, $i = 0, 1, \ldots, d-1$, expressing the coefficients of the polynomial in terms of the elementary symmetric functions in the zeros. (The same method, incidentally, was already used by Weierstrass [1891] in order to prove the Fundamental Theorem of Algebra.) For other global methods, see Werner [1982]. Iterative methods carried out in complex interval arithmetic (based on circular disks or rectangles) are studied in Petković [1989]. From a set of initial complex intervals, each containing a zero of the polynomial, they generate a sequence of complex intervals encapsulating the respective zeros ever more closely. The midpoints of these intervals, taken to approximate the zeros, then come equipped with ready-made error estimates.

§8.1 (a). For Ostrowski's theorem, see Ostrowski [1954], and for literature on the numerical properties of Horner's scheme and more efficient schemes, Gautschi [1975a, §1.5.1(vi)].

The adverse accumulation of error in polynomial deflation can be mitigated somewhat by a more careful deflation algorithm, which depends on the relative magnitude of the zero being removed; see Peters and Wilkinson [1971] and Cohen [1994].

§8.2. The accelerated Newton method is due independently to Kahan and Maehly (cf. Wilkinson [1988, p. 480]). A proof of Theorem 4.8.2 can be found in Stoer and Bulirsch [1993, §5.5].

§9. Among the many other iterative methods for solving systems of nonlinear equations, we mention the nonlinear analogues of the well-known iterative methods for solving linear systems. Thus, for example, the nonlinear Gauss-Seidel method consists of solving the d single equations $f^i(x_{n+1}^1, \dots, x_{n+1}^{i-1}, t, x_n^{i+1}, \dots, x_n^d) = 0$, $i = 1, 2, \dots, d$, for t and letting $x_{n+1}^i = t$. The solution of each of these equations will in turn involve some one-dimensional iterative method, for example, Newton's method, which would constitute "inner iterations" in the "outer iteration" defined by Gauss-Seidel. Evidently, any pair of iterative methods can be so combined. Indeed, the roles can also be reversed, for example, by combining Newton's method for systems of nonlinear equations with the Gauss-Seidel method for solving the linear systems in Newton's method. Newton's method then becomes the outer iteration, and Gauss-Seidel the inner iteration. Still other methods involve homotopies (or continuation), embedding the given system of nonlinear equations in a one-parameter family of equations and approaching the desired solution via a sequence of intermediate solutions corresponding to appropriately chosen values of the parameter. Each of these intermediate solutions serves as an initial approximation to the next solution. Although the basic idea of such methods is simple, many implementational details must be worked out to make it successful; for a discussion of these, see Allgower and Georg [1990]. The application of continuation methods to polynomial systems arising in engineering and scientific problems is considered in Morgan [1987]. Both texts contain computer programs. Also see Watson, Billups, and Morgan [1987] for a software implementation.

§9.1. If one is interested in individual components, rather than just in norms, one can refine the contraction property by defining a map φ to be Γ-contractive on $\mathcal{D}$ if there exists a nonnegative matrix $\Gamma \in \mathbb{R}^{d \times d}$, with spectral radius < 1, such that $|\varphi(x) - \varphi(x^*)| \leq \Gamma |x - x^*|$ (componentwise) for all $x, x^* \in \mathcal{D}$. The contraction mapping principle then extends naturally to Γ-contractive maps (cf. Ortega and Rheinboldt [1970, Ch. 13]). Other generalizations of the contraction mapping principle involve perturbations of the map φ (Ortega and Rheinboldt [1970, Ch. 12, §2]).

§9.2. For quasi-Newton methods (also called modification, or update, methods), see, for example, Ortega and Rheinboldt [1970, Ch. 7, §3], Dennis and Schnabel [1996, Ch. 6]. As with any iterative method, the increment vector (e.g., the modified Newton increment) may be multiplied at each step by an appropriate scalar to ensure that $\|f(x_{n+1})\| < \|f(x_n)\|$. This is particularly advisable during the initial stages of the iteration.

EXERCISES AND MACHINE ASSIGNMENTS TO CHAPTER 4

EXERCISES

1. The following sequences all converge to zero as $n \to \infty$:

$$v_n = n^{-10}, \quad w_n = 10^{-n}, \quad x_n = 10^{-n^2}, \quad y_n = n^{10} \cdot 3^{-n}, \quad z_n = 10^{-3 \cdot 2^n}.$$

Indicate the type of convergence by placing a check mark in the appropriate position in the following table.

Type of Convergence	v	w	x	y	z
sublinear					
linear					
superlinear					
quadratic					
cubic					
none of the above					

2. Suppose a positive sequence $\{\varepsilon_n\}$ converges to zero with order $p > 0$. Does it then also converge to zero with order p' for any $0 < p' < p$?

3. The sequence $\varepsilon_n = e^{-e^n}$, $n = 0, 1, \ldots$, clearly converges to zero as $n \to \infty$. What is the order of convergence?

4. Give an example of a positive sequence $\{\varepsilon_n\}$ converging to zero in such a way that $\lim_{n\to\infty} \frac{\varepsilon_{n+1}}{\varepsilon_n^p} = 0$ for some $p > 1$, but not converging (to zero) with any order $p' > p$.

5. Suppose $\{x_n\}$ converges linearly to α, in the sense that $\lim_{n\to\infty} \frac{x_{n+1}-\alpha}{x_n-\alpha} = c$, $0 < |c| < 1$.

   (a) Define $x_n^* = \frac{1}{2}(x_n + x_{n-1})$, $n = 1, 2, 3, \ldots$. Clearly, $x_n^* \to \alpha$. Does $\{x_n^*\}$ converge appreciably faster than $\{x_n\}$? Explain by determining the asymptotic error constant.

   (b) Do the same for $x_n^* = \sqrt{x_n x_{n-1}}$, assuming $x_n > 0$ for all n, and $\alpha > 0$.

6. Given an iterative method of order p and asymptotic error constant $c \neq 0$, define a new iterative method consisting of m consecutive steps of the given method. Determine the order of this new iterative method and its asymptotic error constant. Hence justify the definition of the efficiency index given near the end of §2.

7. Consider the equation

$$\frac{1}{x-1} + \frac{2}{x+3} + \frac{4}{x-5} - 1 = 0.$$

(a) Discuss graphically the number of real roots and their approximate location.

(b) Are there any complex roots?

8. Consider the equation $x \tan x = 1$.

   (a) Discuss the real roots of this equation: their number, approximate location, and symmetry properties. Use appropriate graphs.

   (b) How many bisections would be required to find the smallest positive root to within an error of $\frac{1}{2} \times 10^{-8}$? (Indicate the initial approximations.) Is your answer valid for all roots?

9. Consider the quadratic equation $x^2 - p = 0$, $p > 0$. Suppose its positive root $\alpha = \sqrt{p}$ is computed by the method of false position starting with two numbers a, b satisfying $0 < a < \alpha < b$. Determine the asymptotic error constant c as a function of b and α. What are the conditions on b for $0 < c < \frac{1}{2}$ to hold, that is, for the method of false position to be (asymptotically) faster than the bisection method?

10. The equation $x^2 - a = 0$ (for the square root $\alpha = \sqrt{a}$) can be written equivalently in the form

$$x = \varphi(x)$$

in many different ways, for example:

$$\varphi(x) = \frac{1}{2}\left(x + \frac{a}{x}\right); \qquad \varphi(x) = \frac{a}{x}\,; \qquad \varphi(x) = 2x - \frac{a}{x}\,.$$

Discuss the convergence (or nonconvergence) behavior of the iteration $x_{n+1} = \varphi(x_n)$, $n = 0, 1, 2, \dots$, for each of these three iteration functions. In case of convergence, determine the order of convergence.

11. Let $\{x_n\}$ be a sequence converging to α. Suppose the errors $e_n = |x_n - \alpha|$ satisfy $e_{n+1} \le M e_n^2 e_{n-1}$ for some constant $M > 0$. What can be said about the order of convergence?

12. (a) Consider the iteration $x_{n+1} = x_n^3$. Give a detailed discussion of the behavior of the sequence $\{x_n\}$ in dependence of x_0.

    (b) Do the same as (a), but for $x_{n+1} = x_n^{1/3}$, $x_0 > 0$.

13. Consider the iteration

$$x_{n+1} = \varphi(x_n), \qquad \varphi(x) = \sqrt{2 + x}.$$

   (a) Show that for any positive x_0 the iterate x_n remains on the same side of $\alpha = 2$ as x_0.

   (b) Show that the iteration converges globally, that is, for any $x_0 > 0$, and not faster than linearly (unless $x_0 = 2$).

(c) If $0 < x_0 < 2$, how many iteration steps are required to obtain α with an error less than 10^{-10}?

(d) If the iteration is perturbed to

$$x_{n+1} = \varphi(x_n)(1 + \varepsilon_n), \quad |\varepsilon_n| \le \varepsilon, \ \ 0 < \varepsilon < 1,$$

how does (a) have to be modified to remain true?

14. Consider the equation $x = \cos x$.

(a) Show graphically that there exists a unique positive root α. Indicate, approximately, where it is located.

(b) Prove local convergence of the iteration $x_{n+1} = \cos x_n$.

(c) For the iteration in (b) prove: if $x_n \in [0, \frac{\pi}{2}]$, then

$$|x_{n+1} - \alpha| < \left(\sin \frac{\alpha + \pi/2}{2}\right) |x_n - \alpha|.$$

In particular, one has global convergence on $[0, \frac{\pi}{2}]$.

(d) Show that Newton's method applied to $f(x) = 0$, $f(x) = x - \cos x$, also converges globally on $[0, \frac{\pi}{2}]$.

15. Consider the equation

$$x = e^{-x}.$$

(a) Show that there is a unique real root α and determine an interval containing it.

(b) Show that the fixed point iteration $x_{n+1} = e^{-x_n}$, $n = 0, 1, 2, \ldots$, converges locally to α and determine the asymptotic error constant.

(c) Illustrate graphically that the iteration in (b) actually converges globally, that is, for arbitrary $x_0 > 0$. Then prove it.

(d) An equivalent equation is

$$x = \ln \tfrac{1}{x}.$$

Does the iteration $x_{n+1} = \ln \frac{1}{x_n}$ also converge locally? Explain.

16. Consider the equation

$$\tan x + \lambda x = 0, \qquad 0 < \lambda < 1.$$

(a) Show graphically, as simply as possible, that there is exactly one root, α, in the interval $[\frac{1}{2}\pi, \pi]$.

(b) Does Newton's method converge to the root $\alpha \in [\frac{1}{2}\pi, \pi]$ if the initial approximation is taken to be $x_0 = \pi$? Justify your answer.

17. Consider the equation

$$f(x) = 0, \quad f(x) = \ln^2 x - x - 1, \quad x > 0.$$

   (a) Graphical considerations suggest that there is exactly one positive root α, and that $0 < \alpha < 1$. Prove this.

   (b) What is the largest positive $b \le 1$ such that Newton's method, started with $x_0 = b$, converges to α?

18. Consider "Kepler's equation"

$$f(x) = 0, \quad f(x) = x - \varepsilon \sin x - \eta, \quad 0 < |\varepsilon| < 1, \quad \eta \in \mathbb{R},$$

where ε, η are parameters.

   (a) Show that for each ε, η there is exactly one real root $\alpha = \alpha(\varepsilon, \eta)$. Furthermore, $\eta - |\varepsilon| \le \alpha(\varepsilon, \eta) \le \eta + |\varepsilon|$.

   (b) Writing the equation in fixed point form

$$x = \varphi(x), \quad \varphi(x) = \varepsilon \sin x + \eta,$$

   show that the fixed point iteration $x_{n+1} = \varphi(x_n)$ converges for arbitrary starting value x_0.

   (c) Let m be an integer such that $m\pi < \eta < (m+1)\pi$. Show that Newton's method with starting value

$$x_0 = \begin{cases} (m+1)\pi & \text{if } (-1)^m \varepsilon > 0, \\ m\pi & \text{otherwise} \end{cases}$$

   is guaranteed to converge (monotonically) to $\alpha(\varepsilon, \eta)$.

   (d) Estimate the asymptotic error constant c of Newton's method.

19. (a) Devise an iterative scheme, using only addition and multiplication, for computing the reciprocal $\frac{1}{a}$ of some positive number a. {*Hint*: Use Newton's method.}

   (b) For what starting values x_0 does the algorithm in (a) converge? What happens if $x_0 < 0$?

   (c) Since in (binary) floating-point arithmetic it suffices to find the reciprocal of the mantissa, assume $\frac{1}{2} \le a < 1$. Show, in this case, that the iterates x_n satisfy

$$\left| x_{n+1} - \frac{1}{a} \right| < \left| x_n - \frac{1}{a} \right|^2, \quad \text{all } n \ge 0.$$

   (d) Using the result of (c), estimate how many iterations are required, at most, to obtain $1/a$ with an error less than 2^{-48}, if one takes $x_0 = \frac{3}{2}$.

20. (a) If $A > 0$, then $\alpha = \sqrt{A}$ is a root of either equation

$$x^2 - A = 0, \qquad \frac{A}{x^2} - 1 = 0.$$

Explain why Newton's method applied to the first equation converges for arbitrary starting value $x_0 > 0$, whereas the same method applied to the second equation produces positive iterates x_n converging to α only if x_0 is in some interval $0 < x_0 < b$. Determine b.

(b) Do the same as (a), but for the cube root $\sqrt[3]{A}$ and the equations

$$x^3 - A = 0, \qquad \frac{A}{x^3} - 1 = 0.$$

21. (a) Show that Newton's iteration

$$x_{n+1} = \frac{1}{2}\left(x_n + \frac{a}{x_n}\right), \qquad a > 0,$$

for computing the square root $\alpha = \sqrt{a}$ satisfies

$$\frac{x_{n+1} - \alpha}{(x_n - \alpha)^2} = \frac{1}{2x_n}.$$

Hence, directly obtain the asymptotic error constant.

(b) What is the analogous formula for the cube root?

22. Consider Newton's method

$$x_{n+1} = \frac{1}{2}\left(x_n + \frac{a}{x_n}\right), \qquad a > 0,$$

for computing the square root $\alpha = \sqrt{a}$. Let $d_n = x_{n+1} - x_n$.

(a) Show that

$$x_n = \frac{a}{d_n + \sqrt{d_n^2 + a}}.$$

(b) Use (a) to show that

$$|d_n| = \frac{d_{n-1}^2}{2\sqrt{d_{n-1}^2 + a}}, \qquad n = 1, 2, \ldots .$$

Discuss the significance of this result with regard to the overall behavior of Newton's iteration in this case.

23. (a) Derive the iteration that results by applying Newton's method to $f(x) := x^3 - a = 0$ to compute the cube root $\alpha = a^{\frac{1}{3}}$ of $a > 0$.

(b) Consider the equivalent equation $f_\lambda(x) = 0$, where $f_\lambda(x) = x^{3-\lambda} - ax^{-\lambda}$, and determine λ so that Newton's method converges cubically. Write down the resulting iteration in its simplest form.

24. Consider the two (equivalent) equations

$$\text{(A)} \quad x \ln x - 1 = 0 \qquad \text{(B)} \quad \ln x - \frac{1}{x} = 0.$$

(a) Show that there is exactly one positive root and find a rough interval containing it.

(b) For both (A) and (B), determine the largest interval on which Newton's method converges. {*Hint*: Investigate the convexity of the functions involved.}

(c) Which of the two Newton iterations converges asymptotically faster?

25. Prove Theorem 4.6.1.

26. Consider the equation

$$f(x) = 0, \quad \text{where} \quad f(x) = \tan x - cx, \quad 0 < c < 1.$$

(a) Show that the smallest positive root α is in the interval $(\pi, \frac{3}{2}\pi)$.

(b) Show that Newton's method started at $x_0 = \pi$ is guaranteed to converge to α if c is small enough. Exactly how small does c have to be?

27. We saw in §1.1 that the equation

$$(*) \qquad \cos x \cosh x - 1 = 0$$

has exactly two roots $\alpha_n < \beta_n$ in each interval $[-\frac{\pi}{2} + 2n\pi, \frac{\pi}{2} + 2n\pi]$, $n = 1, 2, 3 \dots$. Show that Newton's method applied to (*) converges to α_n when initialized by $x_0 = -\frac{\pi}{2} + 2n\pi$, and to β_n when initialized by $x_0 = \frac{\pi}{2} + 2n\pi$.

28. In the engineering of circular shafts the following equation is important for determining critical angular velocities:

$$f(x) = 0, \quad f(x) = \tan x + \tanh x, \quad x > 0.$$

(a) Show that there are infinitely many positive roots, exactly one, α_n, in each interval $[(n - \frac{1}{2})\pi, n\pi]$, $n = 1, 2, 3, \dots$.

(b) Determine $\lim_{n\to\infty}(n\pi - \alpha_n)$.

(c) Discuss the convergence of Newton's method when started at $x_0 = n\pi$.

29. The equation

$$f(x) := x \tan x - 1 = 0,$$

if written as $\tan x = 1/x$ and each side plotted separately, can be seen to have infinitely many positive roots, one, α_n, in each interval $[n\pi, (n + \frac{1}{2})\pi]$, $n = 0, 1, 2, \dots$.

(a) Show that the smallest positive root α_0 can be obtained by Newton's method started at $x_0 = \frac{\pi}{4}$.

(b) Show that Newton's method started with $x_0 = (n + \frac{1}{4})\pi$ converges monotonically decreasing to α_n if $n \geq 1$.

(c) Expanding α_n (formally) in inverse powers of n,

$$\alpha_n = n\pi + c_0 + c_1 n^{-1} + c_2 n^{-2} + c_3 n^{-3} + \cdots ,$$

determine c_0, c_1, c_2, and c_3.

30. Consider the equation

$$f(x) = 0, \quad f(x) = x \sin x - 1, \ \ 0 \leq x \leq \pi.$$

(a) Show graphically (as simply as possible) that there are exactly two roots in the interval $[0, \pi]$ and determine their approximate locations.

(b) What happens with Newton's method when it is started with $x_0 = \frac{1}{2}\pi$? Does it converge, and if so, to which root? Where do you need to start Newton's method to get the other root?

31. (GREGORY, 1672) For an integer $n \geq 1$, consider the equation

$$f(x) = 0, \quad f(x) = x^{n+1} - b^n x + ab^n, \ \ a > 0, \ \ b > 0.$$

(a) Prove that the equation has exactly two distinct positive roots if and only if

$$a < \frac{n}{(n+1)^{1+\frac{1}{n}}}\, b.$$

{*Hint*: Analyze the convexity of f.}

(b) Assuming that the condition in (a) holds, show that Newton's method converges to the smaller positive root, when started at $x_0 = a$, and to the larger one, when started at $x_0 = b$.

32. Suppose the equation $f(x) = 0$ has the root α with exact multiplicity $m \geq 2$, and Newton's method converges to this root. Show that convergence is linear, and determine the asymptotic error constant.

33. (a) Let α be a double root of the equation $f = 0$, where f is sufficiently smooth near α. Show that the "doubly-relaxed" Newton method

$$x_{n+1} = x_n - 2\,\frac{f(x_n)}{f'(x_n)} ,$$

if it converges to α, does so at least quadratically. Obtain the condition under which the order of convergence is exactly 2, and determine the asymptotic error constant c in this case.

(b) What are the analogous statements in the case of an m-fold root?

34. Consider the equation $x \ln x = a$.

   (a) Show that for each $a > 0$ the equation has a unique positive root, $x = x(a)$.

   (b) Prove that

$$x(a) \sim \frac{a}{\ln a} \quad \text{as} \quad a \to \infty$$

   (i.e., $\lim_{a\to\infty} \frac{x(a)\ln a}{a} = 1$). {*Hint*: Use the rule of Bernoulli-L'Hospital.}

   (c) For large a improve the approximation given in (b) by applying one step of Newton's method.

35. The equation $x^2 - 2 = 0$ can be written as a fixed point problem in different ways, for example,

$$\text{(a)} \quad x = \frac{2}{x} \qquad \text{(b)} \quad x = x^2 + x - 2 \qquad \text{(c)} \quad x = \frac{x+2}{x+1} \, .$$

   How does the fixed point iteration perform in each of these three cases? Be as specific as you can.

36. Show that

$$x_{n+1} = \frac{x_n(x_n^2 + 3a)}{3x_n^2 + a}, \qquad n = 0, 1, 2, \ldots ,$$

   is a method for computing $\alpha = \sqrt{a}$, $a > 0$, which converges cubically to α (for suitable x_0). Determine the asymptotic error constant.

37. Consider the fixed point iteration

$$x_{n+1} = \varphi(x_n), \quad n = 0, 1, 2, \ldots ,$$

   where

$$\varphi(x) = Ax + Bx^2 + Cx^3.$$

   (a) Given a positive number α, determine the constants A, B, C such that the iteration converges locally to $1/\alpha$ with order $p = 3$. {This will give a cubically convergent method for computing the reciprocal $1/\alpha$ of α, which uses only addition, subtraction, and multiplication.}

   (b) Determine the precise condition on the initial error $\varepsilon_0 = x_0 - \frac{1}{\alpha}$ for the iteration to converge.

38. The equation $f(x) := x^2 - 3x + 2 = 0$ has the roots 1 and 2. A rearrangement of it produces the algorithm

$$x_{n+1} = \frac{1}{\omega}(x_n^2 - (3 - \omega)x_n + 2), \quad n = 0, 1, 2, \ldots \quad (\omega \neq 0).$$

(a) Identify as large an ω-interval as possible such that for any ω in this interval the iteration converges to 1 (when $x_0 \neq 1$ is suitably chosen).

(b) Do the same as (a), but for the root 2 (and $x_0 \neq 2$).

(c) For what value(s) of ω does the iteration converge quadratically to 1?

(d) Interpret the algorithm produced in (c) as a Newton iteration for some equation $F(x) = 0$, and exhibit F. Hence discuss for what initial values x_0 Newton's method converges.

39. Let α be a simple zero of f and $f \in C^p$ near α, where $p \geq 3$. Show: if $f''(\alpha) = \cdots = f^{(p-1)}(\alpha) = 0$, $f^{(p)}(\alpha) \neq 0$, then Newton's method applied to $f(x) = 0$ converges to α locally with order p. Determine the asymptotic error constant.

40. The iteration

$$x_{n+1} = x_n - \frac{f(x_n)}{f'(x_n) - \frac{1}{2} f''(x_n) \frac{f(x_n)}{f'(x_n)}}, \qquad n = 0, 1, 2, \dots ,$$

for solving the equation $f(x) = 0$ is known as *Halley's method.*

(a) Interpret Halley's method geometrically as the intersection with the x-axis of a hyperbola with asymptotes parallel to the x- and y-axes that is osculatory to the curve $y = f(x)$ at $x = x_n$ (i.e., is tangent to the curve at this point and has the same curvature there).

(b) Show that the method can, alternatively, be interpreted as applying Newton's method to the equation $g(x) = 0$, $g(x) := f(x)/\sqrt{f'(x)}$.

(c) Assuming α is a simple root of the equation, and $x_n \to \alpha$ as $n \to \infty$, show that convergence is exactly cubic, unless the "Schwarzian derivative"

$$(Sf)(x) := \frac{f'''(x)}{f'(x)} - \frac{3}{2}\left(\frac{f''(x)}{f'(x)}\right)^2$$

vanishes at $x = \alpha$, in which case the order of convergence is larger than three.

(d) Is Halley's method more efficient than Newton's method as measured in terms of the efficiency index?

(e) How does Halley's method look in the case $f(x) = x^\lambda - a$, $a > 0$? (Compare with Ex. 36.)

41. Let $f(x) = x^d + a_{d-1}x^{d-1} + \cdots + a_0$ be a polynomial of degree $d \geq 2$ with real coefficients a_i.

(a) In analogy to (8.2), let

$$f(x) = (x^2 - tx - s)(x^{d-2} + b_{d-1}x^{d-3} + \cdots + b_2) + b_1(x - t) + b_0.$$

Derive a recursive algorithm for computing $b_{d-1}, b_{d-2}, \dots, b_1, b_0$ in this order.

(b) Suppose α is a complex zero of f. How can f be deflated to remove the pair of zeros α, $\overline{\alpha}$?

(c) (Bairstow's method) Devise a method based on the division algorithm of (a) to compute a quadratic factor of f. Use Newton's method for a system of two equations in the two unknowns t and s, and exhibit recurrence formulae for computing the elements of the 2×2 Jacobian matrix of the system.

42. The boundary value problem (1.2) and (1.3) may be discretized by replacing the first and second derivatives by centered difference quotients relative to a grid of equally spaced points $x_k = \frac{k}{n+1}$, $k = 0, 1, \ldots, n, n+1$. Interpret the resulting equations as a fixed point problem in $\mathbb{R}^n$. What method of solution does this suggest?

43. Let $p(t)$ be a monic polynomial of degree n. Let $x \in \mathbb{C}^n$ and define

$$f_\nu(x) = [x_1, x_2, \ldots, x_\nu]\, p, \qquad \nu = 1, 2, \ldots, n$$

to be the divided differences of p relative to the coordinates x_μ of x. Consider the system of equations

$$f(x) = 0, \qquad [f(x)]^T = [f_1(x), f_2(x), \ldots, f_n(x)].$$

(a) Let $\alpha^T = [\alpha_1, \alpha_2, \ldots, \alpha_n]$ be the zeros of p. Show that α is, except for a permutation of the components, the unique solution of $f(x) = 0$. {*Hint*: Use Newton's formula of interpolation.}

(b) Describe the application of Newton's iterative method to the preceding system of nonlinear equations, $f(x) = 0$. {*Hint*: Use Ch. 2, Ex. 52.}

(c) Discuss to what extent the procedure in (a) and (b) is valid for nonpolynomial functions p.

44. For the equation $f(x) = 0$ define

$$\begin{aligned} y^{[0]}(x) &= x, \\ y^{[1]}(x) &= \frac{1}{f'(x)}, \\ &\cdots\cdots\cdots\cdots \\ y^{[m]}(x) &= \frac{1}{f'(x)} \frac{d}{dx} y^{[m-1]}(x), \qquad m = 2, 3, \ldots . \end{aligned}$$

Consider the iteration function

$$\varphi_r(x) := \sum_{m=0}^{r} (-1)^m \frac{y^{[m]}(x)}{m!} [f(x)]^m.$$

(When $r = 1$ this is the iteration function for Newton's method.) Show that $\varphi_r(x)$ defines an iteration $x_{n+1} = \varphi_r(x_n)$, $n = 0, 1, 2, \ldots$, converging locally with exact order $p = r + 1$ to a root α of the equation if $y^{[r+1]}(\alpha) f'(\alpha) \neq 0$.

MACHINE ASSIGNMENTS

1. (a) Write a FORTRAN routine that computes in (IEEE Standard) single-precision floating-point arithmetic the polynomial $p(x) = x^5 - 5x^4 + 10x^3 - 10x^2 + 5x - 1$ (as written). Run the routine to print $p(x)/\text{eps}$ for 200 equally spaced x-values in a small neighborhood of $x = 1$ (say, $.93 \le x \le 1.07$), where eps is the machine precision as computed in Ch. 1, Ex. 5. Use MATLAB to prepare a piecewise linear plot of the results. Explain what you observe. What is the "uncertainty interval" for the numerical root corresponding to the mathematical root $x = 1$?

   (b) Do the same as (a), but for the polynomial $p(x) = x^5 - 100x^4 + 3995x^3 - 79700x^2 + 794004x - 3160088$, which has simple zeros at x=18,19,20,21,22. Examine a small interval around $x = 22$ (say, $21.7 \le x \le 22.2$).

2. (a) Run the routine `bisec` of §3.1 for the polynomial $f = p$ of Ex. 1(a) and for `a` = 0(.1).5, `b` = 1.5(.1)2. Use for `tol` two times the machine precision eps with eps as determined in Ex. 1, and print `ntol,x,a,b,p(x),p(a),p(b)`, where `a,b` are the values of `a,b` upon exit from the routine. Comment on the results.

   (b) Do the same as (a) but with the routine `false` of §4.

   (c) Do the same as (a) but with the routine `secant` of §5.

   (d) Do the same as (a) but with the routine `newton` of §6, printing `ntol,x`. Compare the results with the theoretical ones for infinite-precision calculations.

   Repeat (a) through (d) with the polynomial $f = p$ of Ex. 1(b) and for `a` = 21.6(.1)21.8, `b` = 22.2(.1)22.4. Use for `tol` one hundred times the machine precision.

3. To avoid breakdowns in the secant method, combine it with the bisection method in the following manner. Initialize a, b such that $f(a)f(b) < 0$ and choose $c = a$. In general, let b denote the most recent iterate, a the one preceding it, and c the most recent iterate such that $f(c)f(b) < 0$. The algorithm operates on a and b, producing a' and b', and updates c to c'. Basically, if $f(a)f(b) < 0$, the algorithm executes a secant step. It does the same if $f(a)f(b) > 0$, provided the result comes to lie between b and c; otherwise, it takes the midpoint of $[b, c]$ as the new iterate. Thereafter, c is updated to satisfy the condition imposed on it.

   In pseudocode:

```
a'=b
if(fa*fb<0) then
  b'=s(a,b)
```

```
else
  if(s(a,b) in [b,c]) then
    b'=s(a,b)
  else
    b'=(b+c)/2
  end if
end if
if(fb'*fa'>0) then c'=c else c'=a'
```

Here, `s(a,b)` denotes the result of applying a secant step to `a` and `b`. By `[b,c]` is meant the interval between `b` and `c` regardless of whether `b` $<$ `c` or `b` $>$ `c`. The condition `s(a,b) in [b,c]` has to be checked *implicitly* without computing `s(a,b)` (to avoid a possible breakdown). Explain how. Implement the algorithm in a FORTRAN program and run it on the examples of Ex. 1(a) and (b), similarly as in Ex. 2. Print not only `ntol`, the number of iterations, but also the number of secant steps and bisection steps executed.

4. Consider the equation
$$\tfrac{1}{2}x - \sin x = 0.$$
   (a) Discuss the number of real roots and their approximate location.
   (b) Starting with $x_0 = \pi/2$, $x_1 = \pi$, how many iterations are required in the bisection method to find a root to 10 correct decimal places?
   (c) Use the subroutine `bisec` of §3.1 to carry out the bisection method indicated in (b).

5. For an integer $n \geq 2$, consider the equation
$$\frac{x + x^{-1}}{x^n + x^{-n}} = \frac{1}{n}\,.$$
   (a) Write the equation equivalently as a polynomial equation, $p_n(x) = 0$.
   (b) Use Descartes' rule of sign[3] (applied to $p_n(x) = 0$) to show that there are exactly two positive roots, one in (0,1), the other in $(1,\infty)$. How are they related? Denote the larger of the two roots by α_n (>1). It is known (you do not have to prove this) that
$$1 < \alpha_{n+1} < \alpha_n < 3, \qquad n = 2, 3, 4, \ldots\,.$$
   (c) Write and run a program applying the bisection method to compute α_n, $n = 2, 3, \ldots, 20$, to six correct decimal places after the decimal point, using $[1, 3]$ as initial interval for α_2, and $[1, \alpha_n]$ as initial interval for α_{n+1} ($n \geq 2$). For each n count the number of iterations required.

[3]Descartes' rule of sign says that if a real polynomial has s sign changes in its sequence of nonzero coefficients, then it has s positive zeros or a (nonnegative) even number less.

Similarly, apply Newton's method (to the equation $p_n(x) = 0$) to compute α_n to the same accuracy, using the initial value 3 for α_2 and the initial value α_n for α_{n+1} ($n \geq 2$). (Justify these choices.) Again, for each n, count the number of iterations required. In both cases, print n, α_n, and the number of iterations. Use a `do while` construction to program either method.

6. Consider the equation
$$x = e^{-x}.$$
   (a) Implement the fixed-point iteration $x_{n+1} = e^{-x_n}$ on the computer, starting with $x_0 = 1$ and stopping at the first n for which x_{n+1} agrees with x_n to within the machine precision. Print this value of n and the corresponding x_{n+1}.
   (b) If the equation is multiplied by ω ($\neq 0$ and $\neq -1$) and x is added on both sides, one gets the equivalent equation
$$x = \frac{\omega e^{-x} + x}{1 + \omega}.$$
   Under what condition on ω does the fixed-point iteration for this equation converge faster (ultimately) than the iteration in (a)? {This condition involves the root α of the equation.}
   (c) What is the optimal choice of ω? Verify it by a machine computation in a manner analogous to (a).

7. Consider the boundary value problem
$$y'' + \sin y = 0, \quad 0 \leq x \leq \tfrac{1}{4}\pi,$$
$$y(0) = 0, \quad y(1) = 1,$$
which describes the angular motion of a pendulum.
   (a) Use the MATLAB integrator `ode45` to compute and plot the solution $u(x; s)$ of the associated initial value problem
$$u'' + \sin u = 0, \quad 0 \leq x \leq \tfrac{1}{4}\pi,$$
$$u(0) = 0, \quad u'(0) = s$$
   for $s = .2(.2)2$.
   (b) Write and run a MATLAB script that applies the method of bisection to the equation $f(s) = 0$, $f(s) = u(1; s) - 1$. Use the plots of (a) to choose starting values s_0, s_1 such that $f(s_0) < 0$, $f(s_1) > 0$. Do the integrations with the MATLAB routine `ode45`, using a tolerance of 1E-12 and print the value of s obtained to 12 decimal places. Plot the solution curve.

8. Write a program that computes (in double precision) the recursion coefficients $\alpha_0, \ldots, \alpha_{n-1}$; $\beta_0, \ldots, \beta_{n-1}$ for an s-orthogonal polynomial π_n by solving the system of nonlinear equations (1.15), using $(s+1)n$-point Gauss quadrature to evaluate the integrals. Obtain routines for solving systems of nonlinear equations and generating Gauss quadrature rules from an appropriate software package, for example, the routine `hybrd1` from Minpack, and `dgauss`, `drecur`, `d1mach` from TOMS Algorithm 726 (all available on Netlib). Run your program for some of the classical measures $d\lambda$ (Legendre, Chebyshev, Laguerre, and Hermite), taking $n = 1(1)10$ and $s = 1, 2$. Experiment with different schemes for supplying initial approximations and different error tolerances.

   For the symmetric measures of Legendre, Chebyshev, and Hermite, obtain β_0, β_1 analytically when $n = 2$ (you may take $\alpha_0 = \alpha_1 = 0$; why?). Use the analytical results as a check on your computer results.

CHAPTER 5

INITIAL VALUE PROBLEMS FOR ODEs — ONE-STEP METHODS

Initial value problems for ordinary differential equations (ODEs) occur in almost all the sciences, notably in mechanics (including celestial mechanics), where the motion of particles (resp., planets) is governed by Newton's second law — a system of second-order differential equations. It is no wonder, therefore, that astronomers such as Adams, Moulton, and Cowell were instrumental in developing numerical techniques for their integration.[1] Also in quantum mechanics — the analogue of celestial mechanics in the realm of atomic dimensions — differential equations are fundamental; this time it is Schrödinger's equation that reigns, actually a linear *partial* differential equation involving the Laplace operator. Still, when separated in polar coordinates, it reduces to an ordinary second-order linear differential equation. Such equations are at the heart of the theory of special functions of mathematical physics. Coulomb wave functions, for example, are solutions of Schrödinger's equation when the potential is a Coulomb field.

Within mathematics, ordinary differential equations play an important role in the calculus of variations, where optimal trajectories must satisfy the Euler equations, or in optimal control problems, where they satisfy the Pontryagin maximum principle. In both cases one is led to boundary value problems for ordinary differential equations. This type of problem is discussed in Chapter 7. In the present and next chapter, we concentrate on initial value problems.

We begin with some examples of ordinary differential equations as they arise in the context of numerical analysis.

§0.1. **Examples.** Our first example is rather trivial: suppose we want to compute the integral

$$I = \int_a^b f(t)dt \tag{0.1}$$

[1]In fact, it was by means of computational methods that Le Verrier in 1846 predicted the existence of the eighth planet Neptune, based on observed (and unaccounted for) irregularities in the orbit of the next inner planet. Soon thereafter, the planet was indeed discovered at precisely the predicted location. (Some calculations were done previously by Adams, then an undergraduate at Cambridge, but were not published in time.)

for some given function f on $[a, b]$. Letting $y(x) = \int_a^x f(t)dt$, we get immediately

$$\frac{dy}{dx} = f(x), \quad y(a) = 0. \tag{0.2}$$

This is an initial value problem for a first-order differential equation, but a very special one in which y does not appear on the right-hand side. By solving (0.2) on $[a, b]$, one obtains $I = y(b)$. This example, as elementary as it seems to be, is not entirely without interest, because when we integrate (0.2) by modern techniques of solving ODEs, we can take advantage of step control mechanisms, taking smaller steps where f changes more rapidly, and larger ones elsewhere. This gives rise to what may be called *adaptive integration.*

A more interesting extension of this idea is exemplified by the following integral,

$$I = \int_0^\infty J_0(t^2)e^{-t}dt, \tag{0.3}$$

containing the Bessel function $J_0(x^2)$, one of the special functions of mathematical physics. It satisfies the linear second-order differential equation

$$\frac{d^2y}{dx^2} + \left(2 - \frac{1}{x}\right)\frac{dy}{dx} + 4x^2y = 0, \quad x > 0, \tag{0.4}$$

with initial conditions

$$y(0) = 1, \quad y'(0) = 0. \tag{0.5}$$

As is often the case with special functions arising in physics, the associated differential equation has a singularity at the origin $x = 0$, even though the solution $y(x) = J_0(x^2)$ is perfectly regular at $x = 0$. (The singular term in (0.4) has limit 0 at $x = 0$.) Here we let

$$y_1(x) = \int_0^x J_0(t^2)e^{-t}dt, \quad y_2(x) = J_0(x^2), \quad y_3(x) = y_2'(x) \tag{0.6}$$

and obtain

$$\begin{aligned} \frac{dy_1}{dx} &= e^{-x}y_2, & y_1(0) &= 0, \\ \frac{dy_2}{dx} &= y_3, & y_2(0) &= 1, \\ \frac{dy_3}{dx} &= -\left(2 - \frac{1}{x}\right)y_3 - 4x^2y_2, & y_3(0) &= 0, \end{aligned} \tag{0.7}$$

an initial value problem for a system of three (linear) first-order differential equations. We need to integrate this on $(0,\infty)$ to find $I = y_1(\infty)$. An advantage of this approach is that the Bessel function need not be calculated explicitly; it is calculated implicitly (as component y_2) through the integration of the differential equation that it satisfies. Another advantage over straightforward numerical integration is again the possibility of automatic error control through appropriate step change techniques. We could equally well get rid of the exponential e^{-x} in (0.7) by calling it y_4 and adding the differential equation $dy_4/dx = -y_4$ and initial condition $y_4(0) = 1$. (The difficulty with the singularity at $x = 0$ can be handled, e.g., by introducing initially, say, for $0 \le x \le \frac{1}{3}$, a new dependent variable $\tilde{y}_3$, setting $\tilde{y}_3 = (2 - \frac{1}{x})y_2' = (2 - \frac{1}{x})y_3$. Then $\tilde{y}_3(0) = 0$, and the second and third equations in (0.7) can be replaced by

$$\frac{dy_2}{dx} = -\frac{x}{1-2x}\tilde{y}_3,$$
$$\frac{d\tilde{y}_3}{dx} = -\frac{4(1-x)}{1-2x}\tilde{y}_3 + 4x(1-2x)y_2.$$

Here the coefficients are now well-behaved functions near $x = 0$. Once $x = \frac{1}{3}$ has been reached, one can switch back to (0.7), using for y_3 the initial condition $y_3(\frac{1}{3}) = -\tilde{y}_3(\frac{1}{3})$.)

Another example is the *method of lines* in partial differential equations, where one discretizes partial derivatives with respect to all variables but one, thereby obtaining a system of ordinary differential equations. This may be illustrated by the heat equation on a rectangular domain,

$$\frac{\partial u}{\partial t} = \frac{\partial^2 u}{\partial x^2}, \qquad 0 \le x \le 1, \qquad 0 \le t \le T, \tag{0.8}$$

where the temperature $u = u(x,t)$ is to satisfy the initial condition

$$u(x,0) = \varphi(x), \qquad 0 \le x \le 1, \tag{0.8a}$$

and the boundary conditions

$$u(0,t) = \lambda(t), \qquad u(1,t) = \rho(t), \qquad 0 \le t \le T. \tag{0.8b}$$

Here φ is a given function of x, and λ, ρ given functions of t. We discretize in the x-variable by placing a grid $\{x_n\}_{n=0}^{N+1}$ with $x_n = nh$, $h = \frac{1}{N+1}$, on the interval $0 \le x \le 1$ and approximating the second derivative by the second divided difference (cf. Ch. 2, Eq. (2.68)),

$$\left.\frac{\partial^2 u}{\partial x^2}\right|_{x=x_n} \approx \frac{u_{n+1} - 2u_n + u_{n-1}}{h^2}, \qquad n = 1, 2, \ldots, N, \tag{0.9}$$

where

$$u_n = u_n(t) := u(x_n, t), \qquad n = 0, 1, \ldots, N+1. \tag{0.10}$$

Writing down (0.8) for $x = x_n$, $n = 1, 2, \ldots, N$, and using (0.9), we get approximately

$$\begin{aligned} \frac{du_n}{dt} &= \frac{1}{h^2}(u_{n+1} - 2u_n + u_{n-1}), \\ u_n(0) &= \varphi(x_n), \end{aligned} \qquad n = 1, 2, \ldots, N, \tag{0.11}$$

an initial value problem for a system of N differential equations in the N unknown functions $u_1, u_2, \ldots, u_N$. The boundary functions $\lambda(t)$, $\rho(t)$ enter into (0.11) when reference is made to u_0 or u_{N+1} on the right-hand side of the differential equation. By making the grid finer and finer, hence h smaller, one expects to obtain better and better approximations for $u(x_n, t)$. Unfortunately, this comes at a price: the system (0.11) becomes more and more "stiff" as h decreases, calling for special methods designed especially for stiff equations (cf. §5; Ch. 6, §5).

§0.2. **Types of differential equations**. The standard initial value problem involves a system of first-order differential equations

$$\frac{dy^i}{dx} = f^i(x, y^1, y^2, \ldots, y^d), \qquad i = 1, 2, \ldots, d, \tag{0.12}$$

which is to be solved on an interval $[a, b]$, given the initial values

$$y^i(a) = y_0^i, \qquad i = 1, 2, \ldots, d. \tag{0.13}$$

Here the component functions are indexed by superscripts, and subscripts are reserved to indicate step numbers, the initial step having index 0. We use vector notation throughout by letting

$$y^T = [y^1, y^2, \ldots, y^d], \qquad f^T = [f^1, f^2, \ldots, f^d], \qquad y_0^T = [y_0^1, y_0^2, \ldots, y_0^d]$$

and writing (0.12), (0.13) in the form

$$\frac{dy}{dx} = f(x, y), \qquad a \le x \le b; \qquad y(a) = y_0. \tag{0.14}$$

We are thus seeking a vector-valued function $y(x) \in C^1[a, b]$ that satisfies (0.14) identically on $[a, b]$ and has the starting value y_0 at $x = a$. About $f(x, y)$ we assume that it is defined for $x \in [a, b]$ and all $y \in \mathbb{R}^d$. [2]

We note some important special cases of (0.14).

(i) $d = 1$, $y' = f(x, y)$: a single first-order differential equation.

(ii) $d > 1$, $u^{(d)} = g(x, u, u', \ldots, u^{(d-1)})$: a single dth-order differential equation. The initial conditions here take the form $u^{(i)}(a) = u_0^i$, $i = 0, 1, \ldots, d-1$. This problem is easily brought into the form (0.14) by defining

$$y^i = u^{(i-1)}, \qquad i = 1, 2, \ldots, d.$$

Then

$$\begin{aligned}
\frac{dy^1}{dx} &= y^2, & y^1(a) &= u_0^0, \\
\frac{dy^2}{dx} &= y^3, & y^2(a) &= u_0^1, \\
&\cdots\cdots\cdots & &\cdots\cdots\cdots \\
\frac{dy^{d-1}}{dx} &= y^d, & y^{d-1}(a) &= u_0^{d-2}, \\
\frac{dy^d}{dx} &= g(x, y^1, y^2, \ldots, y^d), & y^d(a) &= u_0^{d-1},
\end{aligned} \tag{0.15}$$

which has the form (0.14) with very special (linear) functions $f^1, f^2, \ldots, f^{d-1}$, and $f^d(x, y) = g(x, y)$.

Although this is the canonical way of transforming a single equation of order d into a system of first-order equations, there are other ways of doing this, which are sometimes more advantageous. Consider, for example, the *Sturm-Liouville equation*

$$\frac{d}{dx}\left(p(x)\,\frac{du}{dx}\right) + q(x)u = 0, \qquad a \le x \le b, \tag{0.16}$$

[2]That is, each component function $f^i(x, y)$ is defined on $[a, b] \times \mathbb{R}^d$. In some problems, $f(x, y)$ is defined only on $[a, b] \times \mathcal{D}$, where $\mathcal{D} \subset \mathbb{R}^d$ is a compact domain. In such cases, the solution $y(x)$ must be required to remain in $\mathcal{D}$ as x varies in $[a, b]$. This causes complications, which we avoid by assuming $\mathcal{D} = \mathbb{R}^d$.

where $p(x) \neq 0$ on $[a, b]$. Here the substitution

$$y^1 = u, \quad y^2 = p(x)\,\frac{du}{dx} \tag{0.17}$$

is more appropriate (and also physically more meaningful), leading to the system

$$\begin{aligned} \frac{dy^1}{dx} &= \frac{1}{p(x)}\,y^2, \\ \frac{dy^2}{dx} &= -q(x)y^1. \end{aligned} \tag{0.18}$$

Mathematically, (0.18) has the advantage over (0.16) of not requiring that p be differentiable on $[a, b]$.

(iii) $dy/dx = f(y)$, $y \in \mathbb{R}^d$: an *autonomous* system of differential equations. Here f does not depend explicitly on x. This can always be trivially achieved by introducing, if necessary, $y^0(x) = x$ and writing (0.14) as

$$\frac{d}{dx}\begin{bmatrix} y \\ y^0 \end{bmatrix} = \begin{bmatrix} f(y^0, y) \\ 1 \end{bmatrix}, \quad a \leq x \leq b; \quad \begin{bmatrix} y \\ y^0 \end{bmatrix}(a) = \begin{bmatrix} y_0 \\ a \end{bmatrix}. \tag{0.14'}$$

Many ODE software packages indeed assume that the system is autonomous.

(iv) Second-order system

$$\frac{d^2u^i}{dx^2} = g^i\left(x, u^1, \ldots, u^d, \frac{du^1}{dx}, \ldots, \frac{du^d}{dx}\right), \quad i = 1, 2, \ldots, d. \tag{0.19}$$

Newton's law in mechanics is of this form, with $d = 3$. The canonical transformation here introduces

$$y^1 = u^1, \ldots, y^d = u^d; \quad y^{d+1} = \frac{du^1}{dx}, \ldots, y^{2d} = \frac{du^d}{dx}$$

and yields a system of $2d$ first-order equations,

$$\begin{aligned}
\frac{dy^1}{dx} &= y^{d+1}, \\
&\cdots\cdots\cdots \\
\frac{dy^d}{dx} &= y^{2d}, \\
\frac{dy^{d+1}}{dx} &= g^1(x, y^1, \dots, y^d, y^{d+1}, \dots, y^{2d}), \\
&\cdots\cdots\cdots\cdots\cdots\cdots \\
\frac{dy^{2d}}{dx} &= g^d(x, y^1, \dots, y^d, y^{d+1}, \dots, y^{2d}).
\end{aligned} \tag{0.20}$$

(v) *Implicit system* of first-order differential equations:

$$F^i\left(x, u, \frac{du}{dx}\right) = 0, \qquad i = 1, 2, \dots, d; \quad u \in \mathbb{R}^d. \tag{0.21}$$

Here $F^i = F^i(x, u, v)$ are given functions of $2d + 1$ variables, which again we combine into a vector $F = [F^i]$. We denote by F_x the partial derivative of F with respect to x, and by F_u, F_v the Jacobian matrices

$$F_u(x, u, v) = \left[\frac{\partial F^i}{\partial u^j}\right], \quad F_v(x, u, v) = \left[\frac{\partial F^i}{\partial v^j}\right].$$

Assuming that these Jacobians exist and that F_v is nonsingular on $[a, b] \times \mathbb{R}^d \times \mathbb{R}^d$, the implicit system (0.21) may be dealt with by differentiating it totally with respect to x. This yields (in vector notation)

$$F_x + F_u \frac{du}{dx} + F_v \frac{du^2}{dx^2} = 0, \tag{0.22}$$

where the arguments in F_x, F_u, F_v are x, u, $u' = du/dx$. By assumption, this can be solved for the second derivative,

$$\frac{du^2}{dx^2} = F_v^{-1}(x, u, u')\left\{-F_x(x, u, u') - F_u(x, u, u')\frac{du}{dx}\right\}, \tag{0.23}$$

yielding an (explicit) system of second-order differential equations (cf. (iv)). If we are to solve the initial value problem for (0.21) on $[a, b]$, with $u(a) = u_0$ prescribed, we need for (0.23) the additional initial data $(du/dx)(a) = u_0'$.

This must be obtained by solving $F(a, u_0, u_0') = 0$ for u_0', which in general is a nonlinear system of equations (unless $F(x, u, v)$ depends linearly on v). But from then on, when integrating (0.23) (or the equivalent first-order system), only linear systems (0.22) need to be solved at each step to compute d^2u/dx^2 for given x, u, u', since the numerical method automatically updates x, u, u' from step to step.

§0.3. **Existence and uniqueness.** We recall from the theory of differential equations the following basic existence and uniqueness theorem.

Theorem 5.0.1. *Assume that $f(x, y)$ is continuous in the first variable for $x \in [a, b]$ and with respect to the second satisfies a uniform Lipschitz condition*

$$\|f(x, y) - f(x, y^*)\| \le L\|y - y^*\|, \quad x \in [a, b], \quad y, y^* \in \mathbb{R}^d, \tag{0.24}$$

where $\|\cdot\|$ is some vector norm. Then the initial value problem (0.14) *has a unique solution $y(x)$, $a \le x \le b$, for arbitrary $y_0 \in \mathbb{R}^d$. Moreover, $y(x)$ depends continuously on x_0 and y_0.*

The Lipschitz condition (0.24) certainly holds if all functions $\frac{\partial f^i}{\partial y^j}(x, y)$, $i, j = 1, 2, \ldots, d$, are continuous in the y-variables and bounded on $[a, b] \times \mathbb{R}^d$. This is the case for *linear systems of differential equations*, where

$$f^i(x, y) = \sum_{j=1}^{d} a_{ij}(x) y^j + b_i(x), \qquad i = 1, 2, \ldots, d, \tag{0.25}$$

and $a_{ij}(x)$, $b_i(x)$ are continuous functions on $[a, b]$. In nonlinear problems it is rarely the case, however, that a Lipschitz condition is valid uniformly in all of $\mathbb{R}^d$; it more often holds in some compact and convex domain $\mathcal{D} \subset \mathbb{R}^d$. In this case one can assert the existence of a unique solution only in some neighborhood of x_0 in which $y(x)$ remains in $\mathcal{D}$. To avoid this complication, we assume in this chapter that $\mathcal{D}$ is so large that $y(x)$ exists on the whole interval $[a, b]$ and that all numerical approximations are also contained in $\mathcal{D}$. Bounds on partial derivatives of $f(x, y)$ are assumed to hold uniformly in $[a, b] \times \mathcal{D}$, if not in $[a, b] \times \mathbb{R}^d$, and it is tacitly understood that the bounds may depend on $\mathcal{D}$ but not on x and y.

§0.4. **Numerical methods.** One can distinguish between *analytic approximation methods* and *discrete-variable methods.* In the former one tries to find approximations $y_a(x) \approx y(x)$ to the exact solution, valid for all $x \in [a, b]$. These usually take the form of a truncated series expansion, either in powers of x, in Chebyshev polynomials, or in some other system of basis functions. In discrete-variable methods, on the other hand, one attempts to find approximations $u_n \in \mathbb{R}^d$ of $y(x_n)$ only at discrete points $x_n \in [a, b]$. The abscissas x_n may be predetermined (e.g., equally spaced on $[a, b]$) or, more likely, are generated dynamically as part of the integration process. If desired, one can then from these discrete approximations $\{u_n\}$ again obtain an approximation $y_a(x)$ defined for all $x \in [a, b]$ either by interpolation or, more naturally, by a continuation mechanism built into the approximation method itself. We are concerned here only with discrete-variable methods.

Depending on how the discrete approximations are generated, one distinguishes between *one-step methods* and *multistep methods.* In the former, u_{n+1} is determined solely from a knowledge of x_n, u_n and the step h to proceed from x_n to $x_{n+1} = x_n + h$. In a k-step method ($k > 1$), knowledge of $k - 1$ additional points $(x_{n-\kappa}, u_{n-\kappa})$, $\kappa = 1, 2, \ldots, k - 1$, is required to advance the solution. The present chapter is devoted to one-step methods; multistep methods are discussed in Chapter 6.

When describing a single step of a one-step method, it suffices to show how one proceeds from a generic point (x, y), $x \in [a, b]$, $y \in \mathbb{R}^d$, to the "next" point $(x + h, y_{\text{next}})$. We refer to this as the *local description* of the one-step method. This also includes a discussion of the local accuracy, that is, how closely y_{next} agrees at $x + h$ with the solution (passing through the point (x, y)) of the differential equation. A one-step method solving the initial value problem (0.14) effectively generates a grid function $\{u_n\}$, $u_n \in \mathbb{R}^d$, on a grid $a = x_0 < x_1 < x_2 < \cdots < x_{N-1} < x_N = b$ covering the interval $[a, b]$, whereby u_n is intended to approximate the exact solution $y(x)$ at $x = x_n$. The point (x_{n+1}, u_{n+1}) is obtained from the point (x_n, u_n) by applying a one-step method with an appropriate step $h_n = x_{n+1} - x_n$. This is referred to as the *global* description of a one-step method. Questions of interest here are the behavior of the global error $u_n - y(x_n)$, in particular stability and convergence, and the choice of h_n to proceed from one grid point x_n to the next, $x_{n+1} = x_n + h_n$. Finally, we address special difficulties arising from the stiffness of the given differential equation problem.

§1. Local Description of One-Step Methods

Given a generic point $x \in [a,b]$, $y \in \mathbb{R}^d$, we define a single step of the one-step method by

$$y_{\text{next}} = y + h\Phi(x, y; h), \qquad h > 0. \tag{1.1}$$

The function Φ: $[a,b] \times \mathbb{R}^d \times \mathbb{R}_+ \to \mathbb{R}^d$ may be thought of as the approximate increment per unit step, or the approximate difference quotient, and it defines the method. Along with (1.1), we consider the solution $u(t)$ of the differential equation (0.14) passing through the point (x, y), that is, the local initial value problem

$$\frac{du}{dt} = f(t, u), \qquad x \le t \le x + h; \qquad u(x) = y. \tag{1.2}$$

We call $u(t)$ the *reference solution.* The vector y_{next} in (1.1) is intended to approximate $u(x+h)$. How successfully this is done is measured by the truncation error defined as follows.

Definition 1.1. The *truncation error* of the method Φ at the point (x, y) is defined by

$$T(x, y; h) = \frac{1}{h}\,[y_{\text{next}} - u(x+h)]. \tag{1.3}$$

The truncation error thus is a vector-valued function of $d+2$ variables. Using (1.1) and (1.2), we can write for it, alternatively,

$$T(x, y; h) = \Phi(x, y; h) - \frac{1}{h}\,[u(x+h) - u(x)], \tag{1.3'}$$

showing that T is the difference between the approximate and exact increment per unit step.

An increasingly finer description of local accuracy is provided by the following definitions, all based on the concept of truncation error.

Definition 1.2. The method Φ is called *consistent* if

$$T(x, y; h) \to 0 \text{ as } h \to 0, \tag{1.4}$$

uniformly for $(x, y) \in [a, b] \times \mathbb{R}^d$. [3]

By (1.3′) and (1.2) we have consistency if and only if

$$\Phi(x, y; 0) = f(x, y), \qquad x \in [a, b], \quad y \in \mathbb{R}^d. \tag{1.5}$$

Definition 1.3. The method Φ is said to have *order p* if, for some vector norm $\| \cdot \|$,

$$\|T(x, y; h)\| \le Ch^p, \tag{1.6}$$

uniformly on $[a, b] \times \mathbb{R}^d$, with a constant C not depending on x, y and h.[4]

We express this property briefly as

$$T(x, y; h) = O(h^p), \qquad h \to 0. \tag{1.6′}$$

Note that $p > 0$ implies consistency. Usually, p is an integer ≥ 1. It is called the *exact order*, if (1.6) does not hold for any larger p.

Definition 1.4. A function τ: $[a, b] \times \mathbb{R}^d \to \mathbb{R}^d$ that satisfies $\tau(x, y) \not\equiv 0$ and

$$T(x, y; h) = \tau(x, y)h^p + O(h^{p+1}), \qquad h \to 0, \tag{1.7}$$

is called the *principal error function.*

The principal error function determines the leading term in the truncation error. The number p in (1.7) is the exact order of the method since $\tau \not\equiv 0$.

All the preceding definitions are made with the idea in mind that $h > 0$ is a small number. Then the larger p, the more accurate the method. This can (and should) always be arranged by a proper scaling of the independent variable x; we tacitly assume that such a scaling has already been made.

---

[3] More realistically, one should require uniformity on $[a, b] \times \mathcal{D}$, where $\mathcal{D} \subset \mathbb{R}^d$ is a sufficiently large compact domain; cf. §0.3.

[4] If uniformity of (1.6) is required only on $[a, b] \times \mathcal{D}$, $\mathcal{D} \subset \mathbb{R}^d$ compact, then C may depend on $\mathcal{D}$; cf. §0.3.

§2. Examples of One-Step Methods

Some of the oldest methods are motivated by simple geometric considerations based on the slope field defined by the right-hand side of the differential equation. These include the Euler and modified Euler methods. More accurate and sophisticated methods are based on Taylor expansion.

§2.1. **Euler's method**. Euler proposed his method in 1768, in the early days of calculus. It consists of simply following the slope at the generic point (x, y) over an interval of length h:

$$y_{\text{next}} = y + hf(x, y). \tag{2.1}$$

Thus, $\Phi(x, y; h) = f(x, y)$ does not depend on h, and by (1.5) the method is evidently consistent. For the truncation error we have by (1.3′)

$$T(x, y; h) = f(x, y) - \frac{1}{h}\,[u(x+h) - u(x)], \tag{2.2}$$

where $u(t)$ is the reference solution defined in (1.2). Since $u'(x) = f(x, u(x)) = f(x, y)$, we can write, using Taylor's theorem,

$$\begin{aligned} T(x, y; h) &= u'(x) - \frac{1}{h}\,[u(x+h) - u(x)] \\ &= u'(x) - \frac{1}{h}\,[u(x) + hu'(x) + \tfrac{1}{2}\,h^2 u''(\xi) - u(x)] \\ &= -\tfrac{1}{2}\,hu''(\xi), \qquad x < \xi < x + h, \end{aligned} \tag{2.3}$$

assuming $u \in C^2[x, x+h]$. This is certainly true if $f \in C^1$ on $[a, b] \times \mathbb{R}^d$, as we assume. Note the slight abuse of notation in the last two equations, where ξ is to be understood to differ from component to component but to be always in the interval shown. We freely use this notation later on without further comment.

Now differentiating (1.2) totally with respect to t and then setting $t = \xi$ yields

$$T(x, y; h) = -\tfrac{1}{2}\,h[f_x + f_y f](\xi, u(\xi)), \tag{2.4}$$

where f_x is the partial derivative of f with respect to x and f_y the Jacobian of f with respect to the y-variables. If, in the spirit of Theorem 5.0.1, we assume that f and all its first partial derivatives are uniformly bounded in $[a, b] \times \mathbb{R}^d$, there exists a constant C independent of x, y, and h such that

$$\|T(x, y; h)\| \leq C \cdot h. \tag{2.5}$$

Thus, Euler's method has order $p = 1$. If we make the same assumption about all second-order partial derivatives of f, we have $u''(\xi) = u''(x) + O(h)$ and, therefore, from (2.3),

$$T(x, y; h) = -\tfrac{1}{2}h\ [f_x + f_y f](x, y) + O(h^2), \qquad h \to 0, \tag{2.6}$$

showing that the principal error function is given by

$$\tau(x, y) = -\tfrac{1}{2}\ [f_x + f_y f](x, y). \tag{2.7}$$

Unless $f_x + f_y f \equiv 0$, the order of Euler's method is exactly $p = 1$.

§2.2. **Method of Taylor expansion.** We have seen that Euler's method basically amounts to truncating the Taylor expansion of the reference solution after its second term. It is a natural idea, already proposed by Euler, to use more terms of the Taylor expansion. This requires the computation of successive "total derivatives" of f,

$$\begin{aligned} &f^{[0]}(x, y) = f(x, y), \\ &f^{[k+1]}(x, y) = f_x^{[k]}(x, y) + f_y^{[k]}(x, y) f(x, y), \qquad k = 0, 1, 2, \ldots , \end{aligned} \tag{2.8}$$

which determine (see Ex. 2) the successive derivatives of the reference solution $u(t)$ of (1.2) by virtue of

$$u^{(k+1)}(t) = f^{[k]}(t, u(t)), \qquad k = 0, 1, 2, \ldots . \tag{2.9}$$

These, for $t = x$, become

$$u^{(k+1)}(x) = f^{[k]}(x, y), \qquad k = 0, 1, 2, \ldots , \tag{2.10}$$

and are used to form the Taylor series approximation according to

$$y_{\text{next}} = y + h\left[f^{[0]}(x, y) + \frac{1}{2}\ h f^{[1]}(x, y) + \cdots + \frac{1}{p!}\ h^{p-1} f^{[p-1]}(x, y)\right]; \tag{2.11}$$

that is,

$$\Phi(x, y; h) = f^{[0]}(x, y) + \frac{1}{2}\, h f^{[1]}(x, y) + \cdots + \frac{1}{p!}\, h^{p-1} f^{[p-1]}(x, y). \tag{2.12}$$

For the truncation error, assuming $f \in C^p$ on $[a, b] \times \mathbb{R}^d$ and using (2.10) and (2.12), we obtain from Taylor's theorem

$$\begin{aligned} T(x, y; h) &= \Phi(x, y; h) - \frac{1}{h}\,[u(x+h) - u(x)] \\ &= \Phi(x, y; h) - \sum_{k=0}^{p-1} u^{(k+1)}(x)\, \frac{h^k}{(k+1)!} - u^{(p+1)}(\xi)\, \frac{h^p}{(p+1)!} \\ &= -u^{(p+1)}(\xi)\, \frac{h^p}{(p+1)!}\,, \qquad x < \xi < x+h, \end{aligned}$$

so that

$$\|T(x, y; h)\| \le \frac{C_p}{(p+1)!}\, h^p, \tag{2.13}$$

where C_p is a bound on the pth total derivative of f. Thus, the method has exact order p (unless $f^{[p]}(x, y) \equiv 0$), and the principal error function is

$$\tau(x, y) = -\frac{1}{(p+1)!}\, f^{[p]}(x, y). \tag{2.14}$$

The necessity of computing many partial derivatives in (2.8) was a discouraging factor in the past, when this had to be done by hand. But nowadays, this labor can be delegated to computers, so that the method has become again a viable option.

§2.3. **Improved Euler methods**. There is too much inertia in Euler's method: one should not follow the same initial slope over the whole interval of length h, since along this line segment the slope defined by the slope field of the differential equation changes. This suggests several alternatives. For example, we may wish to reevaluate the slope halfway through the line segment — retake the pulse of the differential equation, as it were — and then follow this revised slope over the whole interval (cf. Figure 5.2.1). In formula,

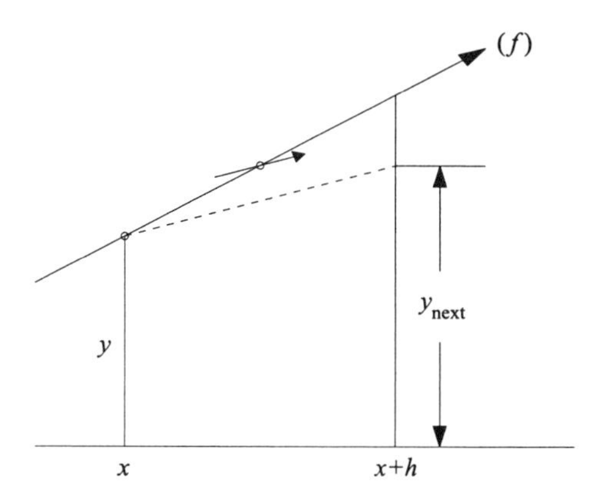

FIGURE 5.2.1. Modified Euler method

$$y_{\text{next}} = y + hf(x + \tfrac{1}{2}h, y + \tfrac{1}{2}hf(x, y)), \tag{2.15}$$

or

$$\Phi(x, y; h) = f(x + \tfrac{1}{2}h, y + \tfrac{1}{2}hf(x, y)). \tag{2.16}$$

Note the characteristic "nesting" of f that is required here. For programming purposes it may be desirable to undo the nesting and write

$$\begin{aligned} k_1(x, y) &= f(x, y), \\ k_2(x, y; h) &= f(x + \tfrac{1}{2}h, y + \tfrac{1}{2}hk_1), \\ y_{\text{next}} &= y + hk_2. \end{aligned} \tag{2.15'}$$

In other words, we are taking two trial slopes, k_1 and k_2, one at the initial point and the other nearby, and then taking the latter as the final slope.

We could equally well take the second trial slope at $(x+h,\ y+hf(x, y))$, but then, having waited too long before reevaluating the slope, take now as the final slope the average of the two slopes:

$$\begin{aligned} k_1(x, y) &= f(x, y), \\ k_2(x, y; h) &= f(x + h, y + hk_1), \\ y_{\text{next}} &= y + \tfrac{1}{2}h(k_1 + k_2). \end{aligned} \tag{2.17}$$

This is sometimes referred to as *Heun's method* or the *trapezoidal rule.*

The effect of both modifications is to raise the order by 1, as is shown in the next section.

§2.4. **Second-order two-stage methods.** We may take a more systematic approach toward modifying Euler's method, by letting

$$\Phi(x, y; h) = \alpha_1 k_1 + \alpha_2 k_2, \tag{2.18}$$

where

$$\begin{aligned} k_1(x, y) &= f(x, y), \\ k_2(x, y; h) &= f(x + \mu h, y + \mu h k_1). \end{aligned} \tag{2.19}$$

We have now three parameters, α_1, α_2, and μ, at our disposal, and we can try to choose them so as to maximize the order p. A systematic way of determining the maximum order p is to expand both $\Phi(x, y; h)$ and $h^{-1}[u(x+h) - u(x)]$ in powers of h and to match as many terms as we can.

To expand Φ, we need Taylor's expansion for (vector-valued) functions of several variables,

$$\begin{aligned} f(x + \Delta x, y + \Delta y) = f + f_x \Delta x + f_y \Delta y + \tfrac{1}{2}\, [f_{xx}(\Delta x)^2 + 2 f_{xy} \Delta x \Delta y \\ + (\Delta y)^T f_{yy} (\Delta y)] + \cdots, \end{aligned} \tag{2.20}$$

where f_y denotes the Jacobian of f and $f_{yy} = [f^i_{yy}]$ the vector of Hessian matrices of f. In (2.20), all functions and partial derivatives are understood to be evaluated at (x, y). Letting $\Delta x = \mu h$, $\Delta y = \mu h f$ then gives

$$k_2(x, y; h) = f + \mu h (f_x + f_y f) + \tfrac{1}{2}\, \mu^2 h^2 (f_{xx} + 2 f_{xy} f + f^T f_{yy} f) + O(h^3). \tag{2.21}$$

Similarly (cf. (2.10)),

$$\frac{1}{h}\, [u(x + h) - u(x)] = u'(x) + \tfrac{1}{2}\, h u''(x) + \tfrac{1}{6}\, h^2 u'''(x) + O(h^3), \tag{2.22}$$

where

$$
\begin{aligned}
u'(x) &= f, \\
u''(x) &= f^{[1]} = f_x + f_y f, \\
u'''(x) &= f^{[2]} = f_x^{[1]} + f_y^{[1]} f \\
&= f_{xx} + f_{xy} f + f_y f_x + (f_{xy} + (f_y f)_y) f \\
&= f_{xx} + 2 f_{xy} f + f^T f_{yy} f + f_y (f_x + f_y f),
\end{aligned}
$$

and where in the last equation we have used (see Ex. 3)

$$(f_y f)_y f = f^T f_{yy} f + f_y^2 f.$$

Now,

$$T(x, y; h) = \alpha_1 k_1 + \alpha_2 k_2 - \frac{1}{h}\,[u(x+h) - u(x)],$$

wherein we substitute the expansions (2.21) and (2.22). We find

$$
\begin{aligned}
T(x, y; h) &= (\alpha_1 + \alpha_2 - 1) f + \left(\alpha_2 \mu - \tfrac{1}{2}\right) h (f_x + f_y f) \\
&+ \tfrac{1}{2} h^2 \left[\left(\alpha_2 \mu^2 - \tfrac{1}{3}\right)(f_{xx} + 2 f_{xy} f + f^T f_{yy} f) - \tfrac{1}{3} f_y (f_x + f_y f)\right] + O(h^3).
\end{aligned}
\tag{2.23}
$$

We can see now that, however we choose the parameters α_1, α_2, μ, we cannot make the coefficient of h^2 equal to zero unless severe restrictions are placed on f (cf. Ex. 7(c)). Thus, the maximum possible order is $p = 2$, and we achieve it by satisfying

$$
\begin{aligned}
\alpha_1 + \alpha_2 &= 1, \\
\alpha_2 \mu &= \tfrac{1}{2}.
\end{aligned}
\tag{2.24}
$$

This has a one-parameter family of solutions,

$$
\begin{aligned}
\alpha_1 &= 1 - \alpha_2, \\
\mu &= \frac{1}{2\alpha_2},
\end{aligned}
\qquad (\alpha_2 \neq 0 \ \text{ arbitrary}).
\tag{2.25}
$$

We recognize the improved Euler method contained therein with $\alpha_2 = 1$, and Heun's method with $\alpha_2 = \frac{1}{2}$. There are other natural choices; one such would be to look at the principal error function

$$\tau(x,y) = \frac{1}{2}\left[\left(\frac{1}{4\alpha_2} - \frac{1}{3}\right)(f_{xx} + 2f_{xy}f + f^T f_{yy} f) - \frac{1}{3}\, f_y(f_x + f_y f)\right] \quad (2.26)$$

and to note that it consists of a linear combination of two aggregates of partial derivatives. We may wish to minimize some norm of the coefficients, say, the sum of their absolute values. In (2.26) this gives trivially $(4\alpha_2)^{-1} - \frac{1}{3} = 0$, that is, $\alpha_2 = \frac{3}{4}$, and hence suggests a method with

$$\alpha_1 = \tfrac{1}{4}, \quad \alpha_2 = \tfrac{3}{4}, \quad \mu = \tfrac{2}{3}. \quad (2.27)$$

§2.5. **Runge-Kutta methods.** Runge-Kutta methods are a straightforward extension of two-stage methods to r-stage methods:

$$\begin{aligned} \Phi(x,y;h) &= \sum_{s=1}^{r} \alpha_s k_s, \\ k_1(x,y) &= f(x,y), \\ k_s(x,y;h) &= f(x + \mu_s h, y + h\sum_{j=1}^{s-1} \lambda_{sj} k_j), \qquad s = 2,3,\ldots,r. \end{aligned} \quad (2.28)$$

It is natural in (2.28) to impose the conditions (cf. Ex. 12 for the first)

$$\mu_s = \sum_{j=1}^{s-1} \lambda_{sj}, \quad s = 2,3,\ldots,r; \qquad \sum_{s=1}^{r} \alpha_s = 1, \quad (2.29)$$

where the last one is nothing but the consistency condition (cf. (1.5)). We call (2.28) an *explicit r-stage Runge-Kutta method*; it requires r evaluations of the right-hand side f of the differential equation. More generally, we can consider *implicit r-stage Runge-Kutta methods*

$$\begin{aligned} \Phi(x,y;h) &= \sum_{s=1}^{r} \alpha_s k_s(x,y;h), \\ k_s &= f(x + \mu_s h, y + h\sum_{j=1}^{r} \lambda_{sj} k_j), \qquad s = 1,2,\ldots,r, \end{aligned} \quad (2.30)$$

in which the last r equations form a system of (in general nonlinear) equations in the unknowns $k_1, k_2, \ldots, k_r$. Since each of these is a vector in $\mathbb{R}^d$, we have a system of rd equations in rd unknowns that must be solved before we can form the approximate increment Φ. Less work is required in *semi-implicit* Runge-Kutta methods where the summation in the formula for k_s extends from $j = 1$ to $j = s$ only. This yields r systems of equations, each having only d unknowns, the components of k_s. The considerable computational expense involved in implicit and semi-implicit methods can only be justified in special circumstances, for example, stiff problems. The reason is that implicit methods not only can be made to have higher order than explicit methods, but also have better stability properties (cf. §5).

Already in the case of *explicit* Runge-Kutta methods, and even more so in implicit methods, we have at our disposal a large number of parameters which we can choose to achieve the maximum possible order for all sufficiently smooth f. The approach is analogous to the one taken in §2.4, only technically much more involved, as the partial-derivative aggregates that are going to appear in the principal error function are becoming much more complicated and numerous as r increases. In fact, a satisfactory solution of this problem has become possible only through the employment of graph-theoretical tools, specifically, the theory of rooted trees. This was systematically developed by J. C. Butcher, the principal researcher in this area. A description of these techniques is beyond the scope of this book. We briefly summarize, however, some of the results that can be obtained.

Denote by $p^*(r)$ the maximum attainable order (for arbitrary sufficiently smooth fs) of an explicit r-stage Runge-Kutta method. Then Kutta[5] has already shown in 1901 that

$$p^*(r) = r \quad \text{for} \quad 1 \le r \le 4, \tag{2.31}$$

and has derived many concrete examples of methods having these orders.

[5]Wilhelm Martin Kutta (1867–1944) was a German applied mathematician, active at the Technical University of Stuttgart from 1911 until his retirement. In addition to his work on the numerical solution of ODEs, he did important work on the application of conformal mapping to hydro- and aerodynamical problems. Best known is his formula for the lift exerted on an airfoil, now known as the Kutta-Joukowski formula. For Runge, see Footnote 4 in Ch. 2, §2.3.

Butcher, in the 1960s, established that

$$\begin{aligned} p^*(r) &= r-1 \quad \text{for} \quad 5 \le r \le 7, \\ p^*(r) &= r-2 \quad \text{for} \quad 8 \le r \le 9, \\ p^*(r) &\le r-2 \quad \text{for} \quad r \ge 10. \end{aligned} \tag{2.32}$$

Specific examples of higher-order Runge-Kutta formulae are used later in connection with error monitoring procedures. Here we mention only the *classical Runge-Kutta formula*[6] of order $p = 4$:

$$\begin{aligned} \Phi(x, y; h) &= \tfrac{1}{6}\,(k_1 + 2k_2 + 2k_3 + k_4), \\ k_1(x, y) &= f(x, y), \\ k_2(x, y; h) &= f\left(x + \tfrac{1}{2}\,h, y + \tfrac{1}{2}\,hk_1\right), \\ k_3(x, y; h) &= f\left(x + \tfrac{1}{2}\,h, y + \tfrac{1}{2}\,hk_2\right), \\ k_4(x, h; h) &= f(x + h, y + hk_3). \end{aligned} \tag{2.33}$$

When f does not depend on y, and thus we are in the case of a numerical quadrature problem (cf. (0.2)), then (2.33) reduces to Simpson's formula.

§3. Global Description of One-Step Methods

We now turn to the numerical solution of the initial value problem (0.14) with the help of one-step formulae such as those developed in §2. The description of such one-step methods is best done in terms of grids and grid functions. A *grid* on the interval $[a, b]$ is a set of points $\{x_n\}_{n=0}^{N}$ such that

$$a = x_0 < x_1 < x_2 < \cdots < x_{N-1} < x_N = b, \tag{3.1}$$

[6]Runge's idea, in 1895, was to generalize Simpson's quadrature formula (cf. Ch. 3, §2.1) to ordinary differential equations. He succeeded only partially, in that the generalization he gave had stage number $r = 4$ but only order $p = 3$. The method (2.33) of order $p = 4$ was discovered in 1901 by Kutta through a systematic search.

with *grid lengths* h_n defined by

$$h_n = x_{n+1} - x_n, \qquad n = 0, 1, \ldots, N-1. \tag{3.2}$$

The *fineness* of the grid is measured by

$$|h| = \max_{0 \le n \le N-1} h_n. \tag{3.3}$$

We often use the letter h to designate the collection of lengths $h = \{h_n\}$. If $h_1 = h_2 = \cdots = h_N = (b-a)/N$, we call (3.1) a *uniform grid*, otherwise a *nonuniform grid*. For uniform grids we use the letter h also to designate the common grid length $h = (b-a)/N$. A vector-valued function $v = \{v_n\}$, $v_n \in \mathbb{R}^d$, defined on the grid (3.1) is called a *grid function*. Thus, v_n is the value of v at the gridpoint x_n. Every function $v(x)$ defined on $[a, b]$ induces a grid function by restriction. We denote the set of grid functions on $[a, b]$ by $\Gamma_h[a, b]$, and for each grid function $v = \{v_n\}$ define its norm by

$$\|v\|_\infty = \max_{0 \le n \le N} \|v_n\|, \qquad v \in \Gamma_h[a, b]. \tag{3.4}$$

A one-step method — indeed, any discrete-variable method — is a method producing a grid function $u = \{u_n\}$ such that $u \approx y$, where $y = \{y_n\}$ is the grid function induced by the exact solution $y(x)$ of the initial value problem (0.14). The grid (3.1) may be predetermined, for example, a uniform grid, or, as is more often the case in practice, be produced dynamically as part of the method (cf. §4.3). The most general scheme involving one-step formulae is a *variable-method variable-step method*. Given a sequence $\{\Phi_n\}$ of one-step formulae, the method proceeds as follows.

$$\begin{aligned} x_{n+1} &= x_n + h_n, \\ u_{n+1} &= u_n + h_n \Phi_n(x_n, u_n; h_n), \end{aligned} \qquad n = 0, 1, \ldots, N-1, \tag{3.5}$$

where $x_0 = a$, $u_0 = y_0$. For simplicity, we only consider *single-method* schemes involving a single method Φ, although the extension to variable-method schemes would not cause any essential difficulties (just more writing).

To bring out the analogy between (0.14) and (3.5), we introduce operators R and R_h acting on $C^1[a, b]$ and $\Gamma_h[a, b]$, respectively. These are the *residual operators*

$$(Rv)(x) := v'(x) - f(x, v(x)), \qquad v \in C^1[a, b], \tag{3.6}$$

$$(R_h v)_n := \frac{1}{h_n}(v_{n+1} - v_n) - \Phi(x_n, v_n; h_n), \quad n = 0, 1, \ldots, N-1; \qquad v = \{v_n\} \in \Gamma_h[a,b]. \tag{3.6$_h$}$$

(The grid function $\{(R_h v)_n\}$ is not defined for $n = N$, but we may arbitrarily set $(R_h v)_N = (R_h v)_{N-1}$.) Then the problem (0.14) and its discrete analogue (3.5) can be written transparently as

$$Ry = 0 \text{ on } [a,b], \qquad y(a) = y_0, \tag{3.7}$$

$$R_h u = 0 \text{ on } [a,b], \qquad u_0 = y_0. \tag{3.8}$$

Note that the discrete residual operator (3.6$_h$) is closely related to the truncation error (1.3′) when we apply the operator at a point $(x_n, y(x_n))$ on the exact solution trajectory. Then indeed the reference solution $u(t)$ coincides with the solution $y(t)$, and

$$(R_h y)_n = \frac{1}{h_n}[y(x_{n+1}) - y(x_n)] - \Phi(x_n, y(x_n); h_n) = -T(x_n, y(x_n); h_n). \tag{3.9}$$

§3.1. **Stability**. Stability is a property of the numerical scheme (3.5) alone and a priori has nothing to do with its approximation power. It characterizes the robustness of the scheme with respect to small perturbations. Nevertheless, stability combined with consistency yields convergence of the numerical solution to the true solution.

We define stability in terms of the discrete residual operators R_h in (3.6$_h$). As usual, we assume $\Phi(x, y; h)$ to be defined on $[a,b] \times \mathbb{R}^d \times [0, h_0]$, where $h_0 > 0$ is some suitable positive number.

Definition 3.1. The method (3.5) is called *stable* on $[a,b]$ if there exists a constant $K > 0$ not depending on h such that for an arbitrary grid h on $[a,b]$, and for arbitrary two grid functions $v, w \in \Gamma_h[a,b]$, there holds

$$\|v - w\|_\infty \le K(\|v_0 - w_0\| + \|R_h v - R_h w\|_\infty), \quad v, w \in \Gamma_h[a,b], \tag{3.10}$$

for all h with $|h|$ sufficiently small. In (3.10), the infinity norm for grid functions is the norm defined in (3.4).

We refer to (3.10) as the *stability inequality*. The motivation for it is as follows. Suppose we have two grid functions u, w satisfying

$$R_h u = 0, \qquad u_0 = y_0, \tag{3.11}$$

$$R_h w = \varepsilon, \qquad w_0 = y_0 + \eta_0, \tag{3.12}$$

where $\varepsilon = \{\varepsilon_n\} \in \Gamma_h[a,b]$ is a grid function with small $\|\varepsilon_n\|$, and $\|\eta_0\|$ is also small. We may interpret $u \in \Gamma_h[a,b]$ as the result of applying the numerical scheme (3.5) in infinite precision, whereas $w \in \Gamma_h[a,b]$ could be the solution of (3.5) in floating-point arithmetic. Then, if stability holds, we have

$$\|u - w\|_\infty \le K(\|\eta_0\| + \|\varepsilon\|_\infty); \tag{3.13}$$

that is, the global change in u is of the same order of magnitude as the local residual errors $\{\varepsilon_n\}$ and initial error η_0. It should be appreciated, however, that the first equation in (3.12) says $w_{n+1} - w_n - h_n\Phi(x_n, w_n; h_n) = h_n\varepsilon_n$, meaning that "rounding errors" must go to zero as $|h| \to 0$.

Interestingly enough, a Lipschitz condition on Φ is all that is required for stability.

Theorem 5.3.1. *If $\Phi(x, y; h)$ satisfies a Lipschitz condition with respect to the y-variables,*

$$\|\Phi(x, y; h) - \Phi(x, y^*; h)\| \le M\|y - y^*\| \quad \text{on} \quad [a,b] \times \mathbb{R}^d \times [0, h_0], \tag{3.14}$$

then the method (3.5) *is stable.*

We precede the proof with the following useful lemma.

Lemma 5.3.1. *Let $\{e_n\}$ be a sequence of numbers $e_n \in \mathbb{R}$ satisfying*

$$e_{n+1} \le a_n e_n + b_n, \qquad n = 0, 1, \ldots, N-1, \tag{3.15}$$

where $a_n > 0$ and $b_n \in \mathbb{R}$. Then

$$e_n \le E_n, \quad E_n = \left(\prod_{k=0}^{n-1} a_k\right) e_0 + \sum_{k=0}^{n-1} \left(\prod_{\ell=k+1}^{n-1} a_\ell\right) b_k, \quad n = 0, 1, \ldots, N. \tag{3.16}$$

We adopt here the usual convention that an empty product has the value 1, and an empty sum the value 0.

Proof of Lemma 5.3.1. It is readily verified that

$$E_{n+1} = a_n E_n + b_n, \qquad n = 0, 1, \ldots, N-1; \qquad E_0 = e_0.$$

Subtracting this from the inequality in (3.15), we get

$$e_{n+1} - E_{n+1} \le a_n(e_n - E_n), \qquad n = 0, 1, \ldots, N-1.$$

Now, $e_0 - E_0 = 0$, so that $e_1 - E_1 \le 0$, since $a_0 > 0$. By induction, more generally, $e_n - E_n \le 0$, since $a_{n-1} > 0$. □

Proof of Theorem 5.3.1. Let $h = \{h_n\}$ be an arbitrary grid on $[a, b]$, and $v, w \in \Gamma_h[a, b]$ two arbitrary (vector-valued) grid functions. By definition of R_h, we can write

$$v_{n+1} = v_n + h_n \Phi(x_n, v_n; h_n) + h_n (R_h v)_n, \qquad n = 0, 1, \ldots, N-1,$$

and similarly for w_{n+1}. Subtraction then gives

$$\begin{aligned} v_{n+1} - w_{n+1} &= v_n - w_n + h_n[\Phi(x_n, v_n; h_n) - \Phi(x_n, w_n; h_n)] \\ &\quad + h_n[(R_h v)_n - (R_h w)_n], \qquad n = 0, 1, \ldots, N-1. \end{aligned} \tag{3.17}$$

Now define

$$e_n = \|v_n - w_n\|, \qquad d_n = \|(R_h v)_n - (R_h w)_n\|, \qquad \delta = \|d\|_\infty. \tag{3.18}$$

Then, using the triangle inequality in (3.17) and the Lipschitz condition (3.14) for Φ, we obtain

$$e_{n+1} \le (1 + h_n M) e_n + h_n \delta, \qquad n = 0, 1, \ldots, N-1. \tag{3.19}$$

This is inequality (3.15) with $a_n = 1 + h_n M$, $b_n = h_n \delta$. Since for $k = 0, 1, \ldots, n-1$, $n \le N$, we have

$$\begin{aligned} \prod_{\ell=k+1}^{n-1} a_\ell \le \prod_{\ell=0}^{N-1} a_\ell = \prod_{\ell=0}^{N-1} (1 + h_\ell M) \le \prod_{\ell=0}^{N-1} e^{h_\ell M} \\ = e^{(h_0 + h_1 + \cdots + h_{N-1})M} = e^{(b-a)M}, \end{aligned}$$

where $1 + x \le e^x$ has been used in the second inequality, we obtain from Lemma 5.3.1 that

$$e_n \le e^{(b-a)M} e_0 + e^{(b-a)M} \sum_{k=0}^{n-1} h_k \delta$$

$$\leq e^{(b-a)M}(e_0 + (b-a)\delta), \qquad n = 0, 1, \ldots, N-1.$$

Therefore,

$$\|e\|_\infty = \|v - w\|_\infty \leq e^{(b-a)M}(\|v_0 - w_0\| + (b-a)\|R_h v - R_h w\|_\infty),$$

which is (3.10) with $K = e^{(b-a)M} \max\{1, b-a\}$. □

We have actually proved stability for *all* $|h| \leq h_0$, not only for $|h|$ sufficiently small. The proof is virtually the same for a variable-method algorithm involving a family of one-step formulae $\{\Phi_n\}$, if we assume a Lipschitz condition for each Φ_n with a constant M independent of n.

All one-step methods used in practice satisfy a Lipschitz condition if f does, and the constant M for Φ can be expressed in terms of the Lipschitz constant L for f. This is obvious for Euler's method, and not difficult to prove for others (see Ex. 15). It is useful to note that Φ need *not* be continuous in x; piecewise continuity suffices, as long as (3.14) holds for all $x \in [a, b]$, taking one-sided limits at points of discontinuity.

For later use we state another application of Lemma 5.3.1, relative to a grid function $v \in \Gamma_h[a, b]$ satisfying

$$v_{n+1} = v_n + h_n(A_n v_n + b_n), \qquad n = 0, 1, \ldots, N-1, \tag{3.20}$$

where $A_n \in \mathbb{R}^{d \times d}$, $b_n \in \mathbb{R}^d$, and $h = \{h_n\}$ is an arbitrary grid on $[a, b]$.

Lemma 5.3.2. *Suppose in* (3.20) *that*

$$\|A_n\| \leq M, \qquad \|b_n\| \leq \delta, \qquad n = 0, 1, \ldots, N-1, \tag{3.21}$$

where the constants M, δ do not depend on h. Then there exists a constant $K > 0$ independent of h, but depending on $\|v_0\|$, such that

$$\|v\|_\infty \leq K. \tag{3.22}$$

Proof. The lemma follows at once by observing that

$$\|v_{n+1}\| \leq (1 + h_n M)\|v_n\| + h_n \delta, \qquad n = 0, 1, \ldots, N-1,$$

which is precisely the inequality (3.19) in the proof of Theorem 5.3.1, hence

$$\|v_n\| \leq e^{(b-a)M}\{\|v_0\| + (b-a)\delta\}. \qquad \square \tag{3.23}$$

§3.2. **Convergence**. Stability is a rather powerful concept; it implies almost immediately convergence, and is also instrumental in deriving asymptotic global error estimates. We begin by defining precisely what we mean by convergence.

Definition 3.2. Let $a = x_0 < x_1 < x_2 < \cdots < x_N = b$ be a grid on $[a, b]$ with grid length $|h| = \max\limits_{1 \le n \le N}(x_n - x_{n-1})$. Let $u = \{u_n\}$ be the grid function defined by applying the method (3.5) on $[a, b]$, and $y = \{y_n\}$ the grid function induced by the exact solution of the initial value problem (0.14). The method (3.5) is said to *converge* on $[a, b]$ if there holds

$$\|u - y\|_\infty \to 0 \quad \text{as} \quad |h| \to 0. \tag{3.24}$$

Theorem 5.3.2. *If the method* (3.5) *is consistent and stable on* $[a, b]$, *then it converges. Moreover, if* Φ *has order* p, *then*

$$\|u - y\|_\infty = O(|h|^p) \quad as \quad |h| \to 0. \tag{3.25}$$

Proof. By the stability inequality (3.10) applied to the grid functions $v = u$ and $w = y$ of Definition 3.2, we have, for $|h|$ sufficiently small,

$$\begin{aligned} \|u - y\|_\infty &\le K(\|u_0 - y(x_0)\| + \|R_h u - R_h y\|_\infty) \\ &= K\|R_h y\|_\infty, \end{aligned} \tag{3.26}$$

since $u_0 = y(x_0)$ and $R_h u = 0$ by (3.5). But, by (3.9),

$$\|R_h y\|_\infty = \|T(\,\cdot\,, y; h)\|_\infty, \tag{3.27}$$

where T is the truncation error of the method Φ. By definition of consistency,

$$\|T(\,\cdot\,, y; h)\|_\infty \to 0 \quad \text{as} \quad |h| \to 0,$$

which proves the first part of the theorem. The second part follows immediately from (3.26) and (3.27), since order p means, by definition, that

$$\|T(\,\cdot\,, y; h)\|_\infty = O(|h|^p) \quad \text{as} \quad |h| \to 0. \qquad \square \tag{3.28}$$

Since, as we already observed, practically all one-step methods are stable and of order $p \geq 1$ (under reasonable smoothness assumptions on f), it follows that they are all convergent as well.

§3.3. **Asymptotics of global error**. Just as the principal error function describes the leading contribution of the local truncation error, it is of interest to identify the leading term in the global error $u_n - y(x_n)$. To simplify matters, we assume a constant grid length h, although it would not be difficult to deal with variable grid lengths of the form $h_n = \vartheta(x_n)h$, where $\vartheta(x)$ is piecewise continuous and $0 < \vartheta(x) \leq \Theta$ for $a \leq x \leq b$. Thus, we consider our one-step method to have the form

$$\begin{aligned} x_{n+1} &= x_n + h, \\ u_{n+1} &= u_n + h\Phi(x_n, u_n; h), \quad n = 0, 1, \ldots, N-1, \\ x_0 &= a, \quad u_0 = y_0, \end{aligned} \tag{3.29}$$

defining a grid function $u = \{u_n\}$ on a uniform grid over $[a, b]$. We are interested in the asymptotic behavior of $u_n - y(x_n)$ as $h \to 0$, where $y(x)$ is the exact solution of the initial value problem

$$\frac{dy}{dx} = f(x, y), \qquad a \leq x \leq b; \qquad y(a) = y_0. \tag{3.30}$$

Theorem 5.3.3. *Assume that*

(1) $\Phi(x, y; h) \in C^2$ *on* $[a, b] \times \mathbb{R}^d \times [0, h_0]$;

(2) Φ *is a method of order* $p \geq 1$ *admitting a principal error function* $\tau(x, y) \in C$ *on* $[a, b] \times \mathbb{R}^d$;

(3) $e(x)$ *is the solution of the linear initial value problem*

$$\begin{aligned} \frac{de}{dx} &= f_y(x, y(x))e + \tau(x, y(x)), \qquad a \leq x \leq b, \\ e(a) &= 0. \end{aligned} \tag{3.31}$$

Then, for $n = 0, 1, \ldots, N$,

$$u_n - y(x_n) = e(x_n)h^p + O(h^{p+1}) \quad as \quad h \to 0. \tag{3.32}$$

Before we prove the theorem, we make the following remarks.

1. The precise meaning of (3.32) is

$$\|u - y - h^p e\|_\infty = O(h^{p+1}) \quad \text{as} \quad h \to 0, \tag{3.22$'$}$$

where u, y, e are the grid functions $u = \{u_n\}$, $y = \{y(x_n)\}$, $e = \{e(x_n)\}$ and $\|\cdot\|_\infty$ is the norm defined in (3.4).

2. Since by consistency $\Phi(x, y; 0) = f(x, y)$, assumption (1) implies $f \in C^2$ on $[a, b] \times \mathbb{R}^d$, which is more than enough to guarantee the existence and uniqueness of the solution $e(x)$ of (3.31) on the whole interval $[a, b]$.

3. The fact that some, but not all, components of $\tau(x, y)$ may vanish identically does *not* imply that the corresponding components of $e(x)$ also vanish, since (3.31) is a *coupled* system of differential equations.

Proof of Theorem 5.3.3. We begin with an auxiliary computation, an estimate for

$$\Phi(x_n, u_n; h) - \Phi(x_n, y(x_n); h). \tag{3.33}$$

By Taylor's theorem (for functions of several variables), applied to the ith component of (3.33), we have

$$\begin{aligned}\Phi^i(x_n, u_n; h) - \Phi^i(x_n, y(x_n); h) &= \sum_{j=1}^{d} \Phi^i_{y^j}(x_n, y(x_n); h)[u_n^j - y^j(x_n)] \\ &+ \frac{1}{2} \sum_{j,k=1}^{d} \Phi^i_{y^j y^k}(x_n, \overline{u}_n; h)[u_n^j - y^j(x_n)][u_n^k - y^k(x_n)],\end{aligned} \tag{3.34}$$

where $\overline{u}_n$ is on the line segment connecting u_n and $y(x_n)$. Using Taylor's theorem once more, in the variable h, we can write

$$\Phi^i_{y^j}(x_n, y(x_n); h) = \Phi^i_{y^j}(x_n, y(x_n); 0) + h\Phi^i_{y^j h}(x_n, y(x_n); \overline{h}),$$

where $0 < \overline{h} < h$. Since by consistency $\Phi(x, y; 0) \equiv f(x, y)$ on $[a, b] \times \mathbb{R}^d$, we have

$$\Phi^i_{y^j}(x, y; 0) = f^i_{y^j}(x, y), \qquad x \in [a, b], \quad y \in \mathbb{R}^d,$$

and assumption (1) allows us to write

$$\Phi^i_{y^j}(x_n, y(x_n); h) = f^i_{y^j}(x_n, y(x_n)) + O(h), \qquad h \to 0. \tag{3.35}$$

Now observing that $u_n - y(x_n) = O(h^p)$ by virtue of Theorem 5.3.2, and using (3.35) in (3.34), we get, again by assumption (1),

$$\begin{aligned}
&\Phi^i(x_n, u_n; h) - \Phi^i(x_n, y(x_n); h) \\
&\quad = \sum_{j=1}^{d} f^i_{y^j}(x_n, y(x_n))[u^j_n - y^j(x_n)] + O(h^{p+1}) + O(h^{2p}).
\end{aligned}$$

But $O(h^{2p})$ is also of order $O(h^{p+1})$, since $p \geq 1$. Thus, in vector notation,

$$\Phi(x_n, u_n; h) - \Phi(x_n, y(x_n); h) = f_y(x_n, y(x_n))[u_n - y(x_n)] + O(h^{p+1}). \tag{3.36}$$

Now, to highlight the leading term in the global error, we define the grid function $r = \{r_n\}$ by

$$r = h^{-p}(u - y). \tag{3.37}$$

Then

$$\begin{aligned}
\frac{1}{h}\,(r_{n+1} - r_n) &= \frac{1}{h}\,[h^{-p}(u_{n+1} - y(x_{n+1})) - h^{-p}(u_n - y(x_n))] \\
&= h^{-p}\left[\frac{1}{h}\,(u_{n+1} - u_n) - \frac{1}{h}\,(y(x_{n+1}) - y(x_n))\right] \\
&= h^{-p}[\Phi(x_n, u_n; h) - \{\Phi(x_n, y(x_n); h) - T(x_n, y(x_n); h)\}],
\end{aligned}$$

where we have used (3.29) and the relation (3.9) for the truncation error T. Therefore, expressing T in terms of the principal error function τ, we get

$$\frac{1}{h}\,(r_{n+1} - r_n) = h^{-p}[\Phi(x_n, u_n; h) - \Phi(x_n, y(x_n); h) + \tau(x_n, y(x_n))h^p + O(h^{p+1})].$$

For the first two terms in brackets we use (3.36) and the definition of r in (3.37) to obtain

$$\begin{aligned}
&\frac{1}{h}\,(r_{n+1} - r_n) = f_y(x_n, y(x_n))r_n + \tau(x_n, y(x_n)) + O(h), \\
&\qquad\qquad n = 0, 1, \ldots, N-1, \\
&r_0 = 0.
\end{aligned} \tag{3.38}$$

Now letting

$$g(x,y) := f_y(x,y(x))y + \tau(x,y(x)), \tag{3.39}$$

we can interpret (3.38) by writing

$$(R_h^{\mathrm{Euler},g} r)_n = \varepsilon_n \quad (n=0,1,\dots,N-1), \quad \varepsilon_n = O(h),$$

where $R_h^{\mathrm{Euler},g}$ is the discrete residual operator (3.6$_h$) that goes with Euler's method applied to $e' = g(x,e)$, $e(a) = 0$. Since Euler's method is stable on $[a,b]$ and g (being linear in y) certainly satisfies a uniform Lipschitz condition, we have by the stability inequality (3.10)

$$\|r - e\|_\infty = O(h),$$

and hence, by (3.37),

$$\|u - y - h^p e\|_\infty = O(h^{p+1}),$$

as was to be shown. □

§4. Error Monitoring and Step Control

Most production codes currently available for solving ODEs monitor local truncation errors and control the step length on the basis of estimates for these errors. Here we attempt to monitor the global error, at least asymptotically, by implementing the asymptotic result of Theorem 5.3.3. This necessitates the evaluation of the Jacobian matrix $f_y(x,y)$ along or near the solution trajectory; but this is only natural, since f_y, in a first approximation, governs the effect of perturbations via the variational differential equation (3.31). This equation is driven by the principal error function evaluated along the trajectory, so that estimates of local truncation errors (more precisely, of the principal error function) are needed also in this approach. For simplicity we again assume constant grid length.

§4.1. **Estimation of global error.** The idea of our estimation is to integrate the "variational equation" (3.31) along with the main equation (3.30). Since we need $e(x_n)$ in (3.32) only to within an accuracy of $O(h)$ (any $O(h)$ error term in $e(x_n)$, multiplied by h^p, being absorbed by the $O(h^{p+1})$ term), we can use Euler's method for that purpose, which will provide the desired approximation $v_n \approx e(x_n)$.

Theorem 5.4.1. *Assume that*

(1) $\Phi(x, y; h) \in C^2$ *on* $[a, b] \times \mathbb{R}^d \times [0, h_0]$;

(2) Φ *is a method of order* $p \geq 1$ *admitting a principal error function* $\tau(x, y) \in C^1$ *on* $[a, b] \times \mathbb{R}^d$;

(3) *an estimate* $r(x, y; h)$ *is available for the principal error function that satisfies*

$$r(x, y; h) = \tau(x, y) + O(h), \qquad h \to 0, \tag{4.1}$$

uniformly on $[a, b] \times \mathbb{R}^d$;

(4) *along with the grid function* $u = \{u_n\}$ *we generate the grid function* $v = \{v_n\}$ *in the following manner .*

$$\begin{aligned} x_{n+1} &= x_n + h, \\ u_{n+1} &= u_n + h\Phi(x_n, u_n; h), \\ v_{n+1} &= v_n + h[f_y(x_n, u_n)v_n + r(x_n, u_n; h)], \\ x_0 &= a, \quad u_0 = y_0, \quad v_0 = 0. \end{aligned} \tag{4.2}$$

Then, for $n = 0, 1, \ldots, N$,

$$u_n - y(x_n) = v_n h^p + O(h^{p+1}) \;\; as \;\; h \to 0. \tag{4.3}$$

Proof. The proof consists of establishing the following estimates.

$$f_y(x_n, u_n) = f_y(x_n, y(x_n)) + O(h), \tag{4.4}$$

$$r(x_n, u_n; h) = \tau(x_n, y(x_n)) + O(h). \tag{4.5}$$

Once this has been done, we can argue as follows. Let (cf. (3.39))

$$g(x, y) := f_y(x, y(x))y + \tau(x, y(x)). \tag{4.6}$$

The equation for v_{n+1} in (4.2) has the form

$$v_{n+1} = v_n + h(A_n v_n + b_n),$$

where A_n are bounded matrices and b_n bounded vectors. By Lemma 5.3.2, we have boundedness of v_n,

$$v_n = O(1), \qquad h \to 0. \tag{4.7}$$

Substituting (4.4) and (4.5) into the equation for v_{n+1}, and noting (4.7), we obtain

$$v_{n+1} = v_n + h[f_y(x_n, y(x_n))v_n + \tau(x_n, y(x_n)) + O(h)]$$

$$= v_n + hg(x_n, v_n) + O(h^2).$$

Thus, in the notation used in the proof of Theorem 5.3.3,

$$(R_h^{\text{Euler},g} v)_n = O(h), \quad v_0 = 0.$$

Since Euler's method is stable, we conclude

$$v_n - e(x_n) = O(h),$$

where $e(x)$ is, as before, the solution of $e' = g(x, e)$, $e(a) = 0$. Therefore, by (3.32),

$$u_n - y(x_n) = e(x_n)h^p + O(h^{p+1}) = v_n h^p + O(h^{p+1}),$$

as was to be shown.

It remains to prove (4.4) and (4.5). From assumption (1) we note, first of all, that $f(x, y) \in C^2$ on $[a, b] \times \mathbb{R}^d$, since by consistency $f(x, y) = \Phi(x, y; 0)$. By virtue of $u_n = y(x_n) + O(h^p)$ (cf. Theorem 5.3.2), we therefore have

$$f_y(x_n, u_n) = f_y(x_n, y(x_n)) + O(h^p),$$

which implies (4.4), since $p \geq 1$.

Next, since $\tau(x, y) \in C^1$ by assumption (2), we have

$$\tau(x_n, u_n) = \tau(x_n, y(x_n)) + \tau_y(x_n, \overline{u}_n)(u_n - y(x_n))$$

$$= \tau(x_n, y(x_n)) + O(h^p),$$

so that by assumption (3),

$$r(x_n, u_n; h) = \tau(x_n, u_n) + O(h) = \tau(x_n, y(x_n)) + O(h^p) + O(h),$$

from which (4.5) follows at once. This completes the proof of Theorem 5.4.1. □

§4.2. **Truncation error estimates.** In order to apply Theorem 5.4.1, we need estimates $r(x, y; h)$ of the principal error function $\tau(x, y)$ which are

$O(h)$ accurate. A number of them, in increasing order of efficiency, are now described.

(1) *Local Richardson extrapolation to zero.* This works for any one-step method Φ, but is usually considered to be too expensive. If Φ has order p, the procedure is as follows.

$$
\begin{aligned}
y_h &= y + h\Phi(x, y; h), \\
y_{h/2} &= y + \tfrac{1}{2}\, h\Phi\left(x, y; \tfrac{1}{2}\, h\right), \\
y_h^* &= y_{h/2} + \tfrac{1}{2}\, h\Phi\left(x + \tfrac{1}{2}\, h, y_{h/2}; \tfrac{1}{2}\, h\right), \\
r(x, y; h) &= \frac{1}{1 - 2^{-p}}\, \frac{1}{h^{p+1}}\, (y_h - y_h^*).
\end{aligned}
\tag{4.8}
$$

Note that y_h^* is the result of applying Φ over two consecutive steps of length $\frac{1}{2}\, h$ each, whereas y_h is the result of one application over the whole step of length h.

We now verify that $r(x, y; h)$ in (4.8) is an acceptable estimator. To do this, we need to assume that $\tau(x, y) \in C^1$ on $[a, b] \times \mathbb{R}^d$. In terms of the reference solution $u(t)$ through (x, y), we have (cf. (1.3′) and (1.7))

$$
\Phi(x, y; h) = \frac{1}{h}\, [u(x + h) - u(x)] + \tau(x, y)h^p + O(h^{p+1}). \tag{4.9}
$$

Furthermore,

$$
\begin{aligned}
\frac{1}{h}\, (y_h - y_h^*) &= \frac{1}{h}\, (y - y_{h/2}) + \Phi(x, y; h) - \frac{1}{2}\, \Phi\left(x + \frac{1}{2}\, h, y_{h/2}; \frac{1}{2}\, h\right) \\
&= \Phi(x, y; h) - \frac{1}{2}\, \Phi\left(x, y; \frac{1}{2}\, h\right) - \frac{1}{2}\, \Phi\left(x + \frac{1}{2}\, h, y_{h/2}; \frac{1}{2}\, h\right).
\end{aligned}
$$

Applying (4.9) to each of the three terms on the right, we find

$$
\begin{aligned}
\frac{1}{h}\, (y_h - y_h^*) &= \frac{1}{h}\, [u(x + h) - u(x)] + \tau(x, y)h^p + O(h^{p+1}) \\
&\quad - \frac{1}{2}\, \frac{1}{h/2} \left[u\left(x + \frac{1}{2}\, h\right) - u(x)\right] - \frac{1}{2}\, \tau(x, y) \left(\frac{1}{2}\, h\right)^p + O(h^{p+1}) \\
&\quad - \frac{1}{2}\, \frac{1}{h/2} \left[u(x + h) - u\left(x + \frac{1}{2}\, h\right)\right] - \frac{1}{2}\, \tau\left(x + \frac{1}{2}\, h, y + O(h)\right) \left(\frac{1}{2}\, h\right)^p \\
&\quad + O(h^{p+1}) = \tau(x, y)(1 - 2^{-p})h^p + O(h^{p+1}).
\end{aligned}
$$

Consequently,

$$\frac{1}{1-2^{-p}} \frac{1}{h} (y_h - y_h^*) = \tau(x,y)h^p + O(h^{p+1}) \tag{4.10}$$

as required.

Subtracting (4.10) from (4.9) shows, incidentally, that

$$\Phi^*(x,y;h) := \Phi(x,y;h) - \frac{1}{1-2^{-p}} \frac{1}{h} (y_h - y_n^*) \tag{4.11}$$

defines a one-step method of order $p+1$.

The procedure in (4.8) is rather expensive. For a fourth-order Runge-Kutta process, it requires a total of 11 evaluations of f per step, almost three times the effort for a single Runge-Kutta step. Therefore, Richardson extrapolation is normally used only after every two steps of Φ; that is, one proceeds according to

$$\begin{aligned} y_h &= y + h\Phi(x,y;h), \\ y_{2h}^* &= y_h + h\Phi(x+h,y_h;h), \\ y_{2h} &= y + 2h\Phi(x,y;2h). \end{aligned} \tag{4.8'}$$

Then (4.10) gives

$$\frac{1}{2(2^p-1)} \frac{1}{h^{p+1}} (y_{2h} - y_{2h}^*) = \tau(x,y) + O(h), \tag{4.12}$$

so that the expression on the left is an acceptable estimator $r(x,y;h)$. If the two steps in (4.8′) yield acceptable accuracy (cf. §4.3), then, again for a fourth-order Runge-Kutta process, the procedure requires only three additional evaluations of f, since y_h and y_{2h}^* would have to be computed anyhow. We show, however, that there are still more efficient schemes.

(2) *Embedded methods.* The basic idea of this approach is very simple: if the given method Φ has order p, take any one-step method Φ^* of order $p^* = p+1$ and define

$$r(x,y;h) = \frac{1}{h^p} [\Phi(x,y;h) - \Phi^*(x,y;h)]. \tag{4.13}$$

This is indeed an acceptable estimator, as follows by subtracting the two relations

$$\Phi(x,y;h) - \frac{1}{h}\,[u(x+h)-u(x)] = \tau(x,y)h^p + O(h^{p+1}),$$
$$\Phi^*(x,y;h) - \frac{1}{h}\,[u(x+h)-u(x)] = O(h^{p+1})$$

and dividing the result by h^p.

The tricky part is making this procedure efficient. Following an idea of Fehlberg, one can try to do this by embedding one Runge-Kutta process (of order p) into another (of order $p+1$). Specifically, let Φ be some explicit r-stage Runge-Kutta method,

$$k_1(x,y) = f(x,y),$$
$$k_s(x,y;h) = f\left(x+\mu_s h, y+h\sum_{j=1}^{s-1}\lambda_{sj}k_j\right), \quad s=2,3,\dots,r,$$
$$\Phi(x,y;h) = \sum_{s=1}^{r}\alpha_s k_s.$$

Then for Φ^* choose a similar r^*-stage process, with $r^* > r$, in such a way that

$$\mu_s^* = \mu_s, \quad \lambda_{sj}^* = \lambda_{sj} \quad \text{for} \quad s=2,3,\dots,r.$$

The estimate (4.13) then costs only r^*-r extra evaluations of f. If $r^*=r+1$, one might even attempt to save the additional evaluation by selecting (if possible)

$$\mu_{r^*} = 1, \quad \lambda_{r^*j} = \alpha_j \quad \text{for} \quad j=1,2,\dots,r^*-1 \quad (r^*=r+1). \tag{4.14}$$

Then indeed, k_{r^*} will be identical with k_1 for the next step.

Pairs of such embedded $(p,p+1)$ Runge-Kutta formulae have been developed in the late 1960s by E. Fehlberg. There is a considerable degree of freedom in choosing the parameters. Fehlberg's choices were guided by an attempt to reduce the magnitude of the coefficients of all the partial derivative aggregates that enter into the principal error function $\tau(x,y)$ of Φ (cf. the end of §2.4 for an elementary example of this technique). He succeeded in obtaining pairs with the following values of parameters p, r, r^*.

p	r	r^*
3	4	5
4	5	6
5	6	8
6	8	10
7	11	13
8	15	17

TABLE 5.4.1. Embedded Runge-Kutta formulae

For the third-order process (and only for that one) one can also arrange for (4.14) to hold (cf. Ex. 17 for a second-order process).

§4.3. **Step control.** Any estimate $r(x, y; h)$ of the principal error function $\tau(x, y)$ implies an estimate

$$h^p r(x, y; h) = T(x, y; h) + O(h^{p+1}) \tag{4.15}$$

for the truncation error, which can be used to monitor the local truncation error during the integration process. However, one has to keep in mind that the local truncation error is quite different from the global error, the error that one really wants to control. To get more insight into the relationship between these two errors, we recall the following theorem, which quantifies the continuity of the solution of an initial value problem with respect to initial values.

Theorem 5.4.2. *Let $f(x, y)$ be continuous in x for $a \le x \le b$ and satisfy a Lipschitz condition uniformly on $[a, b] \times \mathbb{R}^d$ with Lipschitz constant L* (cf. (0.24)). *Then the initial value problem*

$$\begin{aligned} \frac{dy}{dx} &= f(x, y), \qquad a \le x \le b, \\ y(c) &= y_c \end{aligned} \tag{4.16}$$

has a unique solution on $[a, b]$ for any c with $a \le c \le b$ and for any $y_c \in \mathbb{R}^d$. Let $y(x; s)$ and $y(x; s^*)$ be the solutions of (4.16) *corresponding to $y_c = s$ and $y_c = s^*$, respectively. Then, for any vector norm $\|\cdot\|$,*

$$\|y(x; s) - y(x; s^*)\| \le e^{L|x-c|}\|s - s^*\|. \tag{4.17}$$

Solving the given initial value problem (3.30) numerically by a one-step method (not necessarily with constant step) in reality means that one follows a sequence of "solution tracks," whereby at each grid point x_n one jumps from one track to the next by an amount determined by the truncation error at x_n (cf. Figure 5.4.1). This is so by the very definition of truncation error, the reference solution being one of the solution tracks. Specifically, the nth track, $n = 0, 1, \ldots, N$, is given by the solution of the initial value problem

$$\frac{dv_n}{dx} = f(x, v_n), \qquad x_n \le x \le b, \qquad v_n(x_n) = u_n, \tag{4.18}$$

and

$$u_{n+1} = v_n(x_{n+1}) + h_n T(x_n, u_n; h_n), \qquad n = 0, 1, \ldots, N-1. \tag{4.19}$$

Since by (4.18) we have $u_{n+1} = v_{n+1}(x_{n+1})$, we can apply Theorem 5.4.2 to the solutions v_{n+1} and v_n, letting $c = x_{n+1}$, $s = u_{n+1}$, $s^* = u_{n+1} - h_n T(x_n, u_n; h_n)$ (by (4.19)), and thus obtain

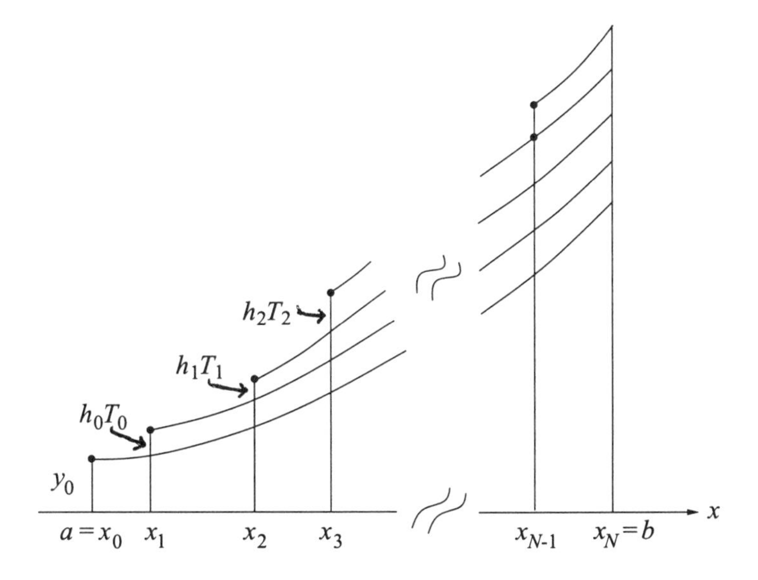

FIGURE 5.4.1. Error accumulation in a one-step method

$$\|v_{n+1}(x) - v_n(x)\| \leq h_n e^{L|x-x_{n+1}|}\|T(x_n, u_n; h_n)\|, \quad n = 0, 1, \ldots, N-1. \tag{4.20}$$

Now

$$\sum_{n=0}^{N-1} [v_{n+1}(x) - v_n(x)] = v_N(x) - v_0(x) = v_N(x) - y(x), \tag{4.21}$$

and since $v_N(x_N) = u_N$, letting $x = x_N$, we get from (4.20) and (4.21) that

$$\begin{aligned} \|u_N - y(x_N)\| &\leq \sum_{n=0}^{N-1} \|v_{n+1}(x_N) - v_n(x_N)\| \\ &\leq \sum_{n=0}^{N-1} h_n e^{L|x_N - x_{n+1}|}\|T(x_n, u_n; h_n)\|. \end{aligned}$$

Therefore, if we make sure that

$$\|T(x_n, u_n; h_n)\| \leq \varepsilon_T, \qquad n = 0, 1, 2, \ldots, N-1, \tag{4.22}$$

then

$$\|u_N - y(x_N)\| \leq \varepsilon_T \sum_{n=0}^{N-1} (x_{n+1} - x_n) e^{L|x_N - x_{n+1}|}.$$

Interpreting the sum on the right as a Riemann sum for a definite integral, we finally obtain, approximately,

$$\|u_N - y(x_N)\| \leq \varepsilon_T \int_a^b e^{L(b-x)} dx = \frac{\varepsilon_T}{L}\,(e^{L(b-a)} - 1).$$

Thus, knowing an estimate for L would allow us to set an appropriate ε_T, namely,

$$\varepsilon_T = \frac{L}{e^{L(b-a)} - 1}\,\varepsilon, \tag{4.23}$$

to guarantee an error $\|u_N - y(x_N)\| \leq \varepsilon$. What holds for the whole grid on $[a, b]$, of course, holds for any grid on a subinterval $[a, x]$, $a < x \leq b$. So, in principle, given the desired accuracy ε for the solution $y(x)$, we can determine a "local tolerance level" ε_T by (4.23) and achieve the desired accuracy by keeping the local truncation error below ε_T (cf. (4.22)). Note that as $L \to 0$ we have $\varepsilon_T \to \varepsilon/(b-a)$. This limit value of ε_T would be appropriate for a quadrature problem but definitely not for a true differential equation problem, where ε_T, in general, has to be chosen considerably smaller than the target error tolerance ε.

Considerations such as these motivate the following *step control* mechanism: each integration step (from x_n to $x_{n+1} = x_n + h_n$) consists of these parts:

1. Estimate h_n.
2. Compute $u_{n+1} = u_n + h_n\Phi(x_n, u_n; h_n)$ and $r(x_n, u_n; h_n)$.
3. Test $h_n^p\|r(x_n, u_n; h_n)\| \le \varepsilon_T$ (cf. (4.15) and (4.22)). If the test passes, proceed with the next step; if not, repeat the step with a smaller h_n, say, half as large, until the test passes.

To estimate h_n, assume first that $n \ge 1$, so that the estimator from the previous step, $r(x_{n-1}, u_{n-1}; h_{n-1})$ (or at least its norm), is available. Then, neglecting terms of $O(h)$,

$$\|\tau(x_{n-1}, u_{n-1})\| \approx \|r(x_{n-1}, u_{n-1}; h_{n-1})\|,$$

and since $\tau(x_n, u_n) \approx \tau(x_{n-1}, u_{n-1})$, likewise

$$\|\tau(x_n, u_n)\| \approx \|r(x_{n-1}, u_{n-1}; h_{n-1})\|.$$

What we want is

$$\|\tau(x_n, u_n)\|h_n^p \approx \theta\varepsilon_T,$$

where θ is "safety factor," say, $\theta = .8$. Eliminating $\tau(x_n, u_n)$, we find

$$h_n \approx \left\{\frac{\theta\varepsilon_T}{\|r(x_{n-1}, u_{n-1}; h_{n-1})\|}\right\}^{1/p}.$$

Note that from the previous step we have $h_{n-1}^p\|r(x_{n-1}, u_{n-1}; h_{n-1})\| \le \varepsilon_T$, so that

$$h_n \ge \theta^{1/p}h_{n-1},$$

and the tendency is toward increasing the step.

If $n = 0$, we proceed similarly, using some initial guess $h_0^{(0)}$ of h_0 and associated $r(x_0, y_0; h_0^{(0)})$ to obtain

$$h_0^{(1)} = \left\{\frac{\theta\varepsilon_T}{\|r(x_0, y_0; h_0^{(0)})\|}\right\}^{1/p}.$$

The process may be repeated once or twice to get the final estimate of h_0 and $\|r(x_0, y_0; h_0)\|$.

§5. Stiff Problems

Although there is no generally accepted definition of stiffness[7] of differential equations, a characteristic feature of stiffness is the presence of rapidly changing transients. This manifests itself mathematically in the Jacobian matrix f_y having eigenvalues with very large negative real parts along with others of normal magnitude. Standard (in particular explicit) numerical ODE methods are unable to cope with such solutions unless they use unrealistically small step lengths. What is called for are methods enjoying a special stability property called A-stability. We introduce this concept in the context of linear homogeneous systems of differential equations with constant coefficient matrix. Padé approximants to the exponential function turn out to be instrumental in constructing A-stable one-step methods.

§5.1. **A-stability**. A model problem exhibiting stiffness is the linear initial value problem

$$\frac{dy}{dx} = Ay, \qquad 0 \le x < \infty; \qquad y(a) = y_0, \tag{5.1}$$

where $A \in \mathbb{R}^{d \times d}$ is a constant matrix of order d having all its eigenvalues in the left half-plane,

$$\operatorname{Re} \lambda_i(A) < 0, \qquad i = 1, 2, \dots, d. \tag{5.2}$$

It is well known that all solutions of the differential system in (5.1) then decay exponentially as $x \to \infty$. Those corresponding to eigenvalues with very large negative parts do so particularly fast, giving rise to the phenomenon of stiffness. In particular, for the solution $y(x)$ of (5.1) we have

$$y(x) \to 0 \quad \text{as} \quad x \to \infty. \tag{5.3}$$

How does a one-step method Φ behave when applied to (5.1)? First of all, a generic step of the one-step method will now have the form

$$y_{\text{next}} = y + h\Phi(x, y; h) = \varphi(hA)y, \tag{5.4}$$

where φ is some function, called the *stability function* of the method. In what follows we assume that the matrix function $\varphi(hA)$ is well defined; minimally,

[7]The word "stiffness" comes from the differential equation governing the oscillation of a "stiff" spring, that is, a spring with a large spring constant.

we require that φ: $\mathbb{C} \to \mathbb{C}$ is analytic in a neighborhood of the origin. Since the reference solution through the point (x, y) is given by $u(t) = e^{A(t-x)}y$, we have for the trunction error of Φ at (x, y) (cf. (1.3))

$$T(x, y; h) = \frac{1}{h}\,[y_{\text{next}} - u(x+h)] = \frac{1}{h}\,[\varphi(hA) - e^{hA}]y. \tag{5.5}$$

In particular, the method Φ in this case has order p if and only if

$$e^z = \varphi(z) + O(z^{p+1}), \qquad z \to 0. \tag{5.6}$$

This shows the relevance of approximations to the exponential function in the context of one-step methods applied to the model problem (5.1).

The approximate solution $u = \{u_n\}$ to the initial value problem (5.1), assuming for simplicity a constant grid length h, is given by

$$u_{n+1} = \varphi(hA)u_n, \qquad n = 0, 1, 2, \dots \;; \qquad u_0 = y_0;$$

hence

$$u_n = [\varphi(hA)]^n y_0, \qquad n = 0, 1, 2, \dots \;. \tag{5.7}$$

This will simulate the behavior (5.3) of the exact solution if and only if

$$\lim_{n\to\infty} [\varphi(hA)]^n = 0. \tag{5.8}$$

A necessary and sufficient condition for (5.8) to hold is that all eigenvalues of the matrix $\varphi(hA)$ be strictly within the unit circle. This in turn is equivalent to

$$|\varphi(h\lambda_i(A))| < 1 \quad \text{for} \quad i = 1, 2, \dots, d, \tag{5.9}$$

where $\lambda_i(A)$ are the eigenvalues of A. In view of (5.2), this gives rise to the following definition.

Definition 5.1. A one-step method Φ is called *A-stable* if the function φ associated with Φ according to (5.4) is defined in the left half of the complex plane and satisfies

$$|\varphi(z)| < 1 \quad \text{for all } z \text{ with } \operatorname{Re} z < 0. \tag{5.10}$$

We are led to the problem of constructing a function φ (and with it, a one-step formula Φ) which is analytic in the left-half plane, approximates well the exponential function near the origin (cf. (5.6)), and satisfies (5.10).

An important tool for this is Padé approximation to the exponential function.

§5.2. **Padé[8] approximation**. For any function $g(z)$ analytic in a neighborhood of $z = 0$, one defines its Padé approximants as follows.

Definition 5.2. The *Padé approximant* $R[n, m](z)$ to the function $g(z)$ is the rational function

$$R[n,m](z) = \frac{P(z)}{Q(z)}, \qquad P \in \mathbb{P}_m, \quad Q \in \mathbb{P}_n, \tag{5.11}$$

satisfying

$$g(z)Q(z) - P(z) = O(z^{n+m+1}) \text{ as } z \to 0. \tag{5.12}$$

Thus, when expanding the left-hand side of (5.12) in powers of z, all initial terms should drop out up to (and including) the one with power z^{n+m}. It is known that the rational function $R[n, m]$ is uniquely determined by this definition, even though in exceptional cases, P and Q may have common factors. If this is not the case, that is, P and Q are irreducible over the complex numbers, we assume without loss of generality that $Q(0) = 1$.

Our interest here is in the function $g(z) = e^z$. In this case, $P = P[n, m]$ and $Q = Q[n, m]$ in (5.11) and (5.12) can be explicitly determined.

Theorem 5.5.1. *The Padé approximant $R[n, m]$ to the exponential function $g(z) = e^z$ is given by*

$$P[n,m](z) = \sum_{k=0}^{m} \frac{m!(n+m-k)!}{(m-k)!(n+k)!} \frac{z^k}{k!}, \tag{5.13}$$

$$Q[n,m](z) = \sum_{k=0}^{n} (-1)^k \frac{n!(n+m-k)!}{(n-k)!(n+m)!} \frac{z^k}{k!}. \tag{5.14}$$

Moreover,

$$e^z - \frac{P[n,m](z)}{Q[n,m](z)} = C_{n,m} z^{n+m+1} + \cdots,$$

[8]Henri Eugène Padé (1863–1953), a French mathematician, was educated partly in Germany and partly in France, where he wrote his thesis under Hermite's supervision. Although much of his time was consumed by high administrative duties, he managed to write many papers on continued fractions and rational approximation. His thesis and related papers became widely known after Borel referred to them in his 1901 book on divergent series.

where

$$C_{n,m} = (-1)^n \, \frac{n!m!}{(n+m)!(n+m+1)!} \, . \tag{5.15}$$

Proof. Let

$$v(t) := t^n(1-t)^m.$$

By Leibniz's rule one finds

$$\begin{aligned} v^{(r)}(t) &= \sum_{k=0}^{r} \binom{r}{k} [t^n]^{(k)}[(1-t)^m]^{(r-k)} \\ &= \sum_{k=0}^{r} \binom{r}{k}\binom{n}{k} k!t^{n-k} \binom{m}{r-k} (r-k)!(1-t)^{m-r+k}(-1)^{r-k}; \end{aligned}$$

hence, in particular,

$$\begin{aligned} v^{(r)}(0) &= (-1)^{r-n}\binom{m}{r-n} r! \ \text{ if } \ r \ge n; \qquad v^{(r)}(0) = 0 \ \text{ if } \ r < n; \\ v^{(r)}(1) &= (-1)^{m}\binom{n}{r-m} r! \ \text{ if } \ r \ge m; \qquad v^{(r)}(1) = 0 \ \text{ if } \ r < m. \end{aligned} \tag{5.16}$$

Given any integer $q \ge 0$, repeated integration by parts yields

$$\begin{aligned} \int_0^1 e^{tz} v(t)dt = e^z \sum_{r=0}^{q} (-1)^r \, \frac{v^{(r)}(1)}{z^{r+1}} - \sum_{r=0}^{q} (-1)^r \, \frac{v^{(r)}(0)}{z^{r+1}} \\ + \frac{(-1)^{q+1}}{z^{q+1}} \int_0^1 e^{tz} v^{(q+1)}(t)dt. \end{aligned} \tag{5.17}$$

Putting here $q = n+m$, so that $v^{(q+1)}(t) \equiv 0$, and multiplying by $(-1)^q z^{q+1}$ gives

$$(-1)^q e^z \sum_{r=0}^{q} (-1)^r v^{(r)}(1) z^{q-r} - (-1)^q \sum_{r=0}^{q} (-1)^r v^{(r)}(0) z^{q-r} + O(z^{n+m+1}), \quad z \to 0,$$

where the O-term comes from multiplying the integral on the left of (5.17) (which is $O(1)$ as $z \to 0$) by $z^{q+1} = z^{n+m+1}$. In the first sum it suffices, by (5.16), to sum over $r \ge m$, and in the second, to sum over $r \ge n$. Using the new variable of summation k defined by $q - r = k$, we find

$$e^z \sum_{k=0}^{n} (-1)^k v^{(n+m-k)}(1) z^k - \sum_{k=0}^{m} (-1)^k v^{(n+m-k)}(0) z^k = O(z^{n+m+1}),$$

which clearly is (5.12) for $g(z) = e^z$. It now suffices to substitute the values (5.16) for the derivatives of v at 0 and 1 to obtain (5.13) and (5.14), after multiplication of numerator and denominator by $(-1)^m/(n+m)!$. Tracing the constants, one readily checks (5.15). □

The Padé approximants to the exponential function have some very useful and important properties. Here are those of interest in connection with A-stability.

(1) $P[n,m](z) = Q[m,n](-z)$: *The numerator polynomial is the denominator polynomial with indices interchanged and z replaced by $-z$.* This reflects the property $1/e^z = e^{-z}$ of the exponential function. The proof follows immediately from (5.13) and (5.14).

(2) *For each $n = 0, 1, 2, \ldots,$ all zeros of $Q[n,n]$ have positive real parts* (hence, by (1), all zeros of $P[n,n]$ have negative real parts). A proof can be given by applying the Routh-Hurwitz criterion[9] for stable polynomials.

(3) *For real $t \in \mathbb{R}$, and $n = 0, 1, 2, \ldots,$ there holds*

$$\left| \frac{P[n,n](it)}{Q[n,n](it)} \right| = 1.$$

Indeed, by property (1), one has $P[n,n](it) = \overline{Q[n,n](it)}$.

(4) *There holds*

$$\left| \frac{P[n+1,n](it)}{Q[n+1,n](it)} \right| < 1 \quad \textit{for} \quad t \in \mathbb{R}, \quad t \neq 0, \quad n = 0, 1, 2, \ldots .$$

The proof follows from the basic property (5.12) of the Padé approximant:

$$e^{it} Q[n+1,n](it) - P[n+1,n](it) = O(|t|^{2n+2}), \qquad t \to 0.$$

Taking absolute values, and using the triangle inequality, gives

$$\big|\, |Q[n+1,n](it)| - |P[n+1,n](it)| \,\big| \leq |e^{it} Q[n+1,n](it) - P[n+1,n](it)| = O(|t|^{2n+2});$$

that is,

$$|Q[n+1,n](it)| - |P[n+1,n](it)| = O(|t|^{2n+2}).$$

[9]The Routh-Hurwitz criterion states that a real polynomial $a_0 x^n + a_1 x^{n-1} + \cdots + a_n$, $a_0 > 0$, has all its zeros in the left half of the complex plane if and only if all leading principal minors of the nth-order Hurwitz matrix H are positive. Here the elements in the ith row of H are $a_{2-i}, a_{4-i}, \ldots, a_{2n-i}$ (where $a_k = 0$ if $k < 0$ or $k > n$).

Multiply this by $|Q[n+1,n](it)| + |P[n+1,n](it)|$ to obtain

$$|Q[n+1,n](it)|^2 - |P[n+1,n](it)|^2 = O(|t|^{2n+2}), \qquad t \to 0, \tag{5.18}$$

where the order term is unaffected since both Q and P have the value 1 at $t = 0$ (cf. (5.13) and (5.14)). But $|P[n+1,n](it)|^2 = P[n+1,n](it) \cdot P[n+1,n](-it)$ is a polynomial of degree n in t^2, and similarly, $|Q[n+1,n](it)|^2$ a polynomial of degree $n+1$ in t^2. Thus, (5.18) can hold only if

$$|Q[n+1,n](it)|^2 - |P[n+1,n](it)|^2 = at^{2n+2},$$

where at^{2n+2} is the leading term in $|Q[n+1,n](it)|^2$, hence $a > 0$. From this the assertion follows immediately.

(5) *For each* $n = 0, 1, 2, \dots,$ *all zeros of* $Q[n+1,n]$ *have positive real parts.* We sketch the proof. From (5.13) and (5.14) one notes that

$$Q[n+1,n](z) + P[n+1,n](-z) = 2Q[n+1,n+1](z).$$

We now use Rouché's theorem to show that $Q[n+1,n]$ and $Q[n+1,n+1]$ have the same number of zeros with negative real part (namely, none according to property (2)). For this, one must show that on the contour C_R: $\{z = it, |t| \leq R, t \neq 0\} \cup \{|z| = R, \text{Re } z < 0\}$ (see Figure 5.5.1) one has, when R

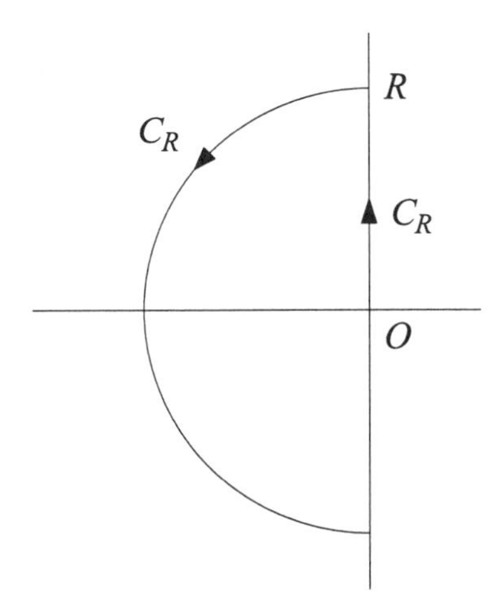

FIGURE 5.5.1. The contour C_R

is sufficiently large,

$$|P[n+1,n](-z)| < |Q[n+1,n](z)|, \qquad Q[n+1,n](0) \neq 0.$$

For $z = it$, since $P[n+1,n](-it) = \overline{P[n+1,n](it)}$, the inequality follows from property (4). For $|z| = R$, the inequality holds for $R \to \infty$, since $\deg P[n+1,n] < \deg Q[n+1,n]$. Finally $Q[n+1,n](0) = 1$.

Note also from (4) that $Q[n+1,n]$ cannot have purely imaginary zeros, since $|Q[n+1,n](it)| > |P[n+1,n](it)| \geq 0$ for $t \neq 0$.

(6) *A rational function R satisfies $|R(z)| < 1$ for $\operatorname{Re} z < 0$ if and only if R is analytic in $\operatorname{Re} z < 0$ and $|R(z)| \leq 1$ for $\operatorname{Re} z = 0$.*

`Necessity`. If $|R(z)| < 1$ in $\operatorname{Re} z < 0$, there can be no pole in $\operatorname{Re} z \leq 0$ or at $z = \infty$. By continuity, therefore, $|R(z)| \leq 1$ on $\operatorname{Re} z = 0$.

`Sufficiency`. R must be analytic in $\operatorname{Re} z \leq 0$ and at $z = \infty$. Clearly, $\lim_{z \to \infty} |R(z)| \leq 1$ and $|R(z)| \leq 1$ on the imaginary axis. Then, by the maximum principle, $|R(z)| < 1$ for $\operatorname{Re} z < 0$.

(7) As a corollary of property (6), we state the following important properties. *For each $n = 0, 1, 2, \ldots,$ there holds*

$$\left| \frac{P[n,n](z)}{Q[n,n](z)} \right| < 1 \quad \textit{for} \quad \operatorname{Re} z < 0, \tag{5.19}$$

$$\left| \frac{P[n+1,n](z)}{Q[n+1,n](z)} \right| < 1 \quad \textit{for} \quad \operatorname{Re} z \leq 0, \quad z \neq 0. \tag{5.20}$$

The first of these inequalities follows from properties (1) through (3), the second from properties (4) and (5).

Property (7) immediately yields the following theorem.

Theorem 5.5.2. *If the function φ associated with the one-step method Φ according to (5.4) is either the Padé approximant $\varphi(z) = R[n,n](z)$ of e^z, or the Padé approximant $\varphi(z) = R[n+1,n](z)$ of e^z, $n = 0, 1, 2, \ldots,$ then the method Φ is A-stable.*

§5.3. Examples of A-stable one-step methods

(1) *Implicit Euler method.* Also called the *backward* Euler method, this is the one-step method defined by

$$u_{n+1} = u_n + hf(x_{n+1}, u_{n+1}). \tag{5.21}$$

It requires, at each step, the solution of a system of (in general) nonlinear equations for $u_{n+1} \in \mathbb{R}^d$. In the case of the model problem (5.1), this becomes $u_{n+1} = u_n + hAu_{n+1}$ and can be solved explicitly: $u_{n+1} = (I - hA)^{-1}u_n$. Thus, the associated function φ here is

$$\varphi(z) = \frac{1}{1-z} = 1 + z + z^2 \dots \,, \tag{5.22}$$

the Padé approximant $R[1,0](z)$ of e^z. Since $\varphi(z) - e^z = O(z^2)$ as $z \to 0$, the method has order $p = 1$ (cf. (5.6)), and by Theorem 5.5.2 is A-stable. (This could easily be confirmed directly.)

(2) *Trapezoidal rule.* Here

$$u_{n+1} = u_n + \tfrac{1}{2}\, h\, [f(x_n, u_n) + f(x_{n+1}, u_{n+1})], \tag{5.23}$$

again a nonlinear equation in u_{n+1}. For (5.1), this becomes $u_{n+1} = \left(I + \frac{1}{2}\, hA\right) \cdot u_n + \frac{1}{2}\, hAu_{n+1}$, hence $u_{n+1} = \left(I - \frac{1}{2}\, hA\right)^{-1} \left(I + \frac{1}{2}\, hA\right) u_n$, and

$$\varphi(z) = \frac{1 + \frac{1}{2}\, z}{1 - \frac{1}{2}\, z} = 1 + z + \frac{1}{2}\, z^2 + \frac{1}{4}\, z^3 + \cdots \,. \tag{5.24}$$

This is the Padé approximant $\varphi(z) = R[1,1](z)$ of e^z and $\varphi(z) - e^z = O(z^3)$, so that again the method is A-stable, but now of order $p = 2$.

(3) *Implicit Runge-Kutta formulae.* As mentioned in §2.5, an r-stage implicit Runge-Kutta formula has the form (cf. (2.30))

$$\begin{aligned} \Phi(x, y; h) &= \sum_{s=1}^{r} \alpha_s k_s(x, y; h), \\ k_s &= f(x + \mu_s h, y + h \sum_{j=1}^{r} \lambda_{sj} k_j), \qquad s = 1, 2, \dots, r. \end{aligned} \tag{5.25}$$

It is possible to show (cf. Notes to §2.5) that (5.25) is a method of order p, $r \le p \le 2r$, if $f \in C^p$ on $[a,b] \times \mathbb{R}^d$ and

$$\sum_{j=1}^{r} \lambda_{sj} \mu_j^k = \frac{\mu_s^{k+1}}{k+1}\,, \qquad k = 0, 1, \dots, r-1; \quad s = 1, 2, \dots, r, \tag{5.26}$$

$$\sum_{s=1}^{r} \alpha_s \mu_s^k = \frac{1}{k+1}\,, \qquad k = 0, 1, \dots, p-1. \tag{5.27}$$

For any set of *distinct* μ_j, and for each $s = 1, 2, \dots, r$, the equations (5.26) represent a system of linear equations for $\{\lambda_{sj}\}_{j=1}^{r}$ whose coefficient matrix is a Vandermonde matrix, hence nonsingular. It thus can be solved uniquely for the $\{\lambda_{sj}\}$. Both conditions (5.26) and (5.27) can be viewed more naturally in terms of quadrature formulae. Indeed, (5.26) is equivalent to

$$\int_0^{\mu_s} p(t)dt = \sum_{j=1}^{r} \lambda_{sj} p(\mu_j), \quad \text{all } \ p \in \mathbb{P}_{r-1}, \tag{5.26$'$}$$

whereas (5.27) means

$$\int_0^1 q(t)dt = \sum_{s=1}^{r} \alpha_s q(\mu_s), \quad \text{all } \ q \in \mathbb{P}_{p-1}. \tag{5.27$'$}$$

We know from Ch. 3, §2.2, that in (5.27′) we can indeed have $r \le p \le 2r$, the extreme values corresponding to Newton-Cotes formulae (with prescribed μ_s) and to the Gauss-Legendre formula on [0,1], where the μ_s are the zeros of the (shifted) Legendre polynomial of degree r. In the latter case we obtain a unique *r-stage Runge-Kutta formula of order $p = 2r$*. We now show that this Runge-Kutta method of maximum order $2r$ is also A-stable.

Instead of the system (5.1), we may as well consider a scalar equation

$$\frac{dy}{dx} = \lambda y, \tag{5.28}$$

to which (5.1) can be reduced by spectral decomposition (i.e., λ represents one of the eigenvalues of A). Applied to (5.28), the k_s in (5.25) must satisfy the linear system:

$$k_s = \lambda \left(y + h \sum_{j=1}^{r} \lambda_{sj} k_j \right);$$

that is (with $z = \lambda h$),

$$k_s - z \sum_{j=1}^{r} \lambda_{sj} k_j = \lambda y, \qquad s = 1, 2, \dots, r.$$

Let

$$d_r(z) = \begin{vmatrix} 1 - z\lambda_{11} & -z\lambda_{12} & \cdots & -z\lambda_{1r} \\ -z\lambda_{21} & 1 - z\lambda_{22} & \cdots & -z\lambda_{2r} \\ \cdots\cdots & \cdots\cdots & \cdot\cdot & \cdots\cdots \\ -z\lambda_{r1} & -z\lambda_{r2} & \cdots & 1 - z\lambda_{rr} \end{vmatrix},$$

$$d_{r,s}(z) = \begin{vmatrix} 1 - z\lambda_{11} & \cdots & 1 & \cdots & -z\lambda_{1r} \\ -z\lambda_{21} & \cdots & 1 & \cdots & -z\lambda_{2r} \\ \cdots\cdots & \cdots & \cdot & \cdots & \cdots\cdots \\ -z\lambda_{r1} & \cdots & 1 & \cdots & 1 - z\lambda_{rr} \end{vmatrix}, \qquad s = 1, 2, \ldots, r,$$

where the column of 1s is the sth column of the determinant. Clearly, d_r and $d_{r,s}$ are polynomials of exact degree r and $r-1$, respectively (if, as we assume, $\lambda_{jj} \neq 0$ for all j). By Cramer's rule,

$$k_s = \frac{d_{r,s}(z)}{d_r(z)}\,\lambda y, \qquad s = 1, 2, \ldots, r,$$

so that

$$y_{\text{next}} = y + h\sum_{s=1}^{r} \alpha_s k_s = \left\{1 + z\sum_{s=1}^{r} \alpha_s \frac{d_{r,s}(z)}{d_r(z)}\right\} y.$$

Thus, the function φ associated with the method Φ of (5.25) is

$$\varphi(z) = \frac{d_r(z) + z\sum_{s=1}^{r} \alpha_s d_{r,s}(z)}{d_r(z)}. \tag{5.29}$$

We see that φ is a rational function of type $[r, r]$ and, the method Φ having order $p = 2r$, we have (cf. (5.6))

$$e^z = \varphi(z) + O(z^{2r+1}), \qquad z \to 0.$$

It follows that φ in (5.29) is the Padé approximant $R[r, r]$ to the exponential function, and hence Φ is A-stable by Theorem 5.5.2.

(4) *Ehle's method.* This is a method involving total derivatives of f (cf. (2.8)),

$$\begin{aligned} \Phi(x, y; h) &= k(x, y; h), \\ k &= \sum_{s=1}^{r} h^{s-1}[\alpha_s f^{[s-1]}(x, y) - \beta_s f^{[s-1]}(x + h, y + hk)]. \end{aligned} \tag{5.30}$$

It is also implicit, as it requires the solution of the second equation in (5.30) for the vector $k \in \mathbb{R}^d$. A little computation (see Ex. 23) will show that the function φ associated with Φ in (5.30) is given by

$$\varphi(z) = \frac{1 + \sum_{s=1}^{r} \alpha_s z^s}{1 + \sum_{s=1}^{r} \beta_s z^s}. \tag{5.31}$$

By choosing this φ to be a Padé approximant to e^z, either $R[r,r]$ or $R[r,r-1]$ (by letting $\alpha_r = 0$), we again obtain two A-stable methods. The latter has the additional property of being *strongly* A-stable (or *L-stable*), in the sense that

$$\varphi(z) \to 0 \quad \text{as} \quad \text{Re } z \to -\infty. \tag{5.32}$$

This means, in view of (5.7), that convergence $u_n \to 0$ as $n \to \infty$ is faster for components corresponding to eigenvalues further to the left in the complex plane.

§5.4. **Regions of absolute stability**. For methods Φ that are *not* A-stable, it is important to know the *region of absolute stability*,

$$\mathcal{D}_A = \{z \in \mathbb{C} : \quad |\varphi(z)| < 1\}. \tag{5.33}$$

If the method Φ applied to the model problem (5.1) is to produce an approximate solution $u = \{u_n\}$ with $\lim_{n\to\infty} u_n = 0$, it is necessary that $h\lambda_i(A) \in \mathcal{D}_A$ for all eigenvalues $\lambda_i(A)$ of A. If some of these have very large negative real parts, then this condition imposes a severe restriction on the step length h, unless $\mathcal{D}_A$ contains a large portion of the left-hand plane. For many classical methods, unfortunately, this is not the case. For Euler's method, for example, we have $\varphi(z) = 1 + z$, hence

$$\mathcal{D}_A = \{z \in \mathbb{C} : \quad |1 + z| < 1\} \qquad \text{(Euler)}, \tag{5.34}$$

and the region of absolute stability is the unit disk in $\mathbb{C}$ centered at -1. More generally, for the Taylor expansion method of order $p \geq 1$, and also for any p-stage explicit Runge-Kutta method of order p, $1 \leq p \leq 4$, one has (see Ex. 20)

$$\varphi(z) = 1 + \frac{1}{1!}\,z + \frac{1}{2!}\,z^2 + \cdots + \frac{1}{p!}\,z^p. \tag{5.35}$$

To compute the contour line $|\varphi(z)| = 1$, which delineates the region $\mathcal{D}_A$, one can find a differential equation for this line and use a one-step method to solve it (see Mach. Ass. 3). That is, we can use a method Φ to analyze its own stability region — a case of self-analysis, as it were. The results for φ in (5.35) and $p = 1, 2, \ldots, 21$ are plotted in Figure 5.5.2. [10]. Because of symmetry, only the parts of the regions in the upper half-plane are shown.

[10]Figure 5.5.2 is reproduced, with permission, from W. Gautschi and J. Waldvogel, *Contour plots of analytic functions*, in *Solving problems in scientific computing using Maple and MATLAB*, Walter Gander and Jiří Hřebíček, eds., 3d ed., Springer, Berlin, 1997, to appear.

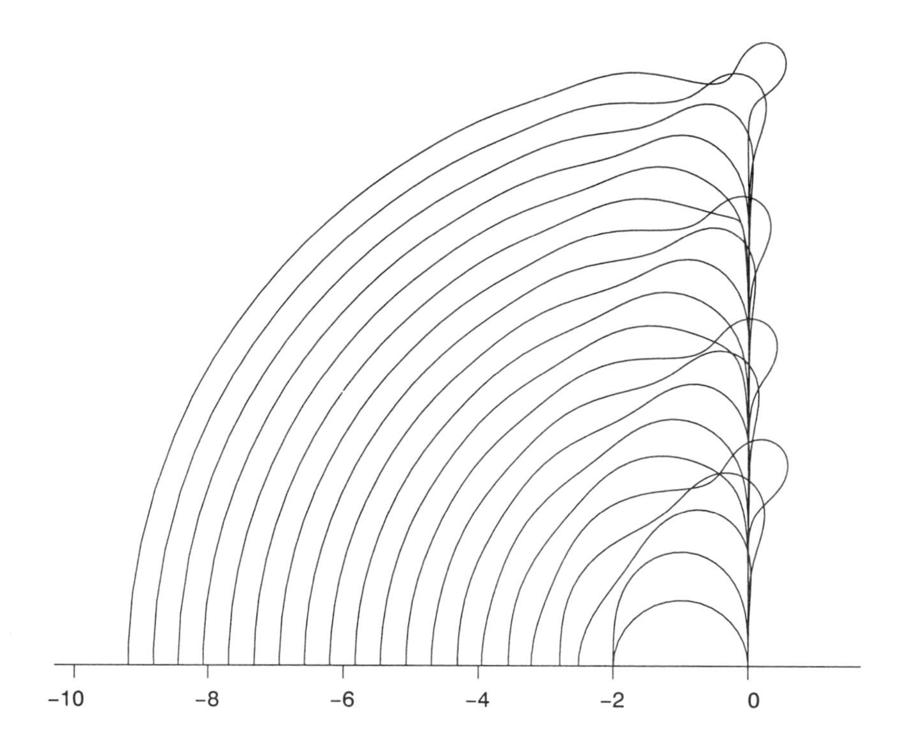

FIGURE 5.5.2. Regions of absolute stability for pth-order methods with φ as in (5.35), $p = 1, 2, \ldots, 21$.

NOTES TO CHAPTER 5

The classic text on the numerical solution of nonstiff ordinary differential equations is Henrici [1962]. It owes much to the pioneering work of Dahlquist [1956] a few years earlier. A number of books have since been written with varying areas of emphasis, but all paying attention to stiff problems. Among those giving an overview of the subject, we mention the older text by Gear [1971a], which has a distinctly pragmatic flavor and contains computer programs that have greatly influenced the subsequent development of software; the very balanced exposition in Lambert [1991]; and the two-volume work of Hairer, Nørsett, and Wanner [1993] and Hairer and Wanner [1996], which is especially rich in historical comments, interesting examples, and numerical experimentation based on FORTRAN codes that are provided in the appendices. The book by Butcher [1987] is an authoritative source on Runge-Kutta methods and contains an extensive bibliography of several thousand items covering the literature up to 1982. Error and stability analyses of Runge-Kutta methods in the context of stiff nonlinear differential equations are the subject of the monograph by Dekker and Verwer [1984]. Texts devoted to general multistep and discretization methods are Henrici [1977] and Stetter [1973]. Shampine and Gordon [1975] is basically a study of the Adams method, complete with many practical details of implementation and computer codes. A more recent general text emphasizing software issues and the practical use of software, and containing many instructive case studies thereof, is Shampine [1994]. Particularly valuable is the survey given in Chapter 3, §5, on quality codes available and the sources from which they can be acquired (cheaply or for free). Another recent text, Iserles [1996], also includes topics in the numerical analysis of partial differential equations.

§0.1. The example (0.3) is from Klopfenstein [1965]. The technique described therein is widely used in codes calculating trajectories of space vehicles. For a detailed presentation of the method of lines, including FORTRAN programs, see Schiesser [1991].

§0.2. There are still other important types of differential equations, for example, singularly perturbed equations and related differential algebraic equations (DAEs), and differential equations with delayed arguments. For the former, we refer to Griepentrog and März [1986], Brenan, Campbell, and Petzold [1996], Hairer, Lubich, and Roche [1989], and Hairer and Wanner [1996, Chs. 6–7], for the latter to Pinney [1958], Bellman and Cooke [1963], Cryer [1972], Driver [1977], and Kuang [1993].

§0.3. For a proof of the existence and uniqueness part of Theorem 5.0.1, see Henrici [1962, §1.2]. The proof is given for a scalar initial value problem, but it extends readily to systems. Continuity with respect to initial data is proved, for example, in Coddington and Levinson [1955, Ch. 1, §7]. Also see Butcher [1987, §112].

A strengthened version of Theorem 5.0.1 involves a one-sided Lipschitz condition,

$$[f(x,y) - f(x,y^*)]^T (y - y^*) \le \lambda \, \|y - y^*\|^2, \quad \text{all } x \ge a, \quad \text{all } y, y^* \in \mathbb{R}^d,$$

where λ is some constant — the one-sided Lipschitz constant. If this holds, and f is continuous in x, then the initial value problem (0.14) has a unique solution on any interval $[a,b]$, $b > a$ (cf. Butcher [1987, §112]).

§0.4. The numerical solution of differential equations can sometimes benefit from a preliminary transformation of variables. Many examples are given in Daniel and Moore [1970, Part 3]; an important example used in celestial mechanics to regularize and linearize the Newtonian equations of motion are the transformations of Levi-Civita, and of Kustaanheimo and Stiefel, for which we refer to Stiefel and Scheifele [1971] for an extensive treatment. The reader may wish to also consult Zwillinger [1992b] for a large number of other analytical tools that may be helpful in the numerical solution of differential equations.

The distinction between one-step and multistep methods may be artificial, as there are theories that allow treating them both in a unified manner; see, for example, Stetter [1973, Ch.5], Butcher [1987, Ch. 4], or Hairer, Nørsett, and Wanner [1993, Ch. 3, §8]. We chose, however, to cover them separately for didactical reasons.

There are many numerical methods in use that are not discussed in our text. Among the more important ones are the extrapolation methods of Gragg and of Gragg, Bulirsch, and Stoer, which extend the ideas in Ch. 3, §2.7, to differential equations. For these, we refer to Stetter [1973, §6.3] and Hairer, Nørsett, and Wanner [1993, Ch. 2, §§8,9]. (Extrapolation methods for stiff problems are discussed in Hairer and Wanner [1996, Ch. 4, §9].) Both these texts also contain accounts of multistep methods involving derivatives, and of Nordsieck-type methods, which carry along not only function values but also derivative values from one step to the next. There are also methods tailored to higher-order systems of differential equations; for second-order systems, for example, see Hairer, Nørsett, and Wanner [1993, Ch. 2, §14]. Very recently, so-called symplectic methods have created a great deal of interest, especially in connection with Hamiltonian systems. These are numerical methods that preserve invariants of the given differential system; cf. Sanz-Serna and Calvo [1994].

§1. A nonstiff initial value problem (0.14) on $[a, b]$, where b is very large, is likely one that also has a very small Lipschitz constant. The problem, in this case, is not properly scaled, and one should transform the independent variable x, for example by letting $x = (1-t)a + tb$, to get an initial value problem on $[0, 1]$, namely, $dz/dt = g(t, z)$, $0 \le t \le 1$, $z(0) = y_0$, where $z(t) = y((1-t)a + tb)$ and $g(t, z) := (b-a)f((1-t)a + tb, z)$. If f has a (very small) Lipschitz constant L, then g has the Lipschitz constant $(b-a)L$, which may well be of more reasonable size.

§2.1. In the spirit of Laplace's exhortation "*Lisez Euler, lisez Euler, c'est notre maître à tous*!," the reader is encouraged to look at Euler's original account of his method in Euler [1768, §650]. Even though it is written in Latin, Euler's use of this language is plain and simple.

§2.2. The method of Taylor expansion was also proposed by Euler [op. cit., §656]. It has long been perceived as being too cumbersome in practice, but recent advances in automatic differentiation helped to revive interest in this method. Codes have been written that carry out the necessary differentiations systematically by recursion (Gibbons [1960] and Barton, Willers, and Zahar [1971]). Combining these techniques with interval arithmetic, as in Moore [1979, §3.4], also provides rigorous error bounds.

§2.5. For the results in (2.32), see Butcher [1965]. Actually, $p^*(10) = 7$, as was shown more recently in Butcher [1985]. The highest order of an explicit Runge-Kutta method ever constructed is $p = 10$ (Hairer [1978]).

It has become customary to associate with the general r-stage Runge-Kutta method (2.30) the array

$$\begin{array}{c|cccc} \mu_1 & \lambda_{11} & \lambda_{12} & \cdots & \lambda_{1r} \\ \mu_2 & \lambda_{21} & \lambda_{22} & \cdots & \lambda_{2r} \\ \vdots & \vdots & \vdots & & \vdots \\ \mu_r & \lambda_{r1} & \lambda_{r2} & \cdots & \lambda_{rr} \\ \hline & \alpha_1 & \alpha_2 & \cdots & \alpha_r \end{array} \qquad \left(\text{in matrix form: } \begin{array}{c|c} \mu & \Lambda \\ \hline & \alpha^{\mathrm{T}} \end{array} \right),$$

called the Butcher array. For an explicit method, $\mu_1 = 0$ and Λ is lower triangular with zeros on the diagonal. With the first r rows of the Butcher array, we may associate the quadrature rules $\int_0^{\mu_s} u(t)dt \approx \sum_{j=1}^r \lambda_{sj} u(\mu_j)$, $s = 1, 2, \ldots, r$, and with the last row the rule $\int_0^1 u(t)dt \approx \sum_{s=1}^r \alpha_s u(\mu_s)$. If the respective degrees of exactness are $d_s = q_s - 1$, $1 \le s \le r+1$ ($d_s = \infty$ if $\mu_s = 0$ and all $\lambda_{sj} = 0$), then by the Peano representation of the error functionals (cf. Ch. 3, Eq. (2.58)) the remainder terms involve derivatives of u of order q_s, and hence, setting $u(t) = y'(x + th)$, one gets

$$\frac{y(x+\mu_s h) - y(x)}{h} - \sum_{j=1}^r \lambda_{sj} y'(x + \mu_j h) = O(h^{q_s}), \quad s = 1, 2, \ldots, r,$$

and

$$\frac{y(x+h) - y(x)}{h} - \sum_{s=1}^r \alpha_s y'(x + \mu_s h) = O(h^{q_{r+1}}).$$

The quantity $q = \min(q_1, q_2, \ldots, q_r)$ is called the stage order of the Runge-Kutta formula, and q_{r+1} the quadrature order.

High-order r-stage implicit Runge-Kutta methods have the property that, when $f(x,y) = f(x)$, they reduce to r-point Gauss-type quadrature formulae, either the Gauss formula proper, or the Gauss-Radau ($\mu_0 = 0$ or $\mu_r = 1$) or Gauss-Lobatto ($\mu_0 = 0$ and $\mu_r = 1$) formula; see, for example, Dekker and Verwer [1984, §3.3], Butcher [1987, §34], Lambert [1991, §5.11], and Hairer and Wanner [1996, Ch. 4, §5]. They can be constructed to have order $2r$, and (in a variety of ways) orders $2r-1$ and $2r-2$, respectively; cf. also §5.3(3). Another interesting way of constructing implicit Runge-Kutta methods is by collocation: define $p \in \mathbb{P}_r$ to be such that $p(x) = y$, $p'(x + \mu_s h) = f(x + \mu_s h, p(x + \mu_s h))$, $s = 1, 2, \ldots, r$ (cf. Ch. 2, Ex. 60), and let $y_{\text{next}} = p(x+h)$. It has been shown by Wright [1970] (also cf. Butcher [1987, §346]) that this indeed is an implicit r-stage method — a collocation method, as it is called. For such methods the stage orders are at least r. This property characterizes collocation methods of orders $\ge r$ (with distinct μ_s); see Hairer, Nørsett, and Wanner [1993, Thm. 7.8, p. 212]. The order of the method is $p = r + k$, $k \ge 0$, if the quadrature order is p (cf. [loc. cit., Ch. 2, Thm. 7.9] and Theorem 3.2.1 in Ch. 3). Some (but not all) of the Gauss-type methods previously mentioned are collocation methods. If the polynomial $p(x + th)$ is determined explicitly (not just $p(x+h)$), it provides a means of computing intermediate approximations for arbitrary t with $0 < t < 1$, giving rise to a "continuous" implicit Runge-Kutta method.

Semi-implicit Runge-Kutta methods with all diagonal elements of Λ in the Butcher array being the same nonzero real number are called DIRK methods

(Diagonally Implicit Runge-Kutta); see Nørsett [1974], Crouzeix [1976], and Alexander [1977]. SIRK methods (Singly-Implicit Runge-Kutta) are fully implicit methods which share with DIRK methods the property that the matrix Λ (though not triangular) has one single real eigenvalue of multiplicity r. These were derived by Nørsett [1976] and Burrage [1978a], [1978b], [1982]. DIRK methods with r stages have maximum order $r+1$, but these are difficult to derive for large r, in contrast to SIRK methods (see Dekker and Verwer [1984, §§3.5 and 3.6]).

The best source for Butcher's theory of Runge-Kutta methods and their attainable orders is Butcher [1987, Ch.3, §§30–34]. A simplified version of this theory can be found in Lambert [1991, Ch. 5] and an alternative approach in Albrecht [1987], [1996]. It may be worth noting that the order conditions for a system of differential equations are not necessarily identical with those for a single differential equation. Indeed, a Runge-Kutta method for a scalar equation may have order $p > 4$, whereas the same method applied to a system has order $< p$. (For $p \leq 4$ this phenomenon does not occur.) Examples of explicit Runge-Kutta formulae of orders 5–8 are given in Butcher [1987, §33].

An informative cross-section of contemporary work on the Runge-Kutta method, as well as historical essays, celebrating the centenary of Runge's 1895 paper, can be found in Butcher [1996].

§3.1. The concept of stability as defined in this section is from Keller [1992, §1.3]. It is also known as zero-stability (relating to $h \to 0$) to distinguish it from other stability concepts used in the context of stiff differential equations; for the latter, see the Notes to §5.1.

§3.2. Theorem 5.3.2 admits a converse if one assumes Φ continuous and satisfying a Lipschitz condition (3.14); that is, consistency is then also necessary for convergence (cf. Henrici [1962, Thm. 3.2]).

§3.3. Theorem 5.3.3 is due independently to Henrici [1962, Thm. 3.4] and Tihonov and Gorbunov [1963], [1964]. Henrici deals also with variable steps in the form alluded to at the beginning of this section, whereas Tihonov and Gorbunov [1964] deal with arbitrary nonuniform grids.

§4.1. Although the idea of getting global error estimates by integrating the variational equation along with the main differential equation has already been expressed by Henrici [1962, p. 81], its precise implementation as in Theorem 5.4.1 is carried out in Gautschi [1975b].

§4.2(2). Fehlberg's embedded (4,5) method with $r^* = 6$ (cf. Fehlberg [1969], [1970]) appears to be a popular method. (It is one of two options provided in MATLAB, the other being a (2,3) pair.) A similar method due to England [1969/1970] has the advantage of possessing coefficients $\alpha_5 = \alpha_6 = 0$, which makes occasional error monitoring more efficient. All of Fehlberg's methods of orders $p \geq 5$ have the (somewhat disturbing) peculiarity of yielding zero error estimates in cases where f does not depend on y. High-order pairs of methods not suffering from this defect have been derived in Verner [1978]. Variable-method codes developed, for example, in Shampine and Wisniewski [1978], use pairs ranging from (3,4) to (7,8). Instead of optimizing the truncation error in the lower-order method of a pair, as was done by Fehlberg, one can do the same with the higher-order method and use the other only for step control. This is the approach taken by Dormand and Prince [1980] and Prince and Dormand [1981]. Their (4,5) and (7,8) pairs appear to be among current state-of-the-art choices (cf. Hairer, Nørsett, and Wanner [1993, Ch. 2, §10] and the appendix of this reference for codes).

§4.3. For a proof of Theorem 5.4.2, see, for example, Butcher [1987, Thm. 112J].

§5. The major text on stiff problems is Hairer and Wanner [1996]. For Runge-Kutta methods, also see Dekker and Verwer [1984].

§5.1. The function $\varphi(z)$ for a general Runge-Kutta method can be expressed in terms of the associated Butcher array as $\varphi(z) = 1 + z\alpha^T(I - z\Lambda)^{-1}e$, where $e^T = [1, 1, \ldots, 1]$ or, alternatively, as $\det(I - z\Lambda + ze\alpha^T)/\det(I - z\Lambda)$; see, for example, Dekker and Verwer [1984, §3.4], Lambert [1991, §5.12], and Hairer and Wanner [1996, Ch. 4, §3]. Thus, φ is a rational function if the method is (semi-) implicit, and a polynomial otherwise.

The concept of A-stability, which was introduced by Dahlquist [1963], can be relaxed by requiring $|\varphi(z)| \le 1$ to hold only in an unbounded subregion S of the left half-plane. Widlund [1967], in the context of multistep methods (cf. Ch. 6, §5.2), for example, takes for S an angular region $|\arg(-z)| \le \alpha$, where $\alpha < \frac{1}{2}\pi$, and speaks of A(α)-stability, whereas Gear [1971a, §11.1] takes the union of some half-plane $\operatorname{Re} z \le \rho < 0$ and a rectangle $\rho \le \operatorname{Re} z \le 0$, $|\operatorname{Im} z| \le \sigma$, and speaks of stiff stability. Other stability concepts relate to more general test problems, some linear and some nonlinear. Of particular interest among the latter are initial value problems (0.14) with f satisfying a one-sided Lipschitz condition with constant $\lambda = 0$. These systems are dissipative in the sense that $\|y(x) - z(x)\|$ is nonincreasing for $x > a$ for any two solutions y, z of the differential equation. Requiring the same to hold for any two numerical solutions u, v generated by the one-step method, that is, requiring that $\|u_{n+1} - v_{n+1}\| \le \|u_n - v_n\|$ for all $n \ge 0$, gives rise to the concept of B-stability (or BN-stability with the "N" standing for "nonautonomous"). The implicit r-stage Runge-Kutta method of order $p = 2r$ (cf. §5.3(3)), for example, is B-stable, and so are some of the other Gauss-type Runge-Kutta methods; see Dekker and Verwer [1984, §4.1], Butcher [1987, §356], and Hairer and Wanner [1996, Ch. 4, §§12 and 13]. Another family of Runge-Kutta methods that are B-stable are the so-called algebraically stable methods, that is, methods which satisfy $D = \operatorname{diag}(\alpha_1, \alpha_2, \ldots, \alpha_r) \ge 0$ and $D\Lambda + \Lambda^T D - \alpha\alpha^T$ nonnegative definite (Dekker and Verwer [1984, §4.2], Butcher [1987, §356], and Hairer and Wanner [1996, Ch. 4, §§12 and 13]). Similar stability concepts can also be associated with test equations satisfying one-sided Lipschitz conditions with constants $\lambda \ne 0$ (Dekker and Verwer [1984, §5.11], Butcher [1987, §357], and Hairer and Wanner [1996, Ch. 4, pp. 193ff]).

§5.2. Standard texts on Padé approximation are Baker [1975] and Baker and Graves-Morris [1996]. The proof of Theorem 5.5.1 follows Perron [1957, §42], who in turn took it from Padé. For a derivation of the Routh-Hurwitz criterion mentioned in (2), see, for example, Marden [1966, Cor. 40,2], and for Rouché's theorem, Henrici [1988, p. 280]. The elegant argument used in the proof of (4) is due to Axelsson [1969].

§5.3. For the study of A-stability it is easier to work with the "relative stability function" $\varphi(z)e^{-z}$ than with $\varphi(z)$ directly. This gives rise to the "order star" theory of Wanner, Hairer, and Nørsett [1978], which, among other things, made it possible to prove that the only Padé approximants to e^z that yield A-stable methods are $\varphi(z) = R[n+k, n](z)$, $n = 0, 1, 2, \ldots$, with $0 \leq k \leq 2$. All Gauss-type Runge-Kutta methods in current use have stability functions given by such Padé approximants and are thus A-stable (Dekker and Verwer [1984, §3.4]). Also see Iserles and Nørsett [1991], and Hairer and Wanner [1996, Ch. 4, §4] for further applications of order stars.

In addition to the implicit Gauss-type Runge-Kutta methods mentioned in (3), there are also A-stable DIRK and SIRK methods; for their construction, see Butcher [1987, §353] and Hairer and Wanner [1996, Ch. 4, §6]. Another class of methods that are A-stable, or nearly so, are basically explicit Runge-Kutta methods that make use of the Jacobian matrix and inverse matrices involving it. They are collectively called Runge-Kutta-Rosenbrock methods; see, for example, Dekker and Verwer [1984, Ch.9] and Hairer and Wanner [1996, Ch. 4, §7].

In all the results previously described it was tacitly assumed that the nonlinear systems of equations to be solved in an implicit Runge-Kutta method have a unique solution. This is not necessarily the case, not even for linear differential equations with constant coefficient matrix. Such questions of existence and uniqueness are considered in Dekker and Verwer [1984, Ch. 5], where one also finds a discussion of, and references to, the efficient implementation of implicit Runge-Kutta methods. Also see Hairer and Wanner [1996, Ch. 4, §§14, 8].

The theory of consistency and convergence developed in §3 for nonstiff problems must be modified when dealing with stiff sytems of nonlinear differential equations. One-sided Lipschitz conditions are then the natural vehicles, and B-consistency and B-convergence the relevant concepts; see Dekker and Verwer [1984, Ch. 7] and Hairer and Wanner [1996, Ch. 4, §15].

§5.4. Regions of absolute stability for the embedded Runge-Kutta pairs 4(5) and the 7(8) pairs of Dormand and Prince (cf. Notes to §4.2(2)), and for the Gragg, Bulirsch, and Stoer extrapolation method (cf. Notes to §0.4), are shown in Hairer and Wanner [1996, Ch. 4, §2]. Attempts to construct explicit Runge-Kutta formulae whose regions of absolute stability, along the negative real axis, extend to the left as far as possible lead to interesting applications of Chebyshev polynomials; see Hairer and Wanner [1996, pp. 31–36].

EXERCISES AND MACHINE ASSIGNMENTS TO CHAPTER 5

EXERCISES

1. Consider the initial value problem

$$\frac{dy}{dx} = \kappa(y + y^3), \quad 0 \le x \le 1; \quad y(0) = s,$$

where $\kappa > 0$ (in fact, $\kappa \gg 1$) and $s > 0$. Under what conditions on s does the solution $y(x) = y(x; s)$ exist on the whole interval $[0, 1]$? {*Hint*: Find y explicitly.}

2. Prove (2.9).

3. Prove

$$(f_y f)_y f = f^T f_{yy} f + f_y^2 f.$$

4. Let

$$f(x, y) = \begin{bmatrix} f^1(x, y) \\ f^2(x, y) \\ \vdots \\ f^d(x, y) \end{bmatrix}$$

be a C^1 map from $[a, b] \times \mathbb{R}^d$ to $\mathbb{R}^d$. Assume that

$$\left| \frac{\partial f^i(x, y)}{\partial y^j} \right| \le M_{ij} \quad \text{on} \quad [a, b] \times \mathbb{R}^d, \quad i, j = 1, 2, \ldots, d,$$

and let $M = [M_{ij}] \in \mathbb{R}_+^{d \times d}$. Determine a Lipschitz constant L of f

(a) in the L_1 vector norm;

(b) in the L_2 vector norm;

(c) in the L_∞ vector norm.

Express L, if possible, in terms of a matrix norm of M.

5. (a) Write the system of differential equations

$$\begin{aligned} u''' &= x^2 u u'' - u v', \\ v'' &= x v v' + 4u' \end{aligned}$$

as a first-order system of differential equations, $y' = f(x, y)$.

(b) Determine the Jacobian matrix $f_y(x, y)$ for the system in (a).

(c) Determine a Lipschitz constant L for f on $[0, 1] \times \mathcal{D}$, where $\mathcal{D} = \{y \in \mathbb{R}^d : \|y\|_1 \le 1\}$, using, respectively, the L_1, L_2, and L_∞ norms (cf. Ex. 4).

6. Consider the surface

$$F(x_1, x_2, \ldots, x_n, y) := y - f(x_1, x_2, \ldots, x_n) = 0$$

in $\mathbb{R}^{n+1}$. Suppose a particle of mass m is placed on the surface, and let it move under the action of the gravitational force (directed along the negative y-axis) and the surface constraint force (in direction of the normal to the surface). Derive a system of differential equations for the projection $x^T = [x_1, x_2, \ldots, x_n]$ into the hyperplane $y = 0$ of the particle's trajectory, assuming Newton's law as the governing law of motion. (Since the solution trajectory started at any point on the surface is expected to pass through a local minimum of f, integrating the system of differential equations provides a method for finding a local minimum.)

7. For the (scalar) differential equation

$$\frac{dy}{dx} = y^\lambda, \quad \lambda > 0,$$

(a) determine the principal error function of the general explicit two-stage Runge-Kutta method;

(b) compare the local accuracy of the modified Euler method with that of Heun's method;

(c) determine a λ-interval such that for each λ in this interval, there is a two-stage explicit Runge-Kutta method of order $p = 3$ having parameters $0 < \alpha_1 < 1$, $0 < \alpha_2 < 1$, and $0 < \mu < 1$.

8. For the implicit Euler method

$$y_{\text{next}} = y + hf(x + h, y_{\text{next}}),$$

(a) state a condition under which y_{next} is uniquely defined;

(b) determine the order and principal error function.

9. Show that any explicit two-stage Runge-Kutta method of order $p = 2$ integrates the special differential equation $dy/dx = f(x)$, $f \in \mathbb{P}_1$, exactly.

10. The (scalar) second-order differential equation

$$\frac{d^2 z}{dx^2} = g(x, z),$$

in which g does not depend on dz/dx, can be written as a first-order system

$$\frac{d}{dx}\begin{bmatrix} y^1 \\ y^2 \end{bmatrix} = \begin{bmatrix} y^2 \\ g(x, y^1) \end{bmatrix}$$

by letting, as usual, $y^1 = z$, $y^2 = dz/dx$. For this system, consider a one-step method $u_{n+1} = u_n + h\Phi(x_n, u_n; h)$ with

$$\Phi(x, y; h) = \begin{bmatrix} y^2 + \frac{1}{2}hk(x, y; h) \\ k(x, y; h) \end{bmatrix}, \quad k = g(x + \mu h, y^1 + \mu h y^2), \quad y = \begin{bmatrix} y^1 \\ y^2 \end{bmatrix}.$$

(Note that this method requires only one evaluation of g per step.)

(a) Can the method be made to have order $p = 2$, and if so, for what value(s) of μ?

(b) Determine the principal error function of any method obtained in (a).

11. Consider the initial value problem

$$y'' = \cos(xy), \quad y(0) = 1, \quad y'(0) = 0, \quad 0 \leq x \leq 1.$$

(a) Does the solution $y(x)$ exist on the whole interval $0 \leq x \leq 1$? Explain.

(b) Describe in detail the generic step of the classical fourth-order Runge-Kutta method applied to this problem.

12. Show that the first condition in (2.29) is equivalent to the condition that

$$k_s(x, y; h) = u'(x + \mu_s h) + O(h^2), \quad s \geq 2,$$

where $u(t)$ is the reference solution through the point (x, y).

13. Suppose that

$$\int_x^{x+h} z(t)dt = h \sum_{k=1}^{\nu} w_k z(x + \vartheta_k h) + ch^{\mu+1} z^{(\mu)}(\xi)$$

is a quadrature formula with $w_k \in \mathbb{R}$, $\vartheta_k \in [0, 1]$, $c \neq 0$, and $\xi \in (x, x+h)$, for z sufficiently smooth. Given increment functions $\overline{\Phi}_k(x, y; h)$ defining methods of order $\overline{p}_k$, $k = 1, 2, \ldots, \nu$, show that the one-step method defined by

$$\Phi(x, y; h) = \sum_{k=1}^{\nu} w_k f(x + \vartheta_k h, y + \vartheta_k h \overline{\Phi}_k(x, y; \vartheta_k h))$$

has order p at least equal to $\min(\mu, \overline{p} + 1)$, where $\overline{p} = \min \overline{p}_k$.

14. Let $g(x, y) = (f_x + f_y f)(x, y)$. Show that the one-step method defined by the increment function

$$\Phi(x, y; h) = f(x, y) + \tfrac{1}{2}hg(x + \tfrac{1}{3}h, y + \tfrac{1}{3}hf(x, y))$$

has order $p = 3$. Express the principal error function in terms of g and its derivatives.

15. Let $f(x, y)$ satisfy a Lipschitz condition in y on $[a, b] \times \mathbb{R}^d$, with Lipschitz constant L.

    (a) Show that the increment function Φ of the second-order Runge-Kutta method

$$\begin{aligned} k_1 &= f(x, y), \\ k_2 &= f(x + h, y + hk_1), \\ \Phi(x, y; h) &= \tfrac{1}{2}(k_1 + k_2) \end{aligned}$$

    also satisfies a Lipschitz condition whenever $x+h \in [a, b]$, and determine a respective Lipschitz constant M.

    (b) What would the result be for the classical Runge-Kutta method?

    (c) What would it be for the general implicit Runge-Kutta method?

16. Describe the application of Newton's method to implement the implicit Runge-Kutta method.

17. Consider the following scheme of constructing an estimator $r(x, y; h)$ for the principal error function $\tau(x, y)$ of Heun's method:

$$\begin{aligned} k_1 &= f(x, y), \\ k_2 &= f(x + h, y + hk_1), \\ y_h &= y + \tfrac{1}{2}h(k_1 + k_2), \\ k_3 &= f(x + h, y_h), \\ k_4 &= f(x + h + \mu h, y_h + \mu h k_3), \\ r(x, y; h) &= h^{-2}(\beta_1 k_1 + \beta_2 k_2 + \beta_3 k_3 + \beta_4 k_4). \end{aligned}$$

    (Note that this scheme requires one additional function evaluation, k_4, beyond what would be required anyhow to carry out Heun's method.) Obtain the conditions on the parameters μ, β_1, β_2, β_3, β_4 in order that

$$r(x, y; h) = \tau(x, y) + O(h).$$

    Show, in particular, that there is a unique set of βs for any μ with $\mu(\mu+1) \neq 0$. What is a good choice of the parameters, and why?

18. Apply the asymptotic error formula (3.32) to the (scalar) initial value problem $dy/dx = \lambda y$, $y(0) = 1$, on $[0, 1]$, when solved by the classical fourth-order Runge-Kutta method. In particular, determine

$$\lim_{h \to 0} h^{-4} \frac{u_N - y(1)}{y(1)},$$

    where u_N is the Runge-Kutta approximation to $y(1)$ obtained with step $h = 1/N$.

19. Consider $y' = \lambda y$ on $[0, \infty)$ for complex λ with $\operatorname{Re}\lambda < 0$. Let $\{u_n\}$ be the approximations to $\{y(x_n)\}$ obtained by the classical fourth-order Runge-Kutta method with the step h held fixed. (That is, $x_n = nh$, $h > 0$, and $n = 0, 1, 2, \dots$.)

    (a) Show that $y(x) \to 0$ as $x \to \infty$, for any initial value y_0.

    (b) Under what condition on h can we assert that $u_n \to 0$ as $n \to \infty$? In particular, what is the condition if λ is real (negative)?

    (c) What is the analogous result for Euler's method?

    (d) Generalize to systems $y' = Ay$, where A is a constant matrix all of whose eigenvalues have negative real parts.

20. Show that any one-step method of order p, which, when applied to the model problem $y' = Ay$, yields

$$y_{\text{next}} = \varphi(hA)y, \qquad \varphi \text{ a polynomial of degree } q \geq p,$$

must have

$$\varphi(z) = 1 + z + \tfrac{1}{2!}z^2 + \cdots + \tfrac{1}{p!}z^p + z^{p+1}\chi(z),$$

where χ is identically zero if $q = p$ and a polynomial of degree $q - p - 1$ otherwise. In particular, show that $\chi \equiv 0$ for a p-stage explicit Runge-Kutta method of order p, $1 \leq p \leq 4$, and for the Taylor expansion method of order $p \geq 1$.

21. Consider the linear homogeneous system

$$(*) \qquad y' = Ay, \qquad y \in \mathbb{R}^d,$$

with constant coefficient matrix $A \in \mathbb{R}^{d \times d}$.

    (a) For Euler's method applied to (*), determine $\varphi(z)$ (cf. (5.4)) and the principal error function.

    (b) Do the same for the classical fourth-order Runge-Kutta method.

22. Consider the model equation

$$\frac{dy}{dx} = a(x)[y - b(x)], \quad 0 \leq x < \infty,$$

where $a(x)$, $b(x)$ are continuous and bounded on $\mathbb{R}_+$, and $a(x)$ negative with $|a(x)|$ large, say,

$$a \leq |a(x)| \leq A \ \text{ on } \mathbb{R}_+, \quad a \gg 1.$$

For the explicit and implicit Euler methods, derive a condition (if any) on the step length h that ensures boundedness of the respective approximations as $x_n = nh \to \infty$ for $h > 0$ fixed.

23. Consider the implicit one-step method

$$\Phi(x, y; h) = k(x, y; h),$$

where $k : [a, b] \times \mathbb{R}^d \times (0, h_0] \to \mathbb{R}^d$ is implicitly defined, in terms of total derivatives of f, by

$$k = \sum_{s=1}^{r} h^{s-1}[\alpha_s f^{[s-1]}(x, y) - \beta_s f^{[s-1]}(x + h, y + hk)],$$

with suitable constants α_s and β_s (Ehle's method; cf. §5.3(4)).

(a) Show how the method works on the model problem $dy/dx = \lambda y$. What is the maximum possible order in this case? Is the resulting method (of maximal order) A-stable?

(b) We may associate with the one-step method the quadrature rule

$$\int_x^{x+h} g(t)dt = \sum_{s=1}^{r} h^s[\alpha_s g^{(s-1)}(x) - \beta_s g^{(s-1)}(x + h)] + E(g).$$

Given any p with $r \leq p \leq 2r$, show that α_s, β_s can be chosen so as to have $E(g) = O(h^{p+1})$ when $g(t) = e^{t-x}$.

(c) With α_s, β_s chosen as in (b), prove that $E(g) = O(h^{p+1})$ for any $g \in C^p$ (not just for $g(t) = e^{t-x}$). {*Hint*: Expand the right-hand side of the quadrature rule in powers of h through h^p inclusive; then specialize to $g(t) = e^{t-x}$ and draw appropriate conclusions.}

(d) With α_s, β_s chosen as in (b), show that the implicit one-step method has order p if $f \in C^p$. {*Hint*: Use the definition of truncation error and Lipschitz conditions on the total derivatives $f^{[s-1]}$.}

(e) Work out the optimal one-step method with $r = 2$ and order $p = 4$.

(f) How can you make the method L-stable and have maximum possible order? Illustrate with $r = 2$.

MACHINE ASSIGNMENTS

1. (a) Write subroutines (in double precision) for Euler's method and the classical fourth-order Runge-Kutta method implementing one basic step $(x, y) \mapsto (x + h, y_{\text{next}})$ of the method. Assume a system of differential equations $y' = f(x, y)$ and let f be one of the variables of the subroutines. (Thus, f has to be declared "external" in FORTRAN.)

(b) Consider the initial value problem

$$(*) \qquad y' = Ay, \quad 0 \le x \le 1, \; y(0) = \mathbf{1},$$

where

$$A = \frac{1}{2}\begin{bmatrix} \lambda_2+\lambda_3 & \lambda_3-\lambda_1 & \lambda_2-\lambda_1 \\ \lambda_3-\lambda_2 & \lambda_1+\lambda_3 & \lambda_1-\lambda_2 \\ \lambda_2-\lambda_3 & \lambda_1-\lambda_3 & \lambda_1+\lambda_2 \end{bmatrix}, \quad \mathbf{1} = \begin{bmatrix} 1 \\ 1 \\ 1 \end{bmatrix}.$$

The exact solution is

$$y(x) = \begin{bmatrix} y^1 \\ y^2 \\ y^3 \end{bmatrix}, \quad \begin{aligned} y^1 &= -e^{\lambda_1 x} + e^{\lambda_2 x} + e^{\lambda_3 x}, \\ y^2 &= e^{\lambda_1 x} - e^{\lambda_2 x} + e^{\lambda_3 x}, \\ y^3 &= e^{\lambda_1 x} + e^{\lambda_2 x} - e^{\lambda_3 x}. \end{aligned}$$

Integrate the initial value problem (*) with constant step length $h = 1/N$ by

(i) Euler's method (order $p = 1$),
(ii) the classical Runge-Kutta method (order $p = 4$),

using the programs written in (a). In each case, along with the approximation vectors $u_n \in \mathbb{R}^3$, $n = 1, 2, \ldots, N$, generate vectors $v_n \in \mathbb{R}^3$, $n = 1, 2, \ldots, N$, that approximate the solution of the variational equation according to Theorem 5.4.1. (For the estimate $r(x, y; h)$ of the principal error function take the true value $r(x, y; h) = \tau(x, y)$ according to Ex. 21.) In this way obtain estimates $\tilde{e}_n = h^p v_n$ (p = order of the method) of the global errors $e_n = u_n - y(x_n)$. Use $N = 5, 10, 20, 40, 80$, and for each N produce the following output:

$h =$ ____

x_n	e_n^1	e_n^2	e_n^3	$\tilde{e}_n^1$	$\tilde{e}_n^2$	$\tilde{e}_n^3$
.2	____	____	____	____	____	____
.4	____	____	____	____	____	____
.6	____	____	____	____	____	____
.8	____	____	____	____	____	____
1.0	____	____	____	____	____	____

(`e11.3` format)

Suggested λ-values:

(i) $\lambda_1 = -1, \; \lambda_2 = 0, \; \lambda_3 = 1$ (to debug the program);
(ii) $\lambda_1 = 0, \; \lambda_2 = -1, \; \lambda_3 = -10$;
(iii) $\lambda_1 = 0, \; \lambda_2 = -1, \; \lambda_3 = -40$;

(iv) $\lambda_1 = 0,\ \lambda_2 = -1,\ \lambda_3 = -160.$

Summarize what you learn from these examples and from others that you may wish to run.

2. On the interval $[2\sqrt{q}, 2\sqrt{q}+1]$, $q \geq 0$ an integer, consider the initial value problem (FEHLBERG, 1968)

$$\frac{d^2c}{dx^2} = -\pi^2 x^2 c - \pi \frac{s}{\sqrt{c^2+s^2}},$$
$$\frac{d^2s}{dx^2} = -\pi^2 x^2 s + \pi \frac{c}{\sqrt{c^2+s^2}},$$

with initial conditions at $x = 2\sqrt{q}$ given by

$$c = 1, \quad \frac{dc}{dx} = 0, \quad s = 0, \quad \frac{ds}{dx} = 2\pi\sqrt{q}.$$

(a) Show that the exact solution is

$$c(x) = \cos(\tfrac{\pi}{2}x^2), \quad s(x) = \sin(\tfrac{\pi}{2}x^2).$$

(b) Write the problem as an initial value problem for a system of first-order differential equations.

(c) Consider the Runge-Kutta-Fehlberg pair Φ, Φ^* of order 3 (resp., 4) given by

$$k_1 = f(x,y),$$
$$k_2 = f(x + \tfrac{2}{7}h, y + \tfrac{2}{7}hk_1),$$
$$k_3 = f(x + \tfrac{7}{15}h, y + \tfrac{77}{900}hk_1 + \tfrac{343}{900}hk_2),$$
$$k_4 = f(x + \tfrac{35}{38}h, y + \tfrac{805}{1444}hk_1 - \tfrac{77175}{54872}hk_2 + \tfrac{97125}{54872}hk_3),$$
$$\Phi(x,y;h) = \tfrac{79}{490}k_1 + \tfrac{2175}{3626}k_3 + \tfrac{2166}{9065}k_4$$

respectively,

$$k_1,\ k_2,\ k_3,\ k_4 \text{ as previously},$$
$$k_5 = f(x+h, y + h\Phi(x,y;h)),$$
$$\Phi^*(x,y;h) = \tfrac{229}{1470}k_1 + \tfrac{1125}{1813}k_3 + \tfrac{13718}{81585}k_4 + \tfrac{1}{18}k_5.$$

Solve the initial value problem in (b) for $q = 0(1)3$ by the method Φ (in double precision), using constant step length $h = .2$. Repeat the integration with half the step length, and keep repeating (and halving the step) until $\max_n \|u_n - y(x_n)\|_\infty \le 10^{-6}$, where u_n, $y(x_n)$ are the approximate (resp., exact) solution vectors at $x_n = 2\sqrt{q}+nh$. For each run print

$$q, \quad h, \quad \max_n \|u_n - y(x_n)\|_\infty, \quad \max_n |c_n^2 + s_n^2 - 1|,$$

where c_n, s_n are the approximate values obtained for $c(x_n)$ (resp., $s(x_n)$) and the maxima are taken over $n = 0, 1, 2, \ldots, N-1$ with N such that $Nh = 1$.

(d) For the same values of q as in (c) and $h = .2, .1, .05, .025, .0125$, print the global and (estimated) local errors,

$$q, \quad h, \quad \|u_n - y(x_n)\|_\infty, \quad h\|\Phi(x_n, u_n; h) - \Phi^*(x_n, u_n; h)\|_\infty,$$

for $x_n = 0(.2).8$.

(e) Implement Theorem 5.4.1 on global error estimation, using the Runge-Kutta-Fehlberg method Φ, Φ^* of (c) and the estimator $r(x, y; h) = h^{-3}[\Phi(x, y; h) - \Phi^*(x, y; h)]$ of the principal error function of Φ. For the same values of q as in (d), and for $h = .05, .025, .0125$, print the exact and estimated global errors,

$$q, \quad h, \quad \|u_n - y(x_n)\|_\infty, \quad h^3\|v_n\|_\infty \quad \text{for} \quad nh = 0(.1).9.$$

3. (a) Let $f(z) = 1 + \frac{1}{1!}z + \frac{1}{2!}z^2 + \cdots + \frac{1}{p!}z^p$. For $p = 1(1)4$ write a MATLAB script, using the `contour` command, to plot the lines along which $|f(z)| = r$, $r = .1(.1)1$ (*level lines* of f) and the lines along which $\arg f(z) = \theta$, $\theta = 0(\frac{1}{8}\pi)2\pi - \frac{1}{8}\pi$ (*phase lines* of f).

(b) For any analytic function f, derive differential equations for the level and phase lines of f. {*Hint*: Write $f(z) = r\exp(i\theta)$ and use θ as the independent variable for the level lines, and r as the independent variable for the phase lines. In each case, introduce arc length as the final independent variable.}

(c) Use the MATLAB function `ode45` (or an ODE solver from some other software package) to compute the level lines $|f(z)| = 1$ of the function f given in (a), for $p = 1(1)21$; these determine the regions of absolute stability of the Taylor expansion method (cf. Ex. 20). {*Hint*: Use initial conditions at the origin. Produce only those parts of the curves that lie in the upper half-plane (why?). To do so in MATLAB, let `ode45` run sufficiently long, interpolate between the first pair of points lying on opposite sides of the real axis to get a point on the axis, and then delete the rest of the data before plotting.}

4. Newton's equations for the motion of a particle on a planar orbit (with eccentricity ε, $0 < \varepsilon < 1$) are

$$x'' = -\frac{x}{r^3}, \quad x(0) = 1 - \varepsilon, \quad x'(0) = 0,$$

$$t \geq 0,$$

$$y'' = -\frac{y}{r^3}, \quad y(0) = 0, \quad y'(0) = \sqrt{\frac{1+\varepsilon}{1-\varepsilon}},$$

where

$$r^2 = x^2 + y^2.$$

(a) Verify that the solution can be written in the form $x(t) = \cos u - \varepsilon$, $y(t) = \sqrt{1-\varepsilon^2}\sin u$, where u is the solution of $u - \varepsilon \sin u - t = 0$.

(b) Reformulate the problem as an initial value problem for a system of first-order differential equations.

(c) Write and run a program (in double precision) for solving the initial value problem in (b) on the interval $[0, 20]$, for $\varepsilon = .3$, $.5$, and $.7$. Print results only at $t = 5, 10, 15, 20$. Use Euler's method, one of the improved Euler methods, and the classical fourth-order Runge-Kutta method, with steps $h = 1, 2^{-1}, 2^{-2}, \ldots, 2^{-5}$. Along with the results $x(t;h)$, $y(t;h)$ (at $t = 5, 10, 15, 20$) also print $e(t) = h^{-p}[x(t;h) - x(t)]$, $f(t) = h^{-p}[y(t;h) - y(t)]$, where p is the order of the method. Compute $x(t)$ for given t by first solving "Kepler's equation" $u = \varepsilon \sin u + t$ for u (cf. Ch. 4, Ex. 18), using the fixed point iteration with starting value t, and then computing $x = \cos u - \varepsilon$. Do this in a similar manner for $y(t)$. Use $\varepsilon = 0$ for debugging purposes. Suggested output is:

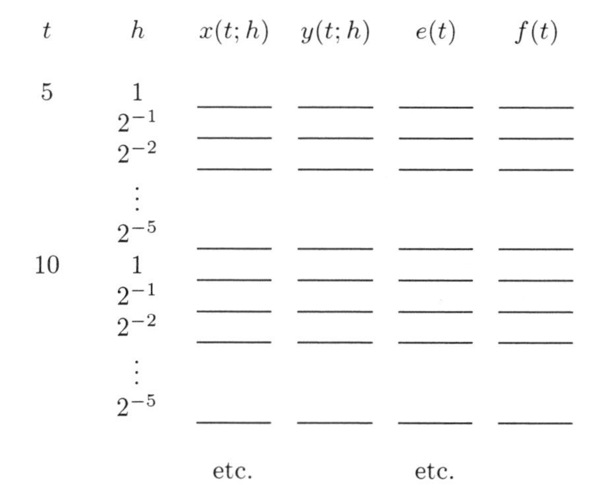

t	h	$x(t;h)$	$y(t;h)$	$e(t)$	$f(t)$
5	1	_____	_____	_____	_____
	2^{-1}	_____	_____	_____	_____
	2^{-2}	_____	_____	_____	_____
	$\vdots$				
	2^{-5}	_____	_____	_____	_____
10	1	_____	_____	_____	_____
	2^{-1}	_____	_____	_____	_____
	2^{-2}	_____	_____	_____	_____
	$\vdots$				
	2^{-5}	_____	_____	_____	_____
		etc.		etc.	

CHAPTER 6

INITIAL VALUE PROBLEMS FOR ODEs — MULTISTEP METHODS

We saw in Chapter 5 that (explicit) one-step methods are increasingly difficult to construct as one upgrades the order requirement. This is no longer true for multistep methods, where an increase in order is straightforward but comes with a price: a potential danger of instability. In addition, there are other complications such as the need for an initialization procedure and considerably more complicated procedures for changing the grid length. Yet, in terms of work involved, multistep methods are still among the most attractive methods. We discuss them along lines similar to one-step methods, beginning with a local description and examples, and proceeding to the global description and problems of stiffness. By the very nature of multistep methods, the discussion of stability is now more extensive.

§1. Local Description of Multistep Methods

§1.1. **Explicit and implicit methods.** We consider as before the initial value problem for a first-order system of differential equations

$$\frac{dy}{dx} = f(x, y), \quad a \le x \le b; \quad y(a) = y_0 \tag{1.1}$$

(cf. Chap. 5, (0.12) through (0.14)). Our task is again to determine a vector-valued grid function $u \in \Gamma_h[a, b]$ (cf. Ch. 5, §3) such that $u_n \approx y(x_n)$ at the nth grid point x_n.

A *k-step method* ($k > 1$) obtains u_{n+k} in terms of k preceding approximations $u_{n+k-1}, u_{n+k-2}, \ldots, u_n$. We call k the *step number* (or *index*) of the method. We consider only *linear* k-step methods, which, in their most general form, but assuming a constant grid length h, can be written as

$$\begin{aligned} &u_{n+k} + \alpha_{k-1} u_{n+k-1} + \cdots + \alpha_0 u_n \\ &= h[\beta_k f_{n+k} + \beta_{k-1} f_{n+k-1} + \cdots + \beta_0 f_n], \quad n = 0, 1, 2, \ldots, N-k, \end{aligned} \tag{1.2}$$

where

$$x_r = a + rh, \quad f_r = f(x_r, u_r), \qquad r = 0, 1, \ldots, N, \tag{1.3}$$

and the αs and βs are given (scalar) coefficients. The relation (1.2) is linear in the function values f_r (in contrast to Runge-Kutta methods); nevertheless, we are still dealing with a nonlinear difference equation for the grid function u.

The definition (1.2) must be supplemented by a *starting procedure* for obtaining the approximations to $y(x_s)$,

$$u_s = u_s(h), \qquad s = 0, 1, \dots, k-1. \tag{1.4}$$

These normally depend on the grid length h; so also may the coefficients α_s, β_s in (1.2). The method (1.2) is called *explicit* if $\beta_k = 0$ and *implicit* otherwise.

Implicit methods require the solution of a system of nonlinear equations,

$$u_{n+k} = h\beta_k f(x_{n+k}, u_{n+k}) + g_n, \tag{1.5}$$

where

$$g_n = h\sum_{s=0}^{k-1} \beta_s f_{n+s} - \sum_{s=0}^{k-1} \alpha_s u_{n+s} \tag{1.6}$$

is a known vector. Fortunately, the nonlinearity in (1.5) is rather weak and in fact disappears in the limit as $h \to 0$. This suggests the use of successive iteration on (1.5),

$$u_{n+k}^{[\nu]} = h\beta_k f(x_{n+k}, u_{n+k}^{[\nu-1]}) + g_n, \qquad \nu = 1, 2, \dots\ , \tag{1.7}$$

where $u_{n+k}^{[0]}$ is a suitable initial approximation for u_{n+k}. By a simple application of the contraction mapping principle (cf. Ch. 4, §9.1), one shows that (1.7) indeed converges as $\nu \to \infty$, for arbitrary initial approximation, provided h is small enough.

Theorem 6.1.1. *Suppose f satisfies a uniform Lipschitz condition on* $[a, b] \times \mathbb{R}^d$ (cf. Ch. 5, §0.3),

$$\|f(x, y) - f(x, y^*)\| \le L\|y - y^*\|, \quad x \in [a, b], \quad y, y^* \in \mathbb{R}^d, \tag{1.8}$$

and assume that

$$\lambda := h|\beta_k|L < 1. \tag{1.9}$$

Then (1.5) *has a unique solution u_{n+k}. Moreover, for arbitrary $u_{n+k}^{[0]}$,*

$$u_{n+k} = \lim_{\nu\to\infty} u_{n+k}^{[\nu]}, \tag{1.10}$$

and

$$\|u_{n+k}^{[\nu]} - u_{n+k}\| \le \frac{\lambda^\nu}{1-\lambda}\|u_{n+k}^{(1)} - u_{n+k}^{(0)}\|, \quad \nu = 1, 2, 3, \ldots . \tag{1.11}$$

Proof. We define the map φ: $\mathbb{R}^d \to \mathbb{R}^d$ by

$$\varphi(y) := h\beta_k f(x_{n+k}, y) + g_n, \qquad y \in \mathbb{R}^d. \tag{1.12}$$

Then, for any y, $y^* \in \mathbb{R}^d$, we have

$$\begin{aligned}\|\varphi(y) - \varphi(y^*)\| &= h|\beta_k|\|f(x_{n+k}, y) - f(x_{n+k}, y^*)\| \\ &\le h|\beta_k|L\|y - y^*\|,\end{aligned}$$

showing, in view of (1.9), that φ is a contraction operator on $\mathbb{R}^d$. By the contraction mapping principle, there is a unique fixed point of φ, that is, a vector $y = u_{n+k}$ satisfying $\varphi(y) = y$. This proves the first part of the theorem. The second part is also a consequence of the contraction mapping principle if one notes that (1.7) is just the fixed point iteration $u_{n+k}^{[\nu]} = \varphi(u_{n+k}^{[\nu-1]})$. □

Strictly speaking, an implicit multistep method requires "iteration to convergence" in (1.7), that is, iteration until the required fixed point is obtained to machine accuracy. This may well entail too many iterations to make the method competitive, since each iteration step costs one evaluation of f. In practice, one often terminates the iteration after the first or second step, having selected the starting value $u_{n+k}^{[0]}$ judiciously (cf. §2.3). It should also be noted that stiff systems, for which L may be quite large, would require unrealistically small steps h to satisfy (1.9). In such cases, Newton's method (see Ex. 1), rather than fixed point iteration, would be preferable.

§1.2. **Local accuracy**. In analogy with one-step methods (cf. Ch. 5, Eq. (3.6) and (3.6$_h$)), we define *residual operators* by

$$(Rv)(x) := v'(x) - f(x, v(x)), \qquad v \in C^1[a, b], \tag{1.13}$$

$$(R_h v)_n := \frac{1}{h}\sum_{s=0}^{k} \alpha_s v_{n+s} - \sum_{s=0}^{k} \beta_s f(x_{n+s}, v_{n+s}), \quad v \in \Gamma_h[a, b], \quad n = 0, 1, \ldots, N-k. \tag{1.13$_h$}$$

(We may arbitrarily define $(R_h v)_N = \cdots = (R_h v)_{N-k+1} = (R_h v)_{N-k}$ to obtain a grid function $R_h v$ defined on the entire grid.) In (1.13_h) and throughout this chapter, we adopt the convention $\alpha_k = 1$. Since there is no longer a natural "generic" point (x, y) in which to define our method, we take the analogue of Ch. 5, Eq. (3.9) (except for sign) as our definition of truncation error:

$$(T_h)_n = (R_h y)_n, \qquad n = 0, 1, \ldots, N, \tag{1.14}$$

where $y(x)$ is the exact solution of (1.1). This defines a grid function T_h on a uniform grid on $[a, b]$. We define consistency, order, and principal error function as before in Ch. 5; that is, the method (1.2) is *consistent* if

$$\|T_h\|_\infty \to 0 \quad \text{as} \quad h \to 0, \tag{1.15}$$

has *order* p if

$$\|T_h\|_\infty = O(h^p) \quad \text{as} \quad h \to 0, \tag{1.16}$$

and admits a *principal error function* $\tau \in C[a, b]$ if

$$\tau(x) \not\equiv 0 \quad \text{and} \quad (T_h)_n = \tau(x_n) h^p + O(h^{p+1}) \quad \text{as} \quad h \to 0 \tag{1.17}$$

in the usual sense that $\|T_h - h^p \tau\|_\infty = O(h^{p+1})$. The infinity norm of grid functions in (1.15) through (1.17) is as defined in Ch. 5, Eq. (3.4).

Note that (1.14) can be written in the simple form

$$\sum_{s=0}^{k} \alpha_s y(x_{n+s}) - h \sum_{s=0}^{k} \beta_s y'(x_{n+s}) = h(T_h)_n, \tag{1.14$'$}$$

since $f(x_{n+s}, y(x_{n+s})) = y'(x_{n+s})$ by virtue of the differential equation (1.1). Although (1.14$'$) is a relation for vector-valued functions, the relationship is exactly the same for each component. This suggests defining a linear operator L_h: $C^1[\mathbb{R}] \to C[\mathbb{R}]$ on scalar functions by letting

$$(L_h z)(x) := \sum_{s=0}^{k} [\alpha_s z(x + sh) - h\beta_s z'(x + sh)], \quad z \in C^1[\mathbb{R}]. \tag{1.18}$$

If $L_h z$ were identically zero for all $z \in C^1[\mathbb{R}]$, then so would be the truncation error, and our method would produce exact answers if started with exact initial values. This is unrealistic, however; nevertheless, we would like L_h to annihiliate as many functions as possible. This motivates the following concept of degree. Given a set of linearly independent "gauge functions" $\{\omega_r(x)\}_{r=0}^{\infty}$ (usually complete on compact intervals), we say that the method (1.2) has *Ω-degree* p if its associated linear operator L_h satisfies

$$L_h\omega \equiv 0 \ \text{ for all } \ \omega \in \Omega_p, \ \text{ all } \ h > 0, \tag{1.19}$$

where Ω_p is the set of functions spanned by the first $p+1$ gauge functions $\omega_0, \omega_1, \dots, \omega_p$; hence

$$\Omega_0 \subset \Omega_1 \subset \Omega_2 \subset \cdots, \quad \dim\, \Omega_m = m+1. \tag{1.20}$$

We say that Ω_m is *closed under translation* if $\omega(x) \in \Omega_m$ implies $\omega(x+c) \in \Omega_m$ for arbitrary real c. Similarly, Ω_m is said to be *closed under scaling* if $\omega(x) \in \Omega_m$ implies $\omega(cx) \in \Omega_m$ for arbitrary real c. For example, algebraic polynomials $\Omega_m = \mathbb{P}_m$ are closed under translation as well as scaling, trigonometric polynomials $\Omega_m = \mathbb{T}_m[0, 2\pi]$ and exponential sums $\Omega_m = \mathbb{E}_m$ (cf. Ch. 2, Examples to Eq. (0.2)) only under translation, and spline functions $\mathbb{S}_m^k(\Delta)$ (for fixed partition) neither under scaling nor under translation.

The following theorem is no more than a simple observation.

Theorem 6.1.2. (a) *If Ω_p is closed under translation, then the method* (1.2) *has Ω-degree p if and only if*

$$(L_h\omega)(0) = 0 \ \textit{ for all } \ \omega \in \Omega_p, \ \textit{ all } \ h > 0. \tag{1.21}$$

(b) *If Ω_p is closed under translation and scaling, then the method* (1.2) *has Ω-degree p if and only if*

$$(L_1\omega)(0) = 0 \ \textit{ for all } \ \omega \in \Omega_p, \tag{1.22}$$

where $L_1 = L_h$ for $h = 1$.

Proof. (a) The necessity of (1.21) is trivial. To prove its sufficiency, it is enough to show that for any $x_0 \in \mathbb{R}$,

$$(L_h\omega)(x_0) = 0, \ \text{ all } \ \omega \in \Omega_p, \ \text{ all } \ h > 0.$$

Take any $\omega \in \Omega_p$ and define $\omega_0(x) = \omega(x + x_0)$. Then, by assumption, $\omega_0 \in \Omega_p$; hence, for all $h > 0$,

$$0 = (L_h\omega_0)(0) = \sum_{s=0}^{k} [\alpha_s \omega(x_0 + sh) - h\beta_s \omega'(x_0 + sh)] = (L_h\omega)(x_0).$$

(b) The necessity of (1.22) is trivial. For the sufficiency, let $\omega_0(x) = \omega(x_0 + xh)$ for any given $\omega \in \Omega_p$. Then, by assumption, $\omega_0 \in \Omega_p$, and

$$\begin{aligned} 0 = (L_1\omega_0)(0) &= \sum_{s=0}^{k} [\alpha_s\omega_0(s) - \beta_s\omega_0'(s)] \\ &= \sum_{s=0}^{k} [\alpha_s\omega(x_0 + sh) - h\beta_s\omega'(x_0 + sh)] = (L_h\omega)(x_0). \end{aligned}$$

Since $x_0 \in \mathbb{R}$ and $h > 0$ are arbitrary, the assertion follows. □

Case (b) of Theorem 6.1.2 suggests the introduction of the *linear functional* L: $C^1[\mathbb{R}] \to \mathbb{R}$ associated with the method (1.2),

$$Lu := \sum_{s=0}^{k} [\alpha_s u(s) - \beta_s u'(s)], \quad u \in C^1[\mathbb{R}\,]. \tag{1.23}$$

For $\Omega_m = \mathbb{P}_m$ we refer to Ω-degree as the *algebraic* (or *polynomial*) *degree*. Thus, (1.2) has algebraic degree p if $Lu = 0$ for all $u \in \mathbb{P}_p$. By linearity, this is equivalent to

$$Lt^r = 0, \quad r = 0, 1, \ldots, p. \tag{1.24}$$

Example. Determine all explicit two-step methods

$$u_{n+2} + \alpha_1 u_{n+1} + \alpha_0 u_n = h(\beta_1 f_{n+1} + \beta_0 f_n) \tag{1.25}$$

having polynomial degree $p = 0, 1, 2, 3$.

Here

$$Lu = u(2) + \alpha_1 u(1) + \alpha_0 u(0) - \beta_1 u'(1) - \beta_0 u'(0).$$

The first four equations in (1.24) are

$$\begin{aligned} 1 + \alpha_1 + \alpha_0 & & &= 0, \\ 2 + \alpha_1 & &-\beta_1 - \beta_0 &= 0, \\ 4 + \alpha_1 & &-2\beta_1 &= 0, \\ 8 + \alpha_1 & &-3\beta_1 &= 0. \end{aligned} \tag{1.26}$$

We have algebraic degree 0, 1, 2, 3 if, respectively, the first, the first two, the first three, and all four equations in (1.26) are satisfied. Thus,

$$\begin{aligned}
&\alpha_1 = -\alpha_0 - 1, \quad \beta_0, \beta_1 \text{ arbitrary} \quad (p = 0),\\
&\alpha_1 = -\alpha_0 - 1,\ \beta_1 = -\alpha_0 - \beta_0 + 1 \quad (p = 1),\\
&\alpha_1 = -\alpha_0 - 1,\ \beta_1 = -\tfrac{1}{2}\,\alpha_0 + \tfrac{3}{2},\ \beta_0 = -\tfrac{1}{2}\,\alpha_0 - \tfrac{1}{2} \quad (p = 2),\\
&\alpha_1 = 4,\ \alpha_0 = -5,\ \beta_1 = 4,\ \beta_0 = 2 \quad (p = 3)
\end{aligned} \tag{1.27}$$

yield $(3-p)$-parameter families of methods of degree p. Since $Lt^4 = 16 + \alpha_1 - 4\beta_1 = 4 \neq 0$, degree $p = 4$ is impossible. This means that the last method in (1.27),

$$u_{n+2} + 4u_{n+1} - 5u_n = 2h(2f_{n+1} + f_n), \tag{1.28}$$

is optimal as far as algebraic degree is concerned. Other special cases include the *midpoint rule*

$$u_{n+2} = u_n + 2hf_{n+1} \quad (\alpha_0 = -1;\ p = 2)$$

and the *Adams-Bashforth* second-order *method* (cf. §2.1),

$$u_{n+2} = u_{n+1} + \tfrac{1}{2}h(3f_{n+1} - f_n) \quad (\alpha_0 = 0;\ p = 2).$$

The "optimal" method (1.28) is nameless — and for good reason! Suppose, indeed, that we apply it to the trivial (scalar) initial value problem

$$y' = 0, \quad y(0) = 0 \ \text{ on } \ 0 \le x \le 1,$$

which has the exact solution $y(x) \equiv 0$. Assume $u_0 = 0$, but $u_1 = \varepsilon$ (to account for a small rounding error in the second starting value). Even if (1.28) is applied with infinite precision, we then get

$$u_n = \tfrac{1}{6}\,\varepsilon[1 + (-1)^{n+1}5^n], \quad n = 0, 1, \ldots, N.$$

Assuming further that $\varepsilon = h^{p+1}$ (pth-order one-step method), we will have, at the end of the interval [0,1],

$$u_N = \tfrac{1}{6}\,h^{p+1}[1 + (-1)^{N+1}5^N] \sim \tfrac{1}{6}\,(-1)^{N+1}\,N^{-p-1}5^N \ \text{ as } \ N \to \infty.$$

Thus, $|u_N| \to \infty$ exponentially fast, and highly oscillating on top of it. We have here an example of "strong instability"; this is analyzed later in more detail (cf. §3).

§1.3. **Polynomial degree vs. order**. We recall from (1.14′) and (1.18) that for the truncation error T_h of (1.2), we have

$$h(T_h)_n = \sum_{s=0}^{k} [\alpha_s y(x_n + sh) - h\beta_s y'(x_n + sh)] = (L_h y)(x_n), \qquad n = 0, 1, 2, \ldots, N-k. \tag{1.29}$$

Let

$$u(t) := y(x_n + th), \quad 0 \le t \le k. \tag{1.30}$$

(More precisely, we should write $u_{n,h}(t)$.) Then

$$h(T_h)_n = \sum_{s=0}^{k} [\alpha_s u(s) - \beta_s u'(s)] = Lu, \tag{1.31}$$

where L on the right is to be applied componentwise to each component of the vector u. If T_h is the truncation error of a method of algebraic degree p, then the linear functional L in (1.31) annihilates all polynomials of degree p and thus, if $u \in C^{p+1}[0,k]$, can be represented in terms of the Peano kernel

$$\lambda_p(\sigma) = L_{(t)}(t-\sigma)_+^p \tag{1.32}$$

by

$$Lu = \frac{1}{p!}\int_0^k \lambda_p(\sigma) u^{(p+1)}(\sigma) d\sigma \tag{1.33}$$

(cf. Ch. 3, §2.6). From the explicit formula

$$\lambda_p(\sigma) = \sum_{s=0}^{k} [\alpha_s (s-\sigma)_+^p - \beta_s p (s-\sigma)_+^{p-1}], \quad p \ge 1, \tag{1.34}$$

it is easily seen that $\lambda_p \in \mathbb{S}_p^{p-2}(\Delta)$, where Δ is the subdivision of $[0,k]$ into subintervals of length 1. Outside the interval $[0,k]$ we have $\lambda_p \equiv 0$. Moreover, if L is definite, then, and only then (cf. Ch. 3, Eq. (2.59) and Ex. 46),

$$Lu = \ell_{p+1} u^{(p+1)}(\overline{\sigma}), \quad 0 < \overline{\sigma} < k; \quad \ell_{p+1} = L\,\frac{t^{p+1}}{(p+1)!}\,. \tag{1.35}$$

The Peano representation (1.33) of L in combination with (1.31) allows us to identify the polynomial degree of a multistep method with its order as defined in §1.2.

Theorem 6.1.3. *A multistep method* (1.2) *of polynomial degree p has order p whenever the exact solution $y(x)$ of* (1.1) *is in the smoothness class $C^{p+1}[a,b]$. If the associated functional L is definite, then*

$$(T_h)_n = \ell_{p+1} y^{(p+1)}(\overline{x}_n) h^p, \qquad x_n < \overline{x}_n < x_{n+k}, \tag{1.36}$$

where ℓ_{p+1} is as given in (1.35). *Moreover, for the principal error function τ of the method, whether definite or not, we have, if $y \in C^{p+2}[a,b]$,*

$$\tau(x) = \ell_{p+1} y^{(p+1)}(x). \tag{1.37}$$

Proof. By (1.30) and (1.31), we have

$$h(T_h)_n = Lu, \quad u(t) = y(x_n + th), \quad n = 0, 1, 2, \ldots, N-k;$$

hence, by (1.33),

$$h(T_h)_n = \frac{1}{p!}\int_0^k \lambda_p(\sigma) u^{(p+1)}(\sigma)d\sigma = \frac{h^{p+1}}{p!}\int_0^k \lambda_p(\sigma) y^{(p+1)}(x_n + \sigma h)d\sigma. \tag{1.38}$$

Therefore,

$$\begin{aligned} \|(T_h)_n\| &= \frac{h^p}{p!}\left\|\int_0^k \lambda_p(\sigma) y^{(p+1)}(x_n + \sigma h)d\sigma\right\| \\ &\le \frac{h^p}{p!}\int_0^k |\lambda_p(\sigma)| \|y^{(p+1)}(x_n + \sigma h)\| d\sigma \\ &\le \frac{h^p}{p!}\|y^{(p+1)}\|_\infty \int_0^k |\lambda_p(\sigma)| d\sigma, \end{aligned}$$

and we see that

$$\|T_h\|_\infty \le Ch^p, \quad C = \frac{1}{p!}\|y^{(p+1)}\|_\infty \int_0^k |\lambda_p(\sigma)| d\sigma. \tag{1.39}$$

This proves the first part of the theorem.

If L is definite, then (1.36) follows directly from (1.31) and from (1.35) applied to the vector-valued function u in (1.30). Finally, (1.37) follows from

(1.38) by noting that $y^{(p+1)}(x_n+\sigma h) = y^{(p+1)}(x_n)+O(h)$ and the fact that $\frac{1}{p!}\int_0^k \lambda_p(\sigma)d\sigma = \ell_{p+1}$ (cf. Ch. 3, Eq. (2.59)). □

The proof of Theorem 6.1.3 also exhibits, in (1.39), an explicit bound on the local truncation error. For methods with definite functional L, there is the even simpler bound (1.39) with

$$C = |\ell_{p+1}|\|y^{(p+1)}\|_\infty \quad (L \text{ definite}), \tag{1.39$'$}$$

which follows from (1.36). It is seen later in §3.4 that for the global discretization error it is not ℓ_{p+1} which is relevant, but rather

$$C_{k,p} = \frac{\ell_{p+1}}{\sum_{s=0}^k \beta_s}, \tag{1.40}$$

which is called the *error constant* of the k-step method (1.2) (of order p). The denominator in (1.40) is positive if the method is stable and consistent (cf. §4.3).

As a simple example, consider the midpoint rule

$$u_{n+2} = u_n + 2hf_n$$

for which

$$Lu = u(2) - u(0) - 2u'(1).$$

We already know that it has order $p = 2$. The Peano kernel is

$$\lambda_2(\sigma) = (2-\sigma)_+^2 - 4(1-\sigma)_+,$$

so that

$$\lambda_2(\sigma) = \begin{cases} (2-\sigma)^2 - 4(1-\sigma) = \sigma^2 & \text{if } 0 < \sigma < 1, \\ (2-\sigma)^2 & \text{if } 1 < \sigma < 2. \end{cases}$$

It consists of two parabolic arcs as shown in Figure 6.1.1. Evidently, L is positive definite; hence by (1.36) and (1.35),

$$(T_h)_n = \ell_3 y^{(3)}(\overline{x}_n)h^2, \quad \ell_3 = L\,\frac{t^3}{6} = \tfrac{1}{3}.$$

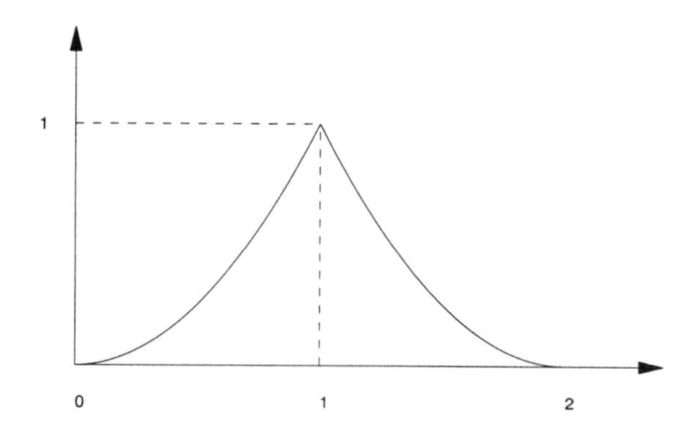

FIGURE 6.1.1. Peano kernel of the midpoint rule

§2. Examples of Multistep Methods

An alternative way of deriving multistep formulae, which does not require the solution of linear algebraic systems (as does the method of §1.2), starts from the fundamental theorem of calculus,

$$y(x_{n+k}) = y(x_{n+k-1}) + \int_{x_{n+k-1}}^{x_{n+k}} y'(x)dx. \tag{2.1}$$

(Instead of x_{n+k-1} on the right, we could take any other grid point $x_{n+\kappa}$, $0 \le \kappa < k-1$, but we limit ourselves to the case shown in (2.1).) A multistep formula results from (2.1) if the integral is expressed (approximately) by a linear combination of derivative values at some grid points selected from the set $\{x_{n+\kappa}\colon\ \kappa = 0, 1, \ldots, k\}$. Those selected may be called the "active" grid points. A simple way to do this is to approximate y' by the unique polynomial interpolating y' at the active grid points. If we carry along the remainder term, we will also get an expression for the truncation error.

We implement this in the two most important cases where the active grid points are $x_n, x_{n+1}, \ldots, x_{n+k-1}$ and $x_{n+1}, x_{n+2}, \ldots, x_{n+k}$, respectively, giving rise to the family of Adams-type multistep methods.

§2.1. **Adams[1]-Bashforth method.** Replace y' in (2.1) by the interpolation polynomial $p_{k-1}(y'; x_n, x_{n+1}, \ldots, x_{n+k-1}; x)$ of degree $\le k-1$

[1]John Couch Adams (1819–1892), son of a tenant farmer, studied at Cambridge University where, in 1859, he became professor of astronomy and geometry and, from 1861 on, director of the observatory. His calculations, while still a student at Cambridge, predicted the existence of the then unknown planet Neptune, based on irregularities in the orbit of

interpolating y' at the grid points $x_n, x_{n+1}, \ldots, x_{n+k-1}$ (cf. Ch. 2, §2.1). If we include the remainder term (cf. Ch. 2, §2.2), assuming that $y \in C^{k+1}$, and make the change of variable $x = x_{n+k-1} + th$ in the integral of (2.1), we obtain

$$y(x_{n+k}) = y(x_{n+k-1}) + h \sum_{s=0}^{k-1} \beta_{k,s} y'(x_{n+s}) + hr_n, \tag{2.2}$$

where

$$\beta_{k,s} = \int_0^1 \prod_{\substack{r=0 \\ r \neq s}}^{k-1} \left(\frac{t+k-1-r}{s-r} \right) dt, \qquad s = 0, 1, \ldots, k-1, \tag{2.3}$$

and

$$r_n = \gamma_k h^k y^{(k+1)}(\overline{x}_n), \quad x_n < \overline{x}_n < x_{n+k-1}; \quad \gamma_k = \int_0^1 \binom{t+k-1}{k} dt. \tag{2.4}$$

The formula (2.2) suggests the multistep method

$$u_{n+k} = u_{n+k-1} + h \sum_{s=0}^{k-1} \beta_{k,s} f(x_{n+s}, u_{n+s}), \tag{2.5}$$

which is called the *kth-order Adams-Bashforth method.* It is called kth-order, since comparison of (2.2) with (1.14′) shows that in fact r_n is the truncation error of (2.5),

$$r_n = (T_h)_n, \tag{2.6}$$

and (2.4) shows that $T_h = O(h^k)$. In view of the form (2.4) of the truncation error, we can infer, as mentioned in §1.3, that the linear functional L associated with (2.5) is definite. We prove later in §4 (cf. Ex. 12) that there is no stable explicit k-step method that has order $p > k$. In this sense, (2.5) with $\beta_{k,s}$ given by (2.3), is optimal.

If $y \in C^{k+2}$, it also follows from (2.4) and (2.6) that

$$(T_h)_n = \gamma_k h^k y^{(k+1)}(x_n) + O(h^{k+1}), \qquad h \to 0;$$

---

the next inner planet Uranus. Unfortunately, publication of his findings was delayed and it was Le Verrier, who did similar calculations (and managed to publish them), and young astronomers in Berlin, who succeeded in locating the planet at the position predicted, who received credit for this historical discovery. Understandably, this led to a prolonged dispute of priority. Adams is also known for his work on lunar theory and magnetism. Later in his life, he turned to computational problems in number theory. An ardent admirer of Newton, he took it upon himself to catalogue a large body of scientific papers left behind by Newton after his death.

that is, the Adams-Bashforth method (2.5) has the principal error function

$$\tau(x) = \gamma_k y^{(k+1)}(x). \tag{2.7}$$

The constant γ_k (defined in (2.4)) is therefore the same as the constant ℓ_{k+1} defined earlier in (1.35).

Had we used Newton's form of the interpolation polynomial (cf. Ch. 2, §2.6) and observed that the divided differences required for equally spaced points are (cf. Ch. 2, Ex. 50)

$$[x_{n+k-1}, x_{n+k-2}, \dots, x_{n+k-s-1}]f = \frac{\nabla^s f_{n+k-1}}{s!h^s}, \tag{2.8}$$

where $\nabla f_{n+k-1} = f_{n+k-1} - f_{n+k-2}$, $\nabla^2 f_{n+k-1} = \nabla(\nabla f_{n+k-1}), \dots$ are ordinary (backward) differences, we would have obtained (2.5) in the form

$$u_{n+k} = u_{n+k-1} + h \sum_{s=0}^{k-1} \gamma_s \nabla^s f_{n+k-1}, \tag{2.5$'$}$$

where

$$\gamma_s = \int_0^1 \binom{t+s-1}{s} dt, \qquad s = 0, 1, 2, \dots . \tag{2.9}$$

The difference form (2.5′) of the Adams-Bashforth method has important practical advantages over the Lagrange form (2.5). For one thing, the coefficients in (2.5′) do not depend on the step number k. Adding more terms in the sum of (2.5′) thus increases the order (and step number) of the method. Related to this is the fact that the first omitted term in the summation of (2.5′) is a good approximation of the truncation error. Indeed, by Ch. 2, Eq. (2.68), we know that

$$y^{(k+1)}(\overline{x}_n) \approx k![x_{n+k-1}, x_{n+k-2}, \dots, x_n, x_{n-1}]y';$$

hence, by (2.4), (2.6), and (2.8),

$$(T_h)_n = \gamma_k h^k y^{(k+1)}(\overline{x}_n) \approx \gamma_k h^k k! \, \frac{\nabla^k f_{n+k-1}}{k!h^k} = \gamma_k \nabla^k f_{n+k-1}.$$

When implementing the method (2.5′), one needs to set up a table of (backward) differences for each component of f. By adding an extra column of differences at the end of the table (the kth differences), we are thus able to monitor the local truncation errors by simply multiplying these differences by γ_k. No such easy procedure is available for the Lagrange form of the method.

It is, therefore, of some importance to have an effective method for calculating the coefficients γ_s, $s = 0, 1, 2, \ldots$. Such a method can be derived from the generating function $\gamma(z) = \sum_{s=0}^{\infty} \gamma_s z^s$ of the coefficients. We have

$$\begin{aligned}
\gamma(z) &= \sum_{s=0}^{\infty} z^s \int_0^1 \binom{t+s-1}{s} dt = \sum_{s=0}^{\infty} z^s (-1)^s \int_0^1 \binom{-t}{s} dt \\
&= \int_0^1 \sum_{s=0}^{\infty} \binom{-t}{s} (-z)^s dt = \int_0^1 (1-z)^{-t} dt \\
&= \int_0^1 e^{-t\ln(1-z)} dt = -\frac{e^{-t\ln(1-z)}}{\ln(1-z)}\bigg|_{t=0}^{t=1} \\
&= -\frac{z}{(1-z)\ln(1-z)} .
\end{aligned}$$

Thus, the γ_s are the coefficients in the Maclaurin expansion of

$$\gamma(z) = -\frac{z}{(1-z)\ln(1-z)} . \tag{2.10}$$

In particular,

$$\frac{z}{1-z} = -\ln(1-z) \sum_{s=0}^{\infty} \gamma_s z^s;$$

that is,

$$z + z^2 + z^3 + \cdots = \left(z + \tfrac{1}{2}\, z^2 + \tfrac{1}{3}\, z^3 + \cdots\right)(\gamma_0 + \gamma_1 z + \gamma_2 z^2 + \cdots),$$

which, on comparing coefficients of like powers on the left and right, yields

$$\begin{aligned}
&\gamma_0 = 1, \\
&\gamma_s = 1 - \tfrac{1}{2}\,\gamma_{s-1} - \tfrac{1}{3}\,\gamma_{s-2} - \cdots - \tfrac{1}{s+1}\,\gamma_0, \\
&\qquad\qquad s = 1, 2, 3, \ldots .
\end{aligned} \tag{2.11}$$

It is, therefore, easy to compute as many of the coefficients γ_s as desired; for example,

$$\gamma_0 = 1, \quad \gamma_1 = \tfrac{1}{2}, \quad \gamma_2 = \tfrac{5}{12}, \quad \gamma_3 = \tfrac{3}{8}, \quad \ldots .$$

Note that they are all positive, as they must be in view of (2.9).

§2.2. **Adams-Moulton[2] method.** This is the implicit analogue of the Adams-Bashforth method; that is, the point x_{n+k} is included among the active grid points. To obtain again a method of order k, we need, as before, k active points and therefore select them to be $x_{n+k}, x_{n+k-1}, \dots, x_{n+1}$. The derivation of the method is then entirely analogous to the one in §2.1, and we limit ourselves to simply stating the results.

The Lagrange form of the method is now

$$u_{n+k} = u_{n+k-1} + h \sum_{s=1}^{k} \beta_{k,s}^{*} f(x_{n+s}, u_{n+s}), \tag{2.12}$$

with

$$\beta_{k,s}^{*} = \int_0^1 \prod_{\substack{r=1 \\ r \neq s}}^{k} \left(\frac{t+k-1-r}{s-r} \right) dt, \qquad s = 1, 2, \dots, k, \tag{2.13}$$

whereas Newton's form becomes

$$u_{n+k} = u_{n+k-1} + h \sum_{s=0}^{k-1} \gamma_s^{*} \nabla^s f_{n+k} \tag{2.12'}$$

with

$$\gamma_s^{*} = \int_{-1}^{0} \binom{t+s-1}{s} dt, \qquad s = 0, 1, 2, \dots . \tag{2.14}$$

The truncation error and principal error function are

$$(T_h^{*})_n = \gamma_k^{*} h^k y^{(k+1)}(\overline{x}_n^{*}), \quad x_{n+1} < \overline{x}_n^{*} < x_{n+k}; \quad \tau^{*}(x) = \gamma_k^{*} y^{(k+1)}(x), \tag{2.15}$$

and the generating function for the γ_s^{*}

$$\gamma^{*}(z) = \sum_{s=0}^{\infty} \gamma_s^{*} z^s = -\frac{z}{\ln(1-z)} \, . \tag{2.16}$$

[2] Forest Ray Moulton (1872–1952) was professor of Astronomy at the University of Chicago and, from 1927 to 1936, director of the Utilities Power and Light Corp. of Chicago. He used his method during World War I and thereafter to integrate the equations of exterior ballistics. He made also contributions to celestial mechanics.

From this, one finds as before,

$$\gamma_0^* = 1,$$

$$\gamma_s^* = -\tfrac{1}{2}\,\gamma_{s-1}^* - \tfrac{1}{3}\,\gamma_{s-2}^* - \cdots - \tfrac{1}{s+1}\,\gamma_0^*, \qquad (2.17)$$

$$s = 1, 2, 3, \dots\ .$$

So, for example,

$$\gamma_0^* = 1, \quad \gamma_1^* = -\tfrac{1}{2}, \quad \gamma_2^* = -\tfrac{1}{12}, \quad \gamma_3^* = -\tfrac{1}{24}, \quad \dots\ .$$

It follows again from (2.15) that the truncation error is approximately the first omitted term in the sum of (2.12′),

$$(T_h^*)_n \approx \gamma_k^*\, \nabla^k f_{n+k}. \qquad (2.18)$$

Since the formula (2.12′) is implicit, however, it has to be solved by iteration. A common procedure is to get a first approximation by assuming the last, $(k-1)$st, difference retained to be constant over the step from x_{n+k-1} to x_{n+k}, which allows one to generate the lower-order differences backward until one obtains (an approximation to) f_{n+k}. We then compute u_{n+k} from (2.12′) and a new f_{n+k} in terms of it. Then we revise all the differences required in (2.12′) and reevaluate u_{n+k}. The process can be repeated until it converges to the desired accuracy. In effect, this is the fixed point iteration of §1.1, with a special choice of the initial approximation.

§2.3. **Predictor-corrector methods**. These are pairs of an explicit and an implicit multistep method, usually of the same order, where the explicit formula is used to predict the next approximation, and the implicit formula to correct it. Suppose we use an explicit k-step method of order k, with coefficients α_s, β_s, for the predictor, and an implicit $(k-1)$-step method of order k with coefficients α_s^*, β_s^*, for the corrector. Assume further, for simplicity, that both methods are definite (in the sense of (1.35)). Then, in Lagrange form, if $\overset{\circ}{u}_{n+k}$ is the predicted approximation, one proceeds as follows.

$$\begin{cases} \overset{\circ}{u}_{n+k} = -\sum_{s=0}^{k-1} \alpha_s u_{n+s} + h \sum_{s=0}^{k-1} \beta_s f_{n+s}, \\ u_{n+k} = -\sum_{s=1}^{k-1} \alpha_s^* u_{n+s} + h\{\beta_k^* f(x_{n+k}, \overset{\circ}{u}_{n+k}) + \sum_{s=1}^{k-1} \beta_s^* f_{n+s}\}, \\ f_{n+k} = f(x_{n+k}, u_{n+k}). \end{cases} \tag{2.19}$$

This requires exactly two evaluations of f per step and is often referred to as a PECE method, where "P" stands for "predict," "E" for "evaluate," and "C" for "correct." One could of course correct once more, and then either quit, or reevaluate f, and so on. Thus there are methods of type $\mathrm{P(EC)}^2$, $\mathrm{P(EC)}^2\mathrm{E}$, and the like. Each additional reevaluation costs another function evaluation and, therefore, the most economic methods are those of type PECE.

Let us analyze the truncation error of the PECE method (2.19). It is natural to define it by

$$(T_h^{\mathrm{PECE}})_n = \frac{1}{h} \sum_{s=1}^{k} \alpha_s^* y(x_{n+s}) - \{\beta_k^* f(x_{n+k}, \overset{\circ}{y}_{n+k}) + \sum_{s=1}^{k-1} \beta_s^* y'(x_{n+s})\},$$

where $\alpha_k^* = 1$ and

$$\overset{\circ}{y}_{n+k} = -\sum_{s=0}^{k-1} \alpha_s y(x_{n+s}) + h \sum_{s=0}^{k-1} \beta_s y'(x_{n+s});$$

that is, we apply (2.19) on exact values $u_{n+s} = y(x_{n+s})$, $s = 0, 1, \ldots, k-1$. We can write

$$\begin{aligned} (T_h^{\mathrm{PECE}})_n &= \frac{1}{h} \sum_{s=1}^{k} \alpha_s^* y(x_{n+s}) - \sum_{s=1}^{k} \beta_s^* y'(x_{n+s}) \\ &\quad + \beta_k^* [y'(x_{n+k}) - f(x_{n+k}, \overset{\circ}{y}_{n+k})] \\ &= \ell_{k+1}^* h^k y^{(k+1)}(\overline{x}_n^*) + \beta_k^* [f(x_{n+k}, y(x_{n+k})) - f(x_{n+k}, \overset{\circ}{y}_{n+k})], \end{aligned} \tag{2.20}$$

having used the truncation error $(T_h^*)_n$ (in (1.36)) of the corrector formula. Since

$$y(x_{n+k}) - \overset{\circ}{y}_{n+k} = \ell_{k+1} h^{k+1} y^{(k+1)}(\overline{x}_n) \tag{2.21}$$

is h-times the truncation error of the predictor formula, the Lipschitz condition on f yields

$$\|f(x_{n+k}, y(x_{n+k})) - f(x_{n+k}, \overset{\circ}{y}_{n+k})\| \leq L|\ell_{k+1}|h^{k+1}\|y^{(k+1)}\|_\infty, \tag{2.22}$$

and hence, from (2.20), we obtain

$$\|T_h^{\text{PECE}}\|_\infty \leq (|\ell_{k+1}^*| + hL|\ell_{k+1}\beta_k^*|)\|y^{(k+1)}\|_\infty h^k \leq Ch^k,$$

where

$$C = (|\ell_{k+1}^*| + (b-a)L|\ell_{k+1}\beta_k^*|)\|y^{(k+1)}\|_\infty.$$

Thus, the PECE method also has order k, and its principal error function is identical with that of the corrector formula, as follows immediately from (2.20) and (2.22).

The local truncation error $(T_h^{\text{PECE}})_n$ can be estimated in terms of the difference between the (locally) predicted approximation $\overset{\circ}{y}_{n+k}$ and the (locally) corrected approximation u_{n+k}. One has, indeed,

$$\begin{aligned} u_{n+k} &= -\sum_{s=1}^{k-1} \alpha_s^* y(x_{n+s}) + h\left\{\beta_k^* f(x_{n+k}, \overset{\circ}{y}_{n+k}) + \sum_{s=1}^{k-1} \beta_s^* y'(x_{n+s})\right\} \\ &= y(x_{n+k}) - h(T_h^{\text{PECE}})_n; \end{aligned}$$

that is,

$$u_{n+k} - y(x_{n+k}) = -\ell_{k+1}^* h^{k+1} y^{(k+1)}(x_n) + O(h^{k+2}),$$

whereas, from (2.21),

$$\overset{\circ}{y}_{n+k} - y(x_{n+k}) = -\ell_{k+1} h^{k+1} y^{(k+1)}(x_n) + O(h^{k+2}),$$

assuming that $y(x) \in C^{k+2}\,[a,b]$. Upon subtraction, one gets

$$u_{n+k} - \overset{\circ}{y}_{n+k} = -(\ell_{k+1}^* - \ell_{k+1}) h^{k+1} y^{(k+1)}(x_n) + O(h^{k+2}),$$

and thus,

$$y^{(k+1)}(x_n) = -\frac{1}{\ell_{k+1}^* - \ell_{k+1}} \frac{1}{h^{k+1}} (u_{n+k} - \overset{\circ}{y}_{n+k}) + O(h).$$

Since, by (2.20) and (2.21),

$$(T_h^{\text{PECE}})_n = \ell_{k+1}^* h^k y^{(k+1)}(x_n) + O(h^{k+1}),$$

we obtain

$$(T_h^{\text{PECE}})_n = -\frac{\ell_{k+1}^*}{\ell_{k+1}^* - \ell_{k+1}} \frac{1}{h} (u_{n+k} - \mathring{y}_{n+k}) + O(h^{k+1}). \tag{2.23}$$

The first term on the right of (2.23) is called the *Milne estimator* of the PECE truncation error.

The most popular choice for the predictor is a kth-order Adams-Bashforth formula, and for the corrector the corresponding Adams-Moulton formula. Here

$$\begin{cases} \mathring{u}_{n+k} = u_{n+k-1} + h \sum_{s=0}^{k-1} \beta_{k,s} f_{n+s}, \\ u_{n+k} = u_{n+k-1} + h \left\{ \beta_{k,k}^* f(x_{n+k}, \mathring{u}_{n+k}) + \sum_{s=1}^{k-1} \beta_{k,s}^* f_{n+s} \right\}, \\ f_{n+k} = f(x_{n+k}, u_{n+k}), \end{cases} \tag{2.24}$$

with coefficients $\beta_{k,s}$, $\beta_{k,s}^*$ as defined in (2.3) and (2.13), respectively.

A predictor-corrector scheme is meaningful only if the corrector formula is more accurate than the predictor formula, not necessarily in terms of order, but in terms of error coefficients. Since the principal error function of the kth-order predictor and corrector are identical except for a multiplicative constant, ℓ_{k+1} in the case of the predictor, and ℓ_{k+1}^* in the case of the corrector, we want $|C_{k,k}^*| < |C_{k,k}|$, which, assuming (3.44), is the same as

$$|\ell_{k+1}^*| < |\ell_{k+1}|. \tag{2.25}$$

For the pair of Adams formulae, we have $\ell_{k+1} = \gamma_k$ and $\ell_{k+1}^* = \gamma_k^*$ (cf. (2.7), (2.15)), and it is easy to show (see Ex. 5) that

$$|\gamma_k^*| < \frac{1}{k-1} \gamma_k, \qquad k \geq 2, \tag{2.26}$$

so that indeed the corrector has a smaller error constant than the predictor. More precisely, it can be shown (see Ex. 6) that

$$\gamma_k \sim \frac{1}{\ln k}, \qquad \gamma_k^* \sim -\frac{1}{k \ln^2 k} \quad \text{as} \quad k \to \infty, \tag{2.27}$$

although both approximations are not very accurate (unless k is extremely large), the relative errors being of $O(1/k)$ and $O(1/\ln k)$, respectively.

§3. Global Description of Multistep Methods

We already commented in §1.1 on the fact that a linear multistep method such as (1.2) represents a system of nonlinear difference equations. To study its properties, one inevitably has to deal with the theory of difference equations. Since nonlinearities are hidden behind a (small) factor h, it turns out that the theory of *linear* difference equations — with constant coefficients at that — will suffice to carry through the analysis. We therefore begin with recalling the basic facts of this theory. We then define stability in a manner similarly to that of Ch. 5, §3.1, and identify a root condition for the characteristic equation of the difference equation as the true source of stability. This, together with consistency, then immediately implies convergence, as for one-step methods.

§3.1. **Linear difference equations.** With notations close to those adopted in §1 we consider a (scalar) linear difference equation of order k,

$$v_{n+k} + \alpha_{k-1}v_{n+k-1} + \cdots + \alpha_0 v_n = \varphi_{n+k}, \quad n = 0, 1, 2, \ldots\ , \tag{3.1}$$

where α_s are given real numbers not depending on n and not necessarily with $\alpha_0 \neq 0$, and $\{\varphi_{n+k}\}_{n=0}^{\infty}$ is a given sequence. Any sequence $\{v_n\}_{n=0}^{\infty}$ satisfying (3.1) is called a *solution* of the difference equation. It is uniquely determined by the *starting values* $v_0, v_1, \ldots, v_{k-1}$. (If $\alpha_0 = \alpha_1 = \cdots = \alpha_{\ell-1} = 0$, $1 \leq \ell < k$, then $\{v_n\}_{n \geq k}$ is not affected by $v_0, v_1, \ldots, v_{\ell-1}$.) The equation (3.1) is called *homogeneous* if $\varphi_{n+k} = 0$ for all $n \geq 0$, and *inhomogeneous* otherwise. It has exact order k if $\alpha_0 \neq 0$.

§3.1.1. *Homogeneous equation.* We begin with the homogeneous equation

$$v_{n+k} + \alpha_{k-1}v_{n+k-1} + \cdots + \alpha_0 v_n = 0, \quad n = 0, 1, 2, \ldots\ . \tag{3.2}$$

We call

$$\alpha(t) = \sum_{s=0}^{k} \alpha_s t^s \qquad (\alpha_k = 1) \tag{3.3}$$

the *characteristic polynomial* of (3.2), and

$$\alpha(t) = 0 \tag{3.4}$$

its *characteristic equation.* If t_s, $s = 1, 2, \ldots, k'$ ($k' \leq k$), denote the distinct roots of (3.4) and m_s their multiplicities, then the general solution of (3.2) is given by

$$v_n = \sum_{s=1}^{k'} \left(\sum_{r=0}^{m_s-1} c_{rs} n^r \right) t_s^n, \qquad n = 0, 1, 2, \dots , \tag{3.5}$$

where c_{rs} are arbitrary (real or complex) constants. There is a one-to-one correspondence between these k constants and the k starting values $v_0, v_1, \dots, v_{k-1}$.

We remark that if $\alpha_0 = 0$, then one of the roots t_s is zero, which contributes an identically vanishing solution (except for $n = 0$, where by convention $t_s^0 = 0^0 = 1$). If additional coefficients $\alpha_1 = \cdots = \alpha_{\ell-1}$, $\ell < k$, are zero, this further restricts the solution manifold. Note also that a complex root $t_s = \rho e^{i\theta}$ contributes a complex solution component in (3.5). However, since the α_s are assumed real, with t_s, also $\bar{t}_s = \rho e^{-i\theta}$ is a root of (3.4), and we can combine the two complex solutions $n^r t_s^n$ and $n^r \bar{t}_s^n$ to form a pair of *real* solutions,

$$\frac{1}{2}\, n^r (t_s^n + \bar{t}_s^n) = n^r \rho^n \cos n\theta, \quad \frac{1}{2i}\, n^r (t_s^n - \bar{t}_s^n) = n^r \rho^n \sin n\theta.$$

If we do this for each complex root t_s and select all coefficients in (3.5) real, we obtain the general solution in real form.

The following is a simple but important observation.

Theorem 6.3.1. *We have $|v_n| \le M$, all $n \ge 0$, for every solution $\{v_n\}$ of the homogeneous equation* (3.2), *with M depending only on the starting values $v_0, v_1, \dots, v_{k-1}$ (but not on n) if and only if*

$$\alpha(t_s) = 0 \ \ implies \ \ \begin{cases} either \ \ |t_s| < 1 \\ or \ \ |t_s| \ = 1, \ m_s = 1. \end{cases} \tag{3.6}$$

Proof. **Sufficiency of** (3.6). If (3.6) holds for every root t_s of (3.4), then every term $n^r t_s^n$ is bounded for all $n \ge 0$ (going to zero as $n \to \infty$ in the first case of (3.6) and being equal to 1 in absolute value in the second case). Since the constants c_{rs} in (3.5) are uniquely determined by the starting values $v_0, v_1, \dots, v_{k-1}$, the assertion $|v_n| \le M$ follows.

Neccesity of (3.6). If $|v_n| \le M$, we cannot have $|t_s| > 1$, since we can always arrange to have $c_{rs} \ne 0$ for a corresponding term in (3.5) and select all other constants to be zero. This singles out an unbounded solution of (3.2). Nor can we have, for the same reason, $|t_s| = 1$ and $m_s > 1$. □

The condition (3.6) is referred to as the *root condition* for the difference equation (3.2) (and also for equation (3.1)).

The representation (3.5) of the general solution is inconvenient insofar as it does not explicitly exhibit its dependence on the starting values. A representation that does, can be obtained by defining k special solutions $\{h_{n,s}\}$, $s = 0, 1, \ldots, k-1$, of (3.2), having as starting values those of the unit matrix; that is,

$$h_{n,s} = \delta_{n,s} \quad \text{for} \quad n = 0, 1, \ldots, k-1, \tag{3.7}$$

with $\delta_{n,s}$ the Kronecker delta. Then indeed, the general solution of (3.2) is

$$v_n = \sum_{s=0}^{k-1} v_s h_{n,s}. \tag{3.8}$$

(Note that if $\alpha_0 = 0$, then $h_{0,0} = 1$ and $h_{n,0} = 0$ for all $n \geq 1$; similarly, if $\alpha_0 = \alpha_1 = 0$, and so on.)

§3.1.2. *Inhomogeneous equation.* To deal with the general inhomogeneous equation (3.1), we define for each $m = k, k+1, k+2, \ldots$ the solution $\{g_{n,m}\}_{n=0}^{\infty}$ of the "initial value problem"

$$\begin{aligned} &\sum_{s=0}^{k} \alpha_s g_{n+s,m} = \delta_{n,m-k}, \quad n = 0, 1, 2, \ldots\ , \\ &g_{0,m} = g_{1,m} = \cdots = g_{k-1,m} = 0. \end{aligned} \tag{3.9}$$

Here the difference equation is a very special case of the equation (3.1), namely, with $\varphi_{n+k} = \delta_{n,m-k} = \begin{cases} 0 & \text{if} \quad n \neq m-k \\ 1 & \text{if} \quad n = m-k \end{cases}$ an "impulse function." We can then superimpose these special solutions to form the solution

$$v_n = \sum_{m=k}^{n} g_{n,m} \varphi_m \tag{3.10}$$

of the initial value problem $v_0 = v_1 = \cdots = v_{k-1} = 0$ for (3.1) (*Duhamel's principle*). This is easily verified by observing first that $g_{n,m} = 0$ for $n < m$, so that (3.10) can be written in the form

$$v_n = \sum_{m=k}^{\infty} g_{n,m} \varphi_m. \tag{3.10$'$}$$

We then have

$$\begin{aligned}\sum_{s=0}^{k} \alpha_s v_{n+s} &= \sum_{s=0}^{k} \alpha_s \sum_{m=k}^{\infty} g_{n+s,m}\varphi_m \\ &= \sum_{m=k}^{\infty} \varphi_m \sum_{s=0}^{k} \alpha_s g_{n+s,m} = \varphi_{n+k},\end{aligned}$$

where the last equation follows from (3.9).

Since effectively (3.9) is a "delayed" initial value problem for a homogeneous difference equation with k starting values $0, 0, \ldots, 0, 1$, we can express $g_{n,m}$ in (3.10) alternatively as $h_{n-m+k-1,k-1}$ (cf. (3.7)). The general solution of the inhomogeneous equation (3.1) is the general solution of the homogeneous equation plus the special solution (3.10) of the inhomogeneous equation; thus, in view of (3.8),

$$v_n = \sum_{s=0}^{k-1} v_s h_{n,s} + \sum_{m=k}^{n} h_{n-m+k-1,k-1}\varphi_m. \tag{3.11}$$

Theorem 6.3.2. *There exists a constant $M > 0$, independent of n, such that*

$$|v_n| \le M \left\{ \max_{0 \le s \le k-1} |v_s| + \sum_{m=k}^{n} |\varphi_m| \right\}, \quad n = 0, 1, 2, \ldots, \tag{3.12}$$

for every solution $\{v_n\}$ of (3.1) and for every $\{\varphi_{n+k}\}$, if and only if the characteristic polynomial $\alpha(t)$ of (3.1) satisfies the root condition (3.6).

Proof. `Sufficiency of` (3.6). By Theorem 6.3.1 the inequality (3.12) follows immediately from (3.11), with M the constant of Theorem 6.3.1.

`Necessity of` (3.6). Take $\varphi_m = 0$, all $m \ge k$, in which case $\{v_n\}$ is a bounded solution of the homogeneous equation, and hence, by Theorem 6.3.1, the root condition must hold. □

As an application of Theorems 6.3.1 and 6.3.2, we take $v_n = h_{n,s}$ (cf. (3.7)) and $v_n = g_{n,m}$ (cf. (3.9)). The former is bounded by M, by Theorem 6.3.1, and so is the latter, by (3.12), since $g_{n,s} = 0$ for $0 \le s \le k-1$ and all φ_m are zero except one, which is 1. Thus, if the root condition is satisfied, then

$$|h_{n,s}| \le M, \quad |g_{n,m}| \le M, \quad \text{all } \ n \ge 0. \tag{3.13}$$

Note that, since $h_{s,s} = 1$, we must have $M \ge 1$.

§3.2. **Stability and root condition.** We now return to the general multistep method (1.2) for solving the initial value problem (1.1). In terms of the residual operator R_h of (1.13_h) we define stability, similarly as for one-step methods (cf. Ch. 5, §3.1), as follows.

Definition 3.1. The method (1.2) is called *stable* on $[a, b]$ if there exists a constant $K > 0$ not depending on h such that for an arbitrary (uniform) grid h on $[a, b]$, and for arbitrary two grid functions v, $w \in \Gamma_h[a, b]$, there holds

$$\|v - w\|_\infty \leq K \left(\max_{0 \leq s \leq k-1} \|v_s - w_s\| + \|R_h v - R_h w\|_\infty \right), \quad v, w \in \Gamma_h[a, b], \tag{3.14}$$

for all h sufficiently small.

The motivation for this "stability inequality" is much the same as for one-step methods (Ch. 5, (3.11) through (3.13)), and we do not repeat it here.

Let $\mathcal{F}$ be the family of functions f satisfying a uniform Lipschitz condition

$$\|f(x, y) - f(x, y^*)\| \leq L\|y - y^*\|, \quad x \in [a, b], \quad y, y^* \in \mathbb{R}^d, \tag{3.15}$$

with Lipschitz constant $L = L_f$ depending on f.

Theorem 6.3.3. *The multistep method* (1.2) *is stable for every* $f \in \mathcal{F}$ *if and only if its characteristic polynomial* (3.3) *satisfies the root condition* (3.6).

Proof. `Necessity` of (3.6). Consider $f(x, y) \equiv 0$, which is certainly in $\mathcal{F}$, and for which

$$(R_h v)_n = \frac{1}{h} \sum_{s=0}^{k} \alpha_s v_{n+s}.$$

Take $v = u$ and $w = 0$ in (3.14), where u is a grid function satisfying

$$\sum_{s=0}^{k} \alpha_s u_{n+s} = 0, \quad n = 0, 1, 2, \ldots . \tag{3.16}$$

Since (3.14) is to hold for arbitrarily fine grids, the integer n in (3.16) can assume arbitrarily large values, and it follows from (3.14) that u is uniformly bounded,

$$\|u\|_\infty \leq K \max_{0 \leq s \leq k-1} \|u_s\|, \tag{3.17}$$

the bound depending only on the starting values. Since u is a solution of the homogeneous difference equation (3.2), its characteristic polynomial must satisfy the root condition by Theorem 6.3.1.

`Sufficiency of` (3.6). Let $f \in \mathcal{F}$ and $v,\ w \in \Gamma_h[a,b]$ be arbitrary grid functions. By definition of R_h, we have

$$\sum_{s=0}^{k} \alpha_s v_{n+s} = h \sum_{s=0}^{k} \beta_s f(x_{n+s}, v_{n+s}) + h(R_h v)_n, \quad n = 0, 1, 2, \ldots, N-k,$$

and similarly for w. Subtraction then gives

$$\sum_{s=0}^{k} \alpha_s (v_{n+s} - w_{n+s}) = \varphi_{n+k}, \quad n = 0, 1, 2, \ldots, N-k,$$

where

$$\varphi_{n+k} = h \sum_{s=0}^{h} \beta_s [f(x_{n+s}, v_{n+s}) - f(x_{n+s}, w_{n+s})] + h[(R_h v)_n - (R_h w)_n]. \tag{3.18}$$

Therefore, $v - w$ is formally a solution of the inhomogeneous difference equation (3.1) (the forcing function φ_{n+k}, though, depending also on v and w), so that by (3.8) and (3.10) we can write

$$v_n - w_n = \sum_{s=0}^{k-1} (v_s - w_s) h_{n,s} + \sum_{m=k}^{n} g_{n,m} \varphi_m.$$

Since the root condition is satisfied, we have by (3.13)

$$|h_{n,s}| \le M, \quad |g_{n,m}| \le M$$

for some constant $M \ge 1$, uniformly in n and m. Therefore,

$$\|v_n - w_n\| \le M \left\{ k \max_{0 \le s \le k-1} \|v_s - w_s\| + \sum_{m=k}^{n} \|\varphi_m\| \right\}, \tag{3.19}$$
$$n = 0, 1, 2, \ldots, N.$$

By (3.18) we can estimate

$$\begin{aligned}\|\varphi_m\| &= \|h\sum_{s=0}^{k}\beta_s[f(x_{m-k+s},v_{m-k+s})-f(x_{m-k+s},w_{m-k+s})]\\ &\qquad + h[(R_hv)_{m-k}-(R_hw)_{m-k}]\| \\ &\le h\beta L\sum_{s=0}^{k}\|v_{m-k+s}-w_{m-k+s}\| + h\|R_hv-R_hw\|_\infty,\end{aligned} \tag{3.20}$$

where

$$\beta = \max_{0\le s\le k}|\beta_s|.$$

Letting $e = v - w$ and $r_h = R_hv - R_hw$, we obtain from (3.19) and (3.20)

$$\|e_n\| \le M\left\{k\max_{0\le s\le k-1}\|e_s\| + h\beta L\sum_{m=k}^{n}\sum_{s=0}^{k}\|e_{m-k+s}\| + Nh\|r_h\|_\infty\right\}.$$

Noting that

$$\begin{aligned}\sum_{m=k}^{n}\sum_{s=0}^{k}\|e_{m-k+s}\| &= \sum_{s=0}^{k}\sum_{m=k}^{n}\|e_{m-k+s}\| \le \sum_{s=0}^{k}\sum_{m=0}^{n}\|e_m\| \\ &= (k+1)\sum_{m=0}^{n}\|e_m\|,\end{aligned}$$

and using $Nh = b - a$, we get

$$\|e_n\| \le M\left\{k\max_{0\le s\le k-1}\|e_s\| + h(k+1)\beta L\sum_{m=0}^{n}\|e_m\| + (b-a)\|r_h\|_\infty\right\}. \tag{3.21}$$

Now let h be so small that

$$1 - h(k+1)\beta LM \ge \tfrac{1}{2}\,.$$

Then, splitting off the term with $\|e_n\|$ on the right of (3.21) and moving it to the left, we obtain

$$\begin{aligned}&(1-h(k+1)\beta LM)\|e_n\| \\ &\le M\left\{k\max_{0\le s\le k-1}\|e_s\| + h(k+1)\beta L\sum_{m=0}^{n-1}\|e_m\| + (b-a)\|r_h\|_\infty\right\},\end{aligned}$$

or

$$\|e_n\| \le 2M\left\{h(k+1)\beta L \sum_{m=0}^{n-1} \|e_m\| + k \max_{0\le s\le k-1} \|e_s\| + (b-a)\|r_h\|_\infty\right\}.$$

Thus,

$$\|e_n\| \le hA \sum_{m=0}^{n-1} \|e_m\| + B, \tag{3.22}$$

where

$$A = 2(k+1)\beta LM, \quad B = 2M\left(k \max_{0\le s\le k-1} \|e_s\| + (b-a)\|r_h\|_\infty\right). \tag{3.23}$$

Consider, along with (3.22), the difference equation

$$E_n = hA \sum_{m=0}^{n-1} E_m + B, \quad E_0 = B. \tag{3.24}$$

It is easily seen by induction that

$$E_n = B(1+hA)^n, \quad n = 0, 1, 2, \ldots \;. \tag{3.25}$$

Subtracting (3.24) from (3.22), we get

$$\|e_n\| - E_n \le hA \sum_{m=0}^{n-1} (\|e_m\| - E_m). \tag{3.26}$$

Clearly, $\|e_0\| \le B = E_0$. Thus, by (3.26), $\|e_1\| - E_1 \le 0$, and by induction on n,

$$\|e_n\| \le E_n, \qquad n = 0, 1, 2, \ldots \;.$$

Thus, by (3.25),

$$\|e_n\| \le B(1+hA)^n \le Be^{nhA} \le Be^{(b-a)A}.$$

Recalling the definition of B in (3.23), we find

$$\|e_n\| \le 2Me^{(b-a)A}\left\{k \max_{0\le s\le k-1} \|e_s\| + (b-a)\|R_hv - R_hw\|_\infty\right\},$$

which is the stability inequality (3.14) with

$$K = 2Me^{(b-a)A}\max\{k, b-a\}. \quad \square$$

Theorem 6.3.3 in particular shows that all Adams methods are stable, since for them

$$\alpha(t) = t^k - t^{k-1} = t^{k-1}(t-1),$$

and the root condition is trivially satisfied.

Theorem 6.3.3 also holds for predictor-corrector methods of the type considered in (2.19), if one defines the residual operator R_h^{PECE} in the definition of stability in the obvious way:

$$(R_h^{\text{PECE}} v)_n = \frac{1}{h} \sum_{s=1}^{k} \alpha_s^* v_{n+s} - \left\{ \beta_k^* f(x_{n+k}, \overset{\circ}{v}_{n+k}) + \sum_{s=1}^{k-1} \beta_s^* f(x_{n+s}, v_{n+s}) \right\}, \tag{3.27}$$

with

$$\overset{\circ}{v}_{n+k} = -\sum_{s=0}^{k-1} \alpha_s v_{n+s} + h \sum_{s=0}^{k-1} \beta_s f(x_{n+s}, v_{n+s}), \tag{3.28}$$

and considers the characteristic polynomial to be that of the corrector formula (see Ex. 8).

The problem of constructing stable multistep methods of maximum order is considered later in §4.

§3.3. **Convergence**. With the powerful property of stability at hand, the convergence of multistep methods follows almost immediately as a corollary. We first define what we mean by convergence.

Definition 3.2. Consider a uniform grid on $[a,b]$ with grid length h. Let $u = \{u_n\}$ be the grid function obtained by applying the multistep method (1.2) on $[a,b]$, with starting approximations $u_s(h)$ as in (1.4). Let $y = \{y_n\}$ be the grid function induced by the exact solution of the initial value problem on $[a,b]$. The method (1.2) is said to converge on $[a,b]$ if there holds

$$\|u - y\|_\infty \to 0 \quad \text{as} \quad h \to 0 \tag{3.29}$$

whenever

$$u_s(h) \to y_0 \quad \text{as} \quad h \to 0, \quad s = 0, 1, \ldots, k-1. \tag{3.30}$$

Theorem 6.3.4. *The multistep method* (1.2) *converges for all* $f \in \mathcal{F}$ (cf. (3.15)) *if and only if it is consistent and stable. If, in addition,* (1.2) *has order* p *and* $u_s(h) - y(x_s) = O(h^p)$, $s = 0, 1, \ldots, k-1$, *then*

$$\|u - y\|_\infty = O(h^p) \quad \textit{as} \quad h \to 0. \tag{3.31}$$

Proof. `Necessity`. Let $f \equiv 0$ (which is certainly in $\mathcal{F}$) and $y_0 = 0$. Then $y(x) \equiv 0$ and (1.2) reduces to (3.16). Since the same relations hold for each component of y and u, we may as well consider a scalar problem. (This holds for the rest of the necessity part of the proof.) Assume first, by way of contradiction, that (1.2) is not stable. Then, by Theorem 6.3.3, there is a root t_s of the characteristic equation $\alpha(t) = 0$ for which either $|t_s| > 1$, or $|t_s| = 1$ and $m_s > 1$. In the first case, (3.16) has a solution $u_n = ht_s^n$ for which $|u_n| = \frac{b-a}{n}|t_s|^n$. Clearly, the starting values $u_0, u_1, \ldots, u_{k-1}$ all tend to $y_0 = 0$ as $h \to 0$, but $|u_n| \to \infty$ as $n \to \infty$. Since for h sufficiently small we can have n arbitrarily large, this contradicts convergence of $\{u_n\}$ to the solution $\{y_n\}$, $y_n \equiv 0$. The same argument applies in the second case if we consider (say) $u_n = h^{\frac{1}{2}} n^{m_s - 1} t_s^n$, where now $|t_s| = 1$, $m_s > 1$. This proves the necessity of stability.

To prove consistency, we must show that $\alpha(1) = 0$ and $\alpha'(1) = \beta(1)$, where (anticipating (4.1)) we define $\beta(t) = \sum_{s=0}^k \beta_s t^s$. For the former, we consider $f \equiv 0$, $y_0 = 1$, which has the exact solution $y(x) \equiv 1$ and the numerical solution still satisfying (3.16). If we take $u_s = 1$, $s = 0, 1, \ldots, k-1$, the assumed convergence implies that $u_{n+s} \to 1$ as $h \to 0$, hence $0 = \sum_{s=0}^k \alpha_s u_{n+s} \to \alpha(1)$. For the latter, we consider $f \equiv 1$ and $y_0 = 0$; that is, $y(x) \equiv x - a$. The multistep method now generates a grid function $u = \{u_n\}$ satisfying

$$\sum_{s=0}^{k} \alpha_s u_{n+s} - \beta(1)h = 0.$$

A particular solution is given by $u_n = \frac{\beta(1)}{\alpha'(1)} nh$. Indeed, since $\alpha(1) = 0$, as already shown, we have

$$\begin{aligned}
\sum_{s=0}^{k} \alpha_s u_{n+s} - \beta(1)h &= \frac{\beta(1)}{\alpha'(1)} h \sum_{s=0}^{k} \alpha_s (n+s) - \beta(1)h \\
&= \frac{\beta(1)}{\alpha'(1)} h[n\alpha(1) + \alpha'(1)] - \beta(1)h \\
&= \frac{\beta(1)}{\alpha'(1)} h\alpha'(1) - \beta(1)h = 0.
\end{aligned}$$

Since also $u_s \to y_0 = 0$ for $h \to 0$, $s = 0, 1, \ldots, k-1$, and since by assumption $\{u_n\}$ converges to $\{y_n\}$, $y_n = nh$, we must have $\frac{\beta(1)}{\alpha'(1)} = 1$; that is, $\alpha'(1) = \beta(1)$.

`Sufficiency`. Letting $v = u$ and $w = y$ in the stability inequality (3.14), and noting that $R_h u = 0$ and $R_h y = T_h$, the grid function $\{(T_h)_n\}$ of the truncation errors (cf. (1.14)), we get

$$\|u - y\|_\infty \le K \left\{ \max_{0 \le s \le k-1} \|u_s - y(x_s)\| + \|T_h\|_\infty \right\}. \tag{3.32}$$

If the method (1.2) is consistent, the second term in braces tends to zero and, therefore, also the term on the left, if (3.30) holds, that is, if $u_s - y(x_s) \to 0$ as $h \to 0$. This completes the proof of the first part of the theorem. The second part follows likewise, since by assumption both terms between braces in (3.32) are of $O(h^p)$. □

In view of the remark near the end of §3.2, Theorem 6.3.4 holds also for the kth-order predictor-corrector method (2.19) with $p = k$. The proof is the same, since for the truncation error, $\|T_h^{\text{PECE}}\|_\infty = O(h^k)$, as was shown in §2.3.

§3.4. **Asymptotics of global error**. A refinement of Theorem 6.3.4, Eq. (3.31), exhibiting the leading term in the global discretization error, is given by the following theorem.

Theorem 6.3.5. *Assume that*

(1) $f(x, y) \in C^2$ *on* $[a, b] \times \mathbb{R}^d$;

(2) *the multistep* (1.2) *is stable* (i.e., satisfies the root condition) *and has order* $p \ge 1$;

(3) *the exact solution* $y(x)$ *of* (1.1) *is of class* $C^{p+2}[a, b]$;

(4) *the starting approximations* (1.4) *satisfy*

$$u_s - y(x_s) = O(h^{p+1}) \quad as \quad h \to 0, \quad s = 0, 1, \ldots, k-1;$$

(5) $e(x)$ *is the solution of the linear initial value problem*

$$\frac{de}{dx} = f_y(x, y(x))e - y^{(p+1)}(x), \quad e(a) = 0. \tag{3.33}$$

Then, for $n = 0, 1, 2, \ldots, N$,

$$u_n - y(x_n) = C_{k,p} h^p e(x_n) + O(h^{p+1}) \quad as \quad h \to 0, \tag{3.34}$$

where $C_{k,p}$ is the error constant of (1.2); *that is,*

$$C_{k,p} = \frac{\ell_{p+1}}{\sum_{s=0}^{k} \beta_s}\,, \qquad \ell_{p+1} = L\,\frac{t^{p+1}}{(p+1)!}\,. \tag{3.35}$$

Before proving the theorem, we make the following remarks.

1. Under the assumptions made in (1) and (3), the solution e of (3.33) exists uniquely on $[a,b]$. It is the same for all multistep methods of order p.

2. The error constant $C_{k,p}$ depends only (through the coefficients α_s, β_s) on the multistep method.

3. For a given differential system (1.1), the asymptotically best k-step method of order p would be one for which $|C_{k,p}|$ is smallest. Unfortunately, as we see later in §4.3, the minimum of $|C_{k,p}|$ over all *stable* k-step methods of order p cannot generally be attained.

4. Stability and $p \geq 1$ implies $\sum\limits_{s=0}^{k} \beta_s \neq 0$. In fact, $\alpha(1) = \sum\limits_{s=0}^{k} \alpha_s = 0$, since $L1 = 0$, and $Lt = \sum\limits_{s=0}^{k} s\alpha_s - \sum\limits_{k=0}^{k} \beta_s = 0$ since $p \geq 1$. Consequently,

$$\sum_{s=0}^{k} \beta_s = \sum_{s=0}^{k} s\alpha_s = \alpha'(1) \neq 0 \tag{3.36}$$

by the root condition. Actually, $\sum\limits_{s=0}^{k} \beta_s > 0$, as we show later in §4.3.

Proof of Theorem 6.3.5. Define the grid function $r = \{r_n\}$ by

$$r = h^{-p}(u - y). \tag{3.37}$$

Then

$$\frac{1}{h}\sum_{s=0}^{k} \alpha_s r_{n+s} = h^{-p}\left[\frac{1}{h}\sum_{s=0}^{k} \alpha_s u_{n+s} - \frac{1}{h}\sum_{s=0}^{k} \alpha_s y(x_{n+s})\right]$$

$$= h^{-p}\left[\sum_{s=0}^{k} \beta_s f(x_{n+s}, u_{n+s}) - \sum_{s=0}^{k} \beta_s y'(x_{n+s}) - (T_h)_n\right],$$

where T_h is the trunction error defined in (1.14′). Expanding f about the exact solution trajectory and noting the form of the principal error function given in Theorem 6.1.3, we obtain

$$\frac{1}{h}\sum_{s=0}^{k}\alpha_s r_{n+s} = h^{-p}\left[\sum_{s=0}^{k}\beta_s\{f(x_{n+s}, y(x_{n+s})) + f_y(x_{n+s}, y(x_{n+s}))(u_{n+s} - y(x_{n+s}))\right.$$

$$\left.+O(h^{2p})\} - \sum_{s=0}^{k}\beta_s y'(x_{n+s}) - \ell_{p+1}y^{(p+1)}(x_n)h^p + O(h^{p+1})\right],$$

having used Assumption (1) and the fact that $u_{n+s} - y(x_{n+s}) = O(h^p)$ by Theorem 6.3.4. Now the sums over f and y' cancel, since y is the solution of the differential system (1.1). Furthermore, $O(h^{2p})$ is of $O(h^{p+1})$ since $p \geq 1$, and making use of the definition (3.37) of r, we can simplify the preceding to

$$\frac{1}{h}\sum_{s=0}^{k}\alpha_s r_{n+s} = h^{-p}\left[\sum_{s=0}^{k}\beta_s f_y(x_{n+s}, y(x_{n+s}))h^p r_{n+s} - \ell_{p+1}y^{(p+1)}(x_n)h^p + O(h^{p+1})\right]$$

$$= \sum_{s=0}^{k}\beta_s f_y(x_{n+s}, y(x_{n+s}))r_{n+s} - \ell_{p+1}y^{(p+1)}(x_n) + O(h).$$

Now

$$\sum_{s=0}^{k}\beta_s y^{(p+1)}(x_{n+s}) = \sum_{s=0}^{k}\beta_s[y^{(p+1)}(x_n) + O(h)] = \left(\sum_{s=0}^{k}\beta_s\right)y^{(p+1)}(x_n) + O(h),$$

so that

$$\begin{aligned}\ell_{p+1}y^{(p+1)}(x_n) &= \frac{\ell_{p+1}}{\sum_{s=0}^{k}\beta_s}\sum_{s=0}^{k}\beta_s y^{(p+1)}(x_{n+s}) + O(h)\\ &= C_{k,p}\sum_{s=0}^{k}\beta_s y^{(p+1)}(x_{n+s}) + O(h),\end{aligned}$$

by the definition (3.35) of the error constant $C_{k,p}$. Thus,

$$\frac{1}{h}\sum_{s=0}^{k}\alpha_s r_{n+s} - \sum_{s=0}^{k}\beta_s[f_y(x_{n+s}, y(x_{n+s}))r_{n+s} - C_{k,p}y^{(p+1)}(x_{n+s})] = O(h).$$

Defining

$$\mathring{r} = \frac{1}{C_{k,p}}\, r, \tag{3.38}$$

we finally get

$$\frac{1}{h}\sum_{s=0}^{k}\alpha_s\overset{\circ}{r}_{n+s}-\sum_{s=0}^{k}\beta_s[f_y(x_{n+s},y(x_{n+s}))\overset{\circ}{r}_{n+s}-y^{(p+1)}(x_{n+s})]=O(h). \tag{3.39}$$

The left-hand side can now be recognized as the residual operator R_h^g of the multistep method (1.2) applied to the grid function $\overset{\circ}{r}$, however, not for the original differential system (1.1), but for the linear system (3.33) with right-hand side

$$g(x,e):=f_y(x,y(x))e-y^{(p+1)}(x),\quad a\le x\le b,\quad e\in\mathbb{R}^d. \tag{3.40}$$

In other words,

$$\|R_h^g\overset{\circ}{r}\|_\infty=O(h). \tag{3.41}$$

For the exact solution $e(x)$ of (3.33), we have likewise

$$\|R_h^g e\|_\infty=O(h), \tag{3.42}$$

since by (1.14) (applied to the linear system $e'=g(x,e)$) R_h^g is the truncation error of the multistep method (1.2), and its order is $p\ge 1$. Since by Assumption (2) the multistep method (1.2) is stable, we can apply the stability inequality (3.14) (for the system $e'=g(x,e)$) to the two grid functions $v=\overset{\circ}{r}$ and $w=e$, giving

$$\begin{aligned}\|\overset{\circ}{r}-e\|_\infty&\le K\left(\max_{0\le s\le k-1}\|\overset{\circ}{r}_s-e(x_s)\|+\|R_h^g\overset{\circ}{r}-R_h^g e\|_\infty\right)\\&=K\left(\max_{0\le s\le k-1}\|\overset{\circ}{r}_s-e(x_s)\|+O(h)\right).\end{aligned} \tag{3.43}$$

It remains to observe that, for $0\le s\le k-1$,

$$\overset{\circ}{r}_s-e(x_s)=\frac{1}{C_{k,p}}\,r_s-e(x_s)=\frac{1}{C_{k,p}}\,h^{-p}[u_s-y(x_s)]-e(x_s),$$

and hence, by Assumption (4) and $e(a)=0$,

$$\overset{\circ}{r}_s-e(x_s)=O(h)-[e(a)+she'(\xi_s)]=O(h),$$

to conclude

$$\max_{0\le s\le k-1}\|\overset{\circ}{r}_s-e(x_s)\|=O(h)$$

and, therefore, by (3.43), (3.38), and (3.37),

$$\|u - y - C_{k,p}h^p e\|_\infty = O(h^{p+1}),$$

as was to be shown. □

The proof of Theorem 6.3.5 applies to the predictor-corrector method (2.19) with $C_{k,p}$ replaced by $C^*_{k,k} = \ell^*_{k+1} / \sum_{s=0}^k \beta^*_s$, the error constant for the corrector formula, once one has shown that $\overset{\circ}{u}_{n+k} - u_{n+k} = O(h^{k+1})$ in (2.19), and $\overset{\circ}{y}_{n+k} - y(x_{n+k}) = O(h^{k+1})$. The first relation is true for special predictor-corrector schemes (cf. (3.44), (3.47)), whereas the second says that h times the truncation error of the predictor formula has the order shown.

§3.5. **Estimation of global error**. It is natural to try, similarly as with one-step methods (cf. Ch. 5, §4.1), to estimate the leading term in the asymptotic formula (3.34) of the global error by integrating with Euler's method the "variational equation" (3.33) along with the multistep integration of the principal equation (1.1). The main technical difficulty is to correctly estimate the driving function $y^{(p+1)}$ in the linear system (3.33). It turns out, however, that Milne's procedure (cf. (2.23)) for estimating the local truncation error in predictor-corrector schemes can be extended to estimate the global error as well, provided the predictor and corrector formulae have the same characteristic polynomial, more precisely, if in the predictor-corrector scheme (2.19) there holds

$$\alpha^*_s = \alpha_s \quad \text{for} \quad s = 1, 2, \ldots, k; \quad \alpha_0 = 0. \tag{3.44}$$

This is true, in particular, for the Adams predictor-corrector scheme (2.24). We formulate the procedure in the following theorem.

Theorem 6.3.6. *Assume that*

(1) $f(x, y) \in C^2$ *on* $[a, b] \times \mathbb{R}^d$;

(2) *the predictor-corrector scheme* (2.19) *is based on a pair of kth-order formulae* ($k \geq 1$) *satisfying* (3.44) *and having local error constants* ℓ_{k+1}, ℓ^*_{k+1} *for the predictor and corrector, respectively*;

(3) *the exact solution* $y(x)$ *of* (1.1) *is of class* $C^{k+2}[a, b]$;

(4) *the starting approximations* (1.4) *satisfy*

$$u_s - y(x_s) = O(h^{k+1}) \quad \textit{as} \quad h \to 0, \quad s = 0, 1, \ldots, k-1;$$

(5) *along with the grid function* $u = \{u_n\}$ *constructed by the predictor-corrector scheme, we generate the grid function* $v = \{v_n\}$ *in the following manner* (where $\overset{\circ}{u}_{n+k}$ is defined as in (2.19)) :

$$v_s = 0, \qquad s = 0, 1, \dots, k-1;$$
$$v_{n+k} = v_{n+k-1} + h\left\{ f_y(x_n, u_n)v_n + \frac{h^{-(k+1)}}{\ell^*_{k+1} - \ell_{k+1}} (\overset{\circ}{u}_{n+k} - u_{n+k}) \right\},$$
$$n = 0, 1, 2, \dots, N-k. \tag{3.45}$$

Then, for $n = 0, 1, \dots, N$,

$$u_n - y(x_n) = C^*_{k,k} h^k v_n + O(h^{k+1}) \ \textit{ as } \ h \to 0, \tag{3.46}$$

where $C^*_{k,k}$ *is the* (global) *error constant for the corrector formula.*

Proof. The proof is the same as that of Theorem 5.4.1, once it has been shown, in place of Ch. 5, Eq. (4.5), that

$$\frac{h^{-(k+1)}}{\ell^*_{k+1} - \ell_{k+1}} (\overset{\circ}{u}_{n+k} - u_{n+k}) = y^{(k+1)}(x_n) + O(h). \tag{3.47}$$

We now proceed to establish (3.47).

By (2.19), we hve

$$\overset{\circ}{u}_{n+k} - u_{n+k} = \sum_{s=1}^{k-1} \alpha^*_s u_{n+s} - \sum_{s=0}^{k-1} \alpha_s u_{n+s}$$
$$+ h\left\{ \sum_{s=0}^{k-1} \beta_s f(x_{n+s}, u_{n+s}) - \beta^*_k f(x_{n+k}, \overset{\circ}{u}_{n+k}) - \sum_{s=1}^{k-1} \beta^*_s f(x_{n+s}, u_{n+s}) \right\}.$$

The first two sums on the right cancel because of (3.44). In the expression between braces, we expand each f about the exact solution trajectory to obtain

$$
\begin{aligned}
\{\cdots\} = &\sum_{s=0}^{k-1} \beta_s[f(x_{n+s}, y(x_{n+s})) + f_y(x_{n+s}, y(x_{n+s}))(u_{n+s} - y(x_{n+s})) + O(h^{2k})] \\
&- \beta_k^*[f(x_{n+k}, y(x_{n+k})) + f_y(x_{n+k}, y(x_{n+k}))(\overset{\circ}{u}_{n+k} - y(x_{n+k})) + O(h^{2k})] \\
&- \sum_{s=1}^{k-1} \beta_s^*[f(x_{n+s}, y(x_{n+s})) + f_y(x_{n+s}, y(x_{n+s}))(u_{n+s} - y(x_{n+s})) + O(h^{2k})] \\
&= \sum_{s=0}^{k-1} \beta_s y'(x_{n+s}) - \sum_{s=1}^{k} \beta_s^* y'(x_{n+s}) \\
&\quad + \sum_{s=0}^{k-1} \beta_s f_y(x_{n+s}, y(x_{n+s}))(u_{n+s} - y(x_{n+s})) \\
&\quad - \beta_k^* f_y(x_{n+k}, y(x_{n+k}))(\overset{\circ}{u}_{n+k} - y(x_{n+k})) \\
&\quad - \sum_{s=1}^{k-1} \beta_s^* f_y(x_{n+s}, y(x_{n+s}))(u_{n+s} - y(x_{n+s})) + O(h^{k+1}),
\end{aligned}
\tag{3.48}
$$

since $O(h^{2k})$ is of $O(h^{k+1})$ when $k \geq 1$.

Now from the definition of truncation error, we have for the predictor and corrector formulae

$$
\frac{1}{h}\left[y(x_{n+k}) + \sum_{s=0}^{k-1} \alpha_s y(x_{n+s})\right] - \sum_{s=0}^{k-1} \beta_s y'(x_{n+s}) = (T_h)_n,
$$

$$
\frac{1}{h}\left[y(x_{n+k}) + \sum_{s=1}^{k-1} \alpha_s^* y(x_{n+s})\right] - \sum_{s=1}^{k} \beta_s^* y'(x_{n+s}) = (T_h^*)_n.
$$

Upon subtraction, and using (3.44), we find

$$
\begin{aligned}
\sum_{s=0}^{k-1} \beta_s y'(x_{n+s}) - \sum_{s=1}^{k} \beta_s^* y'(x_{n+s}) &= (T_h^*)_n - (T_h)_n \\
&= (\ell_{k+1}^* - \ell_{k+1}) h^k y^{(k+1)}(x_n) + O(h^{k+1}).
\end{aligned}
$$

It suffices to show that the remaining terms in (3.48) together are $O(h^{k+1})$. This we do by expanding f_y about the point $(x_n, y(x_n))$ and by using, from (3.34) and (3.35),

$$
u_{n+s} - y(x_{n+s}) = C_{k,k}^* h^k e(x_n) + O(h^{k+1})
$$

as well as

$$\overset{\circ}{u}_{n+k} - y(x_{n+k}) = -\sum_{s=0}^{k-1} \alpha_s(u_{n+s} - y(x_{n+s})) + h \sum_{s=0}^{k-1} \beta_s[f(x_{n+s}, u_{n+s}) - f(x_{n+s}, y(x_{n+s}))] - h(T_h)_n$$

$$= -\sum_{s=0}^{k-1} \alpha_s[C^*_{k,k} h^k e(x_{n+s}) + O(h^{k+1})] + O(h^{k+1})$$
$$= -\left(\sum_{s=0}^{k-1} \alpha_s\right) C^*_{k,k} h^k e(x_n) + O(h^{k+1}) = C^*_{k,k} h^k e(x_n) + O(h^{k+1}),$$

where in the last equation we have used that $\sum_{s=0}^{k} \alpha_s = 1 + \sum_{s=0}^{k-1} \alpha_s = 0$. We see that the terms in question add up to

$$h^k C^*_{k,k} f_y(x_n, y(x_n)) e(x_n) \left\{ \sum_{s=0}^{k-1} \beta_s - \beta^*_k - \sum_{s=1}^{k-1} \beta^*_s \right\} + O(h^{k+1}).$$

Since $k \geq 1$, we have $Lt = L^*t = 0$ for the functionals L and L^* associated with the predictor and corrector formula and, therefore,

$$\sum_{s=0}^{k-1} \beta_s - \sum_{s=1}^{k} \beta^*_s = \sum_{s=0}^{k} s\alpha_s - \sum_{s=1}^{k} s\alpha^*_s = \sum_{s=1}^{k} s(\alpha_s - \alpha^*_s) = 0,$$

again by (3.44). This completes the proof of Theorem 6.3.6. □

The formula (3.47) can also be used to estimate the local trunction error in connection with step control procedures such as those discussed for one-step methods in Ch. 5, §4.3. Changing the grid length in multistep methods, however, is more complicated than in one-step methods, and we refer for this to the specialized literature.

§4. Analytic Theory of Order and Stability

We now turn our attention to the following problems.

(1) Construct a multistep formula of maximum algebraic degree, given its characteristic polynomial $\alpha(t)$. Normally, the latter is chosen to satisfy the root condition.

(2) Determine the maximum algebraic degree among all k-step methods whose characteristic polynomials $\alpha(t)$ satisfy the root condition.

Once we have solved problem (1), it is in principle straightforward to solve (2). We let $\alpha(t)$ vary over all polynomials of degree k satisfying the root condition and for each α construct the multistep formula of maximum degree. We then simply observe the maximum order so attainable.

To deal with problem (1), it is useful to begin with an analytic characterization of algebraic degree — or order — of a multistep formula.

§4.1. **Analytic characterization of order**. With the k-step method (1.2) we associate two polynomials,

$$\alpha(t) = \sum_{s=0}^{k} \alpha_s t^s, \quad \beta(t) = \sum_{s=0}^{k} \beta_s t^s \quad (\alpha_k = 1), \tag{4.1}$$

the first being the characteristic polynomial already introduced in (3.3). We define

$$\delta(\zeta) = \frac{\alpha(\zeta)}{\ln \zeta} - \beta(\zeta), \qquad \zeta \in \mathbb{C}, \tag{4.2}$$

which, since $\alpha(1) = 0$, is a function holomorphic in the disk $|\zeta - 1| < 1$.

Theorem 6.4.1. *The multistep method* (1.2) *has* (*exact*) *polynomial degree p if and only if $\delta(\zeta)$ has a zero of* (*exact*) *multiplicity p at $\zeta = 1$.*

Proof. In terms of the linear functional

$$Lu = \sum_{s=0}^{k} [\alpha_s u(s) - \beta_s u'(s)], \tag{4.3}$$

the method (1.2) has exact polynomial degree p if and only if (cf. (1.33))

$$Lu = \frac{1}{p!} \int_0^k \lambda_p(\sigma) u^{(p+1)}(\sigma) d\sigma, \quad \frac{1}{p!} \int_0^k \lambda_p(\sigma) d\sigma = \ell_{p+1} \neq 0, \tag{4.4}$$

for every $u \in C^{p+1}[0,k]$. Choose $u(t) = e^{tz}$, where z is a complex parameter. Then (4.4) implies

$$\sum_{s=0}^{k} [\alpha_s e^{sz} - \beta_s z e^{sz}] = \frac{z^{p+1}}{p!} \int_0^k \lambda_p(\sigma) e^{\sigma z} d\sigma,$$

that is,

$$\frac{\alpha(e^z)}{z} - \beta(e^z) = \frac{z^p}{p!} \int_0^k \lambda_p(\sigma) e^{\sigma z} d\sigma. \tag{4.5}$$

Since the coefficient of z^p on the right, when $z = 0$, equals $\ell_{p+1} \neq 0$, the function on the left — an entire function — has a zero of exact multiplicity p at $z = 0$. Thus, exact polynomial degree p of L implies that the function $\delta(e^z)$ has a zero of exact multiplicity p at $z = 0$. The converse is also true, since otherwise, (4.4) and (4.5) would hold with a different value of p. The theorem now follows readily by applying the conformal map

$$\zeta = e^z, \qquad z = \ln \zeta, \tag{4.6}$$

and by observing that the multiplicity of a zero remains unchanged under such a map. □

Based on Theorem 6.4.1, the first problem mentioned now allows an easy solution. Suppose we are given the characteristic polynomial $\alpha(t)$ of degree k, and we want to find $\beta(t)$ of degree k' ($\leq k$) such that the method (1.2) has maximum order. (Typically, $k' = k - 1$ for an explicit method, and $k' = k$ for an implicit one.) We simply expand in (4.2) the first term of $\delta(\zeta)$ in a power series about $\zeta = 1$,

$$\frac{\alpha(\zeta)}{\ln \zeta} = c_0 + c_1(\zeta - 1) + c_2(\zeta - 1)^2 + \cdots , \tag{4.7}$$

and then have no other choice for β than to take

$$\beta(\zeta) = c_0 + c_1(\zeta - 1) + \cdots + c_{k'}(\zeta - 1)^{k'}. \tag{4.8}$$

In this way, $\delta(\zeta)$ has a zero of maximum multiplicity at $\zeta = 1$, given $\alpha(t)$ and the degree k' of β. In fact,

$$\delta(\zeta) = c_{k'+1}(\zeta - 1)^{k'+1} + \cdots ,$$

and we get order $p \geq k' + 1$. The order could be larger than $k' + 1$ if by chance $c_{k'+1} = 0$. If p is the exact order so attained, then

$$\delta(\zeta) = c_p(\zeta - 1)^p + \cdots , \quad c_p \neq 0, \quad p \geq k' + 1. \tag{4.9}$$

It is interesting to compare this with (4.5):

$$\begin{aligned}
\delta(\zeta) &= \frac{(\ln \zeta)^p}{p!} \int_0^k \lambda_p(\sigma)\zeta^\sigma d\sigma \\
&= \frac{[\zeta - 1 - \frac{1}{2}(\zeta - 1)^2 + \cdots]^p}{p!} \int_0^k \lambda_p(\sigma) \left[1 + \binom{\sigma}{1}(\zeta - 1) + \cdots \right] d\sigma \\
&= \left(\frac{1}{p!} \int_0^k \lambda_p(\sigma) d\sigma\right) (\zeta - 1)^p + \cdots \\
&= \ell_{p+1}(\zeta - 1)^p + \cdots .
\end{aligned}$$

Thus,

$$\ell_{p+1} = c_p. \tag{4.10}$$

Similarly, if the method is stable, then

$$C_{k,p} = \frac{\ell_{p+1}}{\sum_{s=0}^{k} \beta_s} = \frac{\ell_{p+1}}{\beta(1)} = \frac{c_p}{c_0}\,. \tag{4.11}$$

We see that the local and global error constants can be found directly from the expansion (4.7). It must be observed, however, that (4.10) and (4.11) hold only for the k-step methods of *maximum* degree. If the degree p is not maximal, then

$$\ell_{p+1} = d_p \tag{4.10$'$}$$

and

$$C_{k,p} = \frac{d_p}{c_0}\,, \tag{4.11$'$}$$

where

$$\delta(\zeta) = d_p(\zeta - 1)^p + \cdots\,, \qquad d_p \neq 0. \tag{4.12}$$

It seems appropriate, at this point, to observe that if $\alpha(\zeta)$ and $\beta(\zeta)$ have a common factor $\omega(\zeta)$,

$$\alpha(\zeta) = \omega(\zeta)\alpha_0(\zeta), \qquad \beta(\zeta) = \omega(\zeta)\beta_0(\zeta),$$

and $\omega(1) \neq 0$, then

$$\delta(\zeta) = \omega(\zeta)\delta_0(\zeta), \qquad \delta_0(\zeta) = \frac{\alpha_0(\zeta)}{\ln \zeta} - \beta_0(\zeta),$$

and $\delta_0(\zeta)$ vanishes at $\zeta = 1$ with the same order as $\delta(\zeta)$. The multistep method $\{\alpha, \beta\}$ and the "reduced" multistep method $\{\alpha_0, \beta_0\}$ therefore have the same order and indeed the same error constants (4.10$'$) and (4.11$'$). On the other hand, $\omega(1) = 0$ would imply $\beta(1) = 0$, and the method $\{\alpha, \beta\}$ would not be stable (cf. (3.36)). Since, on top of that, a solution of the difference equation (1.2) corresponding to $\{\alpha_0, \beta_0\}$ is also a solution of (1.2) for $\{\alpha, \beta\}$ (cf. Ex. 9), it would be pointless to consider the method $\{\alpha, \beta\}$. For these reasons it is no restriction to assume that the polynomials $\alpha(\zeta)$, $\beta(\zeta)$ are irreducible.

Example 4.1. Construct all stable implicit two-step methods of maximum order.

Here

$$\alpha(\zeta) = (\zeta - 1)(\zeta - \lambda), \qquad -1 \le \lambda < 1,$$

since 1 is always a zero of $\alpha(\zeta)$ and the second zero λ must satisfy the root condition. For the expansion (4.7) we have

$$\frac{\alpha(\zeta)}{\ln \zeta} = \frac{(\zeta-1)^2 + (1-\lambda)(\zeta-1)}{\zeta - 1 - \frac{1}{2}(\zeta-1)^2 + \cdots} = \frac{1-\lambda+(\zeta-1)}{1-\frac{1}{2}(\zeta-1)+\frac{1}{3}(\zeta-1)^2 - \cdots}$$
$$= c_0 + c_1(\zeta-1) + c_2(\zeta-1)^2 + \cdots .$$

An easy calculation gives

$$c_0 = 1-\lambda, \quad c_1 = \tfrac{1}{2}(3-\lambda), \quad c_2 = \tfrac{1}{12}(5+\lambda),$$
$$c_3 = -\tfrac{1}{24}(1+\lambda), \quad c_4 = \tfrac{1}{720}(11+19\lambda) .$$

Thus,

$$\beta(\zeta) = c_0 + c_1(\zeta-1) + c_2(\zeta-1)^2 = \frac{5+\lambda}{12}\,\zeta^2 + \frac{2-2\lambda}{3}\,\zeta - \frac{1+5\lambda}{12},$$

giving the desired method

$$u_{n+2} - (1+\lambda)u_{n+1} + \lambda u_n = h\left\{\frac{5+\lambda}{12}\,f_{n+2} + \frac{2-2\lambda}{3}\,f_{n+1} - \frac{1+5\lambda}{12}\,f_n\right\}.$$

If $c_3 \neq 0$ (i.e., $\lambda \neq -1$), the order is exactly $p = 3$, and the error constant is

$$C_{2,3} = \frac{c_3}{c_0} = -\frac{1}{24}\,\frac{1+\lambda}{1-\lambda} \qquad (\lambda \neq -1).$$

The case $\lambda = -1$ is exceptional, giving exact order $p = 4$ (since $c_4 = -\frac{1}{90} \neq 0$); in fact,

$$u_{n+2} - u_n = \frac{h}{3}\,(f_{n+2} + 4f_{n+1} + f_n) \qquad (\lambda = -1)$$

is precisely Simpson's rule. This is an example of an "optimal" method — a stable k-step method of order $k+2$ for k even (cf. Theorem 6.4).

Example 4.2. Construct a pair of two-step methods, one explicit, the other implicit, both having $\alpha(\zeta) = \zeta^2 - \zeta$ and order $p = 2$, but global error constants that are equal in modulus and opposite in sign.

Let $\beta(\zeta)$ and $\beta^*(\zeta)$ be the β-polynomials for the explicit and implicit formula, respectively, and $C_{2,2}$, $C^*_{2,2}$ the corresponding error constants. We have

$$\frac{\alpha(\zeta)}{\ln \zeta} = \frac{\zeta(\zeta-1)}{\zeta - 1 - \frac{1}{2}(\zeta-1)^2 + \cdots} = 1 + \tfrac{3}{2}(\zeta-1) + \tfrac{5}{12}(\zeta-1)^2 + \cdots .$$

Thus,

$$\beta(\zeta) = 1 + \tfrac{3}{2}(\zeta-1) = \tfrac{3}{2}\zeta - \tfrac{1}{2},$$

giving

$$C_{2,2} = \frac{\frac{5}{12}}{1} = \frac{5}{12}.$$

For β^*, we try

$$\beta^*(\zeta) = 1 + \tfrac{3}{2}(\zeta-1) + b(\zeta-1)^2.$$

As we are not aiming for optimal degree, we must use (4.11′) and (4.12) and find

$$C^*_{2,2} = \frac{\frac{5}{12} - b}{1} = \frac{5}{12} - b.$$

Since we want $C^*_{2,2} = -C_{2,2}$, we get

$$\tfrac{5}{12} - b = -\tfrac{5}{12}, \qquad b = \tfrac{5}{6},$$

and so,

$$\beta^*(\zeta) = 1 + \tfrac{3}{2}(\zeta-1) + \tfrac{5}{6}(\zeta-1)^2 = \tfrac{5}{6}\zeta^2 - \tfrac{1}{6}\zeta + \tfrac{1}{3}.$$

The desired pair of methods, therefore, is

$$\begin{cases} u_{n+2} = u_{n+1} + \dfrac{h}{2}(3f_{n+1} - f_n), \\[2ex] u^*_{n+2} = u^*_{n+1} + \dfrac{h}{6}(5f^*_{n+2} - f^*_{n+1} + 2f^*_n). \end{cases} \tag{4.13}$$

The interest in such pairs of formulae is rather evident: if both formulae are used independently (i.e., *not* in a predictor-corrector mode, but the corrector formula being iterated to convergence), then by Theorem 6.3.5, Eq. (3.34), we have

$$\begin{aligned} u_n - y(x_n) &= C_{2,2}h^2 e(x_n) + O(h^3), \\ u_n^* - y(x_n) &= -C_{2,2}h^2 e(x_n) + O(h^3); \end{aligned} \tag{4.14}$$

that is, asymptotically for $h \to 0$, the exact solution is halfway between u_n and u_n^*. This generates upper and lower bounds for each solution component and built-in error bounds $\frac{1}{2}\,|u_n^* - u_n|$ (absolute value taken componentwise). A break-down occurs in the ith component if $e^i(x_n) \approx 0$; if $e^i(x)$ changes sign across x_n, the bounds switch from an upper to a lower one and vice versa.

Naturally, it would not be difficult to generate such "equilibrated" pairs of formulae having orders much larger than 2 (cf. Ex. 10).

Example 4.3. Given $\alpha(t)$, construct an explicit k-step method of maximum order in Newton's form (involving backward differences).

Here we want L in the form

$$Lu = \sum_{s=0}^{k} \alpha_s u(s) - \sum_{s=0}^{k-1} \gamma_s \nabla^s u'(k-1).$$

The mapping $\zeta = e^z$ used previously in (4.6) is no longer appropriate here, since we do not want the coefficients of $\beta(\zeta)$. The principle used in the proof of Theorem 6.4.1, however, remains the same: we want $\frac{1}{z}\, L_{(t)}e^{tz}$ to vanish at $z = 0$ with multiplicity as large as possible. Now

$$\frac{1}{z}\, L_{(t)}e^{tz} = \frac{1}{z} \sum_{s=0}^{k} \alpha_s e^{sz} - \sum_{s=0}^{k-1} \gamma_s [\nabla_{(t)}^s e^{tz}]_{t=k-1}$$

and

$$\nabla_{(t)} e^{tz} = e^{tz} - e^{(t-1)z} = e^{tz}\left(\frac{e^z - 1}{e^z}\right),$$

. . . .

$$\nabla_{(t)}^s e^{tz} = e^{tz}\left(\frac{e^z - 1}{e^z}\right)^s.$$

Therefore,

$$\frac{1}{z}\, L_{(t)}e^{tz} = \frac{\alpha(e^z)}{z} - e^{(k-1)z} \sum_{s=0}^{k-1} \gamma_s \left(\frac{e^z - 1}{e^z}\right)^s. \tag{4.15}$$

This suggests the mapping

$$\zeta = \frac{e^z - 1}{e^z}, \qquad z = -\ln(1-\zeta),$$

which maps a neighborhood of $z = 0$ conformally onto a neighborhood of $\zeta = 0$. Thus, (4.15) has a zero at $z = 0$ of maximal multiplicity if and only if

$$\frac{\alpha\left(\dfrac{1}{1-\zeta}\right)}{-\ln(1-\zeta)} - \frac{1}{(1-\zeta)^{k-1}} \sum_{s=0}^{k-1} \gamma_s \zeta^s = \frac{1}{(1-\zeta)^{k-1}} \left\{ \frac{(1-\zeta)^{k-1}\alpha\left(\dfrac{1}{1-\zeta}\right)}{-\ln(1-\zeta)} - \sum_{s=0}^{k-1} \gamma_s \zeta^s \right\}$$

has a zero at $\zeta = 0$ of maximal multiplicity. Thus, we have to expand

$$\frac{(1-\zeta)^{k-1}\alpha\left(\dfrac{1}{1-\zeta}\right)}{-\ln(1-\zeta)} = \sum_{s=0}^{\infty} \gamma_{ks} \zeta^s \tag{4.16}$$

and take

$$\gamma_s = \gamma_{ks}, \qquad s = 0, 1, \dots, k-1. \tag{4.17}$$

To illustrate, for Adams-Bashforth methods we have

$$\alpha(t) = t^k - t^{k-1},$$

hence

$$\frac{(1-\zeta)^{k-1}\alpha\left(\dfrac{1}{1-\zeta}\right)}{-\ln(1-\zeta)} = \frac{\zeta}{-(1-\zeta)\ \ln(1-\zeta)} =: \gamma(\zeta),$$

which is the generating function

$$\gamma(\zeta) = \sum_{s=0}^{\infty} \gamma_s \zeta^s$$

obtained earlier in §2.1, Eq. (2.10). We now see more clearly why the coefficients γ_s are independent of k.

Example 4.4. Given $\alpha(t)$, construct an implicit k-step method of maximum order in Newton's form.

Here

$$Lu = \sum_{s=0}^{k} [\alpha_s u(s) - \gamma_s^* \nabla^s u'(k)],$$

and a calculation similar to the one in Example 4.3 yields

$$\gamma_s^* = \gamma_{ks}^*, \qquad s = 0, 1, \ldots, k, \tag{4.18}$$

where

$$\frac{(1-\zeta)^k \alpha\left(\dfrac{1}{1-\zeta}\right)}{-\ln(1-\zeta)} = \sum_{s=0}^{\infty} \gamma_{ks}^* \zeta^s. \tag{4.19}$$

Again, for Adams-Moulton methods,

$$\frac{(1-\zeta)^k \alpha\left(\dfrac{1}{1-\zeta}\right)}{-\ln(1-\zeta)} = \frac{\zeta}{-\ln(1-\zeta)} =: \gamma^*(\zeta),$$

with

$$\gamma^*(\zeta) = \sum_{s=0}^{\infty} \gamma_s^* \zeta^s$$

the generating function found earlier.

Example 4.5. Given $\beta(t)$, construct a k-step method of maximum order in Newton's form.

In all previous examples we were given $\alpha(t)$ and could thus choose it to satisfy the root condition. There is no intrinsic reason why we should not start with $\beta(t)$ and determine $\alpha(t)$ so as to maximize the order. It then needs to be checked, of course, whether the $\alpha(t)$ thus found satisfies the root condition. Thus, in the example at hand,

$$Lu = \sum_{s=0}^{k} \gamma_s \nabla^s u(k) - \sum_{s=0}^{k} \beta_s u'(s). \tag{4.20}$$

Following the procedure of Examples 4.3 and 4.4, we get

$$\frac{1}{z} L_{(t)} e^{tz} = \frac{e^{kz}}{z} \sum_{s=0}^{k} \gamma_s \left(\frac{e^z - 1}{e^z}\right)^s - \beta(e^z)$$

$$= \frac{1}{-(1-\zeta)^k \ln(1-\zeta)} \sum_{s=0}^{k} \gamma_s \zeta^s - \beta\left(\frac{1}{1-\zeta}\right),$$

and we want this to vanish at $\zeta = 0$ with maximum order. Clearly, $\gamma_0 = 0$, and if

$$-(1-\zeta)^k \ln(1-\zeta)\,\beta\left(\frac{1}{1-\zeta}\right) = \sum_{s=1}^{\infty} d_{ks}\zeta^s, \tag{4.21}$$

we must take the remaining coefficients to be

$$\gamma_s = d_{ks}, \qquad s = 1, 2, \ldots, k, \tag{4.22}$$

to achieve order $p \geq k$.

A particularly simple example obtains if $\beta(t) = t^k$, in which case

$$-(1-\zeta)^k \ln(1-\zeta)\,\beta\left(\frac{1}{1-\zeta}\right) = -\ln(1-\zeta) = \zeta + \tfrac{1}{2}\,\zeta^2 + \tfrac{1}{3}\,\zeta^3 + \cdots\,,$$

so that

$$\gamma_0 = 0, \qquad \gamma_s = \frac{1}{s}\,, \qquad s = 1, 2, \ldots, k. \tag{4.23}$$

The method (4.20) of order k so obtained is called the *backward differentiation method* and is of some interest in connection with stiff problems (cf. §5.2). Its characteristic polynomial $\alpha(t)$ is easily obtained from (4.20) by noting that $\nabla^s u(k) = \sum_{r=0}^{s} (-1)^r \binom{s}{r} u(k-r)$. One finds

$$\alpha(t) = \sum_{s=0}^{k} \alpha_s t^s, \qquad \beta(t) = t^k,$$

$$\alpha_k = \sum_{r=1}^{k} \frac{1}{r}\,, \quad \alpha_s = (-1)^{k-s} \sum_{r=0}^{s} \binom{k-s+r}{k-s} \frac{1}{k-s+r}\,, \quad s = 0, 1, \ldots, k-1. \tag{4.24}$$

Note here that $\alpha_k \neq 1$, but we can normalize (4.20) by dividing both sides by α_k. It turns out that $\alpha(t)$ in (4.24) satisfies the root condition for $k = 1, 2, \ldots, 6$ but not for $k = 7$, in which case one root of α lies outside the unit circle (cf. Ex. 11).

§4.2. **Stable methods of maximum order**. We now give an answer to problem (2) stated at the beginning of this section.

Theorem 6.4.2. (a) *If k is odd, then every stable k-step method has order $p \leq k+1$.*

(b) *If k is even, then every stable k-step method has order $p \leq k+2$, the order being $k+2$ if and only if $\alpha(t)$ has all its zeros on the circumference of the unit circle.*

Before we prove this theorem, we make the following remarks.

1. In case (a), we can attain order $p = k+1$ for any given $\alpha(t)$ satisfying the root condition; cf. (4.9) with $k' = k$.

2. Since $\alpha(t)$ is a real polynomial, all complex zeros of α occur in conjugate pairs. It follows from part (b) of Theorem 6.4.2 that $p = k+2$ if and only if $\alpha(t)$ has zeros at $t = 1$ and $t = -1$, and all other zeros (if any) are located on $|t| = 1$ in conjugate complex pairs.

3. The maximum order among *all* k-step methods (stable or not) is known to be $p = 2k$. The stability requirement thus reduces this maximum possible order to roughly one-half.

Proof of Theorem 6.4.2. We want to determine $\alpha(t)$, subject to the root condition, such that

$$\delta(\zeta) = \frac{\alpha(\zeta)}{\ln \zeta} - \beta(\zeta) \tag{4.25}$$

has a zero at $\zeta = 1$ of maximum multiplicity (cf. Theorem 6.4.1). We map the unit disk $|\zeta| \leq 1$ conformally onto the left half-plane $\mathrm{Re}\, z \leq 0$ (which is easier to deal with) by means of

$$\zeta = \frac{1+z}{1-z}, \qquad z = \frac{\zeta - 1}{\zeta + 1}. \tag{4.26}$$

This maps the point $\zeta = 1$ to the origin $z = 0$ and preserves multiplicities of zeros at these points. The function $\delta(\zeta)$ in (4.25) is transformed to

$$d(z) = \left\{ \frac{\alpha\left(\dfrac{1+z}{1-z}\right)}{\ln \dfrac{1+z}{1-z}} - \beta\left(\frac{1+z}{1-z}\right) \right\} \left(\frac{1-z}{2}\right)^k,$$

except for the factor $[(1-z)/2]^k$ which, however, does not vanish at $z = 0$. We write

$$d(z) = \frac{a(z)}{\ln \dfrac{1+z}{1-z}} - b(z), \tag{4.27}$$

where

$$a(z) = \left(\frac{1-z}{2}\right)^k \alpha\left(\frac{1+z}{1-z}\right), \quad b(z) = \left(\frac{1-z}{2}\right)^k \beta\left(\frac{1+z}{1-z}\right) \tag{4.28}$$

are both polynomials of degree $\leq k$.

Our problem thus reduces to the following purely analytical problem. *How many initial terms of the Maclaurin expansion of* $d(z)$ *in* (4.27) *can be made to vanish if* $a(z)$ *is to have all its zeros in* $\operatorname{Re} z \le 0$, *and those with* $\operatorname{Re} z = 0$ *are to be simple?*

To solve this problem, we need some preliminary facts:

(i) We have $a(1) = 1$. This follows trivially from (4.28), since $\alpha(t)$ has leading coefficient 1.

(ii) The polynomial $a(z)$ has exact degree k, unless $\zeta = -1$ is a zero of $\alpha(\zeta)$, in which case $a(z)$ has degree $k - \mu$, where μ is the multiplicity of the zero $\zeta = -1$. (If the root condition is satisfied, then, of course, $\mu = 1$.) This also follows straightforwardly from (4.28).

(iii) Let

$$a(z) = a_1 z + a_2 z^2 + \cdots + a_k z^k,$$

where $a_1 \neq 0$ by the root condition. If $a(z)$ has all its zeros in $\operatorname{Re} z \le 0$, then $a_s \ge 0$ for $s = 1, 2, \ldots, k$. (The converse is *not* true.) This is easily seen if we factor $a(z)$ with respect to its real and complex zeros,

$$a(z) = a_\ell z \prod_\rho (z - r_\rho) \prod_\sigma [z - (x_\sigma + iy_\sigma)][z - (x_\sigma - iy_\sigma)], \quad a_\ell \neq 0.$$

Simplifying, we can write

$$a(z) = a_\ell z \prod_\rho (z - r_\rho) \prod_\sigma [(z - x_\sigma)^2 + y_\sigma^2].$$

Since by assumption, $r_\rho \le 0$, $x_\sigma \le 0$, all nonzero coefficients of $a(z)$ have the sign of a_ℓ, and $a_\ell > 0$ since $a(1) = 1$.

(iv) Let

$$\frac{z}{\ln \dfrac{1+z}{1-z}} = \lambda_0 + \lambda_2 z^2 + \lambda_4 z^4 + \cdots .$$

Then

$$\lambda_0 = \tfrac{1}{2}, \qquad \lambda_{2\nu} < 0 \ \text{ for } \ \nu = 1, 2, 3, \ldots .$$

The proof of this is deferred to the end of this section.

Suppose now that

$$\frac{z}{\ln \dfrac{1+z}{1-z}} \, \frac{a(z)}{z} = b_0 + b_1 z + b_2 z^2 + \cdots \, . \tag{4.29}$$

For maximum order p, we must take

$$b(z) = b_0 + b_1 z + \cdots + b_k z^k \tag{4.30}$$

in (4.27). The preceding question then amounts to determining how many of the coefficients $b_{k+1}, b_{k+2}, \ldots$ in (4.29) can vanish simultaneously if $a(z)$ is restricted to have all its zeros in $\operatorname{Re} z \le 0$ and only simple zeros on $\operatorname{Re} z = 0$.

The expansion (4.29), in view of (iv), is equivalent to

$$(\lambda_0 + \lambda_2 z^2 + \lambda_4 z^4 + \cdots)(a_1 + a_2 z + a_3 z^2 + \cdots + a_k z^{k-1})$$
$$= b_0 + b_1 z + b_2 z^2 + \cdots \, .$$

Comparing coefficients of like power on the right and left, we get

$$\begin{aligned} b_0 &= \lambda_0 a_1, \\ b_1 &= \lambda_0 a_2, \\ b_{2\nu} &= \lambda_0 a_{2\nu+1} + \lambda_2 a_{2\nu-1} + \cdots + \lambda_{2\nu} a_1 \\ b_{2\nu+1} &= \lambda_0 a_{2\nu+2} + \lambda_2 a_{2\nu} + \cdots + \lambda_{2\nu} a_2 \end{aligned} \Bigg\} \quad \nu = 1, 2, 3, \ldots \, , \tag{4.31}$$

where for convenience we assume $a_\mu = 0$ if $\mu > k$. We distinguish two cases.

Case 1: k is odd. Then by (4.31),

$$b_{k+1} = \lambda_0 a_{k+2} + \lambda_2 a_k + \lambda_4 a_{k-2} + \cdots + \lambda_{k+1} a_1 .$$

Since $a_{k+2} = 0$ by convention, $\lambda_{2\nu} < 0$ by (iv), and $a_s \ge 0$, $a_1 > 0$ by (iii), it follows that $b_{k+1} < 0$. Thus, $p = k + 1$ is the maximum possible order.

Case 2: k is even. Here (4.31) gives

$$b_{k+1} = \lambda_0 a_{k+2} + \lambda_2 a_k + \lambda_4 a_{k-2} + \cdots + \lambda_k a_2 .$$

Again, $a_{k+2} = 0$, $\lambda_{2\nu} < 0$, and $a_s \geq 0$, so that $b_{k+1} \leq 0$. We have $b_{k+1} = 0$ if and only if $a_2 = a_4 = \cdots = a_k = 0$; that is,

$$a(-z) = -a(z). \tag{4.32}$$

Since $a_k = 0$, we conclude from (ii) that $\alpha(\zeta)$ has a zero at $\zeta = -1$. A trivial zero is $\zeta = 1$. In view of (4.32), the polynomial $a(z)$ cannot have zeros off the imaginary axis without violating the root condition. Therefore, $a(z)$ has all its zeros on $\mathrm{Re}\, z = 0$, hence $\alpha(\zeta)$ all its zeros on $|\zeta| = 1$. Conversely, if $\alpha(\zeta)$ has zeros at $\zeta = \pm 1$ and all other zeros on $|\zeta| = 1$, then $a_k = 0$, and

$$a(z) = a_{k-1} z \prod_{\gamma} [(z - iy_\gamma)(z + iy_\gamma)] = a_{k-1} z \prod_{\gamma} (z^2 + y_\gamma^2);$$

that is, $a(z)$ is an odd polynomial and, therefore, $b_{k+1} = 0$.

This proves the second half of part (b) of Theorem 6.4.2. To complete the proof, we have to show that $b_{k+2} = 0$ is impossible. This follows again from (4.31), if we note that (for k even)

$$\begin{aligned} b_{k+2} &= \lambda_0 a_{k+3} + \lambda_2 a_{k+1} + \lambda_4 a_{k-1} + \cdots + \lambda_{k+2} a_1 \\ &= \lambda_4 a_{k-1} + \cdots + \lambda_{k+2} a_1 < 0 \end{aligned}$$

for the same reason as in Case 1. □

It remains to prove the crucial property (iv). Let

$$f(z) := \frac{z}{\ln \dfrac{1+z}{1-z}} = \lambda_0 + \lambda_2 z^2 + \lambda_4 z^4 + \cdots . \tag{4.33}$$

The fact that $\lambda_0 = \frac{1}{2}$ follows easily by taking the limit of $f(z)$ as $z \to 0$. By Cauchy's formula,

$$\lambda_{2\nu} = \frac{1}{2\pi i} \oint_C \frac{f(z)dz}{z^{2\nu+1}} = \frac{1}{2\pi i} \oint_C \frac{dz}{z^{2\nu} \ln \dfrac{1+z}{1-z}}, \quad \nu > 1,$$

where C is a contour encircling the origin in the positive sense of direction anywhere in the complex plane cut along $(-\infty, -1)$ and $(1, \infty)$ (where f in (4.33) is one-valued analytic). To get a negativity result for the $\lambda_{2\nu}$ one would like to push the contour as close to the cuts as possible. This is much easier to do after a change of variables according to

$$u = \frac{1}{z}\ .$$

Then

$$\lambda_{2\nu} = -\frac{1}{2\pi i} \oint_{\Gamma} u^{2\nu-2} \left[\ln \frac{u+1}{u-1}\right]^{-1} du, \tag{4.34}$$

where the cut in the u-plane now runs from -1 to 1, and Γ is a contour encircling this cut in the negative sense of direction. By letting Γ shrink onto the cut and noting that

$$\ln \frac{u+1}{u-1} \to \ln \frac{x+1}{1-x} - i\pi \quad \text{as} \quad u \to x + i0, \quad -1 < x < 1,$$

whereas

$$\ln \frac{u+1}{u-1} \to \ln \frac{x+1}{1-x} + i\pi \quad \text{as} \quad u \to x - i0, \quad -1 < x < 1,$$

we find in the limit

$$\begin{aligned} \lambda_{2\nu} &= -\frac{1}{2\pi i} \left\{ \int_{-1}^{1} x^{2\nu-2} \frac{dx}{\ln \dfrac{x+1}{1-x} - i\pi} + \int_{1}^{-1} x^{2\nu-2} \frac{dx}{\ln \dfrac{x+1}{1-x} + i\pi} \right\} \\ &= -\int_{-1}^{1} \frac{x^{2\nu-2}}{\pi^2 + \ln^2 \dfrac{x+1}{1-x}}\, dx < 0, \qquad \nu \geq 1, \end{aligned} \tag{4.35}$$

as was to be shown.

We also note from (4.35) that

$$|\lambda_{2\nu}| < \frac{1}{\pi^2} \int_{-1}^{1} x^{2\nu-2} dx = \frac{2}{\pi^2} \frac{1}{2\nu - 1}, \qquad \nu = 1, 2, 3, \ldots\ ;$$

that is,

$$\lambda_{2\nu} = O\left(\frac{1}{\nu}\right) \quad \text{as} \quad \nu \to \infty.$$

We now recall a theorem of Littlewood, which says that if

$$f(z) = \sum_{n=0}^{\infty} \lambda_n z^n$$

is convergent in $|z| < 1$ and satisfies

$$f(x) \to s \quad \text{as} \quad x \uparrow 1, \quad \lambda_n = O\left(\frac{1}{n}\right) \quad \text{as} \quad n \to \infty,$$

then

$$\sum_{n=0}^{\infty} \lambda_n = s.$$

In our case,

$$f(x) = \frac{x}{\ln \dfrac{1+x}{1-x}} \to 0 \quad \text{as} \quad x \uparrow 1,$$

and so,

$$\sum_{\nu=0}^{\infty} \lambda_{2\nu} = 0; \quad \text{that is,} \quad \sum_{\nu=1}^{\infty} \lambda_{2\nu} = -\tfrac{1}{2}. \tag{4.36}$$

For an application of (4.36), see Ex. 12(b).

§4.3. **Applications**. Theorem 6.4.2 and its proof technique have a number of interesting consequences, which we now discuss.

Theorem 6.4.3. *For every stable k-step method of maximum order p ($= k+1$ or $k+2$), it is true that*

$$\ell_{p+1} < 0, \quad C_{k,p} = \frac{\ell_{p+1}}{\sum_{s=0}^{k} \beta_s} < 0. \tag{4.37}$$

Proof. We recall from (4.10) and (4.11) that

$$\ell_{p+1} = c_p, \qquad C_{k,p} = \frac{c_p}{c_0},$$

where $c_0 = \beta(1)$ and

$$\delta(\zeta) = \frac{\alpha(\zeta)}{\ln \zeta} - \beta(\zeta) = c_p(\zeta - 1)^p + \cdots, \qquad c_p \neq 0.$$

With the transformation

$$\zeta = \frac{1+z}{1-z}, \quad \zeta - 1 = \frac{2z}{1-z} = 2z + \cdots$$

used in the proof of Theorem 6.4.2, we get for the function $d(z)$ in (4.27)

$$\left(\frac{1-z}{2}\right)^{-k} d(z) = 2^p c_p z^p + \cdots ,$$

or

$$d(z) = 2^{p-k} c_p z^p + \cdots .$$

On the other hand, by (4.29) and (4.30),

$$d(z) = b_p z^p + \cdots .$$

Therefore, $c_p = 2^{k-p} b_p$, and

$$\ell_{p+1} = 2^{k-p} b_p, \qquad C_{k,p} = 2^{k-p} \frac{b_p}{\beta(1)} . \tag{4.38}$$

From the proof of Theorem 6.4.2, we know that $b_p < 0$. This proves the first relation in (4.37).

To prove the second, we must show that

$$\beta(1) > 0. \tag{4.39}$$

Since $p \geq 1$, we have

$$\beta(1) = \sum_{s=0}^{k} \beta_s = \sum_{s=0}^{k} s\alpha_s = \alpha'(1).$$

By the root condition, $\alpha'(1) \neq 0$. If we had $\alpha'(1) < 0$, then $\alpha(1) = 0$ in conjunction with $\alpha(t) \sim t^k$ as $t \to \infty$ would imply that $\alpha(t)$ vanishes for some $t > 1$, contradicting the root condition. Thus, $\alpha'(1) > 0$, proving (4.39). □

We note, incidentally, from (4.38), since $\beta(1) = 2^k b(0) = 2^k b_0$, that

$$C_{k,p} = 2^{-p} \frac{b_p}{b_0} . \tag{4.40}$$

Theorem 6.4.4. (a) *Let $k \geq 3$ be odd, and*

$$\gamma_k = \inf \; |C_{k,k+1}|, \tag{4.41}$$

where the infimum is taken over all stable k-step methods of (maximum) *order $p = k+1$. If $k \geq 5$, there is no such method for which the infimum* *in* (4.41) *is attained. If* $k = 3$, *all three-step methods of order* $p = 4$ *with* $\alpha(\zeta) = (\zeta - 1)(\zeta + 1)(\zeta - \lambda)$, $-1 < \lambda < 1$, *have* $|C_{3,4}| = \gamma_3 = \frac{1}{180}$.

(b) *Let* $k \geq 2$ *be even, and*

$$\gamma_k^* = \inf \ |C_{k,k+2}|, \tag{4.42}$$

where the infimum is taken over all stable k*-step methods of* (maximum) *order* $p = k + 2$. *If* $k \geq 4$, *there is no such method for which the infimum in* (4.42) *is attained. If* $k = 2$, *Simpson's rule is the only two-step method of order* $p = 4$ *with* $|C_{2,4}| = \gamma_2^* = \frac{1}{180}$.

Proof. (a) By (4.40) we have

$$C_{k,k+1} = 2^{-(k+1)} \ \frac{b_{k+1}}{b_0} \ .$$

From the proof of Theorem 6.4.2 (cf. (4.31)), we have

$$b_0 = \lambda_0 a_1 = \tfrac{1}{2}\, a_1, \quad b_{k+1} = \lambda_2 a_k + \lambda_4 a_{k-2} + \cdots + \lambda_{k+1} a_1 < 0.$$

So,

$$|C_{k,k+1}| = \frac{1}{2^k a_1} \ (|\lambda_2| a_k + |\lambda_4| a_{k-2} + \cdots + |\lambda_{k+1}| a_1) \geq \frac{|\lambda_{k+1}|}{2^k} \ . \tag{4.43}$$

We claim that

$$\gamma_k = \inf \ |C_{k,k+1}| = \frac{|\lambda_{k+1}|}{2^k} \ .$$

Indeed, take

$$a(z) = z \ \frac{(z - x_1)(z - x_2) \cdots (z - x_{k-1})}{(1 - x_1)(1 - x_2) \cdots (1 - x_{k-1})} \ ,$$

where the x_i are distinct negative numbers. Then $a(1) = 1$ (as must be), and $a(z)$ satisfies the root condition. Now let $x_i \to -\infty$ (all i); then $a(z) \to z$; that is,

$$a_1 \to 1, \quad a_s \to 0 \ \text{ for } \ s = 2, 3, \ldots, k.$$

Therefore, by (4.43),

$$|C_{k,k+1}| \to \frac{|\lambda_{k+1}|}{2^k} \ .$$

Now suppose that $|C_{k,k+1}| = \gamma_k$ for some stable method. Then, necessarily,

$$a_k = a_{k-2} = \cdots = a_3 = 0;$$

that is,

$$a(z) = a_1 z + a_2 z^2 + a_4 z^4 + \cdots + a_{k-1} z^{k-1}. \tag{4.44}$$

In particular, $\zeta = -1$ is a zero of $\alpha(\zeta)$ and $a_{k-1} \neq 0$ by the stability requirement. We distinguish two cases.

Case 1: $k = 3$. Here

$$a(z) = a_1 z + a_2 z^2 = z(a_1 + a_2 z).$$

By stability, $a(z)$ has a zero anywhere on the negative real axis and a zero at $z = 0$. Transforming back to ζ in the usual way (cf. (4.26)), this means that

$$\alpha(\zeta) = (\zeta - 1)(\zeta + 1)(\zeta - \lambda), \quad -1 < \lambda < 1.$$

All these methods (of order $p = 4$) have by construction

$$|C_{3,4}| = \gamma_3 = \frac{|\lambda_4|}{8} = \frac{1}{180}.$$

Case 2: $k \geq 5$. If z_i are the zeros of $a(z)$ in (4.44), then by Vieta's rule,

$$\sum_{i=1}^{k-1} z_i = -\frac{a_{k-2}}{a_{k-1}} = 0;$$

hence

$$\sum_{i=1}^{k-1} \operatorname{Re} z_i = 0.$$

Since by stability, $\operatorname{Re} z_i \leq 0$ for all i, we must have $\operatorname{Re} z_i = 0$ for all i. This means that

$$a(z) = \text{const} \cdot z \prod_j (z^2 + y_j^2)$$

is an odd polynomial, so $a_2 = a_4 = \cdots = a_{k-1} = 0$, contradicting stability ($\zeta = -1$ is a multiple zero).

(b) The proof is similar to the one in case (a), and we leave it as an exercise for the reader (see Ex. 14). □

It is interesting to observe that if in the infimum (4.41) of Theorem 6.4.4 we admit only methods whose characteristic polynomials $\alpha(\zeta)$ have the zero $\zeta_1 = 1$ and all other zeros bounded in absolute value by $\gamma < 1$ (and hence are stable), then it can be shown that the infimum is attained precisely when

$$\alpha(\zeta) = (\zeta - 1)(\zeta + \gamma)^{k-1}; \tag{4.45}$$

that is, all zeros other than $\zeta = 1$ are placed at the point $\zeta = -\gamma$ farthest away from $\zeta = 1$. Moreover, there are explicit expressions for the minimum error constant. For example, if k is odd, then (cf. Ex. 13)

$$\min |C_{k,k+1}| = 2^{-k} \left\{ |\lambda_{k+1}| + \binom{k-1}{2} |\lambda_{k-1}|\omega^2 + \binom{k-1}{4} |\lambda_{k-3}|\omega^4 + \cdots + |\lambda_2|\omega^{k-1} \right\}, \tag{4.46}$$

where

$$\omega = \frac{1-\gamma}{1+\gamma} \tag{4.47}$$

and $\lambda_2, \lambda_4, \ldots$ are the expansion coefficients in (iv) of the proof of Theorem 6.4.2. If $\gamma = 0$, we of course obtain the Adams-Moulton formulae.

§5. Stiff Problems

In §4 we were concerned with multistep methods applied to problems on a finite interval; however, the presence of "stiffness" (i.e., of rapidly decaying solutions), requires consideration of infinite intervals and related stability concepts. These again, as in the case of one-step methods, are developed in connection with a simple model problem exhibiting exponentially decaying solutions. The relevant concepts of stability then describe to what extent multistep methods are able to simulate such solutions, especially those decaying at a rapid rate. It turns out that multistep methods are much more limited in their ability to effectively deal with such solutions than are one-step methods. This is particularly so if one requires A-stability, as defined in the next section, and to a lesser extent if one weakens the stability requirement in a manner described briefly in §5.2.

§5.1. **A-stability**. For simplicity we consider the scalar model problem (cf. Ch. 5, Eq. (5.28))

$$\frac{dy}{dx} = \lambda y, \quad 0 \le x < \infty, \quad \operatorname{Re} \lambda < 0, \tag{5.1}$$

all of whose solutions decay exponentially at infinity. In particular,

$$y(x) \to 0 \quad \text{as} \quad x \to \infty \tag{5.2}$$

for every solution of (5.1).

Definition 5.1. A multistep method (1.2) is called *A-stable* if, when applied to (5.1), it produces a grid function $\{u_n\}_{n=0}^{\infty}$ satisfying

$$u_n \to 0 \quad \text{as} \quad n \to \infty, \tag{5.3}$$

regardless of the choice of starting values (1.4). (It is assumed that the method is applied with constant grid length $h > 0$.)

We may assume (cf. §4.1) that the multistep method is irreducible; that is, the polynomials $\alpha(t)$ and $\beta(t)$ defined in (4.1) have no common zeros.

Application of (1.2) to (5.1) yields

$$\sum_{s=0}^{k} \alpha_s u_{n+s} - h\lambda \sum_{s=0}^{k} \beta_s u_{n+s} = 0, \tag{5.4}$$

a constant-coefficient difference equation of order k whose characteristic polynomial is (cf. §3.1)

$$\tilde{\alpha}(t) = \alpha(t) - \tilde{h}\beta(t), \qquad \tilde{h} = h\lambda \in \mathbb{C}. \tag{5.5}$$

All solutions of (5.4) will tend to zero as $n \to \infty$ if the zeros of $\tilde{\alpha}$ are all strictly less than 1 in absolute value. The multistep method, therefore, is A-stable if and only if

$$\{\tilde{\alpha}(\zeta) = 0, \quad \operatorname{Re}\lambda < 0\} \text{ implies } |\zeta| < 1.$$

This is the same as saying that

$$\{\tilde{\alpha}(\zeta) = 0, \quad |\zeta| \geq 1\} \text{ implies } \operatorname{Re}\lambda \geq 0.$$

But $\tilde{\alpha}(\zeta) = 0$ implies $\beta(\zeta) \neq 0$ (since otherwise $\alpha(\zeta) = \beta(\zeta) = 0$, contrary to the assumed irreducibility of the method). Thus $\tilde{\alpha}(\zeta) = 0$ implies

$$\tilde{h} = h\lambda = \frac{\alpha(\zeta)}{\beta(\zeta)},$$

and A-stability is characterized by the condition

$$\operatorname{Re}\frac{\alpha(\zeta)}{\beta(\zeta)} \geq 0 \ \text{ if } \ |\zeta| \geq 1. \tag{5.6}$$

Theorem 6.5.1. *If the multistep method (1.2) is A-stable, then it has order $p = 2$ and error constant $C_{k,p} \leq -\frac{1}{12}$. The trapezoidal rule is the only A-stable method for which $p = 2$ and $C_{k,p} = -\frac{1}{12}$.*

Proof. From (4.12) and (4.11′) we have for any k-step method of order p

$$\frac{\alpha(\zeta)}{\ln \zeta} - \beta(\zeta) = c_0 C_{k,p}(\zeta - 1)^p + \cdots ,$$

which, after division by $\alpha(\zeta) = \alpha'(1)(\zeta - 1) + \cdots = \beta(1)(\zeta - 1) + \cdots = c_0(\zeta - 1) + \cdots$ gives

$$\frac{1}{\ln \zeta} - \frac{\beta(\zeta)}{\alpha(\zeta)} = C_{k,p}(\zeta - 1)^{p-1} + \cdots . \tag{5.7}$$

For the trapezoidal rule, having $\alpha_T(\zeta) = \zeta - 1$ and $\beta_T(\zeta) = \frac{1}{2}(\zeta + 1)$, one easily finds

$$\frac{1}{\ln \zeta} - \frac{\beta_T(\zeta)}{\alpha_T(\zeta)} = -\tfrac{1}{12}\,(\zeta - 1) + \cdots . \tag{5.8}$$

Letting

$$\Delta(\zeta) = \frac{\beta(\zeta)}{\alpha(\zeta)} - \frac{\beta_T(\zeta)}{\alpha_T(\zeta)}, \tag{5.9}$$

one obtains from (5.7) and (5.8) by subtraction

$$\Delta(\zeta) = -\left(c + \tfrac{1}{12}\right)(\zeta - 1) + \cdots , \tag{5.10}$$

where

$$c = \begin{cases} C_{k,p} & \text{if} \quad p = 2, \\ 0 & \text{if} \quad p > 2. \end{cases} \tag{5.11}$$

Given that our method is A-stable, we have from (5.6) that $\operatorname{Re}[\alpha(\zeta)/\beta(\zeta)] \geq 0$ if $|\zeta| \geq 1$, or equivalently,

$$\operatorname{Re}\frac{\beta(\zeta)}{\alpha(\zeta)} \geq 0 \ \text{ if } \ |\zeta| \geq 1.$$

On the other hand,

$$\operatorname{Re}\frac{\beta_T(\zeta)}{\alpha_T(\zeta)} = 0 \quad \text{if} \quad |\zeta| = 1.$$

It follows from (5.9) that $\operatorname{Re}\Delta(\zeta) \geq 0$ on $|\zeta| = 1$, and by the maximum principle applied to the real part of $\Delta(\zeta)$, since $\Delta(\zeta)$ is analytic in $|\zeta| > 1$ (there are no zeros of $\alpha(\zeta)$ outside the unit circle), that

$$\operatorname{Re}\Delta(\zeta) \geq 0 \quad \text{for} \quad |\zeta| > 1. \tag{5.12}$$

Now putting $\zeta = 1 + \varepsilon$, $\operatorname{Re}\varepsilon > 0$, we have that $|\zeta| > 1$, and therefore, by (5.10) and (5.12), for $|\varepsilon|$ sufficiently small, that

$$c + \tfrac{1}{12} \leq 0.$$

If $p > 2$, this is clearly impossible in view of (5.11), whereas for $p = 2$ we must have $C_{k,p} \leq -\frac{1}{12}$. This proves the first part of Theorem 6.5.1. To prove the second part, we note that $p = 2$ and $C_{k,p} = -\frac{1}{12}$ imply $\Delta(\zeta) = O((\zeta - 1)^2)$, and taking $\zeta = 1 + \varepsilon$ as previously, the real part of $(\zeta - 1)^2$, and in fact of any power $(\zeta - 1)^q$, $q \geq 2$, can take on either sign if $\operatorname{Re}\varepsilon > 0$. Consequently, $\Delta(\zeta) \equiv 0$ and, therefore, the method being irreducible, it follows that $\alpha(\zeta) = \alpha_T(\zeta)$ and $\beta(\zeta) = \beta_T(\zeta)$. □

§5.2. **A(α)-stability**. According to Theorem 6.5.1, asking for A-stability puts multistep methods into a straitjacket. One can loosen it by weakening the demands on the *region of absolute stability*, that is, the region

$$\mathcal{D}_A = \{\tilde{h} \in \mathbb{C}: \quad \tilde{\alpha}(\zeta) = 0 \quad \text{implies} \quad |\zeta| < 1\}. \tag{5.13}$$

A-stability requires the left half-plane $\operatorname{Re}\tilde{h} < 0$ to be contained in $\mathcal{D}_A$. In many applications, however, it is sufficient that only part of the left half-plane be contained in $\mathcal{D}_A$, for example, the wedge-like region

$$W_\alpha = \{\tilde{h} \in \mathbb{C}: \quad |\arg(-\tilde{h})| < \alpha, \quad \tilde{h} \neq 0\}, \quad 0 < \alpha < \tfrac{1}{2}\pi. \tag{5.14}$$

This gives rise to the following.

Definition 5.2. The multistep method (1.2) is said to be *$A(\alpha)$-stable*, $0 < \alpha < \frac{1}{2}\pi$, if $W_\alpha \subset \mathcal{D}_A$.

There are now multistep methods that have order $p > 2$ and are $A(\alpha)$-stable for suitable α, the best known being the kth-order backward differentiation methods of §4.1, Example 4.5. These are known (cf. Ex. 11) to be stable in the sense of §3.2 for $1 \leq k \leq 6$ (but not for $k > 6$) and for these values of k turn out to be also $A(\alpha)$-stable with angles α as shown in Table 6.5.1. They can therefore be effectively employed for problems whose stiffness is such that the responsible eigenvalues are relatively close to the negative real axis.

k	1	2	3	4	5	6
α	90°	90°	86.03°	73.35°	51.84°	17.84°

TABLE 6.5.1. $A(\alpha)$-stable kth-order backward differentiation methods

In principle, for any given $\alpha < \frac{1}{2}\pi$, there exist $A(\alpha)$-stable k-step methods of order k for every $k = 1, 2, 3, \ldots$, but their error constants may be so large as to make them practically useless.

NOTES TO CHAPTER 6

Most texts mentioned in the Notes to Chapter 5 also contain discussions of multistep methods. A book specifically devoted to the convergence and stability theory of multistep methods is Henrici [1977]. It also applies the statistical theory of roundoff errors to such methods, something that is rarely found elsewhere in this context. A detailed treatment of variable-step/variable-order Adams-type methods is given in the book by Shampine and Gordon [1975].

§1.2. The choice of exponential sums as gauge functions in Ω_p is made in Brock and Murray [1952]; trigonometric polynomials are considered in Quade [1951] and Gautschi [1961], and products respectively sums of ordinary and trigonometric polynomials in Stiefel and Bettis [1969] and Bettis [1969/1970] (also see Stiefel and Scheifele [1971, Ch. 7, §24].

§2.1. Adams described his method in a chapter of the book Bashforth and Adams [1883] on capillary action. He not only derived both the explicit and implicit formulae (2.5′) and (2.12′), but also proposed a scheme of prediction and correction. He did not predict $\overset{\circ}{u}_{n+k}$ by the first formula in (2.19) but rather by the backtracking scheme described at the end of §2.2. He then solved the implicit equation by what amounts to Newton's method, the preferred method nowadays of dealing with (mildly) stiff problems.

§2.2. Moulton describes his predictor-corrector method in Moulton [1926]. He predicts exactly as Adams did, and then iterates on the corrector formula as described in the text, preferably only once, by taking the step sufficiently small. No reference is made to Adams, but then, according to the preface, the history of the subject is not one of the concerns of the book.

§2.3. Milne's suggestion of estimating the truncation error in terms of the difference of the corrected and predicted value is made in Milne [1926].

The first asymptotic formula in (2.27) is due to Spital and Trench, the second to Steffensen; for literature and an elementary derivation, see Gautschi [1976b].

§3.1. A basic and elegant text on linear difference equations is Miller [1968]. Scalar difference equations such as those encountered in this paragraph are viewed there as special cases of first-order systems of difference equations. A more extensive account of difference equations is Agarwal [1992]. Readers who wish to learn about practical aspects of difference equations, including applications to numerical analysis and the sciences, are referred to Wimp [1984], Goldberg [1986], Lakshmikantham and Trigiante [1988], and Kelley and Peterson [1991].

§3.2. As in the case of one-step methods, the stability concept of Definition 3.1, taken from Keller [1992, §1.3], is also referred to as zero-stability (i.e., for $h \to 0$), to distinguish it from other stability concepts used in connection with stiff problems (cf. §5). The phenomenon of (strong) instability (i.e., lack of zero-stability) was first noted by Todd [1950]. Simple roots $t_s \neq 1$ of the characteristic equation (3.6) with $|t_s| = 1$, although not violating the root condition, may give rise to "weak" stability (cf. Henrici [1962, p. 242]). This was first observed by Rutishauser [1952] in connection with Milne's method — the implicit fourth-order method with $\alpha(t) = t^2 - 1$ (Simpson's rule) — although Dahlquist was aware of it independently; see the interesting historical account of Dahlquist [1985] on the evolution of concepts of numerical (in)stability.

Theorem 6.3.3 is due to Dahlquist [1956]. The proof given here follows, with minor deviations, the presentation in Hull and Luxemburg [1960].

§3.3. There is a stability and convergence theory also for linear multistep methods on nonuniform grids. The coefficients $\alpha_s = \alpha_{s,n}$, $\beta_s = \beta_{s,n}$ in the k-step method (1.2) then depend on the grid size ratios $\omega_i = h_i/h_{i-1}$, $i = n+k-1, n+k-2, \dots,$ $n+1$. If the coefficients $\alpha_{s,n}$, $\beta_{s,n}$ are uniformly bounded, the basic stability and convergence results of §§3.2 and 3.3 carry over to grids that are quasi-uniform in the sense of having ratios h_n/h_{n-1} uniformly bounded and bounded away from zero; see, for example, Hairer, Nørsett, and Wanner [1993, Ch. 3, §5].

§3.4. Theorem 6.3.5 is due to Salihov [1962] and, independently, with a refined form of the $O(h^{p+1})$ term in (3.34), to Henrici [1962, Thm. 5.12], [1977, Thm. 4.2].

The result (3.34) may be interpreted as providing the first term in an asymptotic expansion of the error in powers of h. The existence of a full expansion has been investigated by Gragg [1964]. In Gragg [1965], a modified midpoint rule is defined which has an expansion in even powers of h and serves as a basis of Gragg's extrapolation method.

§3.5. The global validity of the Milne estimator, under the assumptions stated, is proved by Henrici [1962, Thm. 5.13]. The idea of Theorem 6.3.6(5) is also suggested there, but not carried out.

Practical codes based on Adams predictor-corrector type methods use variable steps and variable order in an attempt to obtain the solution to a prescribed accuracy with a minimum amount of computational work. In addition to sound strategies of when and how to change the step and the order of the method, this requires either interpolation to provide the necessary values of f relative to the new step, or extensions of Adams multistep formulae to nonuniform grids. The former was already suggested by both Adams and Moulton. It is the latter approach that is usually adopted nowadays. A detailed discussion of the many technical and practical issues involved in it can be found in Chapters 5 through 7, and well-documented FORTRAN programs in Chapter 10, of Shampine and Gordon [1975]. Such codes are "self-starting" in the sense that they start off with Euler's method and a very small step, and from then on let the control mechanisms take over to arrive at a proper order and proper step size. A more recent version of Shampine and Gordon's code is described in Hairer, Nørsett, and Wanner [1993, Ch. 3, §7]. So are two other codes, available on Netlib, both based on Nordsieck's formulation (cf. Notes to Chapter 5, §0.4) of the Adams method, one originally due to Brown, Byrne, and Hindmarsh [1989] and the other to Gear [1971c].

§4.1. Pairs of multistep methods, such as those considered in Example 4.2, having the same global error constants except for sign, are called polar pairs in Rakitskiĭ [1961, §2]. They are studied further in Salihov [1962]. The backward differentiation formula derived in Example 4.5, with $k = 1$ or $k = 2$, was proposed by Curtiss and Hirschfelder [1952] to integrate stiff equations. The method with values of k up to 6, for which it is stiffly stable (Gear [1969]), is implemented as part of the code DIFSUB in Gear [1971b,c] and subsequently in the "Livermore solver" LSODE of Hindmarsh [1980]. For a variable-step version of the backward differentiation formulae, see Hairer, Nørsett, and Wanner [1993, p. 400].

§4.2. Theorem 6.4.2 is a celebrated result of Dahlquist proved in Dahlquist [1956] and extended to higher-order systems of differential equations in his thesis, Dahlquist [1959]. The k-step method of maximum order $p = 2k$ is derived in Dahlquist [1956, §2.4]. The proof of (iv) based on Cauchy's formula is from Dahlquist [1956, p. 51]. A more algebraic proof is given in Henrici [1962, p. 233]. For the theorem of Littlewood, see Titchmarsh [1939, §7.66].

§4.3. The result of Theorem 6.4.4(a) was announced in Gautschi [1963]; the case (b) follows from Problem 37 in Henrici [1962, p. 286]. For the remark at the end of this paragraph, see Gautschi and Montrone [1980].

§5. The standard text on multistep methods for stiff problems is again Hairer and Wanner [1996, Ch. 5]. It contains, in Ch. 5, §5, references to, and numerical experiments with, a number of multistep codes.

§5.1. Theorem 6.5.1 is due to Dahlquist [1963]. The proof follows Hairer and Wanner [1996, Ch. 5, §1]. There are basically two ways to get around the severe limitations imposed by Theorem 6.5.1. One is to to weaken the stability requirement, the other to strengthen the method. An example of the former is A(α)-stability defined in §5.2; various possibilities for the latter are discussed in Hairer and Wanner [1996, Ch. 5, §3]. Order star theory [ibid., §4], this time on Riemann surfaces, is again an indispensable tool for studying the attainable order, subject to A-stability.

§5.2. The concept of A(α)-stability was introduced by Widlund [1967]. For Table 6.5.1 as well as for the remark at the end of the paragraph, see Hairer and Wanner [1996, Ch. 5, §2].

The regions $\mathcal{D}_A$ of absolute stability for explicit and implicit k-step Adams methods, as well as for the respective predictor-corrector methods, are depicted for $k = 1, 2, \ldots, 6$ in Hairer and Wanner [1996, Ch. 5, §1]. As one would expect, they become rapidly small, more so for the explicit method than for the others. More favorable are the stability domains for the backward differentiation methods, which are also shown in the cited reference and nicely illustrate the results of Table 6.5.1.

As in the case of one-step methods, there is a theory of nonlinear stability and convergence also for multistep methods and their "one-legged" companions $\sum_{s=0}^{k} \alpha_s u_{n+s} = hf\left(\sum_{s=0}^{k} \beta_s x_{n+s}, \sum_{s=0}^{k} \beta_s u_{n+s}\right)$. The theory is extremely rich in technical results, involving yet another concept of stability — G-stability, for which we refer again to Hairer and Wanner [1996, Ch. 5, §§6 through 9].

EXERCISES AND MACHINE ASSIGNMENTS TO CHAPTER 6

EXERCISES

1. Describe how Newton's method is applied to solve the system of nonlinear equations

$$u_{n+k} = h\beta_k f(x_{n+k}, u_{n+k}) + g_n, \quad g_n = h\sum_{s=0}^{k-1} \beta_s f_{n+s} - \sum_{s=0}^{k-1} \alpha_s u_{n+s}$$

for the next approximation, u_{n+k}.

2. The system of nonlinear equations

$$u_{n+k} = h\beta_k f(x_{n+k}, u_{n+k}) + g_n, \quad \beta_k \neq 0,$$

arising in each step of an implicit multistep method (cf. (1.5)) may be solved by

- Newton's method;
- the modified Newton method (with the Jacobian held fixed at its value at the initial approximation);
- the method of successive approximations (fixed point iteration).

Assume that f has continuous second partial derivatives with respect to the u-variables and the initial approximation $u_{n+k}^{[0]}$ satisfies $u_{n+k}^{[0]} = u_{n+k} + O(h^g)$ for some $g > 0$.

(a) Show that the νth iterate $u_{n+k}^{[\nu]}$ in Newton's method has the property that $u_{n+k}^{[\nu]} - u_{n+k} = O(h^{r_\nu})$, where $r_\nu = 2^\nu(g+1) - 1$. Derive analogous statements for the other two methods.

(b) Show that $g = 1$, if one takes $u_{n+k}^{[0]} = u_{n+k-1}$.

(c) Suppose one adopts the following stopping criterion: quit the iteration after μ iterations, where μ is the smallest integer ν such that $r_\nu > p + 1$. For each of the three iterations, determine μ for $p = 2, 3, \ldots, 10$. (Assume $g = 1$ for simplicity.)

(d) If $g = p$, what would μ be in (c)?

3. (a) Consider a multistep method of the form

$$u_{n+2} - u_{n-2} + \alpha(u_{n+1} - u_{n-1}) = h[\beta(f_{n+1} + f_{n-1}) + \gamma f_n].$$

Show that the parameters α, β, γ can be chosen uniquely so that the method has order $p = 6$. {*Hint*: To preserve symmetry, and thus algebraic simplicity, define the associated linear functional on the interval $[-2, 2]$ rather than $[0, 4]$ as in §1.2. Why is this permissible?}

(b) Discuss the stability properties of the method obtained in (a).

4. (a) Compute the coefficients $\{\beta_{k,s}\}_{s=0}^{k-1}$ of the kth-order Adams-Bashforth method (2.5) for k=1(1)6.

(b) Do the same as (a), but for the coefficients $\{\beta_{k,s}^*\}_{s=1}^{k}$ of the kth-order Adams-Moulton method (2.12).

5. For the local error constants γ_k, γ_k^* of, respectively, the Adams-Bashforth and Adams-Moulton method, prove that

$$|\gamma_k^*| < \frac{1}{k-1}\gamma_k \quad \text{for } k \geq 2.$$

6. For the local error constants γ_k, γ_k^* of, respectively, the Adams-Bashforth and Adams-Moulton method, show that, as $k \to \infty$,

$$\gamma_k = \frac{1}{\ln k}\left[1 + O\left(\frac{1}{k}\right)\right], \quad \gamma_k^* = -\frac{1}{k\ln^2 k}\left[1 + O\left(\frac{1}{\ln k}\right)\right].$$

{*Hint*: Express the constants in terms of the gamma function, use

$$\frac{\Gamma(k+t)}{\Gamma(k+1)} = k^{t-1}\left[1 + O\left(\frac{1}{k}\right)\right], \quad k \to \infty,$$

and integrate by parts.}

7. Consider the predictor-corrector method using the Adams-Bashforth formula as predictor and the Adams-Moulton formula (once) as corrector, both in difference form:

$$\mathring{u}_{n+k} = u_{n+k-1} + h\sum_{s=0}^{k-1}\gamma_s \nabla^s f_{n+k-1},$$
$$u_{n+k} = u_{n+k-1} + h\sum_{s=0}^{k-1}\gamma_s^* \nabla^s \mathring{f}_{n+k},$$

$$f_{n+k} = f(x_{n+k}, u_{n+k}),$$

where $\mathring{f}_{n+k} = f(x_{n+k}, \mathring{u}_{n+k})$, $\nabla\mathring{f}_{n+k} = \mathring{f}_{n+k} - f_{n+k-1}$, and so on.

(a) Show that

$$u_{n+k} = \mathring{u}_{n+k} + h\gamma_{k-1}\nabla^k \mathring{f}_{n+k}.$$

{*Hint*: First show that $\gamma_s^* = \gamma_s - \gamma_{s-1}$ for $s = 0, 1, 2, \ldots$, where γ_{-1} is defined to be zero.}

(b) Show that

$$\nabla^k \mathring{f}_{n+k} = \mathring{f}_{n+k} - \sum_{s=0}^{k-1}\nabla^s f_{n+k-1}.$$

{*Hint*: Use the binomial identity $\sum_{\sigma=0}^{m}\binom{\sigma+j}{\sigma} = \binom{m+j+1}{j+1}$.}

8. Prove that the predictor-corrector method

$$\begin{cases} \overset{\circ}{u}_{n+k} = -\sum_{s=0}^{k-1} \alpha_s u_{n+s} + h \sum_{s=0}^{k-1} \beta_s f_{n+s}, \\ u_{n+k} = -\sum_{s=1}^{k-1} \alpha_s^* u_{n+s} + h\{\beta_k^* f(x_{n+k}, \overset{\circ}{u}_{n+k}) + \sum_{s=1}^{k-1} \beta_s^* f_{n+s}\}, \\ f_{n+k} = f(x_{n+k}, u_{n+k}) \end{cases}$$

is stable for every $f \in \mathcal{F}$ (cf. (3.15)), if and only if its characteristic polynomial $\alpha^*(t) = \sum_{s=0}^{k} \alpha_s^* t^s$, $\alpha_0^* = 0$, satisfies the root condition.

9. Let $\alpha(\zeta) = \omega(\zeta)\alpha_0(\zeta)$, $\beta(\zeta) = \omega(\zeta)\beta_0(\zeta)$, and suppose $\{u_n\}$ is a solution of the difference equation (1.2) corresponding to $\{\alpha_0, \beta_0\}$. Show that $\{u_n\}$ also satisfies the difference equation (1.2) corresponding to $\{\alpha, \beta\}$.

10. Construct a pair of four-step methods, one explicit, the other implicit, both having $\alpha(\zeta) = \zeta^4 - \zeta^3$ and order $p = 4$, but global error constants that are equal in modulus and opposite in sign.

11. (a) Compute the zeros of the characteristic polynomial $\alpha(t)$ of the k-step backward differentiation method (4.23) and (4.24) for $k = 1(1)7$ and the modulus of the absolutely largest zero other than 1. Hence, confirm the statement made at the end of §4.1.

    (b) Compare the error constant of the k-step backward differentiation method with that of the k-step Adams-Moulton method for $k = 1(1)7$.

12. (a) Show that the polynomial $b(z)$ in (4.28), for an explicit k-step method, must satisfy $b(1) = 0$.

    (b) Use the proof techniques of Theorem 6.4.2 to show that every stable explicit k-step method has order $p \le k$. {*Hint*: Make use of (4.36).}

13. Determine min $|C_{k,k+1}|$, where the minimum is taken over all k-step methods of order $k+1$ whose characteristic polynomials have all their zeros ζ_i (except $\zeta_1 = 1$) in the disk $\Gamma_\gamma = \{z \in \mathbb{C} : \ |z| \le \gamma\}$, where γ is a prescribed number with $0 \le \gamma < 1$. {*Hint*: Use the theory developed in §§4.2 and 4.3.}

14. Prove Theorem 6.4.4(b).

MACHINE ASSIGNMENTS

1. (a) Consider the initial value problem

$$(*) \qquad \frac{dy}{dx} = \frac{1}{1 - \varepsilon \cos y}, \quad y(0) = 0, \quad 0 \le x \le 2\pi, \quad 0 < \varepsilon < 1.$$

Show that the exact solution is $y = y(x)$, the solution of Kepler's equation $y - \varepsilon \sin y - x = 0$ (cf. Ch. 5, Mach. Ass. 4). What is $y(\pi)$? What is $y(2\pi)$?

(b) Write a program for solving (*) with the kth-order Adams-Bashforth method, $k = 1, 2, 3, 4$, using backward differences.

(c) Do the same as (b), but with the kth-order predictor-corrector method, using backward differences (cf. Ex. 7).

In both programs start with "exact" initial values. Store results corresponding to the four argument values $x = \frac{1}{2}\pi, \pi, \frac{3}{2}\pi, 2\pi$, and save results belonging to different grids to allow for a printout in the following form.

	u	$e = u - y$	e/h^k	k_{est}
$N = 40$	______	______	______	______
80	______	______	______	______
160	______	______	______	______
320	______	______	______	______
640	______	______	______	______

y	k
______	______

(One such array for each k and for each of the four x-values specified.) Here k_{est} is an estimate for the order k, defined by

$$k_{\text{est}} = \ln \left| \frac{e_h}{e_{h/2}} \right| / \ln 2,$$

where e_h is the error $u - y$ corresponding to the grid length $h = 2\pi/N$. Use the parameter value $\varepsilon = .2$.

Comment on the results. In particular, verify numerically the remark at the end of §3.4.

2. Consider the (slightly modified) model problem

$$\frac{dy}{dx} = -\omega[y - a(x)], \quad 0 \le x \le 1; \quad y(0) = y_0,$$

where $\omega > 0$ and (i) $a(x) = x^2$, $y_0 = 0$; (ii) $a(x) = e^{-x}$, $y_0 = 1$; (iii) $a(x) = e^x$, $y_0 = 1$.

(a) In each of the cases (i) through (iii), obtain the exact solution $y(x)$.

(b) In each of the cases (i) through (iii), apply (in double precision) the k-step Adams-Moulton method, for $k = 1, 2, 3, 4$, using exact starting values and step lengths $h = \frac{1}{10}, \frac{1}{40}, \frac{1}{160}$. Print the exact values y_n and the absolute errors $u_n - y_n$ for $x_n = \frac{i}{10}$, $i = 1, 2, \ldots, 10$. Try $\omega = 1$, $\omega = 10$, and $\omega = 50$. In each step of the Adams-Moulton method, iterate to machine precision, using as the initial approximation the preceding approximation. Apply Newton's method in one run, and the method of successive approximations in another, and print the number of iterations required. Summarize your experience.

(c) Repeat (b), but using the k-step backward differentiation method.

3. Consider the nonlinear system

$$\begin{aligned} \frac{dy_1}{dx} &= 2y_1(1 - y_2), \quad y_1(0) = 1, \\ & \qquad\qquad\qquad\qquad\qquad\qquad 0 \le x \le 10, \\ \frac{dy_2}{dx} &= -y_2(1 - y_1), \quad y_2(0) = 3, \end{aligned}$$

of interest in population dynamics.

(a) Use MATLAB to plot the solution of the system and obtain an idea of its behavior.

(b) Determine a step length h, or the corresponding number N of steps, in the classical Runge-Kutta method that would produce about eight correct decimal digits. {*Hint*: For $N = 10, 20, 40, 80, \ldots$ compute the solution with N steps and $2N$ steps and stop as soon as the two solutions agree to within eight decimal places at all grid points common to both solutions.}

(c) Apply N steps, $N = 160, 320, 640$, of the pair of fourth-order methods constructed in Ex. 10 to obtain asymptotically upper and lower bounds to the solution. Plot suitably scaled errors $u_n - y_n$, $u_n^* - y_n$, $n = 1(1)N$, where y_n is the solution computed in (b) by the Runge-Kutta method. Use Newton's method to solve the implicit equation for u_n^*.

CHAPTER 7

TWO-POINT BOUNDARY VALUE PROBLEMS FOR ODEs

Many problems in applied mathematics require solutions of differential equations specified by conditions at more than one point of the independent variable. These are called *boundary value problems*; they are considerably more difficult to deal with than initial value problems, largely because of their global nature. Unlike (local) existence and uniqueness theorems known for initial value problems (cf. Theorem 5.0.1), there are no comparably general theorems for boundary value problems. Neither existence nor uniqueness is, in general, guaranteed.

We concentrate here on *two-point boundary value problems*, in which the system of differential equations

$$\frac{dy}{dx} = f(x, y), \qquad f: \ [a, b] \times \mathbb{R}^d \to \mathbb{R}^d, \quad d \geq 2, \tag{0.1}$$

is supplemented by conditions at the two endpoints a and b. In the most general case they take the form

$$g(y(a), y(b)) = 0, \tag{0.2}$$

where g is a nonlinear mapping g: $\mathbb{R}^d \times \mathbb{R}^d \to \mathbb{R}^d$. Often, however, they are linear and even of the very special kind in which some components of y are prescribed at one endpoint, and some (other or the same) components at the other endpoint, the total number of conditions being equal to the dimension d of the system.

There are other important problems, such as eigenvalue problems and problems with free boundary, that can be transformed to two-point boundary value problems and, therefore, also solved numerically in this manner.

An *eigenvalue problem* is an overdetermined problem containing a parameter λ, say,

$$\frac{dy}{dx} = f(x, y; \lambda), \qquad a \leq x \leq b, \tag{0.3}$$

with f as in (0.1), but depending on an additional scalar parameter λ, and $d + 1$ boundary conditions (instead of d) of the form

$$g(y(a), y(b); \lambda) = 0, \qquad g: \mathbb{R}^d \times \mathbb{R}^d \times \mathbb{R} \to \mathbb{R}^{d+1}. \tag{0.4}$$

The best-known example of an eigenvalue problem is the *Sturm-Liouville problem*, where one seeks a nontrivial solution u of

$$\frac{d}{dx}\left(p(x)\frac{du}{dx}\right) + q(x)u = \lambda u, \qquad a \le x \le b, \tag{0.5}$$

subject to homogeneous boundary conditions

$$u(a) = 0, \qquad u(b) = 0. \tag{0.6}$$

Since with any solution of (0.5) and (0.6) every constant multiple of it is also a solution, we may specify, in addition to (0.6), that (for example)

$$u'(a) = 1, \tag{0.6'}$$

and in this way also make sure that $u \not\equiv 0$. The problem then becomes of the form (0.3), (0.4), if (0.5) is written as a system of two first-order equations (cf. (0.16) through (0.18) of Ch. 5). In this particular case, f in (0.3) is linear homogeneous, and g in (0.4) is also linear and independent of λ. Normally, (0.5) will have no solution satisfying all three boundary conditions (0.6) and (0.6′), except for special values of λ; these are called *eigenvalues* of the problem (0.5) through (0.6′). Similarly, there will be exceptional values of λ — again called eigenvalues — for which (0.3) and (0.4) admit a solution.

To write (0.3) and (0.4) as a two-point boundary value problem, we introduce an additional component and associated (trivial) differential equation,

$$y^{d+1} = \lambda, \qquad \frac{dy^{d+1}}{dx} = 0$$

and simply adjoin this to (0.3). That is, we let

$$Y = \begin{bmatrix} y \\ y^{d+1} \end{bmatrix} \in \mathbb{R}^{d+1}$$

and write (0.3) and (0.4) in the form

$$\frac{dY}{dx} = F(x, Y), \qquad a \le x \le b; \qquad G(Y(a), Y(b)) = 0, \tag{0.7}$$

where

$$F(x, Y) = \begin{bmatrix} f(x, y; y^{d+1}) \\ 0 \end{bmatrix}, \quad G(Y(a), Y(b)) = g(y(a), y(b); y^{d+1}(a)). \tag{0.8}$$

Thus, for example, the Sturm-Liouville problem (0.5) and (0.6), with

$$y_1 = u, \qquad y_2 = p(x)\,\frac{du}{dx}\,, \qquad y_3 = \lambda,$$

becomes the standard two-point boundary value problem

$$\begin{aligned} \frac{dy_1}{dx} &= \frac{1}{p(x)}\, y_2, \\ \frac{dy_2}{dx} &= -q(x)y_1 + y_3 y_1, \\ \frac{dy_3}{dx} &= 0, \end{aligned} \tag{0.5'}$$

subject to

$$y_1(a) = 0, \qquad y_1(b) = 0, \qquad y_2(a) = p(a). \tag{0.6''}$$

If one of the boundary points, say, b, is unknown, then the problem (0.1) and (0.2), where now $g : \mathbb{R}^d \times \mathbb{R}^d \to \mathbb{R}^{d+1}$, is a *problem with free boundary.* This too can be reduced to an ordinary two-point boundary value problem if one sets

$$z^{d+1} = b - a,$$

which is a constant as far as dependence on x is concerned, and introduces a new independent variable t by

$$x = a + tz^{d+1}, \qquad 0 \le t \le 1.$$

Letting then

$$z(t) = y(a + tz^{d+1}), \qquad Z(t) = \begin{bmatrix} z(t) \\ z^{d+1} \end{bmatrix},$$

gives

$$\begin{aligned} \frac{dZ}{dt} &= \begin{bmatrix} z^{d+1} f(a + tz^{d+1}, z) \\ 0 \end{bmatrix}, \qquad 0 \le t \le 1, \\ g(z(0), z(1)) &= 0, \end{aligned} \tag{0.9}$$

a two-point boundary value problem on a fixed interval, [0,1]. Once it is solved, one recovers $b = a + z^{d+1}$ and $y(x) = z((x-a)/z^{d+1})$, $a \le x \le b$.

We begin with the problem of existence and uniqueness, both for linear and nonlinear boundary value problems. We then show how initial value techniques can be employed to solve boundary value problems and discuss some of the practical difficulties associated with that. Approaches relying more on systems of linear or nonlinear equations are those based on finite difference or variational approximations, and we give a brief account of these as well.

§1. Existence and Uniqueness

Before dealing with general considerations on the existence and uniqueness (or nonuniqueness) of solutions to boundary value problems, it may be useful to look at some very simple but instructive examples.

§1.1. **Examples.** For linear problems the breakdown of uniqueness or existence is exceptional and occurs, if at all, only for some "critical" intervals, often denumerably infinite in number. For nonlinear problems, the situation can be more complex.

Example 1.1. $y'' - y = 0$, $y(0) = 0$, $y(b) = \beta$.

The general solution here is made up of the hyperbolic cosine and sine. Since the hyperbolic cosine is ruled out by the first boundary condition, one obtains from the second boundary condition uniquely

$$y(x) = \beta \, \frac{\sinh x}{\sinh b}\,, \qquad 0 \leq x \leq b. \tag{1.1}$$

There are no exceptional (critical) intervals here.

Example 1.2. $y'' + y = 0$, $y(0) = 0$, $y(b) = \beta$.

Although this problem differs only slightly from the one in Example 1.1, the structure of the solution is fundamentally different because of the oscillatory nature of the general solution, consisting of the trigonometric cosine and sine. If b is not an integer multiple of π, there is a unique solution as before,

$$y(x) = \beta \, \frac{\sin x}{\sin b}\,, \qquad b \neq n\pi \;\; (n = 1, 2, 3, \ldots). \tag{1.2}$$

If, however, $b = n\pi$, then there are infinitely many solutions, or none, accordingly as $\beta = 0$ or $\beta \neq 0$. In the former case, all solutions have the form $y(x) = c \sin x$, with c an arbitrary constant. In the latter case, the second boundary condition cannot be satisfied since every solution candidate must necessarily vanish at b. In either case, $b = n\pi$ is a critical point.

We now minimally modify these examples to make them nonlinear.

Example 1.3. $y'' + |y| = 0$, $y(0) = 0$, $y(b) = \beta$.

As a preliminary consideration, suppose $y(x_0) = 0$ for a solution of the differential equation. What can we say about $y(x)$ for $x > x_0$?

We distinguish three cases: (1) $y'(x_0) < 0$. In this case, y becomes negative to the right of x_0, and hence Example 1.1 applies: the solution becomes, and remains, a negative hyperbolic sine, $y(x) = c \sinh(x - x_0)$, $c < 0$. (2) $y'(x_0) = 0$. By the uniqueness of the initial value problem (the differential equation satisfies a uniform Lipschitz condition with Lipschitz constant 1), we get $y(x) \equiv 0$ for $x > x_0$. (3) $y'(x_0) > 0$. Now y is positive on a right neighborhood of x_0, so that by Example 1.2, $y(x) = c \sin(x - x_0)$, $c > 0$, for $x_0 \le x \le x_0 + \pi$. At $x = x_0 + \pi$, however, $y(x_0 + \pi) = 0$, $y'(x_0 + \pi) = -c < 0$, and by what was said in Case (1), from then on we have $y(x) = -c \sinh(x - x_0 - \pi)$ (which ensures continuity of the first derivative of y at $x = x_0 + \pi$). Thus, in this third case, the solution $y(x)$, $x > x_0$, consists of two arcs, a trigonometric sine arc followed by a hyperbolic sine arc.

To discuss the solution of the boundary value problem in Example 1.3, we distinguish again three cases. In each case (and subcase) one arrives at the solution by considering all three possibilities $y'(0) < 0$, $y'(0) = 0$, $y'(0) > 0$, and eliminating all but one.

Case I: $b < \pi$. Here we have the unique solution

$$y(x) = \begin{cases} 0 & \text{if} \quad \beta = 0, \\ \beta \dfrac{\sin x}{\sin b} & \text{if} \quad \beta > 0, \\ \beta \dfrac{\sinh x}{\sinh b} & \text{if} \quad \beta < 0. \end{cases} \tag{1.3}$$

Case II: $b = \pi$.

(a) $\beta = 0$: *infinitely* many solutions $y(x) = c \sin x$, $c \ge 0$ arbitrary.

(b) $\beta > 0$: *no* solution.

(c) $\beta < 0$: *unique* solution $y(x) = \beta \dfrac{\sinh x}{\sinh \pi}$.

Case III: $b > \pi$.

(a) $\beta = 0$: *unique* solution $y(x) \equiv 0$.

(b) $\beta > 0$: *no* solution.

(c) $\beta < 0$: *exactly two* solutions,

$$y_1(x) = \beta \frac{\sinh x}{\sinh b}, \qquad 0 \le x \le b, \tag{1.4$_1$}$$

$$y_2(x) = \begin{cases} -\beta \dfrac{\sin x}{\sinh(b-\pi)}, & 0 \le x \le \pi, \\ \beta \dfrac{\sinh(x-\pi)}{\sinh(b-\pi)}, & \pi \le x \le b. \end{cases} \tag{1.4$_2$}$$

In summary, we indicate the number of solutions in Table 7.1.1. It is rather remarkable how the seemingly innocuous modification of changing y to $|y|$ produces such a profound change in the qualitative behavior of the solution.

b \ β	> 0	$=0$	< 0
$< \pi$	1	1	1
$= \pi$	0	∞	1
$> \pi$	0	1	2

TABLE 7.1.1. Number of solutions of the boundary value problem in Example 1.3

§1.2. **A scalar boundary value problem**. A problem of some importance is the two-point boundary value problem for a scalar nonlinear second-order differential equation

$$y'' = f(x, y, y'), \qquad a \le x \le b, \tag{1.5}$$

with linear boundary conditions

$$\begin{aligned} a_0 y(a) - a_1 y'(a) &= \alpha, \\ b_0 y(b) + b_1 y'(b) &= \beta, \end{aligned} \tag{1.6}$$

where we assume, of course, that not both a_0 and a_1 are zero, and similarly for b_0 and b_1. We further assume that f is continuous on $[a, b] \times \mathbb{R} \times \mathbb{R}$ and satisfies uniform Lipschitz conditions

$$\begin{aligned} |f(x, u_1^*, u_2) - f(x, u_1, u_2)| &\leq L_1 |u_1^* - u_1|, \\ |f(x, u_1, u_2^*) - f(x, u_1, u_2)| &\leq L_2 |u_2^* - u_2| \end{aligned} \tag{1.7}$$

for all $x \in [a, b]$ and all real u_1, u_2, u_1^* , u_2^*. These assumptions are sufficient to ensure that each initial value problem for (1.5) has a unique solution on the whole interval $[a, b]$ (cf. Theorem 5.0.1).

We associate with (1.5) and (1.6) the initial value problem

$$u'' = f(x, u, u'), \qquad a \leq x \leq b, \tag{1.8}$$

subject to

$$\begin{aligned} a_0 u(a) - a_1 u'(a) &= \alpha, \\ c_0 u(a) - c_1 u'(a) &= s. \end{aligned} \tag{1.9}$$

For the two initial conditions in (1.9) to be linearly independent, we must assume that

$$\det \begin{bmatrix} a_0 & -a_1 \\ c_0 & -c_1 \end{bmatrix} \neq 0; \quad \text{that is,} \quad c_0 a_1 - c_1 a_0 \neq 0.$$

Since we are otherwise free to choose the constants c_0, c_1 as we please, we may as well take them to satisfy

$$c_0 a_1 - c_1 a_0 = 1. \tag{1.10}$$

Then the initial conditions become

$$\begin{aligned} u(a) &= a_1 s - c_1 \alpha, \\ u'(a) &= a_0 s - c_0 \alpha. \end{aligned} \tag{1.11}$$

We consider c_0, c_1 to be fixed from now on, and s a parameter to be determined.

The solution of the initial value problem (1.8) and (1.11) is denoted by $u(x; s)$. If it is to solve the boundary value problem (1.5) and (1.6), we must have

$$\phi(s) = 0, \qquad \phi(s) := b_0 u(b; s) + b_1 u'(b; s) - \beta. \tag{1.12}$$

Here and in the following, the prime in $u'(x;s)$ always indicates differentiation with respect to the first variable, x. Clearly, (1.12) is a nonlinear equation in the unknown s (cf. Ch. 4).

Theorem 7.1.1. *The boundary value problem* (1.5) *and* (1.6) *has as many distinct solutions as* $\phi(s)$ *has distinct zeros.*

Proof. (a) If $\phi(s_1) = 0$, then clearly $u(x;s_1)$ is a solution of the boundary value problem (1.5) and (1.6). If $s_2 \neq s_1$ is another zero of $\phi(s)$, then by (1.11) either $u(a;s_2) \neq u(a,s_1)$ (if $a_1 \neq 0$) or $u'(a;s_2) \neq u'(a;s_1)$ (if $a_0 \neq 0$); that is, $u(x;s_2) \not\equiv u(x;s_1)$. Thus, to two distinct zeros of $\phi(s)$ there correspond two distinct solutions of (1.5) and (1.6).

(b) If $y(x)$ is a solution of the boundary value problem (1.5) and (1.6), then defining $s := c_0 y(a) - c_1 y'(a)$, we have that $y(x) = u(x;s)$; hence $\phi(s) = 0$. Thus, to every solution of (1.5) and (1.6) there corresponds a zero of ϕ. □

Theorem 7.1.1 is the basis for solving (1.5) and (1.6) numerically. Solve $\phi(s) = 0$ by any of the standard methods for solving nonlinear equations. We discuss this in more detail in §2.

For a certain class of boundary value problems (1.5) and (1.6) one can show that $\phi(s) = 0$ has exactly one solution.

Theorem 7.1.2. *Assume that*

(1) $f(x, u_1, u_2)$ *is continuous on* $[a,b] \times \mathbb{R} \times \mathbb{R}$;

(2) *both* f_{u_1} *and* f_{u_2} *are continuous and satisfy*

$$0 < f_{u_1}(x, u_1, u_2) \leq L_1, \quad |f_{u_2}(x, u_1, u_2)| \leq L_2 \text{ on } [a,b] \times \mathbb{R} \times \mathbb{R};$$

(3) $a_0 a_1 \geq 0, \quad b_0 b_1 \geq 0, \quad |a_0| + |b_0| > 0.$

Then the boundary value problem (1.5) *and* (1.6) *has a unique solution.*

Note that Assumption (2) implies (1.7), hence the unique solvability on $[a,b]$ of initial value problems for (1.5). Assumption (3) requires that a_0 and a_1 be of the same sign, as well as b_0 and b_1, and that not both a_0 and b_0 vanish. We may assume, by multiplying one or both of the boundary conditions (1.6) by -1 if necessary, that

$$(3') \qquad a_0 \geq 0, a_1 \geq 0; \quad b_0 \geq 0, b_1 \geq 0; \quad a_0 + b_0 > 0.$$

Proof of Theorem 7.1.2. The idea of the proof is to show that

$$\phi'(s) \geq c > 0 \quad \text{for all} \quad s \in \mathbb{R}. \tag{1.13}$$

The function $\phi(s)$ then increases monotonically from $-\infty$ to $+\infty$, and hence vanishes for exactly one value of s.

We have

$$\phi'(s) = b_0 \frac{\partial}{\partial s} u(b; s) + b_1 \frac{\partial}{\partial s} u'(b; s).$$

It is convenient to denote

$$v(x) = \frac{\partial}{\partial s} u(x; s),$$

where the dependence on s is suppressed in the notation for v. Since differentiation with respect to x and s may be interchanged under the assumptions made, we can write

$$\phi'(s) = b_0 v(b) + b_1 v'(b). \tag{1.14}$$

Furthermore, $u(x; s)$ satisfies, identically in s,

$$u''(x; s) = f(x, u(x; s), u'(x; s)), \quad a \leq x \leq b,$$

$$u(a; s) = a_1 s - c_1 \alpha, \qquad u'(a; s) = a_0 s - c_0 \alpha,$$

from which, by differentiation with respect to s and interchange with differentiation in x where necessary, one gets

$$\begin{gathered} v''(x) = f_{u_1}(x, u(x; s), u'(x; s))v(x) + f_{u_2}(x, u(x; s), u'(x; s))v'(x), \\ v(a) = a_1, \qquad v'(a) = a_0. \end{gathered} \tag{1.15}$$

Thus, v is the solution of a "linear" boundary value problem,

$$\begin{gathered} v'' = p(x)v' + q(x)v, \qquad a \leq x \leq b, \\ v(a) = a_1, \qquad v'(a) = a_0, \end{gathered} \tag{1.16}$$

where

$$|p(x)| \leq L_2, \qquad 0 < q(x) \leq L_1 \qquad \text{on } [a, b]. \tag{1.17}$$

We are going to show that, on $a \le x \le b$,

$$v(x) > a_1 + a_0 \frac{1 - e^{-L_2(x-a)}}{L_2}, \quad v'(x) > a_0 e^{-L_2(x-a)}. \tag{1.18}$$

From this, (1.13) will follow. Indeed, since not both a_0 and a_1 can vanish and by (3′) at least one is positive, it follows from (1.18) that $v(b) > 0$. If $b_0 > 0$, then (1.14) shows, since $b_1 \ge 0$ and $v'(b) > 0$ by (1.18), that $\phi'(s)$ is positive and bounded away from zero (as a function of s). The same conclusion follows if $b_0 = 0$, since then $b_1 > 0$ and $\phi'(s) = b_1 v'(b) > 0$ in (1.14).

To prove (1.18), we first show that $v(x) > 0$ for $a < x \le b$. This is certainly true in a small right neighborhood of a, since by (1.16) either $v(a) > 0$ or $v(a) = 0$ and $v'(a) > 0$. If the assertion were false, we would therefore have $v(x_0) = 0$ for some x_0 in $(a, b]$. But then v must have a local maximum at some x_1 with $a < x_1 < x_0$. This is clear in the cases where $v(a) = 0$, $v'(a) > 0$, and $v(a) > 0$, $v'(a) > 0$. In the remaining case $v(a) > 0$, $v'(a) = 0$, it follows from the fact that then $v''(a) > 0$ by virtue of the differential equation in (1.16) and the positivity of q (cf. (1.17)). Thus,

$$v(x_1) > 0, \qquad v'(x_1) = 0, \qquad v''(x_1) < 0.$$

But this contradicts the differential equation (1.16) at $x = x_1$, since $q(x_1) > 0$. This establishes the positivity of v on $(a, b]$. We thus have, using again the positivity of q,

$$v''(x) - p(x)v'(x) > 0 \quad \text{for} \quad a < x \le b.$$

Multiplication by the "integrating factor" $\exp\left(-\int_a^x p(t)dt\right)$ yields

$$\frac{d}{dx}\left[e^{-\int_a^x p(t)dt} v'(x)\right] > 0,$$

and upon integration from a to x,

$$e^{-\int_a^x p(t)dt} v'(x) - v'(a) > 0.$$

This, in turn, by the second initial condition in (1.16), gives

$$v'(x) > a_0 e^{\int_a^x p(t)dt},$$

from which the second inequality in (1.18) follows by virtue of $p(t) \geq -L_2$ (cf. (1.17)). The first inequality in (1.18) follows by integrating the second from a to x. □

Theorem 7.1.2 has an immediate application to the Sturm-Liouville problem

$$\mathcal{L}y = r(x), \quad a \leq x \leq b; \quad \mathcal{B}_a y = \alpha, \quad \mathcal{B}_b y = \beta, \tag{1.19}$$

where

$$\begin{aligned} \mathcal{L}y &:= -y'' + p(x)y' + q(x)y, \\ \mathcal{B}_a y = a_0 y(a) - a_1 y'(a), &\qquad \mathcal{B}_b y = b_0 y(b) + b_1 y'(b). \end{aligned} \tag{1.20}$$

Corollary 7.1.1. *If p, q, and r are continuous on $[a,b]$ with*

$$q(x) > 0 \quad for \quad a \leq x \leq b, \tag{1.21}$$

and if a_0, a_1, b_0, b_1 satisfy the condition (3) *of Theorem* 7.1.2, *then* (1.19) *has a unique solution.*

We remark that the differential equation in (1.19) can be written equivalently in "self-adjoint form" if we multiply both sides by $P(x) = \exp(-\int p dx)$. This yields

$$-\frac{d}{dx}\left(P(x)\frac{dy}{dx}\right) + Q(x)y = R(x), \qquad a \leq x \leq b, \tag{1.19$'$}$$

with

$$Q(x) = P(x)q(x), \quad R(x) = P(x)r(x).$$

Note that P in (1.19′) is positive and not only continuous, but also continuously differentiable on $[a,b]$. Furthermore, the positivity of q is equivalent to the positivity of Q.

The following is a result of a somewhat different nature, providing an alternative.

Theorem 7.1.3 (Alternative Theorem). *The boundary value problem* (1.19) *has a unique solution for arbitrary α and β if and only if the corresponding homogeneous problem (with $r \equiv 0$, $\alpha = \beta = 0$) has only the trivial solution $y \equiv 0$.*

Proof. Define u_1 and u_2 by

$$\mathcal{L}u_1 = r(x), \quad a \leq x \leq b; \quad u_1(a) = -c_1\alpha, \ u_1'(a) = -c_0\alpha,$$

and

$$\mathcal{L}u_2 = 0, \quad a \le x \le b; \quad u_2(a) = a_1, \; u_2'(a) = a_0,$$

with c_0, c_1 as defined in (1.10). Then one easily verifies that $\mathcal{B}_a u_1 = \alpha$, $\mathcal{B}_a u_2 = 0$, so that

$$u(x) = u_1(x) + s u_2(x) \tag{1.22}$$

satisfies both the inhomogeneous differential equation and the first boundary condition of (1.19). The inhomogeneous boundary value problem therefore has a unique solution if and only if

$$\mathcal{B}_b u_2 \neq 0, \tag{1.23}$$

so that the second boundary condition $\mathcal{B}_b u = \beta$ can be solved uniquely for s in (1.22). On the other hand, for the homogeneous boundary value problem to have only the trivial solution, we must have (1.23), since otherwise $\mathcal{L}u_2 = 0$, $\mathcal{B}_a u_2 = 0$, and $\mathcal{B}_b u_2 = 0$, whereas one of $u_2(a)$ and $u_2'(a)$ must be different from zero, since not both a_1 and a_0 can vanish. □

§1.3. **General linear and nonlinear systems**. The two-point boundary value problem for the general nonlinear system (0.1), with linear boundary conditions, takes the form

$$\begin{gathered} \frac{dy}{dx} = f(x, y), \qquad a \le x \le b, \\ Ay(a) + By(b) = \gamma, \end{gathered} \tag{1.24}$$

where A, B are square matrices of order d with constant elements, and γ is a given d-vector. For linear independence and consistency we assume that

$$\text{rank } [A, B] = d. \tag{1.25}$$

In this general form, the associated initial value problem is

$$\begin{gathered} \frac{du}{dx} = f(x, u), \qquad a \le x \le b, \\ u(a) = s, \end{gathered} \tag{1.26}$$

where $s \in \mathbb{R}^d$ is a "trial" initial vector. If we denote by $u(x; s)$ the solution of (1.26) and assume it to exist on $[a, b]$, then (1.24) is equivalent to a problem of solving a nonlinear system of equations,

$$\phi(s) = 0, \qquad \phi(s) = As + Bu(b; s) - \gamma. \tag{1.27}$$

Again, the boundary value problem (1.24) has as many distinct solutions as the nonlinear system (1.27) has distinct solution vectors. By imposing sufficiently strong — but often unrealistic — conditions on f, A, and B, it is possible to prove that (1.27), and hence (1.24), has a unique solution, but we do not pursue this any further.

For linear systems, we have

$$f(x, y) = C(x)y + d(x), \qquad a \le x \le b, \tag{1.28}$$

in which case the initial value problem (1.26) is known to have the solution

$$u(x) = Y(x)s + v(x), \tag{1.29}$$

where $Y(x) \in \mathbb{R}^{d\times d}$ is a fundamental solution of the homogeneous system $dY/dx = C(x)Y$ with initial value $Y(a) = I$, and $v(x)$ a particular solution of the inhomogeneous system $dv/dx = C(x)v + d(x)$ satisfying $v(a) = 0$. The boundary value problem (1.24) is then equivalent to the system of linear algebraic equations

$$[A + BY(b)]s = \gamma - Bv(b) \tag{1.30}$$

and has a unique solution if and only if the matrix of this system is nonsingular.

We remark that if some components of $y(a)$ are prescribed as part of the boundary conditions in (1.24), then of course they are incorporated in the vector s, and one obtains a smaller system of nonlinear (resp., linear) equations in the remaining (unknown) components of s.

§2. Initial Value Techniques

The techniques used in §§1.2 and 1.3 are also of computational interest in that they lend themselves to the application of numerical methods for solving nonlinear equations or systems of equations. We show, for example, how Newton's method can be used in this context, first for a scalar second-order boundary value problem, and then for a general problem involving a first-order system of differential equations.

§2.1. **Shooting method for a scalar boundary value problem.** We have seen in §1.2 that the boundary value problem (1.5) and (1.6) leads to the nonlinear equation (1.12) via the initial value problem (1.8) and (1.11).

Solving the initial value problem is referred to, in this context, as "shooting." One aims by means of trial initial conditions to satisfy the second boundary condition, which is the "target." A mechanism of readjusting the aim based on the amount by which the target has been missed, is provided by Newton's method. Specifically, one starts with an initial approximation $s^{(0)}$ for s in (1.12), and then iterates according to

$$s^{(\nu+1)} = s^{(\nu)} - \frac{\phi(s^{(\nu)})}{\phi'(s^{(\nu)})}, \qquad \nu = 0, 1, 2, \dots, \tag{2.1}$$

until, it is hoped, $s^{(\nu)} \to s_\infty$ as $\nu \to \infty$. If that occurs, then $y(x) = u(x; s_\infty)$ will be a solution of the boundary value problem. If there is more than one solution, the process (2.1) needs to be repeated, perhaps several times, with different starting values $s^{(0)}$.

For any given s, the values of $\phi(s)$ and $\phi'(s)$ needed in (2.1) are computed simultaneously by "shooting," that is, by solving the initial value problem (1.8) and (1.11) together with the one in (1.15) obtained by differentiation with respect to s. If both are written as first-order systems, by letting

$$y_1(x) = u(x; s), \quad y_2(x) = u'(x; s), \quad y_3(x) = v(x), \quad y_4(x) = v'(x),$$

one solves on $[a, b]$ the initial value problem

$$\begin{aligned}
\frac{dy_1}{dx} &= y_2, & y_1(a) &= a_1 s - c_1\alpha, \\
\frac{dy_2}{dx} &= f(x, y_1, y_2), & y_2(a) &= a_0 s - c_0\alpha, \\
\frac{dy_3}{dx} &= y_4, & y_3(a) &= a_1, \\
\frac{dy_4}{dx} &= f_{u_1}(x, y_1, y_2)y_3 + f_{u_2}(x, y_1, y_2)y_4, & y_4(a) &= a_0,
\end{aligned} \tag{2.2}$$

with c_0, c_1 as chosen in (1.10), and then computes

$$\phi(s) = b_0 y_1(b) + b_1 y_2(b) - \beta, \quad \phi'(s) = b_0 y_3(b) + b_1 y_4(b). \tag{2.3}$$

Thus, each Newton step (2.1) requires the solution on $[a, b]$ of an initial value problem (2.2) with $s = s^{(\nu)}$.

Example 2.1. $y'' = -e^{-y}$, $0 \le x \le 1$, $y(0) = y(1) = 0$.

We first show that this problem has a unique solution. To do this, we "embed" the problem into the following problem:

$$y'' = f(y), \; 0 \le x \le 1; \;\; y(0) = y(1) = 0, \tag{2.4}$$

where

$$f(y) = \begin{cases} -e^{-y} & \text{if} \;\; y \ge 0, \\ e^{y} - 2 & \text{if} \;\; y \le 0. \end{cases} \tag{2.5}$$

Then $f_y(y) = e^{-y}$ if $y \ge 0$ and $f_y(y) = e^{y}$ if $y \le 0$, so that $0 < f_y(y) \le 1$ for all real y. Thus, Assumption (2) of Theorem 7.1.2 is satisfied, and so are (trivially) the other two assumptions. It follows that (2.4) has a unique solution, and since clearly this solution cannot become negative (the second derivative being necessarily negative), it also solves the problem in Example 2.1.

Since $a_0 = b_0 = 1$, $a_1 = b_1 = \alpha = \beta = 0$ in this example, the system (2.2) becomes, for $0 \le x \le 1$,

$$\begin{aligned} \frac{dy_1}{dx} &= y_2, & y_1(a) &= 0, \\ \frac{dy_2}{dx} &= -e^{-y_1}, & y_2(0) &= s, \\ \frac{dy_3}{dx} &= y_4, & y_3(0) &= 0, \\ \frac{dy_4}{dx} &= e^{-y_1} y_3, & y_4(0) &= 1, \end{aligned}$$

and (2.3) simplifies to

$$\phi(s) = y_1(1), \qquad \phi'(s) = y_3(1).$$

Newton's method (2.1), of course, has to be started with a positive initial approximation $s^{(0)}$. If we use $s^{(0)} = 1$, the MATLAB routine ode45 (with tolerance 1.0×10^{-12}) gives the results shown in Table 7.2.1.

ν	$s^{(\nu)}$
0	1.
1	0.45
2	0.463629
3	0.463632591724

TABLE 7.2.1. Numerical results for Example 2.1

Example 2.2. $y'' = \lambda \sinh(\lambda y)$, $0 \le x \le 1$, $y(0) = y(1) = 0$.

If $y'(0) = s$, $s \neq 0$, the solution y of the differential equation has the sign of s in a right neighborhood of $x = 0$, and so does y''. Thus, $|y|$ is monotonically increasing and cannot attain the value zero at $x = 1$. It follows that $y(x) \equiv 0$ is the unique solution of the boundary value problem. Moreover, it can be shown (see Ex. 7) that for $y'(0) = s$ the modulus $|y|$ of the solution tends to infinity at a finite value x_∞ of x, where

$$x_\infty \sim \frac{1}{\lambda} \ln \frac{8}{|s|}, \qquad s \to 0. \tag{2.6}$$

Thus, if we apply shooting with $y'(0) = s$, we can reach the endpoint $x = 1$ without the solution blowing up on us only if $x_\infty > 1$, that is, approximately, if

$$|s| < 8e^{-\lambda}. \tag{2.7}$$

This places a severe restriction on the permissible initial slope; for example, $|s| < 3.63\ldots \times 10^{-4}$ if $\lambda = 10$, and $|s| < 1.64\ldots \times 10^{-8}$ if $\lambda = 20$. It is indeed one of the limitations of "ordinary" shooting that rather accurate initial data need to be known in order to succeed.

§2.2. **Linear and nonlinear systems**. For linear systems, the shooting method amounts to solving the linear system of algebraic equations (1.30), which requires the numerical solution of $d + 1$ initial value problems to obtain $Y(b)$ and $v(b)$, and possibly one more final integration with the starting vector found to determine the solution $y(x)$ at the desired values of x. It is strictly a superposition method and no iteration is required, but the procedure often suffers from ill-conditioning of the matrix involved.

For the general nonlinear boundary value problem (1.24), there is no difficulty, formally, in defining a shooting method. One simply has to solve the system of nonlinear equations (1.27), for example, by Newton's method,

$$\left.\begin{aligned} s^{(\nu+1)} &= s^{(\nu)} + \Delta s^{(\nu)} \\ \frac{\partial \phi}{\partial s}(s^{(\nu)})\Delta s^{(\nu)} &= -\phi(s^{(\nu)}) \end{aligned}\right\} \quad \nu = 0, 1, 2, \ldots, \tag{2.8}$$

where $\partial\phi/\partial s$ is the Jacobian of ϕ,

$$\frac{\partial \phi}{\partial s}(s) = A + B\,\frac{\partial u(b; s)}{\partial s}.$$

With $u(x;s)$ denoting, as before, the solution of (1.26), we let $V(x)$ be its Jacobian (with respect to s),

$$V(x) = \frac{\partial u(x;s)}{\partial s}, \qquad a \le x \le b.$$

Then, in order to simultaneously compute $\phi(s)$ and $(\partial\phi/\partial s)(s)$, we can integrate the initial value problem (1.26) adjoined with that for its Jacobian,

$$\left.\begin{aligned} &\frac{du}{dx} = f(x,u) \\ &\frac{dV}{dx} = f_y(x,u)V \\ &u(a) = s, \quad V(a) = I \end{aligned}\right\} \quad a \le x \le b, \tag{2.9}$$

and get

$$\phi(s) = As + Bu(b;s) - \gamma, \quad \frac{\partial\phi}{\partial s}(s) = A + BV(b). \tag{2.10}$$

Although the procedure is formally straightforward, it is difficult to implement. For one thing, we may not be able to integrate (2.9) on the whole interval $[a,b]$; some component may blow up before we reach the endpoint b (cf. Example 2.2). Another problem has to do with the convergence of Newton's method (2.8). Typically, this requires s in (2.9) to be very close to s_∞ — the true initial vector for one of the possible solutions of the boundary value problem. A good illustration of these difficulties is provided by the following example.

Example 2.3.

$$\left.\begin{aligned} \frac{dy_1}{dx} &= \frac{y_1^2}{y_2} \\ \frac{dy_2}{dx} &= \frac{y_2^2}{y_1} \end{aligned}\right\} \quad 0 \le x \le 1, \tag{2.11}$$

$$y_1(0) = 1, \quad y_1(1) = e \; (= 2.718\ldots).$$

This is really a linear system in disguise, namely, the one for the reciprocal functions y_1^{-1} and y_2^{-1}. Hence it is easily seen that an exact solution to (2.11) is

$$y_1(x) = y_2(x) = e^x, \qquad 0 \le x \le 1. \tag{2.12}$$

We write the system in this complicated nonlinear form to bring out the difficulties inherent in the shooting method.

Since y_1 is known at $x = 0$, we have only one unknown, $s = y_2(0)$. We thus define $u_1(x; s)$, $u_2(x; s)$ to be the solution of the initial value problem

$$\frac{d}{dx}\begin{bmatrix} u_1 \\ u_2 \end{bmatrix} = \begin{bmatrix} \dfrac{u_1^2}{u_2} \\ \dfrac{u_2^2}{u_1} \end{bmatrix}, \quad 0 \le x \le 1; \quad \begin{bmatrix} u_1 \\ u_2 \end{bmatrix}(0) = \begin{bmatrix} 1 \\ s \end{bmatrix}. \tag{2.13}$$

The equation to be solved then is

$$\phi(s) = 0, \quad \phi(s) = u_1(1; s) - e. \tag{2.14}$$

We denote $\frac{\partial u_1(x;s)}{\partial s} = v_1(x)$, $\frac{\partial u_2(x;s)}{\partial s} = v_2(x)$, differentiate (2.13) with respect to s, and note that the Jacobian of (2.13) is

$$f_y(u_1, u_2) = \begin{bmatrix} \dfrac{2u_1}{u_2} & -\dfrac{u_1^2}{u_2^2} \\ -\dfrac{u_2^2}{u_1^2} & \dfrac{2u_2}{u_1} \end{bmatrix}, \quad y = \begin{bmatrix} u_1 \\ u_2 \end{bmatrix}.$$

If we append the differentiated system to the original one, we get

$$\begin{aligned} \frac{du_1}{dx} &= \frac{u_1^2}{u_2}, & u_1(0) &= 1, \\ \frac{du_2}{dx} &= \frac{u_2^2}{u_1}, & u_2(0) &= s, \\ \frac{dv_1}{dx} &= \frac{2u_1}{u_2}\, v_1 - \frac{u_1^2}{u_2^2}\, v_2, & v_1(0) &= 0, \\ \frac{dv_2}{dx} &= -\frac{u_2^2}{u_1^2}\, v_1 + \frac{2u_2}{u_1}\, v_2, & v_2(0) &= 1. \end{aligned} \tag{2.15}$$

Assuming this can be solved on [0,1], we will have

$$\phi(s) = u_1(1; s) - e, \quad \phi'(s) = v_1(1), \tag{2.16}$$

and can thus apply Newton's method,

$$s^{(\nu+1)} = s^{(\nu)} - \frac{\phi(s^{(\nu)})}{\phi'(s^{(\nu)})}, \tag{2.17}$$

taking $s = s^{(\nu)}$ in (2.15).

Because of the elementary nature of the system (2.13), we can solve it in closed form,

$$u_1(x;s) = \frac{s}{s\cosh x - \sinh x}\,, \quad u_2(x;s) = \frac{s}{\cosh x - s\sinh x}\,. \tag{2.18}$$

It is convenient to look at the solution of (2.13) in the phase plane (u_1, u_2), where the exact solution of (2.13) is given by $u_1 = u_2$.

Clearly, s has to be positive in (2.13), since otherwise u_1 would initially decrease and, to meet the condition at $x = 1$, would have to turn around and have a vanishing derivative at some point in (0,1). That would cause a problem in (2.13), since either $u_1 = 0$ or $u_2 = \infty$ at that point.

For u_1 to remain bounded on [0,1], it then follows from the first relation in (2.18) that we must have $s > \tanh x$ for $0 \le x \le 1$; that is,

$$s > \tanh 1 = .76159\ldots\,.$$

At $s = \tanh 1$, we have

$$\lim_{x\to 1} u_1(x;\tanh 1) = \infty, \quad \lim_{x\to 1} u_2(x;\tanh 1) = \sinh 1 = 1.1752\ldots\,.$$

This solution, in the phase plane, has a horizontal asymptote.

Similarly, for u_2 to remain bounded on [0,1], we must have

$$s < \coth 1 = 1.3130\ldots\,.$$

When $s = \coth 1$, then

$$\lim_{x\to 1} u_1(x;\coth 1) = \cosh 1 = 1.5430\ldots\,, \quad \lim_{x\to 1} u_2(x;\coth 1) = \infty,$$

giving a solution with a vertical asymptote. The locus of points $[u_1(1;s), u_2(1;s)]$ as $\tanh 1 < s < \coth 1$ is easily found to be the hyperbola

$$u_2 = \frac{u_1 \sinh 1}{u_1 - \cosh 1}\,.$$

From this, we get a complete picture in the phase plane of all solutions of (2.13); see Figure 7.2.1. Thus, only in a relatively small s-interval,

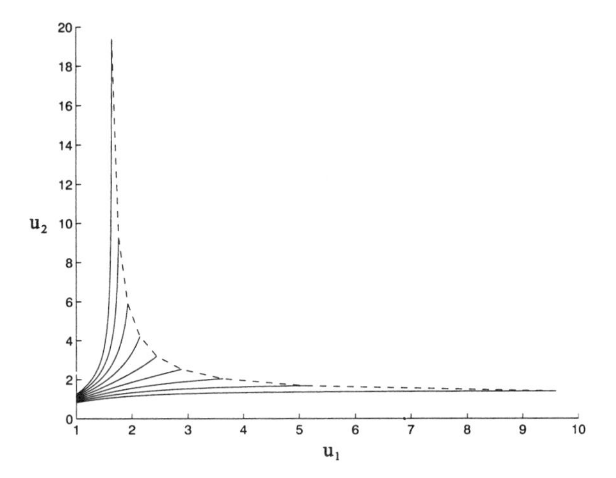

FIGURE 7.2.1. The solutions of (2.13) in the phase plane. The solid lines are solutions for fixed s in $\tanh 1 \leq s \leq \coth 1$ and x varying between 0 and 1. The dashed line is the locus of points $[u_1(1;s), u_2(1;s)]$, $\tanh 1 < s < \coth 1$.

$.76159\ldots < s < 1.3130\ldots$, is it possible to shoot from one endpoint to the other without one component of the solution blowing up in between.

What about the convergence of Newton's method (2.17)? The equation to be solved is

$$\phi(s) = 0, \quad \phi(s) = \frac{s}{s\cosh 1 - \sinh 1} - e.$$

From the graph of $\phi(s)$ (see Figure 7.2.2), in particular, the convexity of ϕ, it follows that Newton's method converges precisely if

$$\tanh 1 < s < s^0, \tag{2.19}$$

where s^0 is such that

$$s^0 - \frac{\phi(s^0)}{\phi'(s^0)} = \tanh 1. \tag{2.20}$$

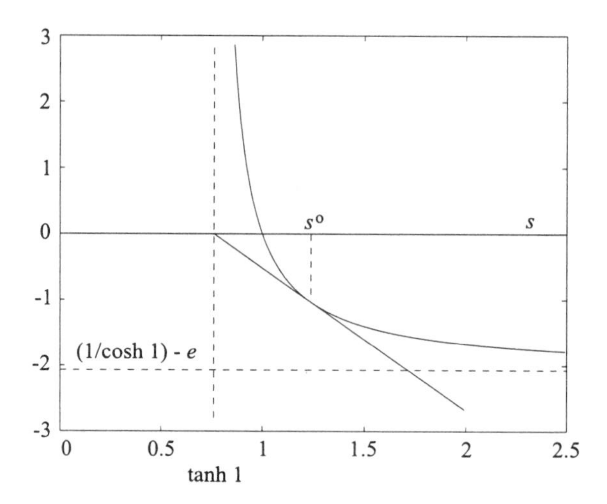

FIGURE 7.2.2. The graph of $\phi(s)$ and convergence of Newton's method

It can be shown (see Ex. 4) that $s^0 < \coth 1$, so that the convergence of Newton's method imposes an additional restriction on the choice of s.

§2.3. **Parallel shooting.** To circumvent the difficulties inherent in ordinary shooting, as indicated in Examples 2.2 and 2.3, one may divide the interval $[a, b]$ into N subintervals,

$$a = x_0 < x_1 < x_2 < \cdots < x_{N-1} < x_N = b, \tag{2.21}$$

and apply shooting concurrently on each subinterval. The hope is that if the intervals are sufficiently small, not only does the appropriate boundary value problem have a unique solution, but also the solution is not given a chance to grow excessively. We say "appropriate" boundary value problem since in addition to the two boundary conditions, there are now also continuity conditions at the interior subdivision points. This, of course, enlarges considerably the problem size. To enhance the prospects of success, it is advisable to generate the subdivision (2.21) dynamically, as described further on, rather than to choose it artificially without regard to the particular features of the problem at hand.

To describe the procedure in more detail, consider the boundary value problem for a general nonlinear system,

$$\frac{dy}{dx} = f(x, y), \quad a \le x \le b; \qquad Ay(a) + By(b) = \gamma. \tag{2.22}$$

Let $h_n = x_n - x_{n-1}$, $n = 1, 2, \ldots, N$, and define

$$y_n(t) = y(x_{n-1} + th_n), \quad 0 \le t \le 1. \tag{2.23}$$

Clearly,

$$\frac{dy_n}{dt} = h_n y'(x_{n-1} + th_n) = h_n f(x_{n-1} + th_n, y_n(t)), \quad 0 \le t \le 1.$$

Thus, by letting

$$f_n(t, z) = h_n f(x_{n-1} + th_n, z), \tag{2.24}$$

we have for the (vector-valued) functions y_n the following system of, as yet uncoupled, differential equations

$$\frac{dy_n}{dt} = f_n(t, y_n), \quad n = 1, 2, \ldots, N; \quad 0 \le t \le 1. \tag{2.25}$$

The coupling comes about through the boundary and interface (continuity) conditions

$$\begin{aligned} &Ay_1(0) + By_N(1) = \gamma, \\ &y_{n+1}(0) - y_n(1) = 0, \quad n = 1, 2, \ldots, N-1. \end{aligned} \tag{2.26}$$

Introducing "supervectors" and "supermatrices,"

$$Y(t) = \begin{bmatrix} y_1(t) \\ \vdots \\ y_N(t) \end{bmatrix}, \quad F(t, Y) = \begin{bmatrix} f_1(t, y_1) \\ \vdots \\ f_N(t, y_N) \end{bmatrix}, \quad \Gamma = \begin{bmatrix} \gamma \\ 0 \\ \vdots \\ 0 \end{bmatrix},$$

$$P = \begin{bmatrix} A & 0 & 0 & \cdots & 0 \\ 0 & I & 0 & \cdots & 0 \\ 0 & 0 & I & \cdots & 0 \\ \vdots & \vdots & \vdots & \ddots & \vdots \\ 0 & 0 & 0 & \cdots & I \end{bmatrix}, \quad Q = \begin{bmatrix} 0 & 0 & \cdots & 0 & B \\ -I & 0 & \cdots & 0 & 0 \\ 0 & -I & \cdots & 0 & 0 \\ \vdots & \vdots & \ddots & \vdots & \vdots \\ 0 & 0 & \cdots & -I & 0 \end{bmatrix}, \tag{2.27}$$

we can write (2.25) and (2.26) compactly as

$$\frac{dY}{dt} = F(t, Y), \quad 0 \le t \le 1; \quad PY(0) + QY(1) = \Gamma; \tag{2.28}$$

this has the same form as (2.22) but is much bigger in size. Parallel shooting consists of applying ordinary shooting to the big system (2.28). Thus, we solve on $0 \le t \le 1$

$$\frac{dU}{dt} = F(t, U), \quad U(0) = S, \tag{2.29}$$

to obtain $U(t) = U(t; S)$ and try to determine the vector $S \in \mathbb{R}^{Nd}$ such that

$$\Phi(S) = PS + QU(1; S) - \Gamma = 0. \tag{2.30}$$

If we use Newton's method, this is done by the iteration

$$\left.\begin{aligned} S^{(\nu+1)} &= S^{(\nu)} + \Delta S^{(\nu)}, \\ [P + QV(1; S^{(\nu)})]\Delta S^{(\nu)} &= -\Phi(S^{(\nu)}) \end{aligned}\right\} \quad \nu = 0, 1, 2, \dots , \tag{2.31}$$

where

$$V(t; S) = \frac{\partial U}{\partial S}(t; S), \qquad 0 \le t \le 1. \tag{2.32}$$

If we partition $U^T = [u_1^T, \dots, u_N^T]$, $S^T = [s_1^T, \dots, s_N^T]$ in accordance with (2.25), then, since (2.25) by itself is uncoupled, we find that $u_n = u_n(t; s_n)$ depends only on s_n. As a consequence, the "big" Jacobian V in (2.32) is block diagonal,

$$V = \begin{bmatrix} v_1 & 0 & \cdots & 0 \\ 0 & v_2 & \cdots & 0 \\ \vdots & \vdots & \dots & \vdots \\ 0 & 0 & \cdots & v_N \end{bmatrix}, \quad v_n(t; s_n) = \frac{\partial u_n}{\partial s_n}(t; s_n), \quad n = 1, 2, \dots, N,$$

and so is the Jacobian $F_Y(t, U)$ of F in (2.29). This means that U in (2.29) and V in (2.32) can be computed by solving uncoupled systems of initial value problems on $0 \le t \le 1$,

$$\left.\begin{aligned} \frac{du_n}{dt} &= f_n(t, u_n), \quad u_n(0) = s_n \\ \frac{dv_n}{dt} &= \frac{\partial f_n}{\partial y_n}(t, u_n)v_n, \quad v_n(0) = I_n \end{aligned}\right\} \quad n = 1, 2, \dots, N. \tag{2.33}$$

This can be done in parallel — hence the name "parallel shooting."

The procedure may be summarized schematically as in Figure 7.2.3. Alternatively, if N is even, one may shoot both forward and backward, as indicated in Figure 7.2.4 for $N = 4$. This reduces the size of the big system by one-half.

Even though multiple shooting can be quite effective, there are many practical problems associated with it. Perhaps the major problems are related to obtaining good initial approximations (recall, we have to choose a reasonable vector S in (2.29)) and to constructing a natural subdivision (2.21). With regard to the latter, suppose we have some rough

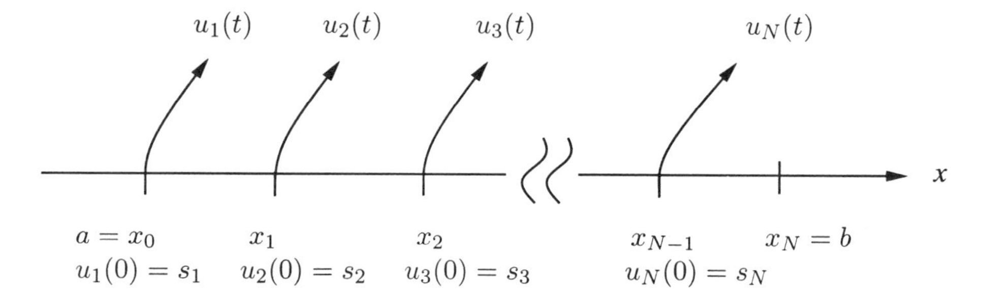

FIGURE 7.2.3. Parallel shooting (one-sided)

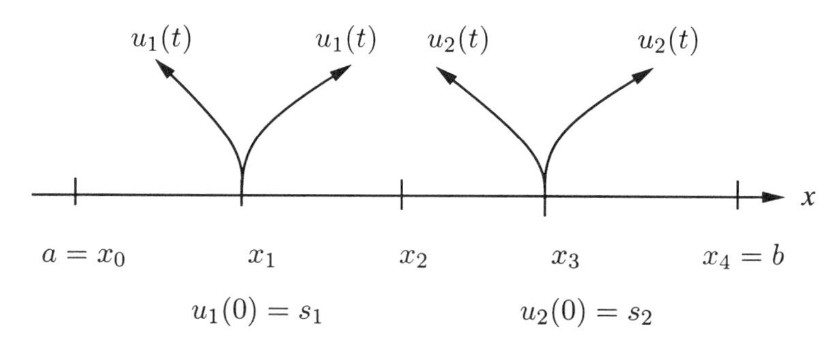

FIGURE 7.2.4. Parallel shooting (two-sided)

approximation $\eta(x) \approx y(x)$ on $a \leq x \leq b$. Then, taking $x_0 = a$, we may construct $x_1, x_2, \ldots$ recursively as follows. For $i = 0, 1, 2, \ldots$ solve the initial value problem

$$\frac{dz_i}{dx} = f(x, z_i), \quad z_i(x_i) = \eta(x_i), \quad x \geq x_i, \tag{2.34}$$

and take for x_{i+1} the smallest $x > x_i$ such that (say) $\|z_i(x)\| \geq 2\|\eta(x)\|$. In other words, we do not allow the solution of (2.34) to increase more than twice in size; see Figure 7.2.5. Thus, (2.34) are strictly auxiliary integrations, whose sole purpose it is to produce an appropriate subdivision of $[a, b]$.

There are circumstances in which reasonable initial approximations may be readily available, for example, if one solves the given boundary value problem (2.22) by a homotopy method. Basically, this means that the problem is embedded in a family of problems,

$$P_\omega : \frac{dy}{dx} = f_\omega(x, y), \quad a \leq x \leq b; \quad Ay(a) + By(b) = \gamma,$$

where ω is a (usually physically meaningful) parameter, say, in the interval $0 \leq \omega \leq 1$. This is done in such a way that for $\omega = 0$ the solution of P_ω is easy, and for $\omega = 1$, we have $f_\omega(x, y) = f(x, y)$. One then solves a sequence

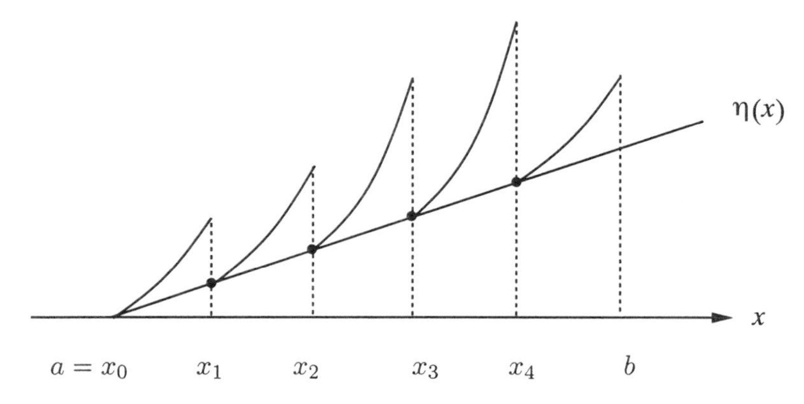

FIGURE 7.2.5. Construction of subdivision

of boundary value problems P_{ω_μ} corresponding to parameter values $0 = \omega_0 < \omega_1 < \omega_2 < \cdots < \omega_m = 1$ which are chosen sufficiently close to one another so that the solution of P_{ω_μ} differs relatively little from the solution of $P_{\omega_{\mu-1}}$. When $\omega = \omega_1$, one takes for $\eta(x)$ the easy solution of P_{ω_0}, constructs the appropriate subdivision as before, and then solves P_{ω_1} by parallel shooting based on the subdivision generated. One next solves P_{ω_2}, using for $\eta(x)$ the solution of P_{ω_1}, and proceeds as with P_{ω_1}. Continuing in this way, we will eventually have solved $P_{\omega_m} = P_1$, the given problem (2.22). Although the procedure is rather labor-intensive, it has the potential of providing accurate solutions to very difficult problems.

§3. Finite Difference Methods

A more static approach toward solving boundary value problems is via direct discretization. One puts a grid on the interval of interest, replaces derivatives by finite difference expressions, and requires the discrete version of the problem to hold at all interior grid points. This gives rise to a system of linear or nonlinear equations for the unknown values of the solution at the grid points.

We consider and analyze only the simplest finite difference schemes. We assume throughout a uniform grid, say,

$$a = x_0 < x_1 < x_2 < \cdots < x_N < x_{N+1} = b, \quad x_n = a + nh, \quad h = \frac{b-a}{N+1}, \tag{3.1}$$

and we continue to use the terminology of grid functions introduced in Ch. 5, §3.

§3.1. **Linear second-order equations**. We consider the Sturm-Liouville problem (cf. (1.19) and (1.20))

$$\mathcal{L}y = r(x), \quad a \le x \le b, \tag{3.2}$$

where

$$\mathcal{L}y := -y'' + p(x)y' + q(x)y, \tag{3.3}$$

with the simplest boundary conditions

$$y(a) = \alpha, \quad y(b) = \beta. \tag{3.4}$$

If p, q, r are continuous and q positive on $[a,b]$, then (3.3) and (3.4) by Corollary 7.1.1 has a unique solution. Under these assumptions, there are positive constants $\overline{p}$, $\underline{q}$, and $\overline{q}$ such that

$$|p(x)| \le \overline{p}, \quad 0 < \underline{q} \le q(x) \le \overline{q} \ \text{ for } \ a \le x \le b. \tag{3.5}$$

A simple finite difference operator, acting on a grid function $u \in \Gamma_h[a,b]$, which approximates the operator $\mathcal{L}$ in (3.3) is

$$(\mathcal{L}_h u)_n = -\frac{u_{n+1} - 2u_n + u_{n-1}}{h^2} + p(x_n)\,\frac{u_{n+1} - u_{n-1}}{2h} + q(x_n)u_n, \quad n = 1, 2, \ldots, N. \tag{3.6}$$

For any smooth function v on $[a, b]$, we define the grid function of the *truncation error* T_h by

$$(T_h v)_n = (\mathcal{L}_h v)_n - (\mathcal{L} v)(x_n), \quad n = 1, 2, \dots, N. \tag{3.7}$$

If $v = y$ is the exact solution of (3.3) and (3.4), this reduces to an earlier definition in Ch. 6, §1.2. By Taylor's formula one easily finds that for $v \in C^4[a, b]$,

$$(T_h v)_n = -\frac{h^2}{12}\,[v^{(4)}(\xi_1) - 2p(x_n)v'''(\xi_2)], \quad \xi_1, \xi_2 \in [x_n - h, x_n + h], \tag{3.8}$$

and more precisely, if $v \in C^6[a, b]$, since $\mathcal{L}_h$ is an even function of h,

$$(T_h v)_n = -\frac{h^2}{12}\,[v^{(4)}(x_n) - 2p(x_n)v'''(x_n)] + O(h^4), \quad h \to 0. \tag{3.9}$$

In analogy to terminology introduced in Ch. 5, we call the difference operator $\mathcal{L}_h$ *stable* if there exists a constant M independent of h such that for h sufficiently small, one has for any grid function $v = \{v_n\}$

$$\|v\|_\infty \le M\{\max(|v_0|, |v_{N+1}|) + \|\mathcal{L}_h v\|_\infty\}, \quad v \in \Gamma_h[a, b], \tag{3.10}$$

where $\|v\|_\infty = \max_{0 \le n \le N+1} |v_n|$ and $\|\mathcal{L}_h v\|_\infty = \max_{1 \le n \le N} |(\mathcal{L}_h v)_n|$. The following theorem gives a sufficient condition for stability.

Theorem 7.3.1. *If* $h\overline{p} \le 2$, *then* $\mathcal{L}_h$ *is stable. Indeed,* (3.10) *holds for* $M = \max(1, 1/\underline{q})$. (Here $\overline{p}, \underline{q}$ are the constants defined in (3.5).)

Proof. From (3.6) one computes

$$\tfrac{1}{2}h^2\,(\mathcal{L}_h v)_n = a_n v_{n-1} + b_n v_n + c_n v_{n+1}, \tag{3.11}$$

where

$$\begin{aligned} a_n &= -\tfrac{1}{2}\left[1 + \tfrac{1}{2}\,hp(x_n)\right], \\ b_n &= 1 + \tfrac{1}{2}\,h^2 q(x_n), \\ c_n &= -\tfrac{1}{2}\left[1 - \tfrac{1}{2}\,hp(x_n)\right]. \end{aligned} \tag{3.12}$$

Since, by assumption, $\frac{1}{2}\,h|p(x_n)| \le \frac{1}{2}\,h\overline{p} \le 1$, we have $a_n \le 0$, $c_n \le 0$, and

$$|a_n| + |c_n| = \tfrac{1}{2}[1 + \tfrac{1}{2}\,hp(x_n)] + \tfrac{1}{2}[1 - \tfrac{1}{2}\,hp(x_n)] = 1. \tag{3.13}$$

Also,

$$b_n \ge 1 + \tfrac{1}{2}\,h^2\underline{q}. \tag{3.14}$$

Now by (3.11), we have

$$b_n v_n = -a_n v_{n-1} - c_n v_{n+1} + \tfrac{1}{2}\,h^2(\mathcal{L}_h v)_n,$$

which, upon taking absolute values and using (3.13) and (3.14) yields

$$(1 + \tfrac{1}{2}\,h^2\underline{q})|v_n| \le \|v\|_\infty + \tfrac{1}{2}\,h^2\|\mathcal{L}_h v\|_\infty, \quad n = 1, 2, \ldots, N. \tag{3.15}$$

We distinguish two cases.

Case I: $\|v\|_\infty = |v_{n_0}|$, $1 \le n_0 \le N$. Here (3.15) gives

$$(1 + \tfrac{1}{2}\,h^2\underline{q})|v_{n_0}| \le |v_{n_0}| + \tfrac{1}{2}\,h^2\|\mathcal{L}_h v\|_\infty;$$

hence

$$|v_{n_0}| \le \frac{1}{\underline{q}}\,\|\mathcal{L}_h v\|_\infty\,,$$

and (3.10) follows since by assumption $\frac{1}{\underline{q}} \le M$.

Case II: $\|v\|_\infty = |v_{n_0}|$, $n_0 = 0$ or $n_0 = N + 1$. In this case, (3.10) is trivial, since $M \ge 1$. □

The method of finite differences now consists of replacing (3.3) and (3.4) by

$$(\mathcal{L}_h u)_n = r(x_n), \quad n = 1, 2, \ldots, N; \quad u_0 = \alpha, \quad u_{N+1} = \beta. \tag{3.16}$$

In view of (3.11) and (3.12), this is the same as

$$a_n u_{n-1} + b_n u_n + c_n u_{n+1} = \tfrac{1}{2}\,h^2 r(x_n), \quad n = 1, 2, \ldots, N,$$

$$u_0 = \alpha, \qquad u_{N+1} = \beta,$$

and gives rise to the system of linear equations

$$\begin{bmatrix} b_1 & c_1 & & & 0 \\ a_2 & b_2 & c_2 & & \\ & \ddots & \ddots & \ddots & \\ & & a_{N-1} & b_{N-1} & c_{N-1} \\ 0 & & & a_N & b_N \end{bmatrix} \begin{bmatrix} u_1 \\ u_2 \\ \vdots \\ u_{N-1} \\ u_N \end{bmatrix} = \frac{1}{2} h^2 \begin{bmatrix} r(x_1) \\ r(x_2) \\ \vdots \\ r(x_{N-1}) \\ r(x_N) \end{bmatrix} - \begin{bmatrix} a_1\alpha \\ 0 \\ \vdots \\ 0 \\ c_N\beta \end{bmatrix}. \tag{3.17}$$

The matrix of the system is tridiagonal and strictly diagonally dominant, since $|a_n| + |c_n| = 1$ and $b_n > 1$ by (3.14). In particular, it is nonsingular, so that the system (3.17) has a unique solution. (Uniqueness follows also from the stability of $\mathcal{L}_h$: the homogeneous system with $r(x_n) = \alpha = \beta = 0$ can have only the trivial solution, since $\mathcal{L}_h u = 0$, $u_0 = u_{N+1} = 0$ implies $\|u\|_\infty = 0$ by (3.10).)

Now that we know that the difference method defines a unique approximation, the next question is: how good is it? An answer to that is given in the next two theorems.

Theorem 7.3.2. *If $h\overline{p} \le 2$, then*

$$\|u - y\|_\infty \le M\|T_h y\|_\infty, \quad M = \max(1, 1/\underline{q}), \tag{3.18}$$

where $u = \{u_n\}$ is the solution of (3.17), $y = \{y_n\}$ the grid function induced by the exact solution $y(x)$ of (3.3) and (3.4), and $T_h y$ the grid function of the truncation errors defined in (3.7), where $v = y$. If $y \in C^4[a,b]$, then

$$\|u - y\|_\infty \le \tfrac{1}{12} h^2 M(\|y^{(4)}\|_\infty + 2\,\overline{p}\,\|y^{(3)}\|_\infty), \tag{3.19}$$

where $\|y^{(k)}\|_\infty = \max_{a\le x\le b} |y^{(k)}(x)|$, $k = 3, 4$.

Proof. From

$$(\mathcal{L}_h u)_n = r(x_n), \qquad u_0 = \alpha, \quad u_{N+1} = \beta,$$

$$(\mathcal{L}y)(x_n) = r(x_n), \quad y(x_0) = \alpha, \quad y(x_{N+1}) = \beta,$$

we obtain, letting $v_n = u_n - y(x_n)$,

$$\begin{aligned}(\mathcal{L}_h v)_n &= (\mathcal{L}_h u)_n - (\mathcal{L}_h y)_n \\ &= r(x_n) - [(\mathcal{L}y)(x_n) + (\mathcal{L}_h y)_n - (\mathcal{L}y)(x_n)] \\ &= r(x_n) - r(x_n) - (T_h y)_n \\ &= -(T_h y)_n,\end{aligned}$$

so that

$$\|\mathcal{L}_h v\|_\infty = \|T_h y\|_\infty. \tag{3.20}$$

By Theorem 7.3.1, $\mathcal{L}_h$ is stable with the stability constant M as defined in (3.18). Since $v_0 = v_{N+1} = 0$, there follows $\|v\|_\infty \le M\|\mathcal{L}_h v\|_\infty$, which in view of (3.20) and the definition of v is (3.18). The second assertion (3.19) follows directly from (3.8). □

In the spirit of Ch. 5, §3.3 and Ch. 6, §3.4, the result (3.19) of Theorem 7.3.2 can be refined as follows.

Theorem 7.3.3. *Let $p, q \in C^2[a,b]$, $y \in C^6[a,b]$, and $h\overline{p} \le 2$. Then*

$$u_n - y(x_n) = h^2 e(x_n) + O(h^4), \quad n = 0, 1, \ldots, N+1, \tag{3.21}$$

where $e(x)$ is the solution of

$$\mathcal{L}e = \theta(x), \quad a \le x \le b; \quad e(a) = 0, \quad e(b) = 0, \tag{3.22}$$

with

$$\theta(x) = \tfrac{1}{12}\,[y^{(4)}(x) - 2p(x)y'''(x)]. \tag{3.23}$$

Proof. We first note that our assumptions are such that $\theta \in C^2[a,b]$, which, by (3.22), implies that $e \in C^4[a,b]$.

Let

$$\overset{\circ}{v}_n = \frac{1}{h^2}\,(u_n - y(x_n)).$$

We want to show that

$$\overset{\circ}{v}_n = e(x_n) + O(h^2). \tag{3.24}$$

As in the proof of Theorem 7.3.2, we have

$$(\mathcal{L}_h \overset{\circ}{v})_n = -\frac{1}{h^2}\,(T_h y)_n.$$

By (3.9) with $v = y$, this gives

$$(\mathcal{L}_h \overset{\circ}{v})_n = \theta(x_n) + O(h^2). \tag{3.25}$$

Furthermore,

$$(\mathcal{L}_h e)_n = (\mathcal{L}e)(x_n) + (\mathcal{L}_h e)_n - (\mathcal{L}e)(x_n) = \theta(x_n) + (T_h e)_n, \tag{3.26}$$

by (3.22) and the definition (3.7) of truncation error. Since $e \in C^4[a,b]$, we have from (3.8) that

$$(T_h e)_n = O(h^2). \tag{3.27}$$

Subtracting (3.26) from (3.25), therefore, yields

$$(\mathcal{L}_h v)_n = O(h^2), \quad \text{where } v_n = \overset{\circ}{v}_n - e(x_n).$$

Since $v_0 = v_{N+1} = 0$ and $\mathcal{L}_h$ is stable, by assumption, there follows from the stability inequality that $|v_n| \le M\|\mathcal{L}_h v\|_\infty = O(h^2)$, which is (3.24). □

Theorem 7.3.3 could be used as a basis for Richardson extrapolation (cf. Ch. 3, §2.7). Another application is the *method of difference correction* due to L. Fox. A "difference correction" is any quantity E_n such that

$$E_n = e(x_n) + O(h^2), \qquad n = 1, 2, \ldots, N. \tag{3.28}$$

It then follows from (3.21) that

$$u_n - h^2 E_n = y(x_n) + O(h^4); \tag{3.29}$$

that is, $\hat{u}_n = u_n - h^2 E_n$ is an improved approximation having order of accuracy $O(h^4)$. Fox's idea is to construct a difference correction E_n by applying the basic difference method to the boundary value problem (3.22) in which $\theta(x_n)$ is replaced by a suitable difference approximation Θ_n:

$$(\mathcal{L}_h E)_n = \Theta_n, \quad n = 1, 2, \ldots, N; \quad E_0 = 0, \quad E_{N+1} = 0. \tag{3.30}$$

Letting $v_n = E_n - e(x_n)$, we then find

$$(\mathcal{L}_h v)_n = (\mathcal{L}_h E)_n - (\mathcal{L}_h e)_n = \Theta_n - \theta(x_n) + O(h^2),$$

by virtue of (3.30), (3.26), and (3.27). Since $v_0 = v_{N+1} = 0$, stability then yields

$$|v_n| = |E_n - e(x_n)| \le M\|\Theta - \theta\|_\infty + O(h^2),$$

so that for (3.28) to hold, all we need is to make sure that

$$\Theta_n - \theta(x_n) = O(h^2), \qquad n = 1, 2, \ldots, N. \tag{3.31}$$

This can be achieved by replacing the derivatives on the right of (3.23) by suitable finite difference approximations (see Ex. 9).

§3.2. **Nonlinear second-order equations**. A natural nonlinear extension of the linear problem (3.3) and (3.4) is

$$\mathcal{K}y = 0, \qquad y(a) = \alpha, \qquad y(b) = \beta, \tag{3.32}$$

where

$$\mathcal{K}y := -y'' + f(x, y, y') \tag{3.33}$$

and $f(x, y, z)$ is a given function of class C^1 defined on $[a, b] \times \mathbb{R} \times \mathbb{R}$, now assumed to be nonlinear in y and/or z. In analogy to (3.5) we make the assumption that

$$|f_z| \le \overline{p}, \qquad 0 < \underline{q} \le f_y \le \overline{q} \quad \text{on } [a, b] \times \mathbb{R} \times \mathbb{R}. \tag{3.34}$$

Then, by Theorem 7.1.2, the problem (3.32) has a unique solution.

We use again the simplest difference approximation $\mathcal{K}_h$ to $\mathcal{K}$,

$$(\mathcal{K}_h u)_n = -\frac{u_{n+1} - 2u_n + u_{n-1}}{h^2} + f\left(x_n, u_n, \frac{u_{n+1} - u_{n-1}}{2h}\right), \tag{3.35}$$

and define the truncation error as before by

$$(T_h v)_n = (\mathcal{K}_h v)_n - (\mathcal{K}v)(x_n), \qquad n = 1, 2, \ldots, N, \tag{3.36}$$

for any smooth function v on $[a, b]$. If $v \in C^4[a, b]$, then by Taylor's theorem, applied at $x = x_n$, $y = v(x_n)$, $z = v'(x_n)$,

$$\begin{aligned}
(T_h v)_n &= -\left[\frac{v(x_n + h) - 2v(x_n) + v(x_n - h)}{h^2} - v''(x_n)\right] \\
&\quad + f\left(x_n, v(x_n), \frac{v(x_n + h) - v(x_n - h)}{2h}\right) - f(x_n, v(x_n), v'(x_n)) \\
&= -\frac{h^2}{12}\, v^{(4)}(\xi_1) + f_z(x_n, v(x_n), \overline{z}_n)\left[\frac{v(x_n + h) - v(x_n - h)}{2h} - v'(x_n)\right] \\
&= -\frac{h^2}{12}\, v^{(4)}(\xi_1) + f_z(x_n, v(x_n), \overline{z}_n)\,\frac{h^2}{6}\, v'''(\xi_2),
\end{aligned}$$

where $\xi_i \in [x_n - h, x_n + h]$, $i = 1, 2$, and $\overline{z}_n$ is between $v'(x_n)$ and $(2h)^{-1}[v(x_n + h) - v(x_n - h)]$. Thus,

$$(T_h v)_n = -\frac{h^2}{12}\,[v^{(4)}(\xi_1) - 2f_z(x_n, v(x_n), \overline{z}_n)v'''(\xi_2)]. \tag{3.37}$$

Since $\mathcal{K}_h$ is nonlinear, the definition of stability needs to be slightly modified. We call $\mathcal{K}_h$ *stable* if for h sufficiently small, and for any two grid functions $v = \{v_n\}$, $w = \{w_n\}$, there is a constant M such that

$$\|v - w\|_\infty \le M\{\max(|v_0 - w_0|,\ |v_{N+1} - w_{N+1}|) + \|\mathcal{K}_h v - \mathcal{K}_h w\|_\infty\},$$
$$v, w \in \Gamma_h[a, b]. \tag{3.38}$$

If $\mathcal{K}_h$ is linear, this reduces to the previous definition (3.10), since $v - w$, just like v, is an arbitrary grid function.

Theorem 7.3.4. *If $h\overline{p} \leq 2$, then $\mathcal{K}_h$ is stable. Indeed,* (3.38) *holds with* $M = \max\,(1,1/\underline{q})$. (Here, $\overline{p}$, $\underline{q}$ are the constants defined in (3.34).)

Proof. Much the same as the proof of Theorem 7.3.1. See Ex. 10. □

The method of finite differences now takes on the following form.

$$(\mathcal{K}_h u)_n = 0, \qquad n = 1, 2, \ldots, N; \qquad u_0 = \alpha, \qquad u_{N+1} = \beta. \tag{3.39}$$

This is a system of N nonlinear equations in the N unknowns $u_1, u_2, \ldots, u_N$. We show shortly that under the assumptions made, the system (3.39) has a unique solution. Its error can be estimated exactly as in Theorem 7.3.2. Indeed, by applying the stability inequality (3.38) with $v = u$ and $w = y$, where u is the grid function satisfying (3.39) and y the grid function induced by the exact solution $y(x)$ of (3.32), we get, with M as defined in Theorem 7.3.4,

$$\begin{aligned}
\|u - y\|_\infty \leq M\|\mathcal{K}_h u - \mathcal{K}_h y\|_\infty &= M\|\mathcal{K}_h y\|_\infty \\
&= M\|\mathcal{K}y + (\mathcal{K}_h y - \mathcal{K}y)\|_\infty \\
&= M\|\mathcal{K}_h y - \mathcal{K}y\|_\infty \\
&= M\|T_h y\|_\infty,
\end{aligned}$$

which is (3.18). The rest of Theorem 7.3.2 then again follows immediately from (3.37) and the first assumption in (3.34).

In order to show that (3.39) has a unique solution, we write the system in fixed point form and apply the contraction mapping principle. It is convenient to introduce a parameter ω in the process — a "relaxation parameter" as it were — by writing (3.39) equivalently in the form

$$\begin{aligned}
&u = g(u), \quad g(u) = u - \tfrac{1}{1+\omega}\,\tfrac{1}{2}h^2\,\mathcal{K}_h u \quad (\omega \neq -1), \\
&u_0 = \alpha, \quad u_{N+1} = \beta.
\end{aligned} \tag{3.40}$$

Here we think of g as a mapping $\mathbb{R}^{N+2} \to \mathbb{R}^{N+2}$, by defining $g_0(u) = \alpha$, $g_{N+1}(u) = \beta$. We want to show that g is a contraction map on $\mathbb{R}^{N+2}$ if h satisfies the condition of Theorem 7.3.4 and ω is suitably chosen. This then will prove (cf. Theorem 4.9.1) the existence and uniqueness of the solution of (3.40), and hence of (3.39).

Given any two grid functions $v = \{v_n\}$, $w = \{w_n\}$, we can write by Taylor's theorem, after a simple calculation,

$$g_n(v) - g_n(w) = \frac{1}{1+\omega}\,[a_n(v_{n-1} - w_{n-1}) + (1 + \omega - b_n)(v_n - w_n) \\ + c_n(v_{n+1} - w_{n+1})], \qquad 1 \le n \le N, \tag{3.41}$$

whereas for $n = 0$ or $n = N + 1$ the difference on the left, of course, is zero; here

$$a_n = \tfrac{1}{2}[1 + \tfrac{1}{2}\,hf_z(z_n, \overline{y}_n, \overline{z}_n)],$$

$$b_n = 1 + \tfrac{1}{2}\,h^2 f_y(x_n, \overline{y}_n, \overline{z}_n),$$

$$c_n = \tfrac{1}{2}[1 - \tfrac{1}{2}\,hf_z(x_n, \overline{y}_n, \overline{z}_n)],$$

with $\overline{y}_n$, $\overline{z}_n$ suitable intermediate values. Since $h\overline{p} \le 2$, we have

$$a_n \ge 0, \qquad c_n \ge 0, \qquad a_n + c_n = 1. \tag{3.42}$$

Assuming, furthermore, that

$$\omega \ge \tfrac{1}{2}\,h^2\overline{q}, \tag{3.43}$$

we have

$$1 + \omega - b_n \ge 1 + \omega - \left(1 + \tfrac{1}{2}\,h^2\overline{q}\right) = \omega - \tfrac{1}{2}\,h^2\overline{q} \ge 0.$$

Hence, all coefficients on the right of (3.41) are nonnegative, and since

$$0 \le 1 + \omega - b_n \le 1 + \omega - \left(1 + \tfrac{1}{2}\,h^2\underline{q}\right) = \omega - \tfrac{1}{2}\,h^2\underline{q},$$

we obtain, upon taking norms and noting (3.42),

$$|g_n(v) - g_n(w)| \le \frac{1}{1+\omega}\left(a_n + \omega - \frac{1}{2}\,h^2\underline{q} + c_n\right)\|v - w\|_\infty \\ = \frac{1}{1+\omega}\left(1 + \omega - \frac{1}{2}\,h^2\underline{q}\right)\|v - w\|_\infty;$$

that is,

$$\|g(v) - g(w)\|_\infty \le \gamma(\omega)\, \|v - w\|_\infty, \quad \gamma(\omega) := 1 - \frac{\frac{1}{2}\, h^2 \underline{q}}{1+\omega}\,. \tag{3.44}$$

Clearly, $\gamma(\omega) < 1$, showing that g is a contraction map on $\mathbb{R}^{N+2}$, as claimed.

In principle, one could apply the method of successive iteration to (3.40), and it would converge for arbitrary initial approximation. Faster convergence can be expected, however, by applying Newton's method directly on (3.39); for details, see Ex. 11.

§4. Variational Methods

Variational methods take advantage of the fact that the solution of important types of boundary value problems satisfy certain extremal properties. This then suggests solving the respective extremal problems — at least approximately — in place of the boundary value problems. This can be done by classical methods. We illustrate the method for a linear second-order boundary value problem with simplified (Dirichlet) boundary conditions.

§4.1. **Variational formulation.** Without restriction of generality (cf. (1.19′)), we can assume that the problem is in self-adjoint form:

$$\mathcal{L}y = r(x), \quad a \le x \le b; \quad y(a) = \alpha, \quad y(b) = \beta, \tag{4.1}$$

where

$$\mathcal{L}y := -\frac{d}{dx}\left(p(x)\,\frac{dy}{dx}\right) + q(x)y, \quad a \le x \le b. \tag{4.2}$$

We assume $p \in C^1[a,b]$ and q, r continuous on $[a,b]$, and

$$p(x) \ge \underline{p} > 0, \quad q(x) > 0 \quad \text{on} \quad [a,b]. \tag{4.3}$$

Under these assumptions, the problem (4.1) has a unique solution (cf. Corollary 7.1.1).

If $\ell(x)$ is a linear function having the same boundary values as y in (4.1), then $z(x) = y(x) - \ell(x)$ satisfies $\mathcal{L}z = r(x) - (\mathcal{L}\ell)(x)$, $z(a) = z(b) = 0$, which is a problem of the same type as (4.1), but with *homogeneous* boundary conditions. We may assume, therefore, that $\alpha = \beta = 0$, and thus consider

$$\mathcal{L}y = r(x), \quad a \le x \le b; \quad y(a) = y(b) = 0. \tag{4.4}$$

Denoting by $C_0^2[a,b]$ the linear space $C_0^2[a,b] = \{u \in C^2[a,b]\colon u(a) = u(b) = 0\}$, we may write (4.4) in operator form as

$$\mathcal{L}y = r, \quad y \in C_0^2[a,b]. \tag{4.4$'$}$$

Note that $\mathcal{L} : C^2[a,b] \to C[a,b]$ is a linear operator. It is convenient to enlarge the space C_0^2 somewhat and define

$$\begin{aligned} V_0 = \{v \in C[a,b] :\ & v' \text{ piecewise continuous} \\ & \text{and bounded on } [a,b],\ v(a) = v(b) = 0\}. \end{aligned}$$

On V_0 we can define the usual inner product

$$(u,v) := \int_a^b u(x)v(x)dx, \quad u,v \in V_0. \tag{4.5}$$

Theorem 7.4.1. *The operator $\mathcal{L}$ in* (4.2) *is symmetric on $C_0^2[a,b]$ relative to the inner product* (4.5); *that is,*

$$(\mathcal{L}u,v) = (u,\mathcal{L}v), \quad \textit{all} \quad u,v \in C_0^2[a,b]. \tag{4.6}$$

Proof. Use integration by parts to obtain

$$\begin{aligned} (\mathcal{L}u,v) &= \int_a^b [-(p(x)u')' + q(x)u]v(x)dx \\ &= -(pu'v)\Big|_a^b + \int_a^b [p(x)u'(x)v'(x) + q(x)u(x)v(x)]dx \\ &= \int_a^b [p(x)u'(x)v'(x) + q(x)u(x)v(x)]dx. \end{aligned}$$

Since the last integral is symmetric in u and v, it is also equal to $(\mathcal{L}v,u)$, which in turn, by the symmetry of $(\cdot\,,\cdot)$, proves the theorem. □

Note that the last integral in the proof of Theorem 7.4.1 is defined not only on $C_0^2[a,b]$, but also on V_0. It suggests an alternative inner product,

$$[u,v] := \int_a^b [p(x)u'(x)v'(x) + q(x)u(x)v(x)]dx, \quad u,v \in V_0, \tag{4.7}$$

and the proof of Theorem 7.4.1 shows that

$$(\mathcal{L}u, v) = [u, v] \quad \text{if} \quad u \in C_0^2[a, b], \quad v \in V_0. \tag{4.8}$$

In particular, if $u = y$ is a solution of (4.4), then

$$[y, v] = (r, v), \quad \text{all} \quad v \in V_0; \tag{4.9}$$

this is the *variational*, or *weak*, form of (4.4).

Theorem 7.4.2. *Under the assumptions made on* p, q, *and* r (cf. (4.3)), *there exist positive constants* $\underline{c}$ *and* $\overline{c}$ *such that*

$$\underline{c}\,\|u\|_\infty^2 \le [u, u] \le \overline{c}\,\|u'\|_\infty^2, \quad \textit{all} \quad u \in V_0. \tag{4.10}$$

In fact,

$$\underline{c} = \frac{\underline{p}}{b-a}, \quad \overline{c} = (b-a)\|p\|_\infty + (b-a)^3\|q\|_\infty. \tag{4.11}$$

Proof. For any $u \in V_0$, since $u(a) = 0$, we have

$$u(x) = \int_a^x u'(t)dt, \quad x \in [a, b].$$

By Schwarz's inequality,

$$u^2(x) \le \int_a^x 1dt \cdot \int_a^x [u'(t)]^2 dt \le (b-a)\int_a^b [u'(t)]^2 dt, \quad x \in [a, b],$$

and, therefore,

$$\|u\|_\infty^2 \le (b-a)\int_a^b [u'(t)]^2 dt \le (b-a)^2\|u'\|_\infty^2. \tag{4.12}$$

Using the assumption (4.3), we get

$$[u, u] = \int_a^b (p(x)[u'(x)]^2 + q(x)u^2(x))dx \ge \underline{p}\int_a^b [u'(x)]^2 dx \ge \frac{\underline{p}}{b-a}\,\|u\|_\infty^2,$$

where the last inequality follows from the left inequality in (4.12). This proves the lower bound in (4.10). The upper bound is obtained by observing that

$$[u, u] \le (b-a)\|p\|_\infty\,\|u'\|_\infty^2 + (b-a)\|q\|_\infty\|u\|_\infty^2 \le \overline{c}\,\|u'\|_\infty^2,$$

where (4.12) has been used in the last step. □

We remark that (4.10) implies the uniqueness of solutions of (4.4). In fact, if

$$\mathcal{L}y = r, \quad \mathcal{L}y^* = r, \quad y, y^* \in C_0^2[a,b],$$

then $\mathcal{L}(y - y^*) = 0$, hence, by (4.8) and (4.10),

$$0 = (\mathcal{L}(y - y^*), y - y^*) = [y - y^*, y - y^*] \geq \underline{c}\, \|y - y^*\|_\infty^2,$$

and it follows that $y \equiv y^*$.

4.2. **The extremal problem**. We define the quadratic functional

$$F(u) := [u, u] - 2(r, u), \quad u \in V_0, \tag{4.13}$$

where r is the right-hand function in (4.4). The extremal property for the solution y of (4.4) is expressed in the following theorem.

Theorem 7.4.3. *Let y be the solution of* (4.4′). *Then*

$$F(u) > F(y), \quad \textit{all} \quad u \in V_0, \quad u \not\equiv y. \tag{4.14}$$

Proof. By (4.9), $(r, u) = [y, u]$, so that

$$\begin{aligned} F(u) &= [u,u] - 2(r,u) = [u,u] - 2[y,u] + [y,y] - [y,y] \\ &= [y - u, y - u] - [y,y] > -[y,y], \end{aligned}$$

where strict inequality holds in view of (4.10) and $y - u \not\equiv 0$. On the other hand, since $[y, y] = (\mathcal{L}y, y) = (r, y)$, by (4.8), we have

$$F(y) = [y,y] - 2(r,y) = (r,y) - 2(r,y) = -(r,y) = -[y,y],$$

which, combined with the previous inequality, proves the theorem. □

Theorem 7.4.3 thus expresses the following extremal property of the solution of (4.4′):

$$F(y) = \min_{u \in V_0} F(u). \tag{4.15}$$

We view (4.15) as an extremal problem for determining y, and in the next section solve it approximately by determining a function u_S from a finite-dimensional subset $S \subset V_0$ that minimizes $F(u)$ on S. In this connection, it is useful to note the identity

$$[y-u, y-u] = F(u) + [y, y], \quad u \in V_0, \tag{4.16}$$

satisfied by the solution y, which was established in the course of the proof of Theorem 7.4.3.

§4.3. **Approximate solution of the extremal problem.** Let $S \subset V_0$ be a finite-dimensional subspace of V_0 and dim $S = n$ its dimension. Let u_1, $u_2, \dots, u_n$ be a basis of S, so that

$$u \in S \ \text{ if and only if } \ u = \sum_{\nu=1}^{n} \xi_\nu u_\nu, \quad \xi_\nu \in \mathbb{R}. \tag{4.17}$$

We approximate the solution y of (4.15) by $u_S \in S$, which satisfies

$$F(u_S) = \min_{u \in S} F(u). \tag{4.18}$$

Before we analyze the quality of the approximation $u_S \approx y$, let us explain the mechanics of the method.

We have, for any $u \in S$,

$$\begin{aligned} F(u) &= \left[\sum_{\nu=1}^{n} \xi_\nu u_\nu, \sum_{\mu=1}^{n} \xi_\mu u_\mu\right] - 2\left(r, \sum_{\nu=1}^{n} \xi_\nu u_\nu\right) \\ &= \sum_{\nu,\mu=1}^{n} [u_\nu, u_\mu]\xi_\nu \xi_\mu - 2\sum_{\nu=1}^{n} (r, u_\nu)\xi_\nu. \end{aligned}$$

Define

$$U = \begin{bmatrix} [u_1, u_1] & [u_1, u_2] & \cdots & [u_1, u_n] \\ [u_2, u_1] & [u_2, u_2] & \cdots & [u_2, u_n] \\ \cdots & \cdots & \cdots & \cdots \\ [u_n, u_1] & [u_n, u_2] & \cdots & [u_n, u_n] \end{bmatrix}, \quad \xi = \begin{bmatrix} \xi_1 \\ \xi_2 \\ \vdots \\ \xi_n \end{bmatrix}, \quad \rho = \begin{bmatrix} (r, u_1) \\ (r, u_2) \\ \vdots \\ (r, u_n) \end{bmatrix}. \tag{4.19}$$

In early applications of the method to structural mechanics, and ever since, the matrix U is called the *stiffness matrix*, and ρ the *load vector*. In terms of these, the functional F can be written in matrix form as

$$F(u) = \xi^T U \xi - 2\rho^T \xi, \quad \xi \in \mathbb{R}^n. \tag{4.20}$$

The matrix U is not only symmetric, but also positive definite, since $\xi^T U \xi = [u, u] > 0$, unless $u \equiv 0$ (i.e., $\xi = 0$).

Our approximate extremal problem (4.18) thus takes the form

$$\begin{aligned} \phi(\xi) &= \min, \\ \phi(\xi) &:= \xi^T U \xi - 2\rho^T \xi, \quad \xi \in \mathbb{R}^n, \end{aligned} \tag{4.21}$$

an unconstrained quadratic minimization problem in $\mathbb{R}^n$. Since U is positive definite, the problem (4.21) has a unique solution $\hat{\xi}$ given by the solution of the linear system

$$U\xi = \rho. \tag{4.22}$$

It is easily verified that

$$\phi(\xi) > \phi(\hat{\xi}), \quad \text{all} \quad \xi \in \mathbb{R}^n, \quad \xi \neq \hat{\xi}; \tag{4.23}$$

indeed, since $\rho = U\hat{\xi}$,

$$\begin{aligned} \phi(\xi) &= \xi^T U \xi - 2\rho^T \xi = \xi^T U \xi - 2\hat{\xi} U \xi \\ &= \xi^T U \xi - 2\hat{\xi} U \xi + \hat{\xi}^T U \hat{\xi} - \hat{\xi}^T U \hat{\xi} \\ &= (\xi - \hat{\xi})^T U (\xi - \hat{\xi}) + \phi(\hat{\xi}), \end{aligned}$$

where $-\hat{\xi}^T U \hat{\xi} = -\hat{\xi}^T \rho = \hat{\xi}^T \rho - 2\rho^T \hat{\xi} = \hat{\xi}^T U \hat{\xi} - 2\rho^T \hat{\xi} = \phi(\hat{\xi})$ has been used in the last step. From this, (4.23) follows immediately. Thus,

$$u_S = \sum_{\nu=1}^{n} \hat{\xi}_\nu u_\nu, \quad \text{where} \quad U\hat{\xi} = \rho. \tag{4.24}$$

In practice, the basis functions of S are chosen to have small support, which results in a matrix U having a band structure.

It is now straightforward to establish the *optimal approximation property* of u_S in the norm $[\cdot, \cdot]$; that is,

$$[y - u_S, y - u_S] = \min_{u \in S} [y - u, y - u]. \tag{4.25}$$

Indeed, by (4.16) and (4.18), the left-hand side is equal to $F(u_S) + [y, y] = \min_{u \in S} \{F(u) + [y, y]\}$, which in turn equals the right-hand side of (4.25).

The approximation property (4.25) gives rise to the following error estimate.

Theorem 7.4.4. *There holds*

$$\|y - u_S\|_\infty \le \sqrt{\overline{c}/\underline{c}}\, \|y' - u'\|_\infty, \quad \textit{all} \quad u \in S, \tag{4.26}$$

where $\overline{c}$ and $\underline{c}$ are the constants defined in (4.11). *In particular,*

$$\|y - u_S\|_\infty \le \sqrt{\overline{c}/\underline{c}}\, \inf_{u \in S} \|y' - u'\|_\infty. \tag{4.27}$$

Proof. By (4.10) and (4.25), we have

$$\underline{c}\, \|y - u_S\|_\infty^2 \le [y - u_S, y - u_S] \le [y - u, y - u] \le \overline{c}\, \|y' - u'\|_\infty^2,$$

from which (4.26) follows. □

The point of Theorem 7.4.4 is that, in order to get a good error bound, we have to use an approximation process $y \approx u$, $u \in S$, which approximates the first derivative of y as well as possible. Note that this approximation process is *independent* of the one yielding u_S; its sole purpose is to provide a good error bound for u_S.

Example 4.1. Let Δ be a subdivision of $[a, b]$, say,

$$a = x_1 < x_2 < x_3 < \cdots < x_{n-1} < x_n = b, \tag{4.28}$$

and take (see Ch. 2, §3.4 for notation)

$$S = \{s \in \mathbb{S}_3^2(\Delta) : \quad s(a) = s(b) = 0\}. \tag{4.29}$$

Here S is a subspace not only of V_0, but even of $C_0^2[a, b]$. Its dimension is easily seen to be n. Given the solution y of (4.4′), there is a unique $s_{\text{compl}} \in S$ such that

$$\begin{aligned} s_{\text{compl}}(x_i) &= y(x_i), \quad i = 1, 2, \ldots, n, \\ s'_{\text{compl}}(a) &= y'(a), \quad s'_{\text{compl}}(b) = y'(b), \end{aligned} \tag{4.30}$$

the "complete cubic spline interpolant" to y (cf. Ch. 2, §3.4 (ii.1)). From Ch. 2, (3.24), we know that

$$\|s'_{\text{compl}} - y'\|_\infty \le \tfrac{1}{24}\, |\Delta|^3 \|y^{(4)}\|_\infty \quad \text{if} \quad y \in C^4[a, b].$$

Combining this with the result of Theorem 7.4.4 (in which $u = s_{\text{compl}}$), we get the error bound

$$\|y - u_S\|_\infty \le \tfrac{1}{24}\,\sqrt{\overline{c}/\underline{c}}\;|\Delta|^3\|y^{(4)}\|_\infty = O(|\Delta|^3), \tag{4.31}$$

which is one order of magnitude better than the one for the ordinary finite difference method (cf. (3.19)). However, there is more work involved in computing the stiffness matrix (many integrals!), and also in solving the linear system (4.22). Even with a basis of S that has small support (extending over at most four consecutive subintervals of Δ), one still has to deal with a banded matrix U having bandwidth 7 (not 3, as in (3.17)).

NOTES TO CHAPTER 7

Background material on the theory of boundary value problems can be found in most textbooks on ordinary differential equations. Specialized texts are Bailey, Shampine, and Waltman [1968], Bernfeld and Lakshmikantham [1974], and Agarwal [1986]; all three, but especially the first, also contain topics on the numerical solution of boundary value problems and applications. An early book strictly devoted to numerical methods for solving boundary value problems is Fox [1990], an eminently practical account still noteworthy for its consistent use of "difference-correction," that is, the incorporation of remainder terms into the solution process. Subsequent books and monographs are Keller [1992], Na [1979], and Ascher, Mattheij, and Russell [1995]. The books by Keller and Na complement each other in that the former gives a mathematically rigorous treatment of the major methods in use, with some applications provided in the final chapter, and the latter a more informal presentation intended for engineers, and containing a wealth of engineering applications. Na's book also discusses methods less familiar to numerical analysts, such as the МЕТОД ПРОГОНКИ developed by Russian scientists, here translated as the "method of chasing," and interesting methods based on transformation of variables. The book by Ascher et al. is currently the major reference work in this area. One of its special features is an extensive discussion of the numerical condition and associated condition numbers for boundary value problems. In its first chapter, it also contains a large sampling of boundary value problems occurring in real-life applications.

Sturm-Liouville eigenvalue problems — both regular and singular — and their numerical solution are given a thorough treatment in Pryce [1993]. A set of 60 test problems is included in one of the appendices, and references to available software in another.

§1.1. Example 1.3 is from Bailey, Shampine, and Waltman [1968, Ch. 1, §4].

§1.2. The exposition in this paragraph, in particular, the proof of Theorem 7.1.2., follows Keller [1992, §1.2].

§1.3. An example of an existence and uniqueness theorem for the general boundary value problem (1.24) is Theorem 1.2.6 in Keller [1992].

§2. In this paragraph we give only a bare outline and some of the key ideas involved in shooting methods. To make shooting a viable method, even for linear boundary value problems, requires attention to many practical and technical details. For these, we must refer to the relevant literature, for example, Roberts and Shipman [1972] or, especially, Ascher, Mattheij, and Russell [1995, Ch. 4]. The latter reference also contains two computer codes, one for linear, the other for nonlinear (nonstiff) boundary value problems.

Shooting basically consists of solving a finite-dimensional system of equations generated by solutions of initial value problems, which is then solved iteratively, for example, by Newton's method. Alternatively, one could apply Newton's method, or more precisely, the Newton-Kantorovich method, directly to the boundary value problem in question, considered as an operator equation in a Banach space of smooth functions on $[a, b]$ satisfying homogeneous boundary conditions. This is the method of quasi-linearization originally proposed by Bellman and Kalaba [1965].

Yet another approach is "invariant imbedding," where the endpoint b of the interval $[a, b]$ is made a variable with respect to which one differentiates to obtain an auxiliary nonlinear initial value problem of the Riccati type. See, for example, Ascher, Mattheij, and Russell [1995, §4.5]. A different view of invariant imbedding based on a system of first-order partial differential equations and associated characteristics is developed in Meyer [1973].

§2.1. Instead of Newton's method (2.1) one could, of course, use other iterative methods for solving $\phi(s) = 0$, for example, a fixed point iteration based on the equivalent equation $s = s - m\phi(s)$, $m \neq 0$. This is analyzed in Keller [1992, §2.2].

For the origin of Example 2.2, see Troesch [1976]. The analytic solution of the associated initial value problem is given, for example, in Stoer and Bulirsch [1993, pp. 514–516]; also cf. Ex. 7.

§2.2. It is relatively straightforward to analyze the effect, in superposition methods, of the errors committed in the numerical integration of the initial value problems involved; see, for example, Keller [1992, §2.1] and Ascher, Mattheij, and Russell [1995, §4.2.2]. More important, and potentially more disastrous, are the effects of rounding errors; see, for example, Ascher, Mattheij, and Russell [1995, §4.2.3]. The remark at the end of §1.3 can be generalized to "partially separated" boundary conditions, which give rise to a "reduced" superposition method; see Ascher, Mattheij, and Russell [1995, §4.2.4].

In place of (2.8), other iterative methods could be used to solve $\phi(s) = 0$, for example, one of the quasi-Newton methods (cf. Ch. 4, Notes to §9.2), or, as in the scalar case, a fixed point iteration based on $s = s - M\phi(s)$, with M a nonsingular matrix chosen such that the map $s \mapsto s - M\phi(s)$ is contractive.

Example 2.3 is from Morrison, Riley, and Zancanaro [1962], where the term "shooting" appears to have been used for the first time in the context of boundary value problems.

§2.3. Parallel shooting is important also for linear boundary value problems of the type (1.28) and (1.29), since without it, numerical linear dependencies may be developing that could render the method of simple shooting useless. There are various versions of parallel shooting, some involving reorthogonalization of solution vectors; see Ascher, Mattheij, and Russell [1995, §§4.3 and 4.4]. For a discussion of homotopy methods, including numerical examples, see Roberts and Shipman [1972, Ch. 7].

§§3.1 and 3.2. The treatment in these paragraphs closely follows Keller [1992, §§3.1 and 3.2]. Maintaining second-order accuracy on nonuniform grids is not entirely straightforward; see, for example, Ascher, Mattheij, and Russell [1995, §5.6.1].

Extensions of the method of finite differences to linear and nonlinear systems of the type (1.24) can be based on the local use of the trapezoidal or midpoint rule. This is discussed in Keller [1992, §3.3] and Ascher, Mattheij, and Russell [1995, §§5.1 and 5.2]. Local use of implicit Runge-Kutta methods afford more accuracy, and so do the methods of extrapolation and "deferred corrections;" for these, see Ascher, Mattheij, and Russell [1995, §§5.3, 5.4, 5.5.2 and 5.5.3].

Boundary value problems for single higher-order differential equations are often solved by collocation methods using spline functions; a discussion of this is given in Ascher, Mattheij, and Russell [1995, §§5.6.2 to 5.6.4].

§4. The treatment of variational methods in this paragraph is along the lines of Stoer and Bulirsch [1993, §7.5]. There are many related methods, collectively called projection methods, whose application to two-point boundary value problems and convergence analysis is the subject of a survey by Reddien [1980] containing extensive references to the literature.

EXERCISES AND MACHINE ASSIGNMENTS TO CHAPTER 7

EXERCISES

1. The following eigenvalue problem arises in the physics of gas discharges. Determine the smallest positive $\lambda > 0$ such that

$$\varphi'' + \frac{1}{r}\varphi' + \lambda^2 \varphi(1-\varphi) = 0, \quad 0 < r \le 1,$$

$$\varphi(0) = a, \quad \varphi'(0) = 0, \quad \varphi(1) = 0,$$

   where a is given, $0 < a < 1$.

   (a) Explain why $\lambda = 0$ cannot be an eigenvalue.

   (b) Reduce the problem to an initial value problem. {*Hint*: Make a change of variables, $x = \lambda r$, $y(x) = \varphi(x/\lambda)$.}

2. Consider the nonlinear boundary value problem (Blasius equation)

$$y''' + \tfrac{1}{2} y y'' = 0, \quad 0 \le x < \infty,$$

$$y(0) = y'(0) = 0, \quad y'(\infty) = 1.$$

   (a) Letting $y''(0) = \lambda$ and $z(t) = \lambda^{-\frac{1}{3}} y(\lambda^{-\frac{1}{3}} t)$ (assuming $\lambda > 0$), derive an initial value problem for z on $0 \le t < \infty$.

   (b) Explain how the solution of the initial value problem in (a) can be used to obtain the solution $y(x)$ of the given boundary value problem.

3. The boundary value problem

$$y'' = -\frac{1}{x} y y', \quad 0 < x \le 1; \quad y(0) = 0, \ y(1) = 1,$$

   although it has a singularity at $x = 0$ and certainly does not satisfy (3.34), has the smooth solution $y(x) = \frac{2x}{1+x}$.

   (a) Determine analytically the s-interval for which the initial value problem

$$u'' = -\frac{1}{x} u u', \quad 0 < x \le 1; \quad u(0) = 0, \ u'(0) = s$$

   has a smooth solution $u(x; s)$ on $0 \le x \le 1$.

   (b) Determine the s-interval for which Newton's method applied to $u(1; s) - 1 = 0$ converges.

4. Let

$$\phi(s) = \frac{s}{s\cosh 1 - \sinh 1} - e$$

and s^0 be the solution of

$$s^0 - \frac{\phi(s^0)}{\phi'(s^0)} = \tanh 1$$

(cf. Example 2.3).

(a) Show that $s^0 < \coth 1$. {*Hint*: With $t^0 = \coth 1$, consider what $s^0 < t^0$ means in terms of one Newton step at t^0.}

(b) Use the bisection method to compute s^0 to six decimal places. What are appropriate initial approximations?

5. Generalizing Example 2.3, let $\nu > 0$ and consider the boundary value problem

$$\left.\begin{aligned} \frac{dy_1}{dx} &= \frac{y_1^{\nu+1}}{y_2^{\nu}} \\ \frac{dy_2}{dx} &= \frac{y_2^{\nu+1}}{y_1^{\nu}} \end{aligned}\right\} \quad 0 \le x \le 1,$$

$$y_1(0) = 1, \quad y_1(1) = e.$$

(a) Determine the exact solution.

(b) Solve the initial value problem

$$\frac{d}{dx}\begin{bmatrix} u_1 \\ u_2 \end{bmatrix} = \begin{bmatrix} u_1^{\nu+1}/u_2^{\nu} \\ u_2^{\nu+1}/u_1^{\nu} \end{bmatrix}, \quad 0 \le x \le 1; \quad \begin{bmatrix} u_1 \\ u_2 \end{bmatrix}(0) = \begin{bmatrix} 1 \\ s \end{bmatrix}$$

in closed form.

(c) Find the conditions on $s > 0$ guaranteeing that $u_1(x)$, $u_2(x)$ both remain positive and finite on $[0, 1]$. In particular, show that, as $\nu \to \infty$, the interval in which s must lie shrinks to the point $s = 1$. What happens when $\nu \to 0$?

6. Suppose Example 2.3 is modified by multiplying the right-hand sides of the differential equation by λ, and by replacing the second boundary condition by $y_1(1) = e^{-\lambda}$, where $\lambda > 0$ is a large parameter.

(a) What is the exact solution?

(b) What are the conditions on s for the associated initial value problem to have positive and bounded solutions? What happens as $\lambda \to \infty$? As $\lambda \to 0$?

7. The Jacobian elliptic functions sn and cn are defined by

$$\operatorname{sn}(u|k) = \sin\varphi, \quad \operatorname{cn}(u|k) = \cos\varphi, \quad 0 < k < 1,$$

where φ is uniquely determined by

$$u = \int_0^{\varphi} \frac{d\theta}{(1 - k^2\sin^2\theta)^{\frac{1}{2}}} \, .$$

(a) Show that $K(k) := \int_0^{\frac{1}{2}\pi}(1 - k^2\sin^2\theta)^{-\frac{1}{2}}\,d\theta$ is the smallest positive zero of cn.

(b) Show that

$$\frac{d}{du}\operatorname{sn}(u|k) = \operatorname{cn}(u|k)\sqrt{1 - k^2[\operatorname{sn}(u|k)]^2} \, ,$$
$$\frac{d}{du}\operatorname{cn}(u|k) = -\operatorname{sn}(u|k)\sqrt{1 - k^2[\operatorname{sn}(u|k)]^2} \, .$$

(c) Show that the initial value problem

$$y'' = \lambda\sinh(\lambda y), \quad y(0) = 0, \; y'(0) = s$$

has the exact solution

$$y(x; s) = \frac{2}{\lambda}\sinh^{-1}\left(\frac{s}{2}\frac{\operatorname{sn}(\lambda x|k)}{\operatorname{cn}(\lambda x|k)}\right), \quad k^2 = 1 - \frac{s^2}{4} \, .$$

Hence show that $y(x; s)$, $x > 0$, becomes singular for the first time when $x = x_\infty$, where

$$x_\infty = \frac{K(k)}{\lambda} \, .$$

(d) From the known expansion

$$K(k) = \ln\frac{4}{\sqrt{1-k^2}} + \frac{1}{4}\left(\ln\frac{4}{\sqrt{1-k^2}} - 1\right)(1 - k^2) + \cdots \, , \quad k \to 1,$$

conclude that

$$x_\infty \sim \frac{1}{\lambda}\ln\frac{8}{|s|} \quad \text{as } s \to 0.$$

8. It has been shown in Example 2.1 that the boundary value problem

$$y'' + e^{-y} = 0, \quad 0 \le x \le 1, \;\; y(0) = y(1) = 0$$

has a unique solution that is nonnegative on $[0, 1]$.

(a) Set up a finite difference method for solving the problem numerically. (Use a uniform grid $x_n = \frac{n}{N+1}$, $n = 0, 1, \ldots, N+1$, and the simplest of finite difference approximations to y''.)

(b) Write the equations for the approximate vector $u^T = [u_1, u_2, \ldots, u_N]$ in fixed point form $u = \varphi(u)$ and find a compact domain $\mathcal{D} \subset \mathbb{R}^N$ such that $\varphi : \mathbb{R}^N \to \mathbb{R}^N$ maps $\mathcal{D}$ into $\mathcal{D}$ and is contractive in $\mathcal{D}$. {*Hint*: Use the fact that the tridiagonal matrix $A = \text{tri}[1, -2, 1]$ has a nonpositive inverse A^{-1} satisfying $\|A^{-1}\|_\infty \le \frac{1}{8}(N+1)^2$.}

(c) Discuss the convergence of the fixed point iteration applied to the system of finite difference equations.

9. Construct finite difference approximations Θ_n for $\theta(x_n)$, where $\theta(x) = \frac{1}{12}[y^{(4)}(x) - 2p(x)y'''(x)]$, such that $\Theta_n - \theta(x_n) = O(h^2)$ (cf. (3.31)). {*Hint*: Distinguish the cases $2 \le n \le N-1$ and $n = 1$ resp., $n = N$.}

10. Prove Theorem 7.3.4.

11. Describe the application of Newton's method for solving the nonlinear finite difference equations $\mathcal{K}_h u = 0$ of (3.39).

12. Let Δ be the subdivision

$$a = x_0 < x_1 < \cdots < x_n < x_{n+1} = b$$

and $S = \{s \in \mathbb{S}_1^0(\Delta) : \; s(a) = s(b) = 0\}$.

(a) Find an expression for $[u_\nu, u_\mu]$ in terms of the basis of hat functions (cf. Ch. 2, Ex. 66) and in terms of the integrals involved; do this in a similar manner for $\rho_\nu = (r, u_\nu)$.

(b) Suppose that each integral is split into a sum of integrals over each subinterval of Δ and the trapezoidal rule is employed to approximate the values of the integrals. Obtain the resulting approximations for the stiffness matrix U and the load vector ρ. Interpret the linear system (4.22) thus obtained as a finite difference method.

13. Apply the approximate variational method of §4.3 to the boundary value problem

$$-y'' = r(x), \quad 0 \le x \le 1; \quad y(0) = y(1) = 0,$$

using for S a space of continuous piecewise quadratic functions. Specifically, take a uniform subdivision

$$\Delta : \quad 0 = x_0 < x_1 < x_2 < \cdots < x_{n-1} < x_n = 1, \quad x_\nu = \nu h,$$

of $[0, 1]$ into n subintervals of length $h = 1/n$ and let $S = \{s \in \mathbb{S}_2^0 : \; s(0) = s(1) = 0\}$.

(a) How many basis functions is S expected to have? Explain.

(b) Construct a basis for S. {*Hint*: For $\nu = 1, 2, \ldots, n$ take $u_\nu = A_{\nu-1}$ to be the quadratic function on $[x_{\nu-1}, x_\nu]$ having values $u_\nu(x_{\nu-1}) = u_\nu(x_\nu) = 0$, $u_\nu(x_{\nu-\frac{1}{2}}) = 1$ and define $A_{\nu-1}$ to be zero outside of $[x_{\nu-1}, x_\nu]$. Add to these functions the basis of hat functions B_ν for $\mathbb{S}_1^0(\Delta)$.}

(c) Compute the stiffness matrix U for the basis constructed in (b).

(d) Interpret the resulting system $U\xi = \rho$ as a finite difference method applied to the given boundary value problem. What are the meanings of the components of ξ?

14. (a) Show that the solution u_S of (4.18) is the orthogonal projection of the exact solution y of (4.15) onto the space S relative to the inner product $[\cdot\,,\cdot]$; that is,

$$[y - u_S, v] = 0 \quad \text{for all } v \in S.$$

(b) With $\|u\|_E$ denoting the energy norm of u (i.e., $\|u\|_E^2 = [u, u]$), show that

$$\|y - u_S\|_E^2 = \|y\|_E^2 - \|u_S\|_E^2.$$

15. Consider the boundary value problem (4.4′) and (4.2). Define the energy norm by $\|u\|_E^2 = [u, u]$. Let Δ_1 and Δ_2 be two subdivisions of $[a, b]$ and $S_i = \{s \in \mathbb{S}_m^k(\Delta_i),\ s(a) = s(b) = 0\}$, $i = 1, 2$, for some integers m, k with $0 \le k < m$.

(a) With y denoting the exact solution of the boundary value problem, and Δ_1 being a refinement of Δ_2, show that

$$\|y - u_{S_1}\|_E \le \|y - u_{S_2}\|_E.$$

(b) Let Δ_2 be an arbitrary subdivision of $[a, b]$ with all grid points (including the endpoints) being rational numbers. Prove that there exists a uniform subdivision Δ_1 of $[a, b]$, with $|\Delta_1| = h$ sufficiently small, such that

$$\|u - u_{S_1}\|_E \le \|y - u_{S_2}\|_E.$$

16. Apply the variational method to the boundary value problem

$$\mathcal{L}y := -py'' + qy = r(x), \quad 0 \le x \le 1;$$
$$y(0) = y(1) = 0,$$

where p and q are constants with $p > 0$, $q \ge 0$. Use approximants from the space $S = \text{span}\{u_\nu(x) = \sin(\nu\pi x),\ \nu = 1, 2, \ldots, n\}$, and interpret $\mathcal{L}u_S$. Find an explicit form for u_S in the case of constant r.

17. Let y be the exact solution of the boundary value problem (4.1) through (4.3) and u_S the approximate solution of the associated extremal problem

with $S = \{s \in \mathbb{S}_1^0(\Delta) : \; s(a) = s(b) = 0\}$ and $\Delta : \; a = x_0 < x_1 < \cdots < x_n < x_{n+1} = b$. Prove that

$$\|y - u_S\|_\infty \leq \tfrac{1}{2}\sqrt{\tfrac{\overline{c}}{\underline{c}}} \; \max_{0\leq\nu\leq n} \operatorname{osc}_{[x_\nu, x_{\nu+1}]}(y'),$$

where $\operatorname{osc}_{[c,d]}(f) := \max_{[c,d]} f - \min_{[c,d]} f$. In particular, show that

$$\|y - u_S\|_\infty \leq \tfrac{1}{2}\sqrt{\tfrac{\overline{c}}{\underline{c}}} \; |\Delta| \|y''\|_\infty.$$

{*Hint*: Apply Theorem 7.4.4, in particular, (4.27).}

18. Consider the boundary value problem (4.1) and (4.4′) with $p(x)$ and $q(x)$ being positive constants,

$$p(x) = p > 0, \quad q(x) = q > 0.$$

Let $S = \{s \in \mathbb{S}_1^0(\Delta) : \; s(a) = s(b) = 0\}$, where the subdivision Δ: $a = x_0 < x_1 < x_2 < \cdots < x_n < x_{n+1} = b$ is assumed to be *quasi-uniform*; that is,

$$\Delta x_\nu := x_{\nu+1} - x_\nu \geq \beta|\Delta|, \quad \nu = 0, 1, \ldots, n,$$

for some positive constant β. (Recall that $|\Delta| := \max_{0\leq\nu\leq n} \Delta x_\nu$.) Let U be the stiffness matrix for the basis $u_\nu = B_\nu$, $\nu = 1, 2, \ldots, n$, of hat functions (cf. Ch. 2, Ex. 66). Write $u(x) = \sum_{\nu=1}^n \xi_\nu u_\nu(x)$ for any $u \in S$, and $\xi^T = [\xi_1, \xi_2, \ldots, \xi_n]$.

(a) Show that $\xi^T U \xi = [u, u]$.

(b) Show that $\|u'\|_{L_2}^2 = \xi^T T_1 \xi$, where T_1 is a symmetric tridiagonal matrix with

$$(T_1)_{\nu,\nu} = \frac{1}{\Delta x_{\nu-1}} + \frac{1}{\Delta x_\nu}, \quad \nu = 1, 2, \ldots, n;$$
$$(T_1)_{\nu+1,\nu} = (T_1)_{\nu,\nu+1} = -\frac{1}{\Delta x_\nu}, \quad \nu = 1, \ldots, n-1.$$

{*Hint*: Use integration by parts, being careful to observe that u' is only piecewise continuous.}

(c) Show that $\|u\|_{L_2}^2 = \xi^T T_0 \xi$, where T_0 is a symmetric tridiagonal matrix with

$$(T_0)_{\nu,\nu} = \tfrac{1}{3}(\Delta x_{\nu-1} + \Delta x_\nu), \quad \nu = 1, 2, \ldots, n;$$
$$(T_0)_{\nu+1,\nu} = (T_0)_{\nu,\nu+1} = \tfrac{1}{6}\Delta x_\nu, \quad \nu = 1, \ldots, n-1.$$

(d) Combine (a) through (c) to compute $[u, u]$ and hence to estimate the Euclidean condition number $\operatorname{cond}_2 U$. {*Hint*: Use Gershgorin's theorem to estimate the eigenvalues of U.}

(e) The analysis in (d) fails if $q = 0$. Show, however, in the case of a *uniform* grid, that when $q = 0$ then $\text{cond}_2 U \le \sin^{-2} \frac{\pi}{2n}$.

(f) Indicate how the argument in (d) can be extended to variable $p(x)$, $q(x)$ satisfying $0 < p(x) \le \bar{p}$, $0 < \underline{q} \le q(x) \le \bar{q}$ on $[a, b]$.

19. The *method of collocation* for solving a boundary value problem

$$\mathcal{L}y = r(x), \quad 0 \le x \le 1; \quad y(0) = y(1) = 0,$$

consists of selecting an n-dimensional subspace $S \subset V_0$ and determining $u_S \in S$ such that $(\mathcal{L}u_S)(x_\mu) = r(x_\mu)$ for a discrete set of points $0 < x_1 < x_2 < \cdots < x_n < 1$. Apply this method to the problem of Ex. 16, with S as defined there. Discuss the solvability of the system of linear equations involved in the method. {*Hint*: Use the known fact that the only trigonometric sine polynomial $\sum_{\nu=1}^{n} \xi_\nu \sin(\nu\pi x)$ of degree n that vanishes at n distinct points in $(0, 1)$ is the one identically zero.}

MACHINE ASSIGNMENTS

1. The shape of an ideal flexible chain of length L, hung from two points $(0, 0)$ and $(1, 1)$, is determined by the solution of the eigenvalue problem

$$y'' = \lambda\sqrt{1 + (y')^2}, \quad y(0) = 0, \ y(1) = 1, \ \int_0^1 \sqrt{1 + (y')^2} dx = L.$$

Strictly speaking, this is not a problem of the form (0.3), (0.4), but nevertheless can be solved analytically as well as numerically.

(a) On physical grounds, what condition must L satisfy for the problem to have a solution?

(b) Derive three equations in three unknowns: the two constants of integration and the eigenvalue λ. Obtain a transcendental equation for λ by eliminating the other two unknowns. Solve the equation numerically and thus find, and plot, the solution for $L = 2, 4, 8$, and 16.

(c) If one attempts to solve the problem by a finite difference method over a uniform grid $x_0 = 0 < x_1 < x_2 < \cdots < x_N < x_{N+1} = 1$, $x_n = \frac{n}{N+1}$, a system of $N + 1$ nonlinear equations in $N + 1$ unknowns results. This can be solved by a homotopy method using L as the homotopy parameter. Since for $L = \sqrt{2}$ the solution is trivial, one selects a sequence of parameter values $L_0 = \sqrt{2} < L_1 < \cdots < L_m$ and solves the finite difference equations for L_i using the solution for L_{i-1} as the initial approximation. Implement this for the values of L given in (b), taking a sequence $\{L_i\}$ which contains these values. Compare the numerical results for the eigenvalues with the analytic ones for $N = 10,\ 20,\ 40$. (Use an appropriate software package for solving systems of nonlinear equations.)

2. Change the boundary value problem of Example 2.1 to

$$y'' = -e^y, \quad 0 \le x \le 1, \quad y(0) = y(1) = 0.$$

Then Theorem 7.1.2 no longer applies (why not?). In fact, it is known that the problem has two solutions. Use MATLAB to compute the respective initial slopes $y'(0)$ to 10 significant digits by Newton's method, as in Example 2.1. {*Hint*: Use approximations $s^{(0)} = 1$ and $s^{(0)} = 15$ to the initial slopes.}

3. Consider the boundary value problem

$$\text{(BVP)} \qquad y'' = y^2, \quad 0 \le x \le b; \quad y(0) = 0,\ y(b) = \beta,$$

and the associated initial value problem

$$\text{(IVP)} \qquad u'' = u^2, \quad u(0) = 0,\ u'(0) = s.$$

Denote the solution of (IVP) by $u(x) = u(x; s)$.

(a) Let $v(x) = u(x; -1)$. Show that

$$v'(x) = -\sqrt{\tfrac{2}{3}v^3(x) + 1},$$

and thus, the function v, being convex (i.e., $v'' > 0$), has a minimum at some $x_0 > 0$ with value $v_{\min} = -(\frac{3}{2})^{\frac{1}{3}} = -1.1447142\ldots$. Show that v is symmetric with respect to the line $x = x_0$.

(b) Compute x_0 numerically in terms of the beta integral. {*Point of information*: The beta integral is $B(p, q) = \int_0^1 t^{p-1}(1-t)^{q-1}dt$ and has the value $\frac{\Gamma(p)\Gamma(q)}{\Gamma(p+q)}$.}

(c) Use MATLAB to compute v by solving the initial value problem

$$v'' = v^2, \quad x_0 \le x \le 3x_0; \quad v(x_0) = v_{\min},\ v'(x_0) = 0.$$

Plot the solution (and its symmetric part) on $-x_0 \le x \le 3x_0$.

(d) In terms of the function v defined in (a), show that

$$u(x; -s^3) = s^2 v(sx), \quad \text{all } s \in \mathbb{R}.$$

As s ranges over all the reals, the solution manifold $\{s^2 v(sx)\}$ thus encompasses all the solutions of (IVP). Prepare a plot of this solution manifold. Note that there exists an envelope of the manifold, located in the lower half-plane. Explain why, in principle, this envelope must be the solution of a first-order differential equation.

(e) Based on the plot obtained in (d), discuss the number of possible solutions to the original boundary value problem (BVP). In particular, determine for what values of b and β there does not exist any solution.

(f) Use the method of finite differences on a uniform grid to compute the two solutions of (BVP) for $b = 3x_0$, $\beta = v_0$, where $v_0 = v(3x_0)$ is a quantity already computed in (c). Solve the systems of nonlinear difference equations by Newton's method. In trying to get the first solution, approximate the solution v of (a) on $0 \le x \le 3x_0$ by a quadratic function $\tilde{v}$ satisfying $\tilde{v}(0) = 0$, $\tilde{v}'(0) = -1$, $\tilde{v}(3x_0) = v_0$, and then use its restriction to the grid as the initial approximation to Newton's method. For the second solution, try the initial approximation obtained from the linear approximation $\overline{v}(x) = v_0 x/(3x_0)$. In both cases, plot initial approximations as well as the solutions to the difference equations. What happens if Newton's method is replaced by the method of successive approximations?

4. The following boundary value problem occurs in soil engineering. Determine $y(r)$, $1 \le r < \infty$, such that

$$\frac{1}{r}\frac{d}{dr}\left(ry\frac{dy}{dr}\right) + \rho(1-y) = 0, \quad y(1) = \eta, \; y(\infty) = 1,$$

where ρ, η are parameters satisfying $\rho > 0$, $0 < \eta < 1$. The quantity of interest is $\sigma = \left.\frac{dy}{dr}\right|_{r=1}$.

(a) Let $z(x) = [y(e^x)]^2$. Derive the boundary value problem and the quantity of interest in terms of z.

(b) Consider the initial value problem associated with the boundary value problem in (a), having initial conditions

$$z(0) = \eta^2, \quad z'(0) = s.$$

Discuss the qualitative behavior of its solutions for real values of s. {*Hint*: Suggested questions may be: admissible domain in the (x, z)-plane, convexity and concavity of the solutions, the role of the line $z = 1$.}

(c) From your analysis in (b), devise an appropriate shooting procedure for solving the boundary value problem numerically and for computing the quantity σ. Run your procedure on the computer (in double precision) for various values of η and ρ. In particular, prepare a 5-decimal table showing the values of s and σ for $\eta = .1(.1).9$ and $\rho = .5, 1, 2, 5, 10$, and plot σ versus η.

References

Abramowitz, Milton [1954]. On the practical evaluation of integrals, *J. Soc. Indust. Appl. Math. 2*, 20–35. (Reprinted in Davis and Rabinowitz [1984, Appendix 1].)

Agarwal, Ravi P. [1986]. *Boundary value problems for higher order differential equations*, World Scientific, Teaneck, NJ.

Agarwal, Ravi P. [1992]. *Difference equations and inequalities: Theory, methods, and applications*, Monographs and Textbooks Pure. Appl. Math., v. 155, Dekker, New York.

Ahlberg, J.H., Nilson, E.N., and Walsh, J.L. [1967]. *The theory of splines and their applications*, Math. Sci. Engrg., v. 38, Academic Press, New York.

Akl, Selim G. and Lyons, Kelly A. [1993]. *Parallel computational geometry*, Prentice-Hall, Englewood Cliffs, NJ.

Albrecht, Peter [1987]. A new theoretical approach to Runge-Kutta methods, *SIAM J. Numer. Anal. 24*, 391–406.

Albrecht, Peter [1996]. The Runge-Kutta theory in a nutshell, *SIAM J. Numer. Anal. 33*, 1712–1735.

Alefeld, Götz and Herzberger, Jürgen [1983]. *Introduction to interval computations*, translated from the German by Jon Rokne, Comput. Sci. Appl. Math., Academic Press, New York.

Alexander, Dan [1996]. *Newton's method — or is it?*, Focus, The Newsletter of the Mathematical Association of America, v. 16, no. 5, 32–33.

Alexander, Roger [1977]. Diagonally implicit Runge-Kutta methods for stiff o.d.e.'s, *SIAM J. Numer. Anal. 14*, 1006–1021.

Allenby, R.B. and Redfern, E.J. [1989]. *Introduction to number theory with computing*, Edward Arnold, London.

Allgower, Eugene L. and Georg, Kurt [1990]. *Numerical continuation methods: An introduction*, Springer Ser. Comput. Math., v. 13, Springer, Berlin.

Ames, William F. [1992]. *Numerical methods for partial differential equations*, 3d ed., Comput. Sci. Sci. Comput., Academic Press, Boston, MA.

Anonymous [1996]. Inquiry board traces Ariane 5 failure to overflow error, *SIAM News 29*, no. 8, pp. 1, 12–13.

Anscombe, Francis John [1981]. *Computing in statistical science through APL*, Springer Ser. Statist., Springer, New York.

Ascher, Uri M., Mattheij, Robert M.M., and Russell, Robert D. [1995]. *Numerical solution of boundary value problems for ordinary differential equations*, corrected reprint of the 1988 original, Classics Appl. Math., v. 13, SIAM, Philadelphia, PA.

Ash, J. Marshall and Jones, Roger L. [1981]. Optimal numerical differentiation using three function evaluations, *Math. Comp. 37*, 159–167.

Ash, J. Marshall, Janson, S., and Jones, R.L. [1984]. Optimal numerical differentiation using N function evaluations, *Calcolo 21*, 151–169.

Ashyralyev, A. and Sobolevskiĭ, P.E. [1994]. *Well-posedness of parabolic difference equations*, translated from the Russian by A. Iacob, Operator Theory: Adv. Appl., v. 69, Birkhäuser, Basel.

Atkinson, Kendall E. [1968]. On the order of convergence of natural cubic spline interpolation, *SIAM J. Numer. Anal. 5*, 89–101.

Atkinson, Kendall E. [1976]. *A survey of numerical methods for the solution of Fredholm integral equations of the second kind*, SIAM, Philadelphia, PA.

Atkinson, Kendall E. [1989]. *An introduction to numerical analysis*, 2d ed., Wiley, New York.

Atteia, Marc [1992]. *Hilbertian kernels and spline functions*, Stud. Comput. Math., v. 4, North-Holland, Amsterdam.

Axelsson, Owe [1969]. A class of A-stable methods, *BIT 9*, 185–199.

Axelsson, O. and Barker, V.A. [1984]. *Finite element solution of boundary value problems: Theory and computation*, Comput. Sci. Appl. Math., Academic Press, Orlando, FL.

Bach, Eric and Shallit, Jeffrey [1996]. *Algorithmic number theory.* v. 1: *Efficient algorithms*, Found. Comput. Ser., MIT Press, Cambridge, MA.

Bailey, Paul B., Shampine, Lawrence F., and Waltman, Paul E. [1968]. *Nonlinear two point boundary value problems*, Math. Sci. Engrg., v. 44, Academic Press, New York.

Baker, Christopher T.H. [1977]. *The numerical treatment of integral equations*, Monographs Numer. Anal., Clarendon Press, Oxford.

Baker, George A., Jr. [1975]. *Essentials of Padé approximants*, Academic Press, New York.

Baker, George A., Jr. and Graves-Morris, Peter [1996]. *Padé approximants*, 2d ed., Encyclopedia Math. Appl., v. 59, Cambridge University Press, Cambridge.

Banerjee, P.K. and Butterfield, R. [1981]. *Boundary element methods in engineering science*, McGraw-Hill, London.

Bartels, Richard H., Beatty, John C., and Barsky, Brian A. [1987]. *An introduction to splines for use in computer graphics and geometric modeling*, with forewords by Pierre Bézier and A. Robin Forrest, Morgan Kaufmann, Palo Alto, CA.

Barton, D., Willers, I.M., and Zahar, R.V.M. [1971]. The automatic solution of systems of ordinary differential equations by the method of Taylor series, *Comput. J. 14*, 243–248.

Bashforth, Francis and Adams, J.C. [1883]. *An attempt to test the theories of capillary action by comparing the theoretical and measured forms of drops of fluid, with an explanation of the method of integration employed in constructing the tables which give the theoretical forms of such drops*, Cambridge University Press, Cambridge.

Bauer, F.L., Rutishauser, H., and Stiefel, E. [1963]. *New aspects in numerical quadrature*, Proc. Sympos. Appl. Math., v. 15, Amer. Math. Soc., Providence, RI, 199–218.

Beatson, R.K. [1986]. On the convergence of some cubic spline interpolation schemes, *SIAM J. Numer. Anal. 23*, 903–912.

Beatson, R.K. and Chacko, E. [1989]. *A quantitative comparison of end conditions for cubic spline interpolation*, in *Approximation theory VI*, v. 1, Charles K. Chui, L.L. Schumaker, and J.D. Ward, eds., Academic Press, Boston, MA, 77–79.

Beatson, R.K. and Chacko, E. [1992]. Which cubic spline should one use?, *SIAM J. Sci. Statist. Comput. 13*, 1009–1024.

Becker, Eric B., Carey, Graham F, and Oden, J. Tinsley [1981]. *Finite elements. v. I: An introduction*, The Texas Finite Element Series, v. 1, Prentice-Hall, Englewood Cliffs, NJ.

Bellman, Richard and Cooke, Kenneth L. [1963]. *Differential-difference equations*, Academic Press, New York, 1963.

Bellman, Richard E. and Kalaba, Robert E. [1965]. *Quasilinearization and nonlinear boundary-value problems*, Modern Analyt. Comput. Methods in Sci. Math., v. 3, American Elsevier, New York.

Bellomo, Nicola and Preziosi, Luigi [1995]. *Modelling mathematical methods and scientific computation*, CRC Math. Modelling Ser., CRC Press, Boca Raton, FL.

Beltzer, A.I. [1990]. *Variational and finite element methods: A symbolic computation approach*, Springer, Berlin.

Bernfeld, Stephen R. and Lakshmikantham, V. [1974]. *An introduction to nonlinear boundary value problems*, Math. Sci. Engrg., v. 109, Academic Press, New York.

Berrut, Jean-Paul [1984]. Baryzentrische Formeln zur trigonometrischen Interpolation I., *Z. Angew. Math. Phys. 35*, 91–105; Baryzentrische Formeln zur trigonometrischen Interpolation II. Stabilität und Anwendung auf die Fourieranalyse bei ungleichabständigen Stützstellen, *ibid.*, 193–205.

Berrut, Jean-Paul [1989]. Barycentric formulae for cardinal (SINC-) interpolants, *Numer. Math. 54*, 703–718. [Erratum: *ibid. 55* (1989), 747.]

Bettis, D.G. [1969/1970]. Numerical integration of products of Fourier and ordinary polynomials, *Numer. Math. 14*, 421–434.

Bini, Dario and Pan, Victor Y. [1994]. *Polynomial and matrix computations*, v. 1: *Fundamental algorithms*, Progr. Theoret. Comput. Sci., Birkhäuser, Boston, MA.

Birkhoff, Garrett and Lynch, Robert E. [1984]. *Numerical solution of elliptic problems*, SIAM Stud. Appl. Math., v. 6, SIAM, Philadelphia, PA.

Björck, Åke [1996]. *Numerical methods for least squares problems*, SIAM, Philadelphia, PA.

Bojanov, B.D., Hakopian, H.A., and Sahakian, A.A. [1993]. *Spline functions and multivariate interpolations*, Math. Appl., v. 248, Kluwer, Dordrecht.

de Boor, Carl [1978]. *A practical guide to splines*, Appl. Math. Sci., v. 27, Springer, New York.

de Boor, C., Höllig, K., and Riemenschneider, S. [1993]. *Box splines*, Applied Math. Sci., v. 98, Springer, New York.

Borwein, Peter and Erdélyi, Tamás [1995]. *Polynomials and polynomial inequalities*, Graduate Texts in Math., v. 161, Springer, New York.

Braess, Dietrich [1986]. *Nonlinear approximation theory*, Springer Ser. Comput. Math., v. 7, Springer, Berlin.

Bramble, James H. [1993]. *Multigrid methods*, Pitman Res. Notes in Math. Ser., v. 294, Wiley, New York.

Branham, Richard L., Jr. [1990]. *Scientific data analysis: An introduction to overdetermined systems*, Springer, New York.

Brass, Helmut [1977]. *Quadraturverfahren*, Studia Mathematica, Skript 3, Vandenhoeck & Ruprecht, Göttingen.

Brebbia, C.A. [1984]. *The boundary element method for engineers*, 2d ed., Pentech Press, London.

Brenan, K.E., Campbell, S.L., and Petzold, L.R. [1996]. *Numerical solution of initial-value problems in differential-algebraic equations*, revised and corrected reprint of the 1989 original, Classics Appl. Math., v. 14, SIAM, Philadelphia, PA.

Brenner, Susanne C. and Scott, L. Ridgway [1994]. *The mathematical theory of finite element methods*, Texts Appl. Math., v. 15, Springer, New York.

Brent, Richard P. [1973]. *Algorithms for minimization without derivatives*, Prentice-Hall Ser. Automat. Comput., Prentice-Hall, Englewood Cliffs, NJ. [Erratum: *Math. Comp. 29* (1975), 1166.]

Brezinski, Claude and Redivo-Zaglia, Michela [1991]. *Extrapolation methods: Theory and practice*, Stud. Comput. Math., v. 2, North-Holland, Amsterdam.

Brezzi, Franco and Fortin, Michel [1991]. *Mixed and hybrid finite element methods*, Springer Ser. Comput. Math., v. 15, Springer, New York.

Briggs, William L. [1987]. *A multigrid tutorial*, SIAM, Philadelphia, PA.

Brock, P. and Murray, F.J. [1952]. The use of exponential sums in step by step integration I., II., *Math. Tables Aids Comput. 6*, 63–78; *ibid.*, 138–150.

Brown, Peter N., Byrne, George D., and Hindmarsh, Alan C. [1989]. VODE: A variable-coefficient ODE solver, *SIAM J. Sci. Statist. Comput. 10*, 1038–1051.

Bruce, J.W., Giblin, P.J., and Rippon, P.J. [1990]. *Microcomputers and mathematics*, Cambridge University Press, Cambridge.

Brunner, H. and van der Houwen, P.J. [1986]. *The numerical solution of Volterra equations*, CWI Monographs, v. 3, North-Holland, Amsterdam.

Brutman, L. [1997a]. Lebesgue functions for polynomial interpolation — a survey, The heritage of P.L. Chebyshev: A Festschrift in honor of the 70th birthday of T.J. Rivlin, *Ann. Numer. Math. 4*, 111–127.

Brutman, L. [1997b]. *Chebyshev quadratures — a survey*, Proc. 4th International Colloquium on Numerical Analysis, Plovdiv, to appear.

Brutman, L. and Passow, E. [1995]. On the divergence of Lagrange interpolation to $|x|$, *J. Approx. Theory 81*, 127–135.

Burnett, David S. [1987]. *Finite element analysis: From concepts to applications*, Addison-Wesley, Reading, MA.

Burrage, Kevin [1978a]. A special family of Runge-Kutta methods for solving stiff differential equations, *BIT 18*, 22–41.

Burrage, Kevin [1978b]. High order algebraically stable Runge-Kutta methods, *BIT 18*, 373–383.

Burrage, Kevin [1982]. Efficiently implementable algebraically stable Runge-Kutta methods, *SIAM J. Numer. Anal. 19*, 245–258.

Butcher, J.C. [1965]. On the attainable order of Runge-Kutta methods, *Math. Comp. 19*, 408–417.

Butcher, J.C. [1985]. The nonexistence of ten-stage eighth order explicit Runge-Kutta methods, *BIT 25*, 521–540.

Butcher, J.C. [1987]. *The numerical analysis of ordinary differential equations: Runge-Kutta and general linear methods*, Wiley-Intersci. Publ., Wiley, Chichester.

Butcher, J.C., guest ed. [1996]. Special issue celebrating the centenary of Runge-Kutta methods, *Appl. Numer. Math. 22*, nos. 1–3.

Canuto, Claudio, Hussaini, M. Yousuff, Quarteroni, Alfio, and Zang, Thomas A. [1988]. *Spectral methods in fluid dynamics*, Springer Ser. Comput. Phys., Springer, New York.

Carey, Graham F. and Oden, J. Tinsley [1983]. *Finite elements.* v. II: *A second course*, The Texas Finite Element Series, v. 2, Prentice-Hall, Englewood Cliffs, NJ.

Carey, Graham F. and Oden, J. Tinsley [1984]. *Finite elements.* v. III: *Computational aspects*, The Texas Finite Element Series, v. 3, Prentice-Hall, Englewood Cliffs, NJ.

Carey, Graham F. and Oden, J. Tinsley [1986]. *Finite elements.* v. VI: *Fluid mechanics*, The Texas Finite Element Series, v. 6, Prentice-Hall, Englewood Cliffs, NJ.

Celia, Michael A. and Gray, William G. [1992]. *Numerical methods for differential equations: Fundamental concepts for scientific and engineering applications*, Prentice-Hall, Englewood Cliffs, NJ.

Chatelin, Françoise [1983]. *Spectral approximation of linear operators*, with a foreword by P. Henrici and solutions to exercises by Mario Ahués, Comput. Sci. Appl. Math., Academic Press, New York.

Chatelin, Françoise [1993]. *Eigenvalues of matrices*, with exercises by Mario Ahués; translated from the French and with additional material by Walter Ledermann, Wiley, Chichester.

Chebyshev, P.L. [1859]. Sur l'interpolation par la méthode des moindres carrés, *Mém. Acad. Impér. Sci. St. Petersbourg (7) 1*, no. 15, 1–24. [Œuvres, v. 1, 473–498.]

Chen, Goong and Zhou, Jian-Xin [1992]. *Boundary element methods*, Comput. Math. Appl., Academic Press, London.

Cheney, E.W. [1966]. *Introduction to approximation theory*, Internat. Ser. Pure Appl. Math., McGraw-Hill, New York.

Cheney, E.W. [1986]. *Multivariate approximation theory: Selected topics*, CBMS-NSF Regional Conf. Ser. in Appl. Math., no. 51, SIAM, Philadelphia, PA.

Cheney, Ward and Kincaid, David [1994]. *Numerical mathematics and computing*, 3d ed., Brooks/Cole, Pacific Grove, CA.

Chihara, T.S. [1978]. *An introduction to orthogonal polynomials*, Math. Appl., v. 13, Gordon and Breach, New York.

Chui, Charles K. [1988]. *Multivariate splines*, with an appendix by Harvey Diamond, CBMS-NSF Regional Conf. Ser. in Appl. Math., no. 54, SIAM, Philadelphia, PA.

Chui, Charles K. [1992]. *An introduction to wavelets*, Wavelet Analysis and Its Applications, v. 1, Academic Press, Boston, MA.

Ciarlet, Phillipe G. [1978]. *The finite element method for elliptic problems*, Stud. Math. Appl., v. 4, North-Holland, Amsterdam.

Ciarlet, Philippe G. [1989]. *Introduction to numerical linear algebra and optimisation*, with the assistance of Bernadette Miara and Jean-Marie Thomas; translated from the French by A. Buttigieg, Cambridge Texts Appl. Math., Cambridge University Press, Cambridge.

Ciarlet, P.G. and Lions, J.-L., eds. [1990, 1991, 1994]. *Handbook of numerical analysis*, vols. 1–3, North-Holland, Amsterdam.

Coddington, Earl A. and Levinson, Norman [1955]. *Theory of ordinary differential equations*, Internat. Ser. Pure Appl. Math., McGraw-Hill, New York.

Coe, Tim, Mathisen, Terje, Moler, Cleve, and Pratt, Vaughan [1995]. Computational aspects of the Pentium affair, *IEEE Comput. Sci. Engrg. 2*, no. 1, 18–30.

Cohen, A.M. [1994]. *Is the polynomial so perfidious?*, Numer. Math. **68**, 225–238.

Cohen, Henri [1993]. *A course in computational algebraic number theory*, Grad. Texts Math., v. 138, Springer, Berlin.

Conte, Samuel D. and de Boor, Carl [1980]. *Elementary numerical analysis: An algorithmic approach*, 3d ed., McGraw-Hill, New York.

Cools, Ronald and Rabinowitz, Philip [1993]. Monomial cubature rules since "Stroud": A compilation, *J. Comput. Appl. Math. 48*, 309–326.

Cormen, Thomas H., Leiserson, Charles E., and Rivest, Ronald L. [1990]. *Introduction to algorithms*, MIT Electr. Engrg. Comput. Sci. Ser., MIT Press, Cambridge, MA.

Cox, David, Little, John, and O'Shea, Donal [1992]. *Ideals, varieties, and algorithms: An introduction to computational algebraic geometry and commutative algebra*, Undergrad. Texts Math., Springer, New York.

Crandall, Richard E. [1994]. *Projects in scientific computation*, Springer, New York.

Crouzeix, Michel [1976]. *Sur les méthodes de Runge-Kutta pour l'approximation des problèmes d'évolution*, Lecture Notes in Econom. and Math. Systems, no. 134, Springer, Berlin, 206–223.

Cryer, Colin W. [1972]. *Numerical methods for functional differential equations*, in *Delay and functional differential equations and their applications*, Klaus Schmitt, ed., Academic Press, New York, 17–101.

Cullum, Jane K. and Willoughby, Ralph A. [1985]. *Lánczos algorithms for large symmetric eigenvalue computations.* v. 1: *Theory*; v. 2: *Programs*, Progr. Sci. Comput., vols. 3, 4, Birkhäuser, Boston, MA.

Curtiss, C.F. and Hirschfelder, J.O. [1952]. Integration of stiff equations, *Proc. Nat. Acad. Sci. U.S.A. 38*, 235–243.

Dahlquist, Germund [1956]. Convergence and stability in the numerical integration of ordinary differential equations, *Math. Scand. 4*, 33–53.

Dahlquist, Germund [1959]. *Stability and error bounds in the numerical integration of ordinary differential equations*, Kungl. Tekn. Högsk. Handl. Stockholm, no. 130.

Dahlquist, Germund G. [1963]. A special stability problem for linear multistep methods, *BIT 3*, 27–43.

Dahlquist, Germund [1985]. 33 years of numerical instability. I., *BIT 25*, 188–204.

Dahlquist, Germund and Björck, Åke [1974]. *Numerical methods*, translated from the Swedish by Ned Anderson, Prentice-Hall Ser. Automat. Comput., Prentice-Hall, Englewood Cliffs, NJ.

Daniel, James W. and Moore, Ramon E. [1970]. *Computation and theory in ordinary differential equations*, Ser. Books Math., Freeman, San Francisco, CA.

Datta, Biswa Nath [1995]. *Numerical linear algebra and applications*, Brooks/Cole, Pacific Grove, CA.

Daubechies, Ingrid [1992]. *Ten lectures on wavelets*, CBMS-NSF Regional Conf. Ser. in Appl. Math., no. 61, SIAM, Philadelphia, PA.

Davenport, J.H., Siret, Y., and Tournier, E. [1993]. *Computer algebra: Systems and algorithms for algebraic computation*, 2d ed., with a preface by Daniel Lazard; translated from the French by A. Davenport and J.H. Davenport; with a foreword by Antony C. Hearn, Academic Press, London.

Davis, Philip J. [1975]. *Interpolation and approximation*, republication, with minor corrections, of the 1963 original, with a new preface and bibliography, Dover, New York.

Davis, Philip J. and Rabinowitz, Philip [1984]. *Methods of numerical integration*, 2d ed., Comput. Sci. Appl. Math., Academic Press, Orlando, FL.

Dekker, K. and Verwer, J.G. [1984]. *Stability of Runge-Kutta methods for stiff nonlinear differential equations*, CWI Monographs, v. 2, North-Holland, Amsterdam.

Delves, L.M. and Mohamed, J.L. [1988]. *Computational methods for integral equations*, Cambridge University Press, Cambridge.

Dennis, J.E., Jr. and Schnabel, Robert B. [1996]. *Numerical methods for unconstrained optimization and nonlinear equations*, corrected reprint of the 1983 original, Classics Appl. Math., v. 16, SIAM, Philadelphia, PA.

Deuflhard, Peter and Hohmann, Andreas [1995]. *Numerical analysis: A first course in scientific computation*, translated from the second (1993) German edition by F.A. Potra and F. Schulz, de Gruyter Textbook, de Gruyter, Berlin.

DeVore, Ronald A. and Lorentz, George G. [1993]. *Constructive approximation*, Grundlehren Math. Wiss., v. 303, Springer, Berlin.

Devroye, Luc [1986]. *Nonuniform random variate generation*, Springer, New York.

Dierckx, Paul [1993]. *Curve and surface fitting with splines*, Monographs Numer. Anal., Oxford Sci. Publ., Oxford University Press, New York.

Dormand, J.R. and Prince, P.J. [1980]. A family of embedded Runge-Kutta formulae, *J. Comput. Appl. Math. 6*, 19–26.

Driver, R.D. [1977]. *Ordinary and delay differential equations*, Appl. Math. Sci., v. 20, Springer, New York.

Duff, I.S., Erisman, A.M., and Reid, J.K. [1989]. *Direct methods for sparse matrices*, 2d ed., Monographs Numer. Anal., Oxford Sci. Publ., Oxford University Press, New York.

Durand, E. [1960, 1961]. *Solutions numériques des équations algébriques.* v. 1: *Équations du type $F(x) = 0$, racines d'un polynôme*; v. 2: *Systèmes de plusieurs équations, valeurs propres des matrices*, Masson, Paris.

Dutka, Jacques [1984]. Richardson extrapolation and Romberg integration, *Historia Math. 11*, 3–21.

Edelman, Alan [preprint]. *The mathematics of the Pentium division bug.*

Edelsbrunner, Herbert [1987]. *Algorithms in combinatorial geometry*, EATCS Monographs Theoret. Comput. Sci., v. 10, Springer, Berlin.

Ehle, Byron L. [1968]. High order A-stable methods for the numerical solution of systems of D.E.'s, *BIT 8*, 276–278.

Ehle, Byron L. [1973]. A-stable methods and Padé approximations to the exponential, *SIAM J. Math. Anal. 4*, 671–680.

Eijgenraam, P. [1981]. *The solution of initial value problems using interval arithmetic: Formulation and analysis of an algorithm*, Mathematical Centre Tracts, v. 144, Mathematisch Centrum, Amsterdam.

Elhay, Sylvan and Kautsky, Jaroslav [1987]. Algorithm 655 — IQPACK: FORTRAN subroutines for the weights of interpolatory quadratures, *ACM Trans. Math. Software 13*, 399–415.

Engels, Hermann [1979]. Zur Geschichte der Richardson-Extrapolation, *Historia Math. 6*, 280–293.

Engels, H. [1980]. *Numerical quadrature and cubature*, Comput. Math. Appl., Academic Press, London.

England, R. [1969/1970]. Error estimates for Runge-Kutta type solutions to systems of ordinary differential equations, *Comput. J. 12*, 166–170.

Epperson, James F. [1987]. On the Runge example, *Amer. Math. Monthly 94*, 329–341.

Erdös, P. and Vértesi, P. [1980]. On the almost everywhere divergence of Lagrange interpolatory polynomials for arbitrary system of nodes, *Acta Math. Acad. Sci. Hungar. 36*, 71–89.

Euler, Leonhard [1768]. *Institutionum calculi integralis*, v. 1, sec. 2, Ch. 7, Impenfis Academiae Imperialis Scientiarum, Petropoli. [Opera Omnia, ser. 1, v. 11, 424–434.]

Evans, Gwynne [1993]. *Practical numerical integration*, Wiley, Chichester.

Evtushenko, Yurij G. [1985]. *Numerical optimization techniques*, translated from the Russian; translation edited and with a foreword by J. Stoer, Transl. Ser. Math. Engrg., Springer, New York.

Fang, K.-T. and Wang, Y. [1994]. *Number-theoretic methods in statistics*, Monographs Statist. Appl. Probab., v. 51, Chapman & Hall, London.

Farin, Gerald E. [1995]. *NURB curves and surfaces: From projective geometry to practical use*, A K Peters, Wellesley, MA.

Farin, Gerald [1997]. *Curves and surfaces for computer aided geometric design: A practical guide*, 4th ed., Chapter 1 by P. Bézier; Chapters 11 and 22 by W. Boehm, Comput. Sci. Sci. Comput., Academic Press, San Diego, CA.

Fehlberg, E. [1969]. Klassische Runge-Kutta-Formeln fünfter und siebenter Ordnung mit Schrittweiten-Kontrolle, *Computing 4*, 93–106. [Corrigendum: *ibid.* *5*, 184.]

Fehlberg, E. [1970]. Klassische Runge-Kutta-Formeln vierter und niedrigerer Ordnung mit Schrittweiten-Kontrolle und ihre Anwendung auf Wärmeleitungsprobleme, *Computing 6*, 61–71.

Feinerman, Robert P. and Newman, Donald J. [1974]. *Polynomial approximation*, Williams & Wikins, Baltimore, MD.

Fejér, L. [1918]. Interpolation und konforme Abbildung, *Göttinger Nachrichten*, 319–331. [Ges. Arbeiten, v. 2, Akadémiai Kiadó, Budapest, 1970, 100–111.]

Fejér, L. [1933]. Mechanische Quadraturen mit positiven Cotesschen Zahlen, *Math. Z. 37*, 287–309. [Ges. Arbeiten, v. 2, Akadémiai Kiadó, Budapest, 1970, 457–478.]

Fletcher, R. [1987]. *Practical methods of optimization*, 2d ed., Wiley-Intersci. Publ., Wiley, Chichester.

Fornberg, Bengt [1981]. Numerical differentiation of analytic functions, *ACM Trans. Math. Software 7*, 512–526; Algorithm 579 — CPSC: Complex power series coefficients, *ibid.* 542–547.

Fornberg, Bengt [1996]. *A practical guide to pseudospectral methods*, Cambridge Monographs Appl. Comput. Math., v. 1, Cambridge University Press, Cambridge.

Förster, Klaus-Jürgen [1993]. *Variance in quadrature — a survey*, in *Numerical integration IV*, H. Brass and G. Hämmerlin, eds., Internat. Ser. Numer. Math., v. 112, Birkhäuser, Basel, 91–110.

Forsythe, George E. [1957]. Generation and use of orthogonal polynomials for data-fitting with a digital computer, *J. Soc. Indust. Appl. Math. 5*, 74–88.

Forsythe, George E. [1958]. Singularity and near singularity in numerical analysis, *Amer. Math. Monthly 65*, 229–240.

Forsythe, George E. [1970]. Pitfalls in computation, or why a math book isn't enough, *Amer. Math. Monthly 77*, 931–956.

Forsythe, George E. and Moler, Cleve B. [1967]. *Computer solution of linear algebraic systems*, Prentice-Hall, Englewood Cliffs, NJ.

Fox, L. [1990]. *The numerical solution of two-point boundary problems in ordinary differential equations*, reprint of the 1957 original, Dover, New York.

Freeman, T.L. and Phillips, C. [1992]. *Parallel numerical algorithms*, Prentice-Hall Internat. Ser. Comput. Sci., Prentice-Hall International, New York.

Freud, Géza [1971]. *Orthogonal polynomials*, translated from the German by I. Földes, Pergamon Press, Oxford.

Fröberg, Carl-Erik [1985]. *Numerical mathematics: Theory and computer applications*, Benjamin/Cummings, Menlo Park, CA.

Gaier, Dieter [1987]. *Lectures on complex approximation*, translated from the German by Renate McLaughlin, Birkhäuser, Boston, MA.

Garfunkel, S. and Steen, L.A., eds. [1988]. *For all practical purposes: Introduction to contemporary mathematics*, Freeman, New York.

Gauss, K.F. [1995]. *Theory of the combination of observations least subject to errors, Part* 1, *Part* 2, *Supplement*, translated from the Latin by G.W. Stewart, Classics Appl. Math., v. 11, SIAM, Philadelphia, PA.

Gautschi, Walter [1961]. Numerical integration of ordinary differential equations based on trigonometric polynomials, *Numer. Math. 3*, 381–397.

Gautschi, Walter [1963]. On error reducing multi-step methods (abstract), *Notices Amer. Math. Soc. 10*, no. 1, Part I, 95.

Gautschi, Walter [1971/1972]. Attenuation factors in practical Fourier analysis, *Numer. Math. 18*, 373–400.

Gautschi, Walter [1973]. On the condition of algebraic equations, *Numer. Math. 21*, 405–424.

Gautschi, Walter [1975a]. *Computational methods in special functions — a survey*, in *Theory and application of special functions*, Richard A. Askey, ed., Math. Res. Center, Univ. Wisconsin Publ., no. 35, Academic Press, New York, 1–98.

Gautschi, Walter [1975b]. Stime dell'errore globale nei metodi "one-step" per equazioni differenziali ordinarie, collection of articles dedicated to Mauro Picone, *Rend. Mat. (6) 8*, 601–617.

Gautschi, Walter [1976a]. *Advances in Chebyshev quadrature*, in *Numerical analysis Dundee 1975*, G.A. Watson, ed., Lecture Notes in Math., v. 506, Springer, Berlin, 100–121.

Gautschi, Walter [1976b]. Comportement asymptotique des coefficients dans les formules d'intégration d'Adams, de Störmer et de Cowell, *C.R. Acad. Sci. Paris Sér. A-B 283*, A787–A788.

Gautschi, Walter [1981]. *A survey of Gauss-Christoffel quadrature formulae*, in *E.B. Christoffel: The influence of his work on mathematics and the physical sciences*, P.L. Butzer and F. Fehér, eds., Birkhäuser, Basel, 72–147.

Gautschi, Walter [1982]. On generating orthogonal polynomials, *SIAM J. Sci. Statist. Comput. 3*, 289–317.

Gautschi, Walter [1983]. On the convergence behavior of continued fractions with real elements, *Math. Comp. 40*, 337–342.

Gautschi, Walter [1984]. *Questions of numerical condition related to polynomials*, MAA Stud. Math., v. 24: *Studies in numerical analysis*, Gene H. Golub, ed., Math. Assoc. America, Washington, DC, 140–177.

Gautschi, Walter [1988]. *Gauss-Kronrod quadrature — a survey*, in *Numerical methods and approximation theory III*, G.V. Milovanović, ed., Univ. Niš, Niš, 39–66.

Gautschi, Walter [1990]. *How (un)stable are Vandermonde systems?*, in *Asymptotic and computational analysis*, R. Wong, ed., Lecture Notes in Pure Appl. Math., v. 124, Dekker, New York, 193–210.

Gautschi, Walter [1993]. *Gauss-type quadrature rules for rational functions*, in *Numerical integration IV*, H. Brass and G. Hämmerlin, eds., Internat. Ser. Numer. Math., v. 112, Birkhäuser, Basel, 111–130.

Gautschi, Walter, ed. [1994a]. *Mathematics of Computation 1943–1993: A half-century of computational mathematics*, Proc. Sympos. Appl. Math., v. 48, Amer. Math. Soc., Providence, RI.

Gautschi, Walter [1994b]. Algorithm 726: ORTHPOL — a package of routines for generating orthogonal polynomials and Gauss-type quadrature rules, *ACM Trans. Math. Software 20*, 21–62.

Gautschi, Walter [1996]. *Orthogonal polynomials: Applications and computation*, in *Acta numerica 1996*, A. Iserles, ed., Cambridge University Press, Cambridge, 45–119.

Gautschi, Walter [1997]. Moments in quadrature problems, *Comput. Math. Appl.*, to appear.

Gautschi, Walter and Milovanović, Gradimir V. [submitted]. *s-Orthogonality and construction of Gauss-Turán type quadrature formulae.*

Gautschi, W. and Montrone, M. [1980]. Metodi multistep con minimo coefficiente dell'errore globale, *Calcolo 17*, 67–75.

Gear, C. W. [1969]. *The automatic integration of stiff ordinary differential equations*, in *Information processing 68.* v. 1: *Mathematics, Software*, A.J.H. Morrell, ed., North-Holland, Amsterdam, 187–193.

Gear, C. William [1971a]. *Numerical initial value problems in ordinary differential equations*, Prentice-Hall Ser. Autom. Comput., Prentice-Hall, Englewood Cliffs, NJ.

Gear, C.W. [1971b]. The automatic integration of ordinary differential equations, *Comm. ACM 14*, 176–179.

Gear, C. W. [1971c]. Algorithm 407 — DIFSUB for solution of ordinary differential equations, *Comm. ACM 14*, 185–190.

Geddes, K.O., Czapor, S.R., and Labahn, G. [1992]. *Algorithms for computer algebra*, Kluwer, Boston, MA.

Gericke, Helmuth [1970]. *Geschichte des Zahlbegriffs*, Bibliographisches Institut, Mannheim.

Ghizzetti, A. and Ossicini, A. [1970]. *Quadrature formulae*, Academic Press, New York.

Gibbons, A. [1960]. A program for the automatic integration of differential equations using the method of Taylor series, *Comput. J. 3*, 108–111.

Gill, Philip E., Murray, Walter, and Wright, Margaret H. [1981]. *Practical optimization*, Academic Press, London.

Gill, Philip E., Murray, Walter, and Wright, Margaret H. [1991]. *Numerical linear algebra and optimization*, v. 1, Addison-Wesley, Redwood City, CA.

Girault, Vivette and Raviart, Pierre-Arnaud [1986]. *Finite element methods for Navier-Stokes equations: Theory and algorithms*, Springer Ser. Comput. Math., v. 5, Springer, Berlin.

Godlewski, Edwige and Raviart, Pierre-Arnaud [1996]. *Numerical approximation of hyperbolic systems of conservation laws*, Appl. Math. Sci., v. 118, Springer, New York.

Godunov, S.K. and Ryaben'kiĭ, V.S. [1987]. *Difference schemes: An introduction to the underlying theory*, translated from the Russian by E.M. Gelbard, Stud. Math. Appl., v. 19, North-Holland, Amsterdam.

Goldberg, Samuel [1986]. *Introduction to difference equations: With illustrative examples from economics, psychology, and sociology*, 2d ed., Dover, New York.

Goldstine, Herman H. [1977]. *A history of numerical analysis from the 16th through the 19th century*, Stud. Hist. Math. Phys. Sci., v. 2, Springer, New York.

Golub, Gene H. [1973]. Some modified matrix eigenvalue problems, *SIAM Rev. 15*, 318–334.

Golub, Gene H. and Ortega, James M. [1992]. *Scientific computing and differential equations: An introduction to numerical methods*, Academic Press, Boston, MA.

Golub, Gene and Ortega, James M. [1993]. *Scientific computing: An introduction with parallel computing*, Academic Press, Boston, MA.

Golub, Gene H. and Van Loan, Charles F. [1996]. *Matrix computations*, 3d ed., Johns Hopkins Ser. in Math. Sci., v. 3, Johns Hopkins University Press, Baltimore, MD.

Golub, Gene H. and Welsch, John H. [1969]. Calculation of Gauss quadrature rules, *Math. Comp. 23*, 221–230. Loose microfiche suppl. A1–A10.

Gottlieb, David and Orszag, Steven A. [1977]. *Numerical analysis of spectral methods: Theory and applications*, CBMS-NSF Regional Conf. Ser. in Appl. Math., no. 26, SIAM, Philadelphia, PA.

Gragg, William B. [1964]. *Repeated extrapolation to the limit in the numerical solution of ordinary differential equations*, Ph.D. thesis, University of California, Los Angeles, CA.

Gragg, William B. [1965]. On extrapolation algorithms for ordinary initial value problems, *J. Soc. Indust. Appl. Math. Ser. B Numer. Anal. 2*, 384–403.

Greenspan, Donald and Casulli, Vincenzo [1988]. *Numerical analysis for applied mathematics, science, and engineering*, Addison-Wesley, Advanced Book Program, Redwood City, CA.

Gregory, Robert T. and Karney, David L. [1978]. *A collection of matrices for testing computational algorithms*, corrected reprint of the 1969 edition, Krieger, Huntington, NY.

Griepentrog, Eberhard and März, Roswitha [1986]. *Differential-algebraic equations and their numerical treatment*, Teubner Texts Math., v. 88, Teubner, Leipzig.

Griewank, Andreas and Corliss, George F., eds. [1991]. *Automatic differentiation of algorithms: Theory, implementation, and application*, SIAM Proc. Ser., SIAM, Philadelphia, PA.

Griewank, Andreas, Juedes, David, and Utke, Jean [1997]. ADOL-C: a package for the automatic differentiation of algorithms written in C/C++, *ACM Trans. Math. Software*, to appear.

Gustafsson, Bertil, Kreiss, Heinz-Otto, and Oliger, Joseph [1995]. *Time dependent problems and difference methods*, Pure Appl. Math., Wiley-Intersci. Publ., Wiley, New York.

Hackbusch, Wolfgang [1985]. *Multigrid methods and applications*, Springer Ser. Comput. Math., v. 4, Springer, Berlin.

Hackbusch, Wolfgang [1994]. *Iterative solution of large sparse systems of equations*, translated and revised from the 1991 German original, Appl. Math. Sci., v. 95, Springer, New York.

Hageman, Louis A. and Young, David M. [1981]. *Applied iterative methods*, Comput. Sci. Appl. Math., Academic Press, New York.

Hairer, E. [1978]. A Runge-Kutta method of order 10, *J. Inst. Math. Appl. 21*, 47–59.

Hairer, E. and Wanner, G. [1996]. *Solving ordinary differential equations. II: Stiff and differential-algebraic problems*, 2d ed., Springer Ser. Comput. Math., v. 14, Springer, Berlin.

Hairer, Ernst, Lubich, Christian, and Roche, Michel [1989]. *The numerical solution of differential-algebraic systems by Runge-Kutta methods*, Lecture Notes in Math., v. 1409, Springer, Berlin.

Hairer, E., Nørsett, S.P., and Wanner, G. [1993]. *Solving ordinary differential equations. I: Nonstiff problems*, 2d ed., Springer Ser. Comput. Math., v. 8, Springer, Berlin.

Hall, Charles A. and Meyer, W.Weston [1976]. Optimal error bounds for cubic spline interpolation, *J. Approx. Theory 16*, 105–122.

Hall, Charles A. and Porsching, Thomas A. [1990]. *Numerical analysis of partial differential equations*, Prentice-Hall, Englewood Cliffs, NJ.

Hall, W.S. [1994]. *The boundary element method*, Solid Mech. Appl., v. 27, Kluwer, Dordrecht.

Hammer, R., Hocks, M., Kulisch, U., and Ratz, D. [1993]. *Numerical toolbox for verified computing. I: Basic numerical problems, theory, algorithms, and Pascal-XSC programs*, Springer Ser. Comput. Math., v. 21, Springer, Berlin.

Hammer, R., Hocks, M., Kulisch, U., and Ratz, D. [1995]. *C++ toolbox for verified computing. I: Basic numerical problems, theory, algorithms, and programs*, Springer, Berlin.

Hämmerlin, Günther and Hoffmann, Karl-Heinz [1991]. *Numerical mathematics*, translated from the German by Larry Schumaker, Undergrad. Texts Math., Readings in Math., Springer, New York.

Hansen, Eldon [1992]. *Global optimization using interval analysis*, Monographs and Textbooks Pure Appl. Math., v. 165, Dekker, New York.

Hartmann, Friedel [1989]. *Introduction to boundary elements: Theory and applications*, with an appendix by Peter Schoepp, Springer, Berlin.

Heath, Michael T. [1997]. *Scientific computing: An introductory survey*, McGraw-Hill Ser. Comput. Sci., McGraw-Hill, New York.

Heck, André [1996]. *Introduction to Maple*, 2d ed., Springer, New York.

Heiberger, Richard M. [1989]. *Computation for the analysis of designed experiments*, Wiley Ser. Probab. Math. Statist.: Appl. Probab. Statist., Wiley, New York.

Henrici, Peter [1962]. *Discrete variable methods in ordinary differential equations*, Wiley, New York.

Henrici, Peter [1977]. *Error propagation for difference methods*, reprint of the 1963 edition, Krieger, Huntington, NY.

Henrici, Peter [1979a]. Fast Fourier methods in computational complex analysis, *SIAM Rev. 21*, 481–527.

Henrici, Peter [1979b]. Barycentric formulas for interpolating trigonometric polynomials and their conjugates, *Numer. Math. 33*, 225–234.

Henrici, Peter [1988, 1991, 1986]. *Applied and computational complex analysis.* v 1: *Power series — integration — conformal mapping — location of zeros*, reprint of the 1974 original, Wiley Classics Lib., Wiley-Intersci. Publ., Wiley, New York; v. 2: *Special functions — integral transforms — asymptotics —*

continued fractions, reprint of the 1977 original, Wiley Classics Lib., Wiley-Intersci. Publ., Wiley, New York; v. 3: *Discrete Fourier analysis — Cauchy integrals — construction of conformal maps — univalent functions*, Pure Appl. Math., Wiley-Intersci. Publ., Wiley, New York.

Hermite, Charles and Stieltjes, Thomas Jan [1905]. *Correspondance d'Hermite et de Stieltjes*, published by B. Baillaud and H. Bourget, with a preface by Emile Picard, vols. 1,2, Gauthier-Villars, Paris. [Partie inédite de la correspondance d'Hermite avec Stieltjes, *Proc. Sem. History Math. 4*, Inst. Henri Poincaré, Paris, 1983, 75–87.]

Hestenes, Magnus Rudolf [1980]. *Conjugate direction methods in optimization*, Appl. Math., v. 12, Springer, New York.

Hewitt, Edwin and Stromberg, Karl [1975]. *Real and abstract analysis: A modern treatment of the theory of functions of a real variable*, 3d printing, Graduate Texts in Math., no. 25, Springer, New York.

Higham, Nicholas J. [1996]. *Accuracy and stability of numerical algorithms*, SIAM, Philadelphia, PA.

Hindmarsh, Allan C. [1980]. LSODE and LSODI, two new initial value ordinary differential equation solvers, *ACM SIGNUM Newsletter 15*, no. 4, 10–11.

Holladay, John C. [1957]. A smoothest curve approximation, *Math. Tables Aids Comput. 11*, 233–243.

Hoschek, Josef and Lasser, Dieter [1993]. *Fundamentals of computer aided geometric design*, translated from the 1992 German edition by Larry L. Schumaker, A K Peters, Wellesley, MA.

Householder, A.S. [1970]. *The numerical treatment of a single nonlinear equation*, Internat. Ser. Pure Appl. Math., McGraw-Hill, New York.

Hu, T.C. [1982]. *Combinatorial algorithms*, Addison-Wesley, Reading, MA.

Hughes, Thomas J.R. [1987]. *The finite element method: Linear static and dynamic finite element analysis*, with the collaboration of Robert M. Ferencz and Arthur M. Raefsky, Prentice-Hall, Englewood Cliffs, NJ.

Hull, T.E. and Luxemburg, W.A.J. [1960]. Numerical methods and existence theorems for ordinary differential equations, *Numer. Math. 2*, 30–41.

IEEE [1985]. *IEEE standard for binary floating-point arithmetic, ANSI/IEEE standard 754-1985*, Institute of Electrical and Electronics Engineers, New York. [Reprinted in *SIGPLAN Notices 22*, no. 2 (1987), 9–25.]

Il'in, V.P. [1992]. *Iterative incomplete factorization methods*, Ser. Soviet East European Math., v. 4, World Scientific, River Edge, NJ.

Imhof, J.P. [1963]. On the method for numerical integration of Clenshaw and Curtis, *Numer. Math. 5*, 138–141.

Isaacson, Eugene and Keller, Herbert Bishop [1994]. *Analysis of numerical methods*, corrected reprint of the 1966 original, Dover, New York.

Iserles, Arieh, ed. [1992–1996]. *Acta numerica*, Cambridge University Press, Cambridge.

Iserles, Arieh [1996]. *A first course in the numerical analysis of differential equations*, Cambridge Texts Appl. Math., Cambridge University Press, Cambridge.

Iserles, A. and Nørsett, S.P. [1991]. *Order stars*, Appl. Math. Math. Comput., v. 2, Chapman & Hall, London.

Jacobi, C.G.J. [1826]. Ueber Gaußs neue Methode, die Werthe der Integrale näherungsweise zu finden, *J. Reine Angew. Math. 1*, 301–308.

Jenkins, M.A. and Traub, J.F. [1970]. A three-stage algorithm for real polynomials using quadratic iteration, *SIAM J. Numer. Anal. 7*, 545–566.

Jennings, Alan and McKeown, J.J. [1992]. *Matrix computation*, 2d ed., Wiley, Chichester.

Johnson, Claes [1987]. *Numerical solution of partial differential equations by the finite element method*, Cambridge University Press, Cambridge.

Joyce, D.C. [1971]. Survey of extrapolation processes in numerical analysis, *SIAM Rev. 13*, 435–490.

Kalos, Malvin H. and Whitlock, Paula A. [1986]. *Monte Carlo methods.* v. 1: *Basics*, Wiley-Intersci. Publ., Wiley, New York.

Kantorovich, L.V. and Akilov, G.P. [1982]. *Functional analysis*, translated from the Russian by Howard L. Silcock, 2d ed., Pergamon Press, Oxford.

Kaucher, Edgar W. and Miranker, Willard L. [1984]. *Self-validating numerics for function space problems: Computation with guarantees for differential and integral equations*, Notes Rep. Comput. Sci. Appl. Math., v. 9, Academic Press, Orlando, FL.

Kautsky, J. and Elhay, S. [1982]. Calculation of the weights of interpolatory quadratures, *Numer. Math. 40*, 407–422.

Kedem, Gershon [1980]. Automatic differentiation of computer programs, *ACM Trans. Math. Software 6*, 150–165.

Keller, Herbert B. [1992]. *Numerical methods for two-point boundary-value problems*, corrected reprint of the 1968 edition, Dover, New York.

Kelley, C.T. [1995]. *Iterative methods for linear and nonlinear equations*, Frontiers Appl. Math., v. 16, SIAM, Philadelphia, PA.

Kelley, Walter G. and Peterson, Allan C. [1991]. *Difference equations: An introduction with applications*, Academic Press, Boston, MA.

Kennedy, William J., Jr. and Gentle, James E. [1980]. *Statistical computing*, Statistics: Textbooks and Monographs, v. 33, Dekker, New York.

Kerner, Immo O. [1966]. Ein Gesamtschrittverfahren zur Berechnung der Nullstellen von Polynomen, *Numer. Math. 8*, 290–294.

Kershaw, D. [1971]. A note on the convergence of interpolatory cubic splines, *SIAM J. Numer. Anal. 8*, 67–74.

Kincaid, David and Cheney, Ward [1996]. *Numerical analysis: Mathematics of scientific computing*, 2d ed., Brooks/Cole, Pacific Grove, CA.

King, J. Thomas and Murio, Diego A. [1986]. Numerical differentiation by finite-dimensional regularization, *IMA J. Numer. Anal. 6*, 65–85.

Klopfenstein, R.W. [1965]. Applications of differential equations in general problem solving, *Comm. ACM 8*, 575–578.

Knuth, Donald E. [1975, 1981, 1973]. *The art of computer programming.* v. 1: *Fundamental algorithms*, 2d printing of the 2d ed.; v. 2: *Seminumerical algorithms*, 2d ed.; v. 3: *Sorting and searching*, Addison-Wesley Ser. Comput. Sci. Inform. Process., Addison-Wesley, Reading, MA.

Ko, Ker-I [1991]. *Complexity theory of real functions*, Progr. Theoret. Comput. Sci., Birkhäuser, Boston, MA.

Köckler, Norbert [1994]. *Numerical methods and scientific computing: Using software libraries for problem solving*, Oxford Sci. Publ., Oxford University Press, New York.

Kollerstrom, Nick [1992]. Thomas Simpson and "Newton's method of approximation": An enduring myth, *British J. Hist. Sci. 25*, 347–354.

Křížek, M. and Neittaanmäki, P. [1990]. *Finite element approximation of variational problems and applications*, Pitman Monographs Surveys Pure Appl. Math., v. 50, Wiley, New York.

Kronsjö, Lydia [1987]. *Algorithms: Their complexity and efficiency*, 2d ed., Wiley Ser. Comput., Wiley, Chichester.

Krylov, Vladimir Ivanovich [1962]. *Approximate calculation of integrals*, translated from the Russian by Arthur H. Stroud, ACM Monograph Ser., Macmillan, New York.

Kuang, Yang [1993]. *Delay differential equations with applications in population dynamics*, Math. Sci. Engrg., v. 191, Academic Press, Boston, MA.

Kulisch, Ulrich W. and Miranker, Willard L. [1981]. *Computer arithmetic in theory and practice*, Comput. Sci. Appl. Math., Academic Press, New York.

Lakshmikantham, V. and Trigiante, D. [1988]. *Theory of difference equations: Numerical methods and applications*, Math. Sci. Engrg., v. 181, Academic Press. Boston, MA.

Lambert, J.D. [1991]. *Numerical methods for ordinary differential systems: The initial value problem*, Wiley, Chichester.

Laurie, D.P. [1997]. Calculation of Gauss-Kronrod quadrature rules, *Math. Comp. 66*, to appear.

Lax, Peter D. [1973]. *Hyperbolic systems of conservation laws and the mathematical theory of shock waves*, CBMS Regional Conf. Ser. in Appl. Math., no. 11, SIAM, Philadelphia, PA.

Legendre, A.M. [1805]. *Nouvelles méthodes pour la détermination des orbites des comètes*, Courcier, Paris.

Lehmer, D.H. [1961]. A machine method for solving polynomial equations, *J. Assoc. Comput. Mach. 8*, 151–162.

LeVeque, Randall J. [1992]. *Numerical methods for conservation laws*, 2d ed., Lectures in Math. ETH Zürich, Birkhäuser, Basel.

Levin, Meishe and Girshovich, Jury [1979]. *Optimal quadrature formulas*, Teubner-Texte Math., Teubner, Leipzig.

Linz, Peter [1985]. *Analytical and numerical methods for Volterra equations*, SIAM Stud. Appl. Math., v. 7, SIAM, Philadelphia, PA.

Lorentz, George G., Golitschek, Manfred v., and Makovoz, Yuly [1996]. *Constructive approximation: Advanced problems*, Grundlehren Math. Wiss., v. 304, Springer, Berlin.

Lorentz, George G., Jetter, Kurt, and Riemenschneider, Sherman D. [1983]. *Birkhoff interpolation*, Encyclopedia Math. Appl., v. 19, Addison-Wesley, Reading, MA.

Lorentz, Rudolph A. [1992]. *Multivariate Birkhoff interpolation*, Lecture Notes in Math., v. 1516, Springer, Berlin.

Lozier, D.W. and Olver, F.W.J. [1994]. *Numerical evaluation of special functions*, in Gautschi, ed. [1994a, 79–125].

Luke, Yudell L. [1975]. *Mathematical functions and their approximations*, Academic Press, New York.

Luke, Yudell L. [1977]. *Algorithms for the computation of mathematical functions*, Academic Press, New York.

Lyness, J.N. [1968]. Differentiation formulas for analytic functions, *Math. Comp. 22*, 352–362.

Lyness, J.N. and Moler, C.B. [1967]. Numerical differentiation of analytic functions, *SIAM J. Numer. Anal. 4*, 202–210.

Lyness, J.N. and Sande, G. [1971]. Algorithm 413 — ENTCAF and ENTCRE: Evaluation of normalized Taylor coefficients of an analytic function, *Comm. ACM 14*, 669–675.

Maas, Christoph [1985]. Was ist das Falsche an der Regula Falsi?, *Mitt. Math. Ges. Hamburg 11*, 311–317.

MacLeod, Allan J. [1996]. Algorithm 757: MISCFUN, a software package to compute uncommon special functions, *ACM Trans. Math. Software 22*, 288–301.

Maindonald, J.H. [1984]. *Statistical computation*, Wiley Ser. Probab. Math. Statist.: Appl. Probab. Statist., Wiley, New York.

Mäntylä, Martti [1988]. *An introduction to solid modeling*, Principles Comput. Sci. Ser., v. 13, Computer Science Press, Rockville, MD.

Marden, Morris [1966]. *Geometry of polynomials*, 2d. ed., Math. Surveys, no. 3, Amer. Math. Soc., Providence, RI.

McCormick, Stephen F. [1989]. *Multilevel adaptive methods for partial differential equations*, Frontiers in Appl. Math., v. 6, SIAM, Philadelphia, PA.

McCormick, Stephen F. [1992]. *Multilevel projection methods for partial differential equations*, CBMS-NSF Regional Conf. Ser. in Appl. Math., no. 62, SIAM, Philadelphia, PA.

Mercier, Bertrand [1989]. *An introduction to the numerical analysis of spectral methods*, translated from the French, edited and with a preface by Nessan Mac Giolla Mhuiris and Mohammed Youssuff Hussaini, Lecture Notes in Phys., v. 318, Springer, Berlin.

Meyer, Arnd [1987]. *Modern algorithms for large sparse eigenvalue problems*, Math. Res., v. 34, Akademie-Verlag, Berlin.

Meyer, Gunter H. [1973]. *Initial value methods for boundary value problems: Theory and application of invariant imbedding*, Math. Sci. Engrg., v. 100, Academic Press, New York.

Miel, George and Mooney, Rose [1985]. On the condition number of Lagrangian numerical differentiation, *Appl. Math. Comput. 16*, 241–252.

Mignotte, Maurice [1992]. *Mathematics for computer algebra*, translated from the French by Catherine Mignotte, Springer, New York.

Miller, Kenneth S. [1968]. *Linear difference equations*, Benjamin, New York.

Milne, W.E. [1926]. Numerical integration of ordinary differential equations, *Amer. Math. Monthly 33*, 455–460.

Milovanović, G.V., Mitrinović, D.S., and Rassias, Th.M. [1994]. *Topics in polynomials: Extremal problems, inequalities, zeros*, World Scientific, River Edge, NJ.

Mishra, Bhubaneswar [1993]. *Algorithmic algebra*, Texts Monographs Comput. Sci., Springer, New York.

Moore, Ramon E. [1966]. *Interval analysis*, Prentice-Hall Ser. Automat. Comput., Prentice-Hall, Englewood Cliffs, NJ.

Moore, Ramon E. [1979]. *Methods and applications of interval analysis*, SIAM Stud. Appl. Math., v. 2, SIAM, Philadelphia, PA.

Moré, Jorge J. and Wright, Stephen J. [1993]. *Optimization software guide*, Frontiers Appl. Math., v. 14, SIAM, Philadelphia, PA.

Morgan, Alexander [1987]. *Solving polynomial systems using continuation for engineering and scientific problems*, Prentice-Hall, Englewood Cliffs, NJ.

Morrison, David D., Riley, James D., and Zancanaro, John F. [1962]. Multiple shooting method for two-point boundary value problems, *Comm. ACM 5*, 613–614.

Morton, K.W. and Mayers, D.F. [1994]. *Numerical solution of partial differential equations: An introduction*, Cambridge University Press, Cambridge.

Moulton, Forest Ray [1926]. *New methods in exterior ballistics*, University of Chicago Press, Chicago, IL.

Muller, David E. [1956]. A method for solving algebraic equations using an automatic computer, *Math. Tables Aids Comput. 10*, 208–215.

Mysovskikh, I.P. [1981]. *Interpolatory cubature formulas* (Russian), Nauka, Moscow.

Na, Tsung Yen [1979]. *Computational methods in engineering boundary value problems*, Math. Sci. Engrg., v. 145, Academic Press, New York.

Nash, Stephen G., ed. [1990]. *A history of scientific computing*, ACM Press History Ser., Addison-Wesley, Reading, MA.

Natanson, I.P. [1964, 1965, 1965]. *Constructive function theory.* v. 1: *Uniform approximation*, translated from the Russian by Alexis N. Obolensky; v. 2: *Approximation in mean*, translated from the Russian by John R. Schulenberger; v. 3: *Interpolation and approximation quadratures*, translated from the Russian by John R. Schulenberger, Ungar, New York.

Nazareth, J.L. [1987]. *Computer solution of linear programs*, Monographs Numer. Anal., Oxford Sci. Publ., Oxford University Press, New York.

Németh, Géza [1992]. *Mathematical approximation of special functions: Ten papers on Chebyshev expansions*, Nova Science, Commack, NY.

Neumaier, Arnold [1990]. *Interval methods for systems of equations*, Encyclopedia Math. Appl., v. 37, Cambridge University Press, Cambridge.

Niederreiter, Harald [1992]. *Random number generation and quasi-Monte Carlo methods*, CBMS-NSF Regional Conf. Ser. in Appl. Math., no. 63, SIAM, Philadelphia, PA.

Nijenhuis, Albert and Wilf, Herbert S. [1978]. *Combinatorial algorithms: For computers and calculators*, 2d ed., Comput. Sci. Appl. Math., Academic Press, New York.

Nikol'skiĭ, S.M. [1988]. *Quadrature formulas* (Russian), 4th ed., with an appendix by N.P. Korneĭchuk and a preface by Nikol'skiĭ and Korneĭchuk, Nauka, Moscow.

Niven, Ivan, Zuckerman, Herbert S., and Montgomery, Hugh L. [1991]. *An introduction to the theory of numbers*, 5th ed., Wiley, New York.

Nørsett, S.P. [1974]. *Semi explicit Runge-Kutta methods*, Rep. No. 6/74, ISBN 82-7151-009-6, Dept. Math., Univ. Trondheim, Norway.

Nørsett, Syvert P. [1976]. Runge-Kutta methods with a multiple real eigenvalue only, *BIT 16*, 388–393.

Notaris, Sotirios E. [1994]. *An overview of results on the existence or nonexistence and the error term of Gauss-Kronrod quadrature formulae*, in *Approximation and computation: A festschrift in honor of Walter Gautschi*, R.V.M. Zahar, ed., Internat. Ser. Numer. Math., v. 119, Birkhäuser, Boston, 485–496.

Nürnberger, Günther [1989]. *Approximation by spline functions*, Springer, Berlin.

Oden, J. Tinsley [1983]. *Finite elements.* v. IV: *Mathematical aspects*, in collaboration with Graham F. Carey, The Texas Finite Element Series, v. 4, Prentice-Hall, Englewood Cliffs, NJ.

Oden, J. Tinsley and Carey, Graham F. [1984]. *Finite elements.* v. V: *Special problems in solid mechanics*, The Texas Finite Element Series, v. 5, Prentice-Hall, Englewood Cliffs, NJ.

Oliver, J. [1980]. Algorithm 017 — An algorithm for numerical differentiation of a function of one real variable, *J. Comp. Appl. Math. 6*, 145–160.

Ortega, James M. [1989]. *Introduction to parallel and vector solution of linear systems*, Frontiers Comput. Sci., Plenum, New York.

Ortega, J.M. and Rheinboldt, W.C. [1970]. *Iterative solution of nonlinear equations in several variables*, Monographs Textbooks Comput. Sci. Appl. Math., Academic Press, New York.

Ortega, James M. and Voigt, Robert G. [1985]. *Solution of partial differential equations on vector and parallel computers*, SIAM, Philadelphia, PA.

Østerby, Ole and Zlatev, Zahari [1983]. *Direct methods for sparse matrices*, Lecture Notes in Comput. Sci., v. 157, Springer, Berlin.

Ostrowski, A. [1954]. *On two problems in abstract algebra connected with Horner's rule*, in *Studies in mathematics and mechanics presented to Richard von Mises*, Academic Press, New York, 40–48. [Collected Math. Papers, v. 2, Birkhäuser, Basel, 1983, 510–518.]

Ostrowski, A.M. [1973]. *Solution of equations in Euclidean and Banach spaces*, 3d ed. of *Solution of equations and systems of equations*, Pure Appl. Math., v. 9, Academic Press, New York.

Panik, Michael J. [1996]. *Linear programming: Mathematics, theory, and algorithms*, Applied Optimization, v. 2, Kluwer, Dordrecht.

Pardalos, P.M., Phillips, A.T., and Rosen, J.B. [1992]. *Topics in parallel computing in mathematical programming*, Appl. Discr. Math. Theoret. Comput. Sci., v. 2, Science Press, New York.

Parlett, Beresford N. [1980]. *The symmetric eigenvalue problem*, Prentice-Hall Ser. in Comput. Math., Prentice-Hall, Englewood Cliffs, NJ.

Paszkowski, Stefan [1983]. *Computational applications of Chebyshev polynomials and series* (Russian), translation from the Polish by S.N. Kiro, Nauka, Moscow.

Perron, Oskar [1957]. *Die Lehre von den Kettenbrüchen*, 3d ed.. v. 2: *Analytisch-funktionentheoretische Kettenbrüche*, Teubner, Stuttgart.

Peters, G. and Wilkinson, J.H. [1971]. Practical problems arising in the solution of polynomial equations, *J. Inst. Math. Appl. 8*, 16–35.

Petković, Miodrag [1989]. *Iterative methods for simultaneous inclusion of polynomial zeros*, Lecture Notes in Math., v. 1387, Springer, Berlin.

Petrushev, P.P. and Popov, V.A. [1987]. *Rational approximation of real functions*, Encyclopedia Math. Appl., v. 28, Cambridge University Press, Cambridge.

Piessens, Robert, de Doncker-Kapenga, Elise, Überhuber, Christoph W., and Kahaner, David K. [1983]. *QUADPACK: A subroutine package for automatic integration*, Springer Ser. Comput. Math., v. 1, Springer, Berlin.

Pinkus, Allan M. [1989]. *On L^1-approximation*, Cambridge Tracts in Math., v. 93, Cambridge University Press, Cambridge.

Pinney, Edmund [1958]. *Ordinary difference-differential equations*, University of California Press, Berkeley, CA.

Plofker, Kim [1996]. An example of the secant method of iterative approximation in a fifteenth-century Sanskrit text, *Historia Math. 23*, 246–256.

Pohst, Michael E. [1993]. *Computational algebraic number theory*, DMV Seminar, v. 21, Birkhäuser, Basel.

Pomerance, Carl, ed. [1990]. *Cryptology and computational number theory*, Proc. Sympos. Appl. Math., v. 42, Amer. Math. Soc., Providence, RI.

Potra, F.A. [1989]. On Q-order and R-order of convergence, *J. Optim. Theory Appl. 63*, 415–431.

Potra, F.-A. and Pták, V. [1984]. *Nondiscrete induction and iterative processes*, Res. Notes Math., v. 103, Pitman, Boston.

Powell, M.J.D. [1981]. *Approximation theory and methods*, Cambridge University Press, Cambridge.

Preparata, Franco P. and Shamos, Michael Ian [1985]. *Computational geometry: An introduction*, Texts Monographs Comput. Sci., Springer, New York.

Prince, P.J. and Dormand, J.R. [1981]. High order embedded Runge-Kutta formulae, *J. Comput. Appl. Math. 7*, 67–75.

Prössdorf, Siegfried and Silbermann, Bernd [1991]. *Numerical analysis for integral and related operator equations*, Operator Theory: Adv. Appl., v. 52, Birkhäuser, Basel.

Pryce, John D. [1993]. *Numerical solution of Sturm-Liouville problems*, Monographs Numer. Anal., Oxford Sci. Publ., Oxford University Press, New York.

Quade, W. [1951]. Numerische Integration von Differentialgleichungen bei Approximation durch trigonometrische Ausdrücke, *Z. Angew. Math. Mech. 31*, 237–238.

Quarteroni, Alfio and Valli, Alberto [1994]. *Numerical approximation of partial differential equations*, Springer Ser. Comput. Math., v. 23, Springer, Berlin.

Rabinowitz, Philip [1992]. Extrapolation methods in numerical integration, *Numer. Algorithms 3*, 17–28.

Rakitskiĭ, Ju. V. [1961]. Some properties of solutions of systems of ordinary differential equations by one-step methods of numerical integration (Russian), *Ž. Vyčisl. Mat. i Mat. Fiz. 1*, 947–962. [Engl. transl. in: *U.S.S.R. Comput. Math. Math. Phys. 1* (1962), 1113–1128.]

Ratschek, H. and Rokne, J. [1984]. *Computer methods for the range of functions*, Ellis Horwood Ser. Math. Appl., Horwood, Chichester.

Ratschek, H. and Rokne, J. [1988]. *New computer methods for global optimization*, Ellis Horwood Ser. Math. Appl., Horwood, Chichester.

Reddien, G.W. [1980]. Projection methods for two-point boundary value problems, *SIAM Rev. 22*, 156–171.

Resnikoff, Howard L. and Wells, Raymond O., Jr. [1997]. *Wavelet analysis and the scalable structure of information*, Springer, New York.

Rice, John R. [1966]. A theory of condition, *SIAM J. Numer. Anal. 3*, 287–310.

Rice, John R. and Boisvert, Ronald F. [1985]. *Solving elliptic problems using ELLPACK*, with appendices by W.R. Dyksen, E.N. Houstis, Rice, J.F. Brophy, C.J. Ribbens, and W.A. Ward, Springer Ser. Comput. Math., v. 2, Springer, Berlin.

Riesel, Hans [1994]. *Prime numbers and computer methods for factorization*, 2d ed., Progr. Math., v. 126, Birkhäuser, Boston, MA.

Rivlin, T.J. [1975]. Optimally stable Lagrangian numerical differentiation, *SIAM J. Numer. Anal. 12*, 712–725.

Rivlin, Theodore J. [1981]. *An introduction to the approximation of functions*, corrected reprint of the 1969 original, Dover Books Adv. Math., Dover, New York.

Rivlin, Theodore J. [1990]. *Chebyshev polynomials: from approximation theory to algebra and number theory*, 2d ed., Pure Appl. Math., Wiley, New York.

Roberts, Sanford M. and Shipman, Jerome S. [1972]. *Two-point boundary value problems: Shooting methods*, Modern Analyt. Comput. Methods in Sci. Math., no. 31, American Elsevier, New York.

Rosen, Kenneth H. [1993]. *Elementary number theory and its applications*, 3d ed., Addison-Wesley, Advanced Book Program, Reading, MA.

Rudin, Walter [1976]. *Principles of mathematical analysis*, 3d ed., Internat. Ser. Pure Appl. Math., McGraw-Hill, New York.

Rutishauser, Heinz [1952]. Über die Instabilität von Methoden zur Integration gewöhnlicher Differentialgleichungen, *Z. Angew. Math. Physik 3*, 65–74.

Rutishauser, Heinz [1957]. *Der Quotienten-Differenzen-Algorithmus*, Mitt. Inst. Angew. Math. Zürich, no. 7.

Rutishauser, Heinz [1990]. *Lectures on numerical mathematics*, edited by M. Gutknecht with the assistance of Peter Henrici, Peter Läuchli, and Hans-Rudolf Schwarz; with a foreword by Gutknecht and a preface by Henrici, Läuchli, and Schwarz; translated from the German and with a preface by Walter Gautschi, Birkhäuser, Boston, MA.

Saad, Youcef [1992]. *Numerical methods for large eigenvalue problems*, Algorithms Architect. Adv. Sci. Comput., Halsted Press, New York.

Saad, Youcef [1996]. *Iterative methods for sparse linear systems*, PWS Publ. Co., Boston, MA.

Salihov, N.P. [1962]. Polar difference methods for solving the Cauchy problem for a system of ordinary differential equations (Russian), *Ž. Vyčisl. Mat. i Mat. Fiz. 2*, 515–528. [Engl. tranl. in: *U.S.S.R. Comput. Math. Math. Phys. 2* (1963), 535–553.]

Salzer, Herbert E. [1949]. Coefficients for facilitating trigonometric interpolation, *J. Math. Physics 27*, 274–278.

Sanz-Serna, J.M. and Calvo, M.P. [1994]. *Numerical Hamiltonian problems*, Appl. Math. Math. Comput., v. 7, Chapman & Hall, London.

Schendel, U. [1984]. *Introduction to numerical methods for parallel computers*, translated from the German by B.W. Conolly, Ellis Horwood Ser. Math. Appl., Halsted Press, New York.

Schiesser, W.E. [1991]. *The numerical method of lines: Integration of partial differential equations*, Academic Press, San Diego, CA.

Schinzinger, Roland and Laura, Patricio A.A. [1991]. *Conformal mapping: Methods and applications*, Elsevier, Amsterdam.

Schoenberg, I.J. [1946]. Contributions to the problem of approximation of equidistant data by analytic functions. Part A: On the problem of smoothing or graduation. A first class of analytic approximation formulae, *Quart. Appl. Math. 4*, 45–99; Part B: On the problem of osculatory interpolation. A second class of analytic approximation formulae, *ibid.*, 112–141.

Schoenberg, I.J. [1973]. *Cardinal spline interpolation*, CBMS Regional Conf. Ser. in Appl. Math., no. 12, SIAM, Philadelphia, PA.

Schumaker, Larry L. [1993]. *Spline functions: Basic theory*, correlated reprint of the 1981 original, Krieger, Malabar, FL.

Schwarz, H.-R. [1988]. *Finite element methods*, translated from the German by Caroline M. Whiteman; translation edited by J.R. Whiteman, Comput. Math. Appl., Academic Press, London.

Schwarz, H.-R. [1989]. *Numerical analysis: A comprehensive introduction*, with a contribution by J. Waldvogel; translated from the German, Wiley, Chichester.

Sehmi, N.S. [1989]. *Large order structural eigenanalysis techniques: Algorithms for finite element systems*, Ellis Horwood Ser. Math. Appl., Halsted Press, New York.

Sewell, Granville [1985]. *Analysis of a finite element method: PDE/PROTRAN*, Springer, New York.

Sewell, Granville [1988]. *The numerical solution of ordinary and partial differential equations*, Academic Press, Boston, MA.

Shaĭdurov, V.V. [1995]. *Multigrid methods for finite elements*, translated from the 1989 Russian original by N.B. Urusova and revised by the author, Math. Appl., v. 318, Kluwer, Dordrecht.

Shampine, L.F. [1975]. Discrete least squares polynomial fits, *Comm. ACM 18*, 179–180.

Shampine, Lawrence F. [1994]. *Numerical solution of ordinary differential equations*, Chapman & Hall, New York.

Shampine, L.F. and Gordon, M.K. [1975]. *Computer solution of ordinary differential equations: The initial value problem*, Freeman, San Francisco, CA.

Shampine, L.F. and Wisniewski, J.A. [1978]. *The variable order Runge-Kutta code RKSW and its performance*, Sandia Rep. SAND78-1347, Sandia National Laboratories, Albuquerque, NM.

Sheynin, Oscar [1993]. On the history of the principle of least squares, *Arch. Hist. Exact Sci. 46*, 39–54.

Sima, Vasile [1996]. *Algorithms for linear-quadratic optimization*, Monographs Textbooks Pure Appl. Math., v. 200, Dekker, New York.

Sloan, I.H. and Joe, S. [1994]. *Lattice methods for multiple integration*, Clarendon Press, Oxford.

Smale, Steve [1987]. *Algorithms for solving equations*, Proc. Internat. Congress Mathematicians, v. 1, Amer. Math. Soc., Providence, RI, 172–195.

Smith, Barry F., Bjørstad, Petter E., and Gropp, William D. [1996]. *Domain decomposition: Parallel multilevel methods for elliptic partial differential equations*, Cambridge University Press, Cambridge.

Smith, B.T., Boyle, J.M., Dongarra, J.J., Garbow, B.S., Ikebe, Y., Klema, V.C., and Moler, C.B. [1976]. *Matrix eigensystem routines — EISPACK guide*, 2d ed., Lecture Notes in Comput. Sci., v. 6, Springer, Berlin.

Sobol', Ilya M. [1994]. *A primer for the Monte-Carlo method*, CRC Press, Boca Raton, FL.

Späth, Helmuth [1992]. *Mathematical algorithms for linear regression*, translated and revised from the 1987 German original by the author, Comput. Sci. Sci. Comput., Academic Press, Boston, MA.

Späth, Helmuth [1995]. *One-dimensional spline interpolation algorithms*, with the collaboration of Jörg Meier; translated from the German by Len Bos, A K Peters, Wellesley, MA.

Steffensen, J.F. [1950]. *Interpolation*, 2d ed., Chelsea, New York.

Stegun, Irene A. and Abramowitz, Milton [1956]. Pitfalls in computation, *J. Soc. Indust. Appl. Math. 4*, 207–219.

Stenger, Frank [1993]. *Numerical methods based on sinc and analytic functions*, Springer Ser. Comput. Math., v. 20, Springer, New York.

Stepleman, R.S. and Winarsky, N.D. [1979]. Adaptive numerical differentiation, *Math. Comp. 33*, 1257–1264.

Sterbenz, Pat H. [1974]. *Floating-point computation*, Prentice-Hall Ser. Automat. Comput., Prentice-Hall, Englewood Cliffs, NJ.

Stetter, Hans J. [1973]. *Analysis of discretization methods for ordinary differential equations*, Springer Tracts Nat. Philos., v. 23, Springer, New York.

Stewart, G.W. [1973]. *Introduction to matrix computations*, Comput. Sci. Appl. Math., Academic Press, New York.

Stewart, William J. [1994]. *Introduction to the numerical solution of Markov chains*, Princeton University Press, Princeton, NJ.

Stiefel, E. and Bettis, D.G. [1969]. Stabilization of Cowell's method, *Numer. Math. 13*, 154–175.

Stiefel, E.L. and Scheifele, G. [1971]. *Linear and regular celestial mechanics: Perturbed two-body motion, numerical methods, canonical theory*, Grundlehren Math. Wiss., v. 174, Springer, New York.

Stoer, J. and Bulirsch, R. [1993]. *Introduction to numerical analysis*, translated from the German by R. Bartels, W. Gautschi, and C. Witzgall, 2d ed., Texts Appl. Math., v. 12, Springer, New York.

Strikwerda, John C. [1989]. *Finite difference schemes and partial differential equations*, The Wadsworth & Brooks/Cole Math. Ser., Wadsworth & Brooks/Cole, Pacific Grove, CA.

Stroud, A.H. [1971]. *Approximate calculation of multiple integrals*, Prentice-Hall Ser. Automat. Comput., Prentice-Hall, Englewood Cliffs, NJ.

Stroud, A.H. and Secrest, Don [1966]. *Gaussian quadrature formulas*, Prentice-Hall, Englewood Cliffs, NJ.

Szabados, J. and Vértesi, P. [1990]. *Interpolation of functions*, World Scientific, Teaneck, NJ.

Szegő, Gabriel [1936]. On some Hermitian forms associated with two given curves of the complex plane, *Trans. Amer. Math. Soc. 40*, 450–461. [Collected Papers, Richard Askey, ed., v. 2, Birkhäuser, Boston, 1982, 666–683.]

Szegö, Gabor [1975]. *Orthogonal polynomials*, 4th ed., Amer. Math. Soc. Colloq. Publ., v. 23, Amer. Math. Soc., Providence, RI.

Taylor, Walter F. [1992]. *The geometry of computer graphics*, The Wadsworth & Brooks/Cole Math. Ser., Wadsworth & Brooks/Cole, Pacific Grove, CA.

Thisted, Ronald A. [1988]. *Elements of statistical computing: Numerical computation*, Chapman & Hall, New York.

Thomas, J.W. [1995]. *Numerical partial differential equations: Finite difference methods*, Texts Appl. Math., v. 22, Springer, New York.

Tihonov, A.N. and Gorbunov, A.D. [1963]. Asymptotic error bounds for the Runge-Kutta method (Russian), *Ž. Vyčisl. Mat. i Mat. Fiz. 3*, 195–197. [Engl. transl. in: *U.S.S.R. Comput. Math. Math. Phys. 3*, no. 1 (1963), 257–261.]

Tihonov, A.N. and Gorbunov, A.D. [1964]. Error estimates for a Runge-Kutta type method and the choice of optimal meshes (Russian), *Ž. Vyčisl. Mat. i Mat. Fiz. 4*, 232–241. [Engl. transl. in: *U.S.S.R. Comput. Math. Math. Phys. 4*, no. 2 (1964), 30–42.]

Titchmarsh, E.C. [1939]. *The theory of functions*, 2d ed., Oxford University Press, Oxford.

Todd, John [1950]. Notes on modern numerical analysis. I. Solution of differential equations by recurrence relations, *Math. Tables Aids Comput. 4*, 39–44.

Todd, John [1954]. *The condition of the finite segments of the Hilbert matrix*, in *Contributions to the solution of systems of linear equations and the determination of eigenvalues*, Olga Taussky, ed., NBS Appl. Math. Ser., v. 39, U.S. Government Printing Office, Washington, DC, 109–116.

Todd, John [1980, 1977]. *Basic numerical mathematics*. v. 1: *Numerical analysis*, Internat. Ser. Numer. Math., v. 14, Birkhäuser, Basel; v. 2: *Numerical algebra*, Internat. Ser. Numer. Math., v. 22, Birkhäuser, Basel.

Traub, J.F. [1964]. *Iterative methods for the solution of equations*, Prentice-Hall Ser. Automat. Comput., Prentice-Hall, Englewood Cliffs, NJ.

Traub, Joe Fred and Woźniakowski, H. [1980]. *A general theory of optimal algorithms*, ACM Monograph Ser., Academic Press, New York.

Traub, Joseph Frederick, Wasilkowski, G.W., and Woźniakowski, H. [1983]. *Information, uncertainty, complexity*, Addison-Wesley, Reading, MA.

Traub, J.F., Wasilkowski, G.W., and Woźniakowski, H. [1988]. *Information-based complexity*, with contributions by A.G. Werschulz and T. Boult, Comput. Sci. Sci. Comput., Academic Press, Boston, MA.

Trefethen, L.N. and Weideman, J.A.C. [1991]. Two results on polynomial interpolation in equally spaced points, *J. Approx. Theory 65*, 247–260.

Troesch, B.A. [1976]. A simple approach to a sensitive two-point boundary value problem, *J. Comput. Phys. 21*, 279–290.

Turán, P. [1950]. On the theory of the mechanical quadrature, *Acta Sci. Math. Szeged 12*, 30–37.

Van Assche, Walter and Vanherwegen, Ingrid [1993]. Quadrature formulas based on rational interpolation, *Math. Comp. 61*, 765–783.

Van de Velde, Eric F. [1994]. *Concurrent scientific computing*, Texts Appl. Math., v. 16, Springer, New York.

van der Laan, C.G. and Temme, N.M. [1984]. *Calculation of special functions: The gamma function, the exponential integrals and error-like functions*, CWI Tract, v. 10, Stichting Mathematisch Centrum, Amsterdam.

Van Loan, Charles [1992]. *Computational frameworks for the fast Fourier transform*, Frontiers Appl. Math., v. 10, SIAM, Philadelphia, PA.

Varga, Richard S. [1962]. *Matrix iterative analysis*, Prentice-Hall, Englewood Cliffs, NJ.

Varga, Richard S. [1990]. *Scientific computation on mathematical problems and conjectures*, CBMS-NSF Regional Conf. Ser. in Appl. Math., no. 60, SIAM, Philadelphia, PA.

Verner, J.H. [1978]. Explicit Runge-Kutta methods with estimates of the local truncation error, *SIAM J. Numer. Anal. 15*, 772–790.

Wait, R. and Mitchell, A.R. [1985]. *Finite element analysis and applications*, Wiley-Intersci. Publ., Wiley, New York.

Walsh, J.L. [1969]. *Interpolation and approximation by rational functions in the complex domain*, 5th ed., Amer. Math. Soc. Colloq. Publ., v. 20, Amer. Math. Soc., Providence, RI.

Walter, Gilbert G. [1994]. *Wavelets and other orthogonal systems with applications*, CRC Press, Boca Raton, FL.

Wang, Ze Ke, Xu, Sen Lin, and Gao, Tang An [1994]. *Algebraic systems of equations and computational complexity theory*, with a preface by H.W. Kuhn, Math. Appl., v. 269, Kluwer, Dordrecht.

Wanner, G., Hairer, E., and Nørsett, S.P. [1978]. Order stars and stability theorems, *BIT 18*, 475–489.

Watkins, David S. [1991]. *Fundamentals of matrix computations*, Wiley, New York.

Watson, G.Alistair [1980]. *Approximation theory and numerical methods*, Wiley-Intersci. Publ., Wiley, Chichester.

Watson, Layne T., Billups, Stephen C., and Morgan, Alexander P. [1987]. Algorithm 652 — HOMPACK: A suite of codes for globally convergent homotopy algorithms, *ACM Trans. Math. Software 13*, 281–310.

Weierstrass, K. [1891]. Neuer Beweis des Satzes, dass jede ganze rationale Function einer Veränderlichen dargestellt werden kann als Product aus linearen Functionen derselben Veränderlichen, *Sitzungsber. Königl. Akad. Wiss.*, 1891. [Math. Werke, v. 3, 251–269.]

Weiss, Rüdiger [1996]. *Parameter-free iterative linear solvers*, Math. Res., v. 97, Akademie Verlag, Berlin.

Werner, Wilhelm [1982]. *On the simultaneous determination of polynomial roots*, in *Iterative solution of nonlinear systems of equations*, R. Ansorge, Th. Meis, and W. Törnig, eds., Lecture Notes in Math., v. 953, Springer, Berlin, 188–202.

Werner, Wilhelm [1984]. Polynomial interpolation: Lagrange versus Newton, *Math. Comp. 43*, 205–217.

Werschulz, Arthur G. [1991]. *The computational complexity of differential and integral equations: An information-based approach*, Oxford Math. Monographs, Oxford Sci. Publ., Oxford University Press, New York.

White, R.E. [1985]. *An introduction to the finite element method with applications to nonlinear problems*, Wiley-Intersci. Publ., Wiley, New York.

Wickerhauser, Mladen Victor [1994]. *Adapted wavelet analysis from theory to software*, A K Peters, Wellesley, MA.

Widlund, Olof B. [1967]. A note on unconditionally stable linear multistep methods, *BIT 7*, 65–70.

Wilkinson, J.H. [1963]. *Rounding errors in algebraic processes*, Prentice-Hall, Englewood Cliffs, NJ.

Wilkinson, James H. [1984]. *The perfidious polynomial*, MAA Stud. Math., v. 24: *Studies in numerical analysis*, Gene H. Golub, ed., Math. Assoc. America, Washington, DC, 1–28.

Wilkinson, J.H. [1988]. *The algebraic eigenvalue problem*, Monographs Numer. Anal., Oxford Sci. Publ., Oxford University Press, New York.

Wimp, Jet [1981]. *Sequence transformations and their applications*, Math. Sci. Engrg., v. 154, Academic Press, New York.

Wimp, Jet [1984]. *Computation with recurrence relations*, Applicable Math. Ser., Pitman, Boston, MA.

Wolfe, Michael Anthony [1978]. *Numerical methods for unconstrained optimization: An introduction*, Van Nostrand Reinhold, New York.

Wright, K. [1970]. Some relationships between implicit Runge-Kutta, collocation Lanczos τ methods, and their stability properties, *BIT 10*, 217–227.

Young, David M. [1971]. *Iterative solution of large linear systems*, Academic Press, New York.

Young, David M. and Gregory, Robert Todd [1988]. *A survey of numerical mathematics*, v. 1, corrected reprint of the 1972 original, Dover, New York.

Ypma, Tjalling J. [1995]. Historical development of the Newton-Raphson method, *SIAM Rev. 37*, 531–551.

Zayed, Ahmed I. [1993]. *Advances in Shannon's sampling theory*, CRC Press, Boca Raton, FL.

Zlatev, Zahari [1991]. *Computational methods for general sparse matrices*, Math. Appl., v. 65, Kluwer, Dordrecht.

Zwillinger, Daniel [1992a]. *Handbook of integration*, Jones and Bartlett, Boston, MA.

Zwillinger, Daniel [1992b]. *Handbook of differential equations*, 2d ed., Academic Press, Boston, MA.

Zygmund, A. [1988]. *Trigonometric series*, vols. 1, 2, reprint of the 1979 edition, Cambridge Math. Lib., Cambridge University Press, Cambridge.

Subject Index